AF876561

Sustainable Potato Production: Global Case Studies

Zhongqi He • Robert Larkin • Wayne Honeycutt
Editors

Sustainable Potato Production: Global Case Studies

Springer

Editors
Zhongqi He
USDA-ARS
Southern Regional Research Center
Robert E. Lee Blvd 1100
New Orleans, LA, USA

Robert P. Larkin
USDA-ARS
New England Plant, Soil,
and Water Laboratory
Orono, ME, USA

Wayne Honeycutt
USDA
Natural Resources Conservation Service
14th and Independence Avenue
SW Rm.5006-S
Washington, DC, USA

ISBN 978-94-007-4103-4 ISBN 978-94-007-4104-1 (eBook)
DOI 10.1007/978-94-007-4104-1
Springer Dordrecht Heidelberg New York London

Library of Congress Control Number: 2012939427

Printed on acid-free paper

Springer is part of Springer Science+Business Media (www.springer.com)

Preface

Potato (*Solanum tuberosum* L.) is grown in over 100 countries throughout the world. As a staple food, potato is the fourth most important crop after rice, wheat, and maize, and has historically contributed to food and nutrition security in the world. Global interest in potato increased sharply in 2008 as world food prices soared, threatening the food security and stability of dozens of low-income countries. Unlike major cereals, potato is not a globally traded commodity, and prices are usually determined by local production costs. Thus, potato is increasingly regarded as a vital food-security crop and as a substitute for costly cereal imports. With such importance, we organized this edited collection of global case studies that address the issues of sustainable potato production.

This book begins with an introduction on sustainable potato production and global food security (Chap. 1). This introductory chapter provides the latest updates on geospatial patterns of potato production world-wide and briefly discusses the potential impacts of climatic change, biotechnology and soil resource management on sustainable potato production.

This book presents eight case studies selected globally and covering different issues relevant to sustainable potato production in both developed and developing countries. Part II is a case study on enhancing potato system sustainability in the Northeast USA. Research in this study case was conducted to identify the constraints to potato system sustainability and develop practices and management strategies to overcome or reduce those constraints. For this purpose, five cropping systems were designed and managed as (a) Status Quo, (b) Soil Conserving, (c) Soil Improving, and (d) Disease Suppressive Systems under both irrigated and rainfed management. Four chapters (i.e. Chaps. 2, 3, 4, and 5) in Part II evaluated the five systems for their impacts on soil physical, chemical, and biological properties; plant growth; plant diseases; nutrient availability; and their interactions. Part III focuses on the case studies of sustainable potato production managements for irrigated agriculture in the Western USA. Chapter 6 discusses research conducted in Colorado to evaluate the effect of different cover crops as a management tool in potato cropping systems. Chapter 7 reports mustard green manure use in eastern Washington State. Chapter 8 reviews the field trial experiments of effects of application of commercial humate products on yields of potato and several other crops conducted in the Western USA. Chapter 9 examines the late blight epidemics in the Columbia Basin of south-central Washington and north-central Oregon.

Part IV presents the case studies of rainfed potato production in eastern Canada with the focus on nitrogen management issues. Chapter 10 evaluates a series of N fertilization strategies and recovery of fertilizer N by the potato crop. The subsequent four chapters address the use of soil- and plant-based test systems to improve fertilizer N recommendations (Chap. 11), N management in organic potato production systems (Chap. 12), and N losses to water (Chap. 13) and to the atmosphere (Chap. 14) in potato production systems in eastern Canada. The five chapters in Part V are the case studies of sustaining potato production in the cool-temperate climate of Tasmania, Australia. Chapter 15 briefly describes Tasmania's geography, the farming systems of which potato production is part, and introduces some of the management challenges facing the industry. Chapters 16, 17, 18, and 19 provide more fully-developed descriptions of these challenges and how recent research and development efforts have helped Tasmanian potato growers to meet them.

Sustainable potato production in developing countries may face greater challenges due to resource limits. Part VI present three chapters covering water-saving potato production research for the semi-arid areas of Northern China. Chapters 20 and 21 examine potato growth and yields affected by dripping irrigation and plastic mulch. Chapter 22 reports the case study of enhanced drought and salinity tolerance of transgenic potato plants with a betaine aldehyde dehydrogenase gene from spinach. Some of the efforts towards increasing sustainability of potato production systems in South America are reported in Part VII. Chapter 23 discusses the relationship between potato yield and nitrogen rates obtained by different mathematical models and how the model chose affects plant nitrogen indices under Brazilian conditions. Chapter 24 examines "deep soil loosening" tillage system in two Brazilian potato producing regions, suggesting it as an alternative to improve potato production in compacted areas, and as a tool to promote the recuperation of soils damaged by compaction. Chapter 25 reports experiences and lessons from two distinct potato production systems of Peru in developing integrated pest management for potato.

Four chapters in Part VIII and IX of this book cover the case studies carried out in Mediterranean and tropical African regions. Chapter 26 reviews the role of green manure and amendments application in soil fertility management in organic potato production with a case study in Tuscany (Central Italy). Chapter 27 focuses on effect of humic substances application on potato tubers yield quantity, quality, and nutrients concentration under Egyptian soil conditions. Chapter 28 examines residual pesticides and heavy metals levels in conventionally and organically farmed potato tubers in Egypt. And, finally, Chap. 29 reviews various management techniques for late blight and bacterial wilt diseases highlighted with examples drawn from research conducted in Sub-Saharan Africa.

Chapter contribution was by invitation only. For each chapter to stand alone, there is occasionally some overlap in literature review, and some experiments have been used as examples in more than one chapter. This book is basically the results of the combined efforts of our distinguished group of contributors. We wish to thank all contributors for their timely contributions. Special thanks go to Drs. Bernie Zebarth and Leigh Sparrow for their coordination of the chapter contributions in Part IV and V, respectively. Finally, we would like to thank all reviewers for their many helpful comments and suggestions which certainly improved the quality of this book.

About the Editors

ZHONGQI HE is a Research Chemist at the United States Department of Agriculture-Agricultural Research Service, Southern Regional Research Center, New Orleans, Louisiana. He was a recipient of the National Research Council postdoctoral fellowship hosted by the United States Air Force Research Laboratory, Tyndall Air Force Base, Florida. The author or co-author of over 120 research articles, patents, proceedings, and book chapters, he has actively pursued basic and applied research in environmental chemistry and biochemistry of agricultural products, byproducts, and plant nutrients. Previously, Dr. He organized and served as editor of three books. He has provided peer review services for more than 40 journals and served in an editorial board. He received the B.S. degree (1982) in applied chemistry from Chongqing University, China, the M.S. degrees (1985 and 1992) in applied chemistry from South China University of Technology, Guangzhou, and in chemistry from the University of Georgia, Athens, USA, and the Ph.D. degree (1996) in biochemistry from the University of Georgia, Athens, USA.

ROBERT P. LARKIN is a Research Plant Pathologist specializing in soil microbiology and soilborne diseases of potato at the United States Department of Agriculture-Agricultural Research Service, New England Plant, Soil and Water Laboratory, Orono, Maine. The author or co-author of over 100 scientific publications, he has actively engaged in basic and applied research dealing with biological and cultural control of soilborne plant diseases, disease-suppressive soils, sustainable cropping systems, and active management of soil microbial communities. He received the (USA) Federal Laboratory Consortium's Regional and National Excellence in Technology Transfer Awards for the Potato System Planner, an interactive decision-making tool for potato growers, and also received the Storkan-Hanes Foundation award for excellence in graduate research on soilborne diseases. He has been an associate and senior editor for the journal Phytopathology, editor for Biological Control, consulting editor for Plant & Soil, and is currently assigning editor for Plant Disease Management Reports, as well as regularly providing reviews for several other publications and granting agencies. Dr. Larkin received the B.S. degree (1978) in Environmental Science-Biology from Bradley University, Peoria

IL, M.S. (1981) in Ecology from the University of Tennessee, Knoxville, and Ph.D. (1990) in Plant Pathology from the University of Florida, Gainesville.

C. WAYNE HONEYCUTT is the Deputy Chief for Science and Technology of the United States Department of Agriculture-Natural Resources Conservation Service, Washington, DC. The author or co-author of over 90 scientific publications, he has actively pursued soil and agronomic research relevant to potato production for more than 20 years. He received the (USA) Federal Laboratory Consortium's Regional and National Excellence in Technology Transfer Awards for the Potato System Planner-an interactive decision-making tool for potato growers. He has been chair of the (USA-Canada) Northeast Potato Technology Forum four times. Dr. Honeycutt served as associate editor of the Agronomy Journal and frequently provides peer review services for soil and agricultural journals. Dr. Honeycutt received the B.S. (1979) and M.S. (1981) degrees in forestry and forest soils, respectively, from University of Kentucky, Lexington, and the Ph.D. degree (1986) in soil genesis from Colorado State University, Fort Collins, USA.

Contents

Part I
Introduction

Chapter 1
Sustainable Potato Production and Global Food Security

Sherri L. DeFauw, Zhongqi He, Robert P. Larkin, and Sameeh A. Mansour

Abstract The potato (*Solanum* spp.) is currently the leading non-grain commodity in the global food system with production exceeding 329 million metric tonnes in 2009. The extraordinary adaptive range of this species complex combined with ease of cultivation and high nutritional content have promoted steady increases in potato consumption especially in developing countries. Recent uncertainties in world food supply and demand have placed the potato in the upper echelon of recommended food security crops. This introductory chapter provides the latest updates on geospatial patterns of potato production world-wide. In addition, the potential impacts of climate change, agrobiodiversity, biotechnology, and soil resource management on sustainable potato production are briefly discussed.

1.1 Introduction

The potato (*Solanum* spp.) has helped sustain humanity for centuries, and now ranks as the leading non-grain commodity in the global food system (FAO 2009a), with production exceeding 329 million tonnes in 2009 (FAOSTAT 2011). The extraordinary adaptive range of this plant combined with relative ease of cultivation (Haverkort 1990) and high nutritional content have promoted steady increases in potato consumption

S.L. DeFauw (✉) • R.P. Larkin
USDA-ARS, New England Plant, Soil, and Water Laboratory, Orono, ME 04469, USA
e-mail: sld42@psu.edu; bob.larkin@ars.usda.gov

Z. He
USDA, SRRC, New Orleans, LA, USA
e-mail: zhongqi.he@ars.usda.gov

S.A. Mansour
Environmental Toxicology Research Unit (ETRU), Pesticide Chemistry Department, National Research Centre, Dokki, Cairo, Egypt
e-mail: samansour@hotmail.com

Z. He et al. (eds.), *Sustainable Potato Production: Global Case Studies*,
DOI 10.1007/978-94-007-4104-1_1, © Springer Science+Business Media B.V. 2012

especially in developing countries which, in turn, account for over half of the total global harvest (FAO 2009a). In fact, the developing world's potato production exceeded that of the developed world for the first time in 2005 (FAO 2010). Millions of farmers depend on potatoes for subsistence and as a local cash crop. Recent uncertainties in world food supply and demand have placed the potato high on the list of recommended food security crops (FAO 2009a, 2010; Lutaladio and Castaldi 2009; Pandey et al. 2005). Potato production potential is exceptionally high as approximately 80% of the plant's biomass constitutes economic yield (Osaki et al. 1996).

Recently, the United Nations (UN) declared the year 2008 as the International Year of the Potato (IYP). The initial resolution, proposed by the Permanent Representative of Peru (at the biennial Conference of the Food and Agriculture Organization (FAO) of the UN convened in November 2005), explained the pivotal role that the potato has served in global diets as well as may serve in achieving international development objectives to further reduce undernourishment. Throughout the celebration of IYP 2008, opportunities were taken to underscore how potatoes could contribute to: (1) improvements in diet and food security; (2) alleviating the income and subsistence challenges for small-scale farming families; and (3) conserving the genetic resources needed to utilize potato biodiversity in order to supply improved varieties in the future (FAO 2009a). In that same year, the Government of Peru created a register of Peruvian native potato varieties (FAO 2009a). The Andes are a center of origin (Vavilov 1992) and diversity for numerous crop species, including the potato (Spooner et al. 2005), with the Huancavelica region of Central Peru recognized as a center of potato genetic diversity (Torres 2001; Huamán 2002; de Haan 2009). Systematic documentation of diversity hotspots such as this one is essential as it deepens our understanding of conservation units (alleles, cultivars, and species mixtures) as well as the socioeconomic scales (at the household-, community-, and regional-levels) associated with improving food security (de Haan et al. 2010).

This introductory chapter highlights the importance of potato production in contributing to global food security and provides maps describing the current geospatial distribution of production areas world-wide. Other inter-related topics briefly reviewed and discussed here as they contribute to strengthening the sustainability of food systems include considerations of the effects climate change, genetic modification of potato, agrobiodiversity reservoirs, and soil conservation strategies. More specific issues on sustainable potato production cultural practices such as control of soilborne pests and pathogens, or improvements in nutrient- and water-use efficiencies that enhance yield are reviewed and discussed in detail in the relevant chapters of the eight case studies.

1.2 The Importance of Potato in Global Food Security

Food webs are central to life, and human-centered food systems are dynamically intertwined with complex changes in the socio-cultural contexts of food production, dispersion of cultures, political alliances, economies and ecosystems

(Eriksen et al. 2009). Increasing demands for food have induced global environmental change (GEC), including soil degradation, loss of biodiversity, rapid proliferation of greenhouse gas emissions, nutrient loading of ground and surface waters, and in some areas critical water shortages. Population and income growth combined with high energy prices, biofuels, science and technology breakthroughs, climate change, globalization, and urbanization are causing drastic changes in food consumption, production, and markets. Adapting to these food security challenges requires integrated food systems approaches (at multiple scales and considering cross-scale dynamics) that engage a broad spectrum of researchers from the social and natural sciences because many other factors, in addition to food production, need to be considered such as food availability, access, utilization and stability (Steffen et al. 2003; Stamoulis and Zezza 2003; Cash et al. 2006; Eriksen et al. 2009). However, further discussion of these highly relevant issues is well beyond the scope of this introductory chapter.

The global agriculture sector is confronting significant challenges within the next four decades. FAO estimates that worldwide agricultural production will need to grow by 70% over an approximated 45-year interval (between 2005–2007 and 2050), and by 100% in developing countries (FAO 2011). By 2050, predictions indicate that the global population will be between 8.0 and 10.4 billion people, with a median estimate of 9.1 billion (Jaggard et al. 2010). Today, more than one in seven people still lack sufficient protein and energy in their diet, and even more have some form of micronutrient malnourishment (FAO 2009b).

Global interest in potato production increased dramatically in 2008 as world food prices soared, creating instabilities in the food security of low-income countries (FAO 2009a, b; Lutaladio and Castaldi 2009). High food prices also tend to worsen poverty and malnutrition (FAO 2011). The nutrient-laden potato yields more food (carbohydrate- and micronutrient-rich, B and C vitamins, protein content comparable to cereal grains, plus dietary antioxidants) (Burlingame et al. 2009) more rapidly on less land than any other major crop as up to 85% of the plant may constitute edible food for humans, compared to only 50% for most cereal grains (FAO 2009a). Potatoes are C3 plants along with wheat, rice, soya, sunflower, oilseed rape, sugar beet and dry bean. Together with key C4 plants that include maize, sugar cane and sorghum, these 11 crops occupy 56% of the world's arable area (Jaggard et al. 2010).

Potatoes are currently grown on an estimated 20 million ha of farmland spanning the subtropical lowlands of India (near sea level) to the Andean highlands of Peru and Bolivia approaching 4,000 m elevation (FAO 2010). Collectively, the top 20 potato-producing countries (Fig. 1.1) harvested close to 257 million metric tonnes from an estimated 13.7 million ha with a crop valuation of close to 30 billion international dollars in 2008 (data compiled from FAOSTAT 2011). These nations accounted for close to 80% of global production. Under irrigation in temperate climates, yields typically vary between 25 and 45 tonnes ha^{-1} with 120–150 d crops requiring from 500 to 700 mm of water; water deficits in the middle to late stages of the growing season generally have the greatest negative impacts on yield (FAO 2009a; FAOSTAT 2011). Subtropical yields tend to be substantially lower, ranging

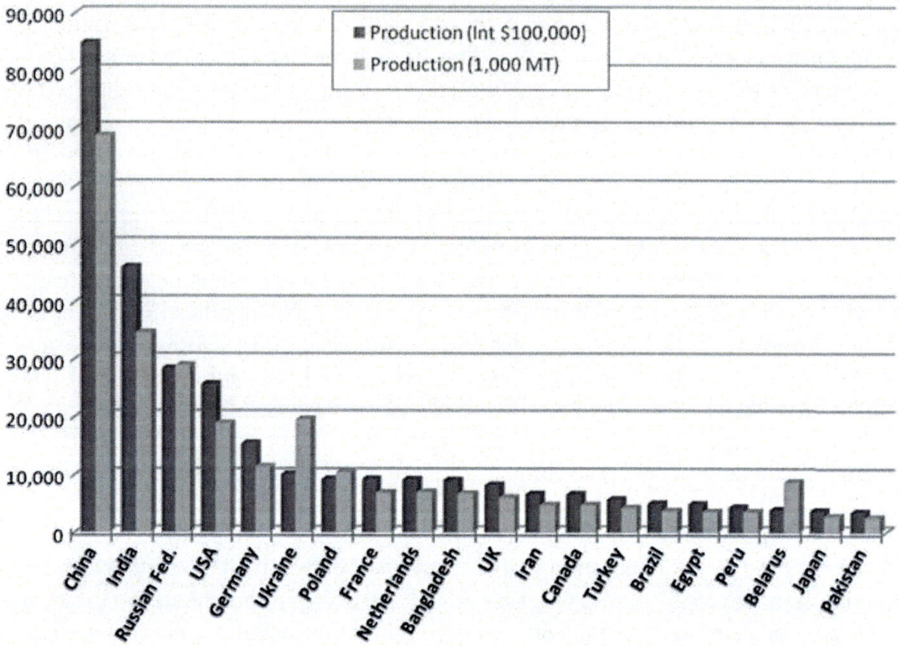

Fig. 1.1 Summary of the top 20 potato-producing nations (2008) comparing valuation in international dollars (Int) with quantity in metric tonnes (MT or Megagrams, Mg) (FAOSTAT 2011)

from 5 to 25 tonnes ha^{-1} (FAO 2009a; FAOSTAT 2011). However, across global landscapes, the versatility of this crop coupled with notable increases in production in most countries over the last two decades is unparalleled. Consumption of fresh potatoes accounts for approximately two-thirds of the harvest, which exceeded 329 million tonnes from an estimated 18.6 million ha in 2009 (FAOSTAT 2011).

Developing countries with high food demands provide case studies on the impact of increased potato production. From 1990 to 2009, 35 countries recorded production increases ranging from approximately 130% to 2,300% (with ten nations profiled in Table 1.1; FAOSTAT 2011). African nations posting some of the largest gains in quantity and harvested areas over this 20-year interval included Angola (2,321%), Nigeria, and Rwanda. Yield averages (for 2009) for these three nations were 8.0, 4.0, and 10.2 Mg ha^{-1}, respectively (FAOSTAT 2011). Other African nations exhibiting substantial growth in potato production (both tonnes and hectares, though not shown in Table 1.1) were Algeria, Cameroon, Mali, Namibia, Niger, Tanzania, and Uganda with average yields (for 2009) ranging from 2.2 to 25.1 Mg ha^{-1} (FAOSTAT 2011). Although consumption is generally rather low for most of these nations (approximately 5–15 kg per capita per year), the potato underpins Rwanda's food security (approximately 125 kg per capita per year; FAO 2009a). Since 1990, the potato has contributed, in part, to reducing the proportion

Table 1.1 Summary of ten selected nations with noteworthy increases in production and harvested area (Based on data compilation from FAOSTAT 2011)

	Production (tonnes)			Harvest (ha)		
Nation	1990	2009	Prod. % diff	1990	2009	Harv% diff
Angola	34,000	823,266	2,321	8,500	103,440	1,117
Bangladesh	1,065,680	5,268,000	394	116,582	395,000	239
China	32,031,189	73,281,890	129	2,829,384	5,083,034	80
Egypt	1,637,810	4,000,000	144	79,663	145,000	82
India	14,770,800	34,391,000	133	940,000	1,828,000	94
Nepal	671,810	2,424,050	261	83,350	181,900	118
Nigeria	54,000	914,778	1,594	7,700	227,519	2,855
Pakistan	830,976	2,941,300	254	79,900	145,000	81
Peru	1,153,980	3,716,700	222	146,435	282,100	93
Rwanda	283,673	1,287,400	354	42,055	126,167	200

of undernourished among the total populations of most of the aforementioned sub-Saharan nations by 33–61% (FAO 2011). Egypt is Africa's top potato producer, and has increased production tonnes by 144% from 1990 to 2009 (Table 1.1; FAOSTAT 2011). Egypt also ranks among the world's top exporters of fresh and frozen potato products directed mostly to European markets (FAO 2009a). In December 2007, a conference was held in Alexandria, Egypt called "Potato, Sweet Potato, and Root Crops Improvement for Facing Poverty and Hunger in Africa", hosted by the African Potato Association. The theme was chosen for the same reason that the UN declared this the International Year of the Potato: "because we realized the food gap problem and we realized that the potato crop can substitute a large quantity of the wheat importation" (Sherk 2008).

Potatoes are grown widely in Nepal, and now serve as this nation's second staple food crop (after rice). For smallholder Nepalese farmers in highland areas (1,800–3,000 m ASL), it is more productive than rice or maize and they also produce seed tubers for sale at lower altitudes (FAO 2009a). The potato has also become a significant source of rural income in Pakistan; most production occurs in the Punjab where spring and autumn crops account for 85% of the harvest. Expansion of irrigated Pakistani land has resulted in substantial increases in production output (up 254% from 1990 to 2009) and area under cultivation (Table 1.1). Annual potato intake in Pakistan is estimated at 11 kg per capita (FAO 2009a). Potato is a highly successful winter crop (October–March) in Bangladesh where it is typically grown for cash sale by smallholder farmers; annual consumption was approximately 24 kg per capita in 2005 (FAO 2009a). The bulk of Peru's potato production occurs on smallholder highland farms (most at 2,500–4,500 m ASL) where it has been a staple food for millennia. Peruvian annual consumption is high estimated at 80 kg per capita (FAO 2009a). From the handful of brief national profiles presented here, potato cropping systems help improve resilience especially among smallholder farmers by providing direct access to nutritious food, increasing household incomes, and reducing their vulnerability to food price volatility.

1.3 Potato Production Areas and Yields: A Global Perspective

The most recent production reports from FAO indicated that the 2009 global harvest exceeded 329 million tonnes from an estimated 18.6 million ha (FAOSTAT 2011). Profiling the relative distribution of production areas revealed that four nations grew well over 1 million ha annually with China recording a harvest area of close to 5.1 million ha (Fig. 1.2); the others included the Russian Federation (about 2.2 million ha), India (over 1.8 million ha) and the Ukraine (over 1.4 million ha). Harvests from Belarus, Bangladesh, the United States and Poland ranged from approximately 0.4–0.5 million ha. Additional nations that exceeded 250,000 ha in harvest area extents for 2009 were Peru, Germany, and Romania.

Aggregated yield data reported from 157 nations ranged from 1.2 to 46.3 Mg ha^{-1} in 2009 (FAOSTAT 2011) with a "global" yield of 18.4 ±10.6 Mg ha^{-1} (mean ± SD). Denmark, France, New Zealand, United Kingdom, Germany, Belgium, the Netherlands, Switzerland and the United States had potato yields greater than 40 Mg ha^{-1} in 2009 (Fig. 1.3) with the Netherlands typically achieving world record average yields (FAO 2011). In rather sharp contrast, of the four nations with the largest harvested production areas, India led with a recorded average yield at 18.8 Mg ha^{-1} and Belarus reported a comparable yield of 18.6 Mg ha^{-1}; whereas, China and the Russian Federation reported average yields of 14.4 and 14.3 Mg ha^{-1}, respectively.

Resolving patterns of 2009 potato productivity revealed that China contributed over 22% of the total global production. India, the Russian Federation, Ukraine, and the United States contributed substantially lesser quantities amounting to 10.4%,

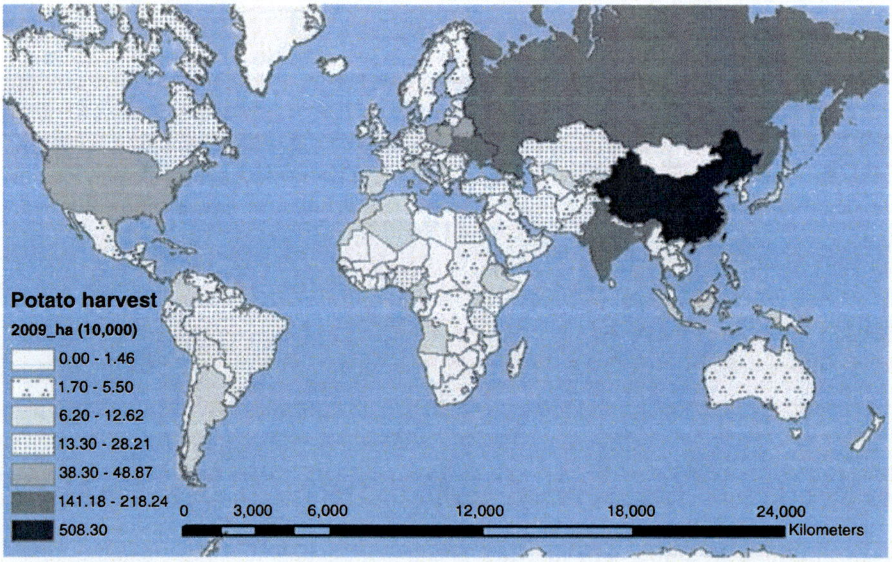

Fig. 1.2 Relative distribution of potato production areas in 2009 with aerial extents reported in 10,000-ha units (Based on data compilation from FAOSTAT 2011)

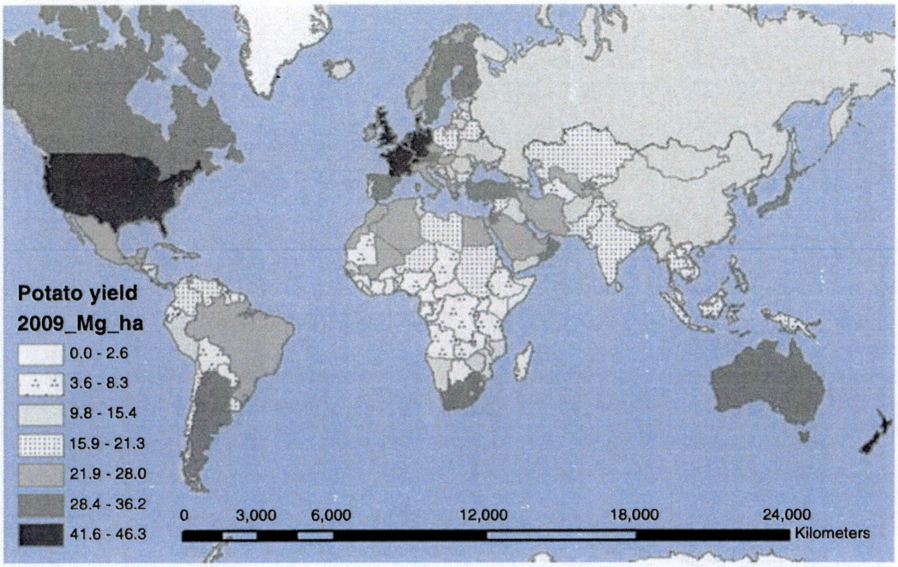

Fig. 1.3 Global distribution of potato yields (Mg ha^{-1}) in 2009 (Based on data compilation from FAOSTAT 2011)

9.4%, 6.0% and 5.9% of the global total, respectively. It is noteworthy that production output in China has more than doubled since 1990 (Table 1.1) and it is becoming a prominent global supplier. Chinese farmers in mountainous areas now rely on potato sales for approximately one-half of their household earnings. In addition, major expansion of potato cultivation is underway in the dry areas which account for approximately 60% of China's arable land (FAO 2009a). Production output in India has also more than doubled from 1990 to 2009 (Table 1.1). There the potato serves as a rural staple as well as a cash crop with production concentrated on the Indo-Gangetic Plain from October through March and some year-round production occurring at higher altitudes in the south. Annual per capita consumption in India is approaching 20 kg (FAO 2009a). The Russian Federation and the Ukraine have very high annual potato intakes compared to other nations; these are estimated at 130–136 kg/year, respectively, although pest and disease pressures result in annual losses of approximately several million tonnes (FAO 2009a). In the USA, over 90% of annual output is for human consumption with approximately 60% processed into frozen products, 33% consumed fresh, and the remainder reserved for seed (FAO 2009a). Other countries commonly ranking in the top 20 of 2009 production output (Fig. 1.1, though Belgium exceeded Pakistan in that year) contributed from 1.0% to 3.5% of global production.

Tracking yields for the top 20 potato-producing nations (shown in Fig. 1.1) for 1990–2009 reveals greater details concerning yield "stability" from year-to-year. Also highlighted are those countries that have experienced rather steady and substantial yield increases over this time interval; in particular, Belarus, Brazil,

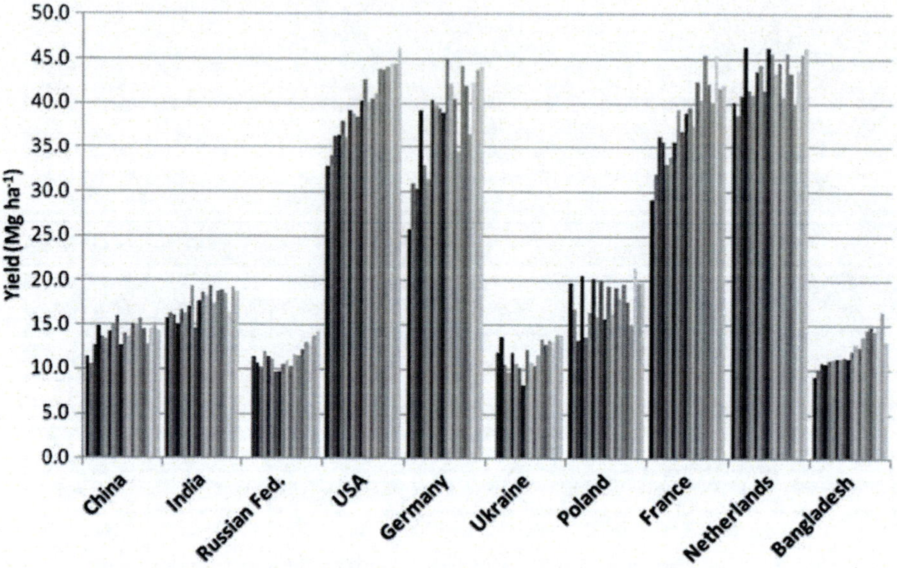

Fig. 1.4 Variations in yield (Mg ha^{-1}) for the past 20 years (1990–2009) for "tier 1" potato-producing nations (Based on data compilation from FAOSTAT 2011)

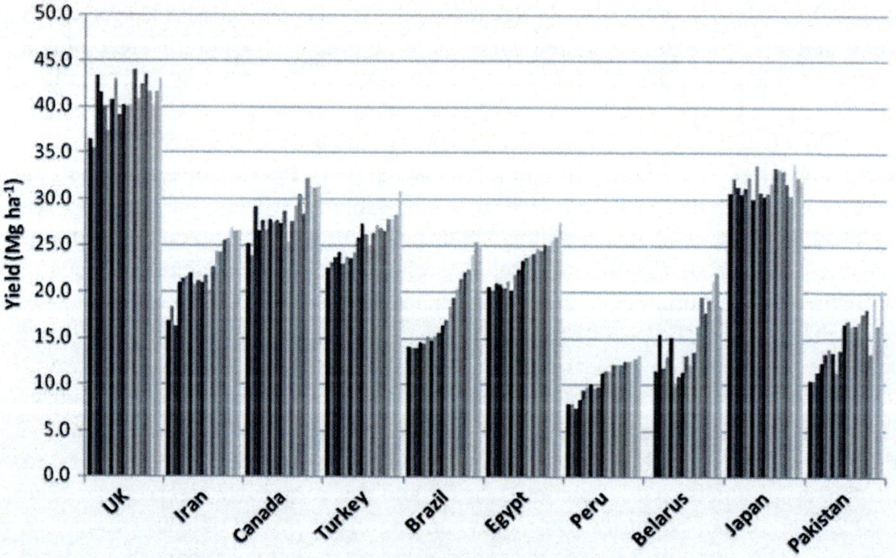

Fig. 1.5 Variations in yield (Mg ha^{-1}) for the past 20 years (1990–2009) for "tier 2" potato-producing nations (Based on data compilation from FAOSTAT 2011)

France, Iran and Pakistan (all experiencing 10+ Mg ha^{-1} increases within a 20-year span) (Figs. 1.4 and 1.5). In addition, these time-series comparisons begin to underscore the interactions of biogeographical and environmental phenomena as well as socioeconomic factors that influence the growth, yield and quality of this drought sensitive crop. France is Europe's leading exporter of fresh potatoes and dedicates approximately 10% of its production area to growing seedlings (FAO 2009a). Belarus ranks 8th among the world potato-producing countries (Fig. 1.1) and exports both fresh and seed potato. Belarusians consume more potatoes per capita than any other country, an estimated 180 kg year^{-1} (FAO 2009a). Iran is the third largest potato producer in Asia (after China and India). This nation is steadily increasing its irrigated lands, and the potato is one of Iran's leading agricultural exports (FAO 2009a). These 20-year yield "snapshots" (Figs. 1.4 and 1.5) may also serve as tracers for potential yield declines resulting from global environmental change (GEC). According to model simulations on the effect of climate change on global potato production presented by Hijmans (2003), Bangladesh, Brazil and the Ukraine were predicted to experience potential yield decreases in excess of 20% for the interval 2040–2069.

1.4 Global Climate Change and Potato Production

The ecology of potato cropping systems in relation to climate as affected by latitude and altitude was reviewed by Haverkort (1990). This author reaffirmed that the potato is a versatile commodity adapted to a wide range of environmental conditions while underscoring the plant's sensitivity to drought stress (dependent on cultivar rooting depth) along with preferences for tuberizing under short-day conditions and best performances in cool temperate climates (Haverkort 1977, 1990). Higher temperatures, for example, promote foliar development, delay tuberization and influence potato quality characteristics such as higher numbers of smaller tubers per plant, and lower specific gravity which is indicative of lower dry matter contents (Haverkort 1988). In addition, water stresses (i.e., either waterlogging or drought conditions) occur to varying degrees dependent on site-specific heterogeneity of soils, complexity of field-scale topography, soil resource management by the farmer, and availability of water for irrigation. Drought events occurring early in the growing season reduce the number of tubers per plant (Haverkort et al. 1990). Furthermore, a single, short-term drought event during tuber bulking can inhibit future bulking of those potatoes set and result in initiation of new tubers; these plant responses not only decrease potato grade (i.e., tuber size and quality) but lower overall yield. High soil moisture conditions prior to harvest are known to negatively affect tuber specific gravity, whereas other in season stressors influence the development of disorders such as internal heat necrosis and hollow heart (Hiller et al. 1985).

Models that couple the biology of crop growth with the physics of climate change help policy-makers and scientists envision these potential changes in patterns of production and galvanize research efforts to mitigate their effects on future food supply. Hijmans (2003) assessed the effect of climate change on global potato production using a simulation model linking temperature and solar radiation datasets (with plant performance based on radiation use efficiency (RUE) algorithms). Climate data inputs included "current climate" (i.e., monthly averages for 1961–1990) as well as 7 scenarios from 5 climate models (CGCM1, CSIRO-Mk2, ECHAM4, GFDL-RI5 and HadCM2); these inputs were used to create two sets of projected climate surfaces for 2010–2039 and 2040–2069. Model runs for each grid cell (1° by 1° in size) included 12 planting times, 5 maturity classes (early to late senescence), and non-heat tolerant versus heat-tolerant potato. Mapped results (for countries with > 100,000 ha of potato area) permitted comparison of average change in potential potato yield (by country) due to climate change with and without adaptation (adaptation was defined as changes in planting month or cultivar maturity class). Climate scenarios for 2040–2069 predicted the increase in global average temperature will be between 2.1°C and 3.2°C, although the predicted temperature increase was smaller (between 1.0°C and 1.4°C) when weighted by the potato area and adaptation of planting time and cultivar choice were allowed. For the 2040–2069 interval, global potential potato yields were forecasted to decrease by 18% to 32% 'without adaptation', and by 9% to 18% 'with adaptation'. In addition, when adaptation was considered for the 2040–2069 scenario, Bangladesh, Brazil, Colombia and the Ukraine were predicted to experience the largest decrease in potential yield (>20%). Argentina, Canada, China, Japan, Peru, the Russian Federation, Spain, the UK, and USA were listed as notable examples where adaptation could mitigate much of the negative effects of global warming – particularly by shifting the location of production with existing potato growing regions. In general, the strongest negative impacts to potato production were predicted for the tropical and subtropical lowlands though these impacts could be ameliorated by the development of heat-tolerant cultivars (Hijmans 2003).

Higher resolution crop modeling entails systematically structuring biotic and abiotic spatiotemporal factors that influence crop development, growth and yield (i.e., genotype*environment*management (g*e*m)). Detailed models permit the potato industry to perform agro-ecological zoning by estimating timing (from planting to crop maturity), yields, hazards, and water-use efficiency (Haverkort 2007). These models also help serve as decision support systems for farmers (for irrigation as well as timing and dosing of nutrients and crop-protection inputs), help guide procurement strategies and aid in price policy establishment (MacKerron and Haverkort 2004). SPUDSIM, for example, is a new explanatory-type potato crop model that upgrades the RUE ('big-leaf') approach with a more detailed biochemical leaf-level model system that better depicts canopy architecture (for both shaded and sunlit leaves), and assimilate allocation to branches, roots and tubers (Fleisher et al. 2010). Comparison of SPUDSIM predictions versus gas exchange and dry mass data indicated the model predicted plant growth accurately (with high indices of agreement (≥0.80)) over a broad range of temperatures (12.6–32.3°C, on a 24-h average basis) except for the 34/29 case study. The model will be suitable for a variety of applications (i.e., farmscape to regional-scales) that involve complex soil-plant-atmosphere-water relationships (Fleisher et al. 2010).

Potato productivity and production in India was estimated under future climate change scenarios (Singh and Lal 2009). The authors concluded that without adaptations potato production under the impact of climate change and global warming may decline by 3% and 14% in the years 2020 and 2050, respectively. Possible adaptations like change in planting time, breeding heat tolerant varieties, efficient agronomic and water management and shifting cultivation to new and suitable agro-climatic zones can significantly arrest the decline in the production.

The "SUBSTOR" model is a mechanistic, process-oriented model for simulating tuber yield and crop development, and was employed to simulate physiological processes and yield of potato production in Egypt (Abdrabbo et al. 2010). Actual measurements of potato production were used to compare present and predicted. The climate change data was used from two general circulation models (GCMs), CSIRO and HadCM3, for the A1 greenhouse gas scenario to 2050 (Pearman 1988). The results of the work indicated that the potato yield in 2050 may be decreased by 11% to 13% compared to 2005/2006. Irrigation at 100% gave the highest tuber yield for different cultivars with the two climate change GCM models.

1.5 Agrobiodiversity and Biotechnology Considerations

The widely-cultivated potato, *Solanum tuberosum* L., along with six other cultivated species grown only in the Andes (Walker et al. 1999) collectively constitute one of the world's principal food crops. Results from recent molecular research indicate that the widest grown species retains but a portion of the actual range of genetic diversity found among the key recognized species, and they are divided into two cultivar groups based on adaptation to day length conditions. The 'Chilotanum' group (also referred to as the *"European"* potato) is now cultivated around the world, whereas the 'Andigenum' group (adapted to short-day conditions) is still primarily grown in the Andes (FAO 2009a). However, breeders in Europe and North American have produced many of the cultivars in current use by drawing on potato germplasm from Chile (Lutaladio and Castaldi 2009).

Farmers in the high Andes recognize potatoes not only by species and variety, but also by the microenvironmental niche where the tubers grow best; in fact, it is customary to find 8–20 cultivars per field at altitudes varying from 3,550 to 4,250 m in Huancavelica where weather extremes are frequent occurrences (de Haan et al. 2010). Natural potato pollination sustains the diversity of Andean farmer-developed, locally-adapted varieties. The systems perspective on planting levels and emergent properties of potato biodiversity documented by de Haan et al. (2010) in this central Peruvian highlands region (latitude 11°59′94″S to 14°7′48″S and longitude −74°16′11″W to −75°48′38″W) is of great importance for ongoing crop conservation efforts that will, in turn, contribute to future global germplasm enhancement. Of all the major crops, potato is arguably one of the most important species groups sustaining mountain agriculture, for the highest levels of genetic diversity are maintained by farmer communities at altitudes well above 3,000 m (Zimmerer 1991). An individual farm household may retain as many as 160 unique cultivars based on long-established culinary preferences as well as intimate site-specific knowledge of the deployment of cultivar diversity to ensure a reasonable harvest return; therefore, food system interventions designed to enhance food security should encourage participant growers to emulate these high altitude farmers by building on agrobiodiversity (de Haan 2009; de Haan et al. 2010).

Genetic modification (GM) technologies and the on-going debate over their acceptance has become highly politicized and polarized, particularly in Europe. Although biotechnology tools (genomics and bioinformatics) are likely to be of pivotal importance (Phillips 2010) in contributing to some rapid advances (i.e., single transgene events that could dramatically alter plant performance), broader-based concerns expressed by agronomists indicate that diverting resources and focus from conventional breeding could slow the rate of yield increases required to feed a rapidly expanding world population (Jaggard et al. 2010). Leading researchers on global food and farming futures (e.g., Godfray et al. 2010, pp 815–816) agree that "genetic modification is a potentially valuable technology whose advantages and disadvantages need to be considered rigorously on an evidential, inclusive, case-by-case basis." These authors also affirm that this technology needs to garner greater public trust and acceptance "before it can be considered as one among a set of technologies that may contribute to improved global food security." Furthermore, new technologies (both GM and non-GM) must be community-directed; those efforts meant to benefit the poorest nations will require innovative alliances of civilians, governments, and businesses (Godfray et al. 2010). As mentioned in the previous section of this chapter, in general, the strongest negative impacts to potato production have been predicted for the heavily-populated subtropical and tropical lowlands (Hijmans 2003); these impacts could be ameliorated, in part, by the development of a broader spectrum of stress-tolerant cultivars. Si and colleagues (Chap. 22 of this volume; Zhang et al. 2011) have developed transgenic potato plants that are resistant to drought and salinity stresses; however, these plants have yet to be grown commercially. Performance trials and biosafety assessments are underway.

Biotechnology has provided a fast alternative for potato crop improvement in the areas of resistance to insects and viruses. Genetic transformation of potato by *Agrobacterium tumefaciens* has been most successful and this method has been particularly efficient in introducing several useful genes into various potato genomes (Kumar et al. 1995). It is well known and documented that the choice of cultivar, type and physiological status of the explants tissue, transformation vector and the *Agrobacterium* strain utilized are major variables in any attempted transformation system (Badr et al. 1998). Potato tuber moth (PTM), *Phthorimaea operrculella,* is a caterpillar insect pest that attacks potato plants in field and storage causing great damage to foliage and tubers. Derivation of genetically modified and PTM-resistant potato using *Bacillus thuringiensis* (Bt) toxin genes is the most suitable way to control PTM and avoid the negative impacts of chemical insecticides. An Egyptian Bt isolate produces a potent Cry1Aa7 toxin that kills the larval stages of PTM more efficiently than other standard Bt toxins (Ibrahim et al. 2001). The transformed tubers were challenged by releasing PTM larvae (1st instar) and emerged adults were scored. The authors' results revealed that the transformed tubers resisted insect attack 30–40% better than their non-transformed counterparts. The benefits of Bt transgenic plants, however, are still the subject of much debate between supporters and opponents of genetic modification (IUPAC 2004; EPA 2004).

More serious than the PTM insect are the Potyviruses which belong to one of the largest plant virus groups. Potyvirus Y (PVY), the most important pathogen of cultivated potato, is responsible for substantial damage to potato production throughout the world and can reduce yield by up to 80%. Saker (2003) reported for the first time the development of transgenic potato plants harboring potato virus Y coat protein gene (CP-PVY), which confers recipient plant resistance against PVY without the use of antibiotic resistance gene as a selectable marker gene. This avoids theoretical environmental risks of horizontal gene flow of antibiotic resistance genes from transgenic plants to enteric bacteria. The obtained results indicated that it is possible to avoid the use of antibiotic and herbicide resistance genes as selectable markers and consequently avoid environmental risks. Moreover, this system may be useful for the transformation of crops known to be recalcitrant for *in vitro* regeneration, as the omitting of antibiotics from the regeneration medium enhances the percentage of shoot recovery. Further studies have to be undertaken to evaluate the efficacy of the transgenic potato clone for resistance to potato virus Y isolates.

1.6 Soil Resource Assessments and Management Strategies

Soil erosion is a major cause of soil degradation in arable landscapes, adversely impacting hydropedologic dynamics (such as infiltration, patterns of subsurface flow, and shallow aquifer recharge) as well as farmscape- to watershed- and regional-level processes including redistribution of nutrients, pesticides and emission of greenhouse gases (e.g., Pennock and Corre 2001; Lin 2003; MacLauchlan 2006). Potato is in the top tier of crops with the highest erosion risk; in some production settings, harvest erosion rates were of the same order of magnitude (almost 10 Mg ha^{-1} $year^{-1}$) as water and tillage erosion on sloping land (Auerswald et al. 2006; Ruysschaert et al. 2006). Tiessen et al. (2007a, b, c) demonstrated that the tillage erosivity of commercial potato production systems in Atlantic Canada (due to primary, secondary and tertiary tillage operations – especially the latter which included planting, hilling and harvesting) was greater than that for the other major cropping systems in Canada. Soil losses varying between 20 and 100 Mg^{-1} ha $year^{-1}$ have been reported from convex landscape positions with the significance of these observations linked to reductions in crop yield (up to 40%); rolling agricultural terrains have been estimated to range between 15% and 30% of arable landscapes (Tiessen et al. 2007d). Investigations that detail crop yields in topographic contexts and assess soil resource risks at finer-scales (preferably from the farmscape-assemblage to watershed or sub-regional levels) for key cropping systems such as potato using geographic information systems (GIS) based approaches are urgently needed for most production areas.

High-resolution GIS-based investigations have been conducted on rainfed commercial potato cropping systems in Prince Edward Island and Québec, Canada (DeHaan et al. 1999; Cambouris et al. 2006). These datasets show that potato production constraints related to soil degradation have developed over decades, and when combined with annual variability in a multitude of environmental factors, the apparent results are five to ten-fold differences in yield at the field- scale. Geospatial integration of extrinsic (temperature and rainfall) as well as intrinsic field-specific (e.g., soil heterogeneity, topography, surficial and subsurface drainage patterns) datasets in potato production settings could help growers resolve field- to farm-scape-level complexities in order to better manage soil and water resources as well as evaluate pest- and pathogen-related risks.

Increasing awareness of the adverse influences of soil degradation and the role of conventional agriculture in potentially accelerating erosion rates an average of 1–2 orders of magnitude greater than rates of soil production (e.g., Montgomery 2007) prompted GIS-based assessments of the soils sustaining potato production systems in Maine (using farmland and erodibility classifiers) in order to help producers, communities, and policy makers gauge future food systems security risks (DeFauw et al. 2011). In addition, DeFauw et al. (2011) examined crop sequences and detected rotational patterns based on 3 years of Cropland Data Layer (CDL 2008–2010) classified imagery released by the USDA, National Agricultural Statistics Service. The 3-year potato production footprint covered approximately 62,000 ha. Zonal assessments of agri-environmental indicators that combined farmland and erodibility classifiers showed that close to 85% of potato production soils in Maine (over 52,000 ha) were either "potentially highly erodible" (PHEL) or "highly erodible" (HEL), therefore, requiring the highest standards in soil conservation practices. These geospatial frameworks help resolve patterns and trends in production environments (at multiple scales) that may, in turn, facilitate the wider adoption of adaptive management strategies which enhance yield, increase whole-farm profitability, and foster sustainable land use (DeFauw et al. 2011). Future refinements to the Maine potato systems geodatabase will include derived layers based on topographic and climatic datasets that will, in turn, facilitate higher resolution modeling (i.e., local- to regional-scales at 30 m resolution) of erosion potential as well as crop yields using SPUDSIM, a new process-level model that has the ability to accurately predict plant growth and yield for different potato varieties (Fleisher et al. 2010).

1.7 Conclusion

The widely-cultivated potato, *S. tuberosum* L., along with six other cultivated species grown only in the Andes collectively constitute one of the world's principal food crops; it ranks fourth after rice, maize and wheat. The global agriculture sector is confronting significant challenges within the next four decades as predictions indicate that the global population will be between 8.0 and 10.4 billion people, with a median estimate of 9.1 billion. Recently released studies estimate that worldwide agricultural production will need to grow by 70% over an approximated 45-year interval (between 2005–2007 and 2050), and by 100% in developing countries.

The highly-adaptable, high-yielding, nutrient-rich potato species complex (as an integral part of diversified cropping systems) has a deep history of helping relieve food insecurities, and has provided millions of smallholder farmers with the means to improve household income.

Some of the greatest challenges to overcome in improving the sustainability of potato production systems, driven by varying economies of scale, are: heterogeneity in soil resources, availability of nutrient and pest control inputs, pest resistance issues, weather-related constraints, demographic changes, and shifts in the availability of arable lands. High resolution geospatial investigations will help detect patterns and trends in commercial production environments (at local to regional scales) that may, in turn, facilitate the wider implementation of adaptive management strategies which enhance yield, increase whole-farm profitability, and foster sustainable land use. In many places, GEC will result in increased temperatures that will, in turn, require manipulation of agronomic practices to improve crop performance; however, it is imperative to ensure that locally-adapted crop and livestock germplasm (agrobiodiversity reservoirs) are protected and not displaced by improved varieties and breeds that may be susceptible in the future. The road ahead is difficult requiring integrated, trans-disciplinary research agendas involving researchers from the natural and social sciences as food systems are inherently multi-scale and multi-level and the adaptive options developed (whether science- or policy-based) must begin to recognize cross-scale and cross-level synergies as well as antagonistic interactions. With the tireless efforts of potato researchers worldwide, we believe locally-to-regionally specified sustainable and environmentally responsible potato production systems will help meet the challenges for long-term and country-driven food security and poverty alleviation.

References

Abdrabbo MAA, Khalil AA, Hassanien MKK, Abou-Hadid AF (2010) Sensitivity of potato yield to climate change. J Appl Sci Res 6:751–755

Auerswald K, Gerl G, Kainz M (2006) Influence of cropping system on harvest erosion under potato. Soil Till Res 89:22–34

Badr A, Ghanem SA, Saker MM, El-Bahr MK, Moursy HA, Salah F (1998) PCR analysis confirms insertion and maintenance of Agrobacterium-transferred herbicide resistance gene in callus cultures of kidney bean. Egypt J Genet Cytol 27:193–203

Burlingame B, Mouillé B, Charrondiére UR (2009) Review: nutrients, bioactive non-nutrients and anti-nutrients in potatoes. J Food Compos Anal 22:494–502

Cambouris AN, Nolin MC, Zebarth BJ, Laverdière MR (2006) Soil management zones delineated by electrical conductivity to characterize spatial and temporal variations in potato yield and in soil properties. Am J Potato Res 83:381–395

Cash DW, Adger WN, Berkes F, Garden P, Lebel L, Olsson P, Prtichard L, Young O (2006) Scale and cross-scale dynamics: governance and information in a multilevel world. Ecol Soc 11(2):8. http://www.ecologyandsocietyorg/vol11/iss2/art8/

DeFauw SL, English PJ, Larkin RP, Halloran JM, Hoshide AK (2011) Potato production systems in Maine: geospatial assessments of agri-environmental indicators. ASA-CSSA-SSSA conference proceedings. http://a-c-s.confex.com/crops/2011am/webprogram/Paper64689.html

DeHaan KR, Vessey GT, Holmstrom DA, MacLeod JA, Sanderson JB, Carter MR (1999) Relating potato yield to the level of soil degradation using a bulk yield monitor and differential global positioning systems. Comput Electron Agric 23:133–143
de Haan S (2009) Potato diversity at height: multiple dimensions of farmer-driven in-situ conservation in the Andes. Dissertation, Wageningen Agricultural University
de Haan S, Núñez J, Bonierbale M, Ghislainet M (2010) Multilevel agrobiodiversity and conservation of Andean potatoes in central Peru species, morphological, genetic, and spatial diversity. Mountain Res Dev 30:222–231
EPA (2004) Biopesticides. URL: http://www.epa.gov/agriculture/tbio.html. Accessed 30 June 2004
Eriksen PJ, Ingram JSI, Liverman DM (2009) Food security and global environ mental change: emerging challenges. Environ Sci Policy 12:373–377
FAO (Food and Agriculture Organization of the United Nations) (2009a) International year of the potato 2008: new light on a hidden treasure. http://www.potato2008.org/en/events/book.html
FAO (Food and Agriculture Organization of the United Nations) (2009b) The state of food insecurity in the world 2009. FAO, Rome. http://www.fao.org/docrep/012/i0876e/i0876e.pdf
FAO (Food and Agriculture Organization of the United Nations) (2010) Strengthening potato value chains: technical and policy options for developing countries. http://www.fao.org/docrep/013/i1710e/i1710e00.pdf
FAO (Food and Agriculture Organization of the United Nations) (2011) The state of food insecurity in the world 2011. FAO, Rome. http://www.fao.org/docrep/014/i2330e/i2330e00.pdf
FAOSTAT (2011) World potato production quantity (tonnes), yields and harvested areas for 2009. http://faostat.fao.org/site/567. Accessed 16 Aug 2011
Fleisher DH, Timlin DJ, Yang Y, Reddy VR (2010) Simulation of potato gas exchange rates using SPUDSIM. Agric Forest Meteorol 150:432–442
Godfray HC, Beddington JR, Crute IR, Haddad L (2010) Food security: the challenge of feeding 9 billion people. Science 327:812–818
Haverkort AJ (1977) Drought tolerance of the potato. The International Potato Center, Lima
Haverkort AJ (1988) Climate and potato crop growth interactions in Africa's continental divide region. Acta Hort 214:137–148
Haverkort AJ (1990) Ecology of potato cropping systems in relation to latitude and altitude. Agric Systems 32:251–272
Haverkort AJ (2007) The canon of potato science: 46. Potato crop modeling. Potato Res 50:399–402
Haverkort AJ, van de Waart M, Bodlaender KBA (1990) The effect of early drought stress on tuber and stolon numbers of potato in controlled and field conditions. Potato Res 33:89–96
Hijmans RJ (2003) The effect of climate change on global potato production. Am J Potato Res 80:271–280
Hiller LK, Koller DC, Thornton RE (1985) Physiological disorders of potato tubers. In: Li PH (ed) Potato physiology. Academic Press, Inc, Orlando, pp 389–455
Huamán Z (2002) Tecnología disponible para reforzar la conservación "in-situ"de los cultivares de papa tradicionales de los Andes. Revista Electrónica de la Red Mundial de Científicos Peruanos 1:1–10
Ibrahim MA, Metry EA, Osman YA, Nasr El-Din TM, Madkour MA (2001) Genetically modified potato (Solanum tuberosum L.) resistant to potato tuber moth (*Phthorimaea operculella*). http://www.acgssr.org/BioTechnology/V5N1January2002/fullpaper/p01.pdf
IUPAC (2004) Impact of transgenic crops on the use of agrochemicals and the environment. URL: http://www.iupac.org/projects/2001/200 1-024-2-600.html
Jaggard KW, Qi A, Ober ES (2010) Possible changes to arable crop yields. Phil Trans R Soc B 365:2835–2851
Kumar A, Miller M, Whitty P, Lyon J, Davie P (1995) Agrobacterium mediated transformation of five wild Solanum species using in vitro micro tubers. Plant Cell Rep 14:324–328
Lin H (2003) Hydropedology: bridging disciplines, scales, and data. Vadose Zone J 2:1–11
Lutaladio N, Castaldi L (2009) Potato: the hidden treasure. J Food Compos Anal 22:491–493
MacKerron DKL, Haverkort AJ (2004) Decision support systems in potato production: bringing models to practice. Wageningen Academic Publishers, Wageningen

MacLauchlan K (2006) The nature and longevity of agricultural impacts on soil carbon and nutrients: a review. Ecosystems 9:1364–1382

Montgomery DR (2007) Soil erosion and agricultural sustainability. Proc Natl Acad Sci U S A 104:13268–13272

Osaki M, Matsumoto M, Shinano T, Tadano T (1996) A root-shoot interaction hypothesis for high productivity of root crops. Soil Sci Plant Nutri 42:289–301

Pandey SK, Singh SV, Sarkar D (2005) Potato (*Solanum tuberosum*) for sustaining food and nutrition security in developing world. Indian J Agric 75:3–18

Pearman GI (1988) Greenhouse gases: evidence for atmospheric changes and anthropogenic cases. In: Pearman GI (ed) Planning for climate change. CSIRO, Division of Atmospheric Research. E.J. Brill, New York

Pennock DJ, Corre MD (2001) Development and application of landform segmentation procedures. Soil Till Res 58:151–162

Phillips RL (2010) Mobilizing science to break yield barriers. Crop Sci 50:99–108. doi:10.2135/cropsci2009.09.0525

Ruysschaert G, Poesen J, Verstraeten G, Govers G (2006) Soil losses due to mechanized potato harvesting. Soil Till Res 86:52–72

Saker MM (2003) Production of biosafe transgenic potato plants with coat protein gene for potato virus Y. Arab J Biotechnol 6(1):125–138

Sherk E (2008) Egypt aims to increase potato production. http://www.africanagricultureblog.com/2008/10/egypt-aims-to-increase-potato.html. Accessed 23 Oct 2008

Singh JP, Lal SS (2009) Climate change and potato production in India. In: Panigrahy S, Ray SS, Parihar JS (eds) ISPRS Ahmedabad 2009 workshop: impact of climate change on agriculture. Space Application Centre (ISRO) Ahmedabad, India, 17–18 Dec 2009, Proceedings, vol XXXVIII, part 8/W3, pp 115–117

Spooner DM, McLean K, Ramsay G, Waugh R, Bryan GJ (2005) A single domestication for potato based on multilocus amplified fragment length polymorphism genotyping. Proc Natl Acad Sci U S A 102:14694–14699

Stamoulis K, Zezza A (2003) A conceptual framework for national agricultural, rural development, and food security strategies and policies. ESA working papers no. 03-17, FAO, Agricultural and Development Economics Division, Rome

Steffen W, Sanderson A, Tyson PD, Jager J, Matson PA, Moore B III, Oldfield F, Richardson K, Schellhuber HJ, Turner BL II, Wasson RJ (eds) (2003) Global change and the earth system: a planet under pressure. Springer, Berlin/New York

Tiessen KHD, Lobb DA, Mehuys GR, Rees HW (2007a) Tillage erosion within potato production in Atlantic Canada: I. Measurement of tillage translocation by implements used in seedbed preparation. Soil Till Res 95:308–319

Tiessen KHD, Lobb DA, Mehuys GR, Rees HW (2007b) Tillage erosion within potato production in Atlantic Canada: II. Erosivity of primary and secondary tillage operations. Soil Till Res 95:320–331

Tiessen KHD, Lobb DA, Mehuys GR, Rees HW (2007c) Tillage translocation and tillage erosivity by planting, hilling and harvesting operations common to potato production in Atlantic Canada. Soil Till Res 97:123–129

Tiessen KHD, Lobb DA, Mehuys GR (2007d) The canon of potato science: 30. Tillage erosion within potato production – soil tillage, earthing up and planting. Potato Res 50:327–330

Torres J (2001) Estrategia y plan de acción de la biodiversidad para el departamento de Huancavelica como base de su desarrollo sostenible. Comunidad Andina, Lima

Vavilov NI (1992) Origin and geography of cultivated plants. Cambridge University Press, Cambridge

Walker TS, Schmiediche PE, Hijamans RJ (1999) World trends and patterns in the potato crop: an economic and geographic survey. Potato Res 42:241–264

Zhang N, Si HJ, Wen G, Du HH, Liu BL, Wang D (2011) Enhanced drought and salinity tolerance in transgenic potato plants with a *BADH* gene from spinach. Plant Biotechnol Rep 5:71–77

Zimmerer KS (1991) The regional biogeography of native potato cultivars in high land Peru. J Biogeograph 18:165–178

Part II
Enhancing Potato System Sustainability in the Northeast USA

Chapter 2
Impacts of Crop Rotation and Irrigation on Soilborne Diseases and Soil Microbial Communities

Robert P. Larkin, C. Wayne Honeycutt, O. Modesto Olanya, John M. Halloran, and Zhongqi He

Abstract Crop rotation provides numerous benefits to crop production, and is essential to reduce the build-up of soilborne plant pathogens and diseases that can devastate potato crops grown in multiple consecutive years. Crop rotations can reduce soilborne diseases through a variety of mechanisms, including changes in soil microbial communities, but different types of rotation crops can have very different effects. Crop rotations may be implemented as full-season harvestable crops, cover crops, or as green manures, with each approach having different impacts on soilborne diseases. In recent research in Maine, full-season rotation crops, such as barley, ryegrass, canola, and rapeseed, in 2-year and 3-year rotations with potato substantially reduced (15–50% reduction) *Rhizoctonia* and other soilborne diseases. Addition of a fall cover crop of winter rye to existing rotations further reduced *Rhizoctonia* and common scab diseases by another 5–20%. Use of specific disease-suppressive rotation crops as green manures can provide even greater reductions in soilborne diseases. In an ongoing large-scale study examining the effects of several different cropping system strategies both with and without irrigation, a disease-suppressive approach (utilizing *Brassica* and sudangrass green manures,

R.P. Larkin (✉) • J.M. Halloran
USDA-ARS, New England Plant, Soil, and Water Laboratory, Orono, ME 04469, USA
e-mail: bob.larkin@ars.usda.gov; john.halloran@ars.usda.gov

C.W. Honeycutt
USDA, Natural Resources Conservation Service, Washington, DC 20250, USA
e-mail: Wayne.Honeycutt@wdc.usda.gov

O.M. Olanya
USDA-ARS, Eastern Regional Research Laboratory, Wyndmoor, PA 19038, USA
e-mail: modesto.olanya@ars.usda.gov

Z. He
USDA-ARS, Southern Regional Research Center, New Orleans, LA 70122, USA
e-mail: zhongqi.he@ars.usda.gov

Z. He et al. (eds.), *Sustainable Potato Production: Global Case Studies*,
DOI 10.1007/978-94-007-4104-1_2,

fall cover crops, and high crop diversity) reduced soilborne diseases better than any other cropping system (25–58% reduction), and both the disease-suppressive and a soil improving (with compost amendments) system substantially increased tuber yield (19–42%). Irrigation also increased yield (~28%) in most systems. Combining the disease-suppressive rotation with irrigation increased yield by 53% relative to non-irrigated continuous potato. Combining effective crop rotations with other compatible components of integrated pest management can provide more effective and sustainable disease management and crop productivity.

2.1 Introduction

Crop rotation has been an important component of potato (*Solanum tuberosum* L.) production since the earliest days of cultivation. Long rotations and fallow periods between potato crops were developed by the Incas in South America to increase soil fertility and reduce soilborne pests and diseases (Bridge 1996; Thurston 1990). Crop rotations, in general, provide numerous benefits to crop production, and serve multiple functions. They can help conserve, maintain, or replenish soil resources, including organic matter, nitrogen and other nutrient inputs, and physical and chemical properties (Ball et al. 2005; Karlen et al. 1992, 2006; Magdoff 2000; Magdoff and van Es 2000; Pankhurst et al. 1997). Crop rotations have been associated with increased soil fertility, increased soil tilth and aggregate stability, improved soil water management, and reduced erosion (Ball et al. 2005; Grandy et al. 2002). Probably most importantly, for potatoes as well as many other crops, rotations are essential to maintain crop productivity and reduce the build-up of soilborne plant pathogens and diseases, which can devastate crops grown in multiple consecutive years (Cook 1986, 2000; Krupinsky et al. 2002; Sumner 1982).

Numerous soilborne diseases are persistent, recurrent problems in potato production, resulting in reduced plant growth and vigor, lower tuber quality, and reduced yield. Soilborne potato diseases of most concern in the Northeast U.S. and other potato-growing regions include: Rhizoctonia canker and black scurf, caused by *Rhizoctonia solani*; common scab, caused by *Streptomyces scabiei*; powdery scab, caused by *Spongospora subterranea* f.sp. *subterranea*; white mold, caused by *Sclerotinia sclerotiorum*; silver scurf, caused by *Helminthosporium solani*; pink rot, caused by *Phytophthora erythroseptica*; and Verticillium wilt, caused by *Verticillium dahliae*. Most of these diseases are difficult to control, and there are few effective control measures readily available.

Potato production systems in the Northeast U.S. are characterized by short rotations, extensive tillage, minimal crop residue return, and minimal crop diversity. The overall productivity of these systems has remained constant for several decades, despite increasing inputs of pesticides, nutrients, and water (Halloran et al. 2005). Potato production faces numerous constraints to productivity and profitability, including the lack of profitable rotation crops, high potential for disease problems, high fertilizer and pesticide requirements, degrading soil quality, and variable rainfall during the growing season. The long-term sustainability of potato production will depend on balancing the physiological production requirements of the crop with overcoming these additional constraints. Improved cropping systems that address the most important constraints should improve production.

Current production practices in the Northeast and many other potato production areas are based on a 2-year rotation with a low maintenance grain crop (such as barley or oats). Although 2-year rotations have been shown to reduce soilborne disease levels compared to continuous potato (Honeycutt et al. 1996; Specht and Leach 1987), longer rotation lengths of 3 or 4 years between potato crops are known to be more effective in controlling soilborne diseases (Carter and Sanderson 2001; Hide and Read 1991; Hoekstra 1989; Peters et al. 2003, 2004; Scholte 1987). The use of crops with known disease-suppressive capabilities, such as *Brassica* spp. and Sudangrass crops, and fall cover crops may provide additional resources for reducing diseases through improved cropping systems, as shown by our previous research (Larkin and Honeycutt 2006; Larkin and Griffin 2007; Larkin et al. 2010, 2011a, b). Such disease-suppressive cropping systems may suppress disease through a combination of multiple potential mechanisms, including biofumigation, manipulation of soil microbial communities, and other mechanisms. Through conserving or replenishing soil resources such as organic matter, effective cropping systems can alter soil chemical, physical, and biological properties (Ball et al. 2005; Karlen et al. 1992, 2006; Magdoff 2000; Magdoff and van Es 2000; Pankhurst et al. 1997) that improve soil water management and reduce erosion (Ball et al. 2005; Grandy et al. 2002). Even larger or more rapid changes in soil fertility, structure, and microbial communities can be observed by adding green manures or other organic amendments, such as compost, which provides much greater biomass and organic matter than is achieved through crop rotation alone (Abdallahi and N'Dayegamiye 2000; Collins et al. 2006; Grandy et al. 2002; Little et al. 2004; MacRae and Mehuys 1985; Stark et al. 2007; Thorup-Kristensen et al. 2003).

In the subsequent sections, recent examples of the effects of rotation, cover, and green manure crops on soilborne pathogens and diseases will be presented, highlighted by results from our own research program to show the potential for improved management of soilborne diseases using different cropping systems.

2.2 Rotation Crop Effects on Soilborne Diseases

In previous research evaluating different rotation crops in 2- and 3-year rotations with potato over several years, Rhizoctonia diseases (stem canker and black scurf) and common scab were the soilborne diseases most commonly and consistently observed in these studies, and rotation crop significantly affected the incidence and severity of these diseases throughout the years of study (Larkin and Honeycutt 2006; Larkin et al. 2010). In the 2-year rotations, although disease levels of black scurf varied somewhat among rotations each year, canola and rapeseed rotations consistently

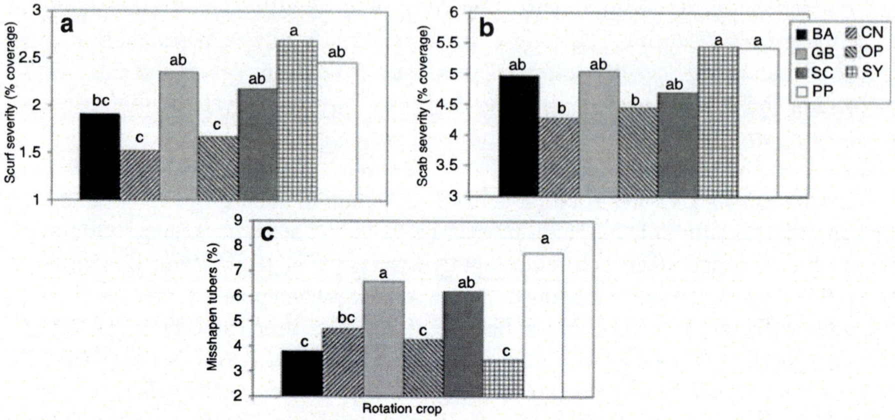

Fig. 2.1 Effects of different rotation crops on soilborne disease symptoms on potato tubers from 2-year crop rotations averaged over a 7-year study period (2000–2006) for (**a**) black scurf, (**b**) common scab, and (**c**) misshapen tubers. *BA* = barley, *CN* = canola, *OP* = millet/rapeseed, *GB* = green bean, *SC* = sweet corn, *SY* = soybean, *PP* = potato (control). Bars topped by the same letter are not significantly different according to Fisher's protected LSD test ($P < 0.05$)

reduced severity of black scurf over a 7-year evaluation period (2000–2006), averaging reductions of 25–32% relative to continuous potato (Fig. 2.1a). However, soybean and green bean rotations resulted in high scurf severity comparable to continuous potato. Barley/clover rotations initially were associated with lower levels of scurf through the first several years of the study (through 2002, or three full rotation cycles), but resulted in higher severity levels in subsequent years (2003–2005), and the 7-year average was not significantly lower than for continuous potato (but was lower than soybean rotations). Significant differences in common scab were also observed among rotations, with average disease severity significantly lower for canola and rapeseed rotations than continuous potato (25–31% reduction) over multiple years of study (Fig. 2.1b). However, incidence and severity of common scab increased substantially in all plots and all rotations from 2002 to 2005. Most rotations, including barley/clover, canola, rapeseed, and soybean reduced the percentage of severely misshapen tubers (55–70% reduction) relative to continuous potato (Fig. 2.1c). Misshapen tubers reduce the market value of the crop and can be an indication of soilborne disease problems. Each rotation crop was also associated with distinct changes in soil microbial community characteristics as determined by fatty acid methyl ester (FAME) and substrate utilization (SU; aka BIOLOG or CLPP) profile analyses (Larkin 2003; Larkin et al. 2010).

In addition, although not present during the early years of the study, Verticillium wilt developed to become a prominent problem in all plots, with incidence of wilt ranging from 60% to 100% in 2005 and 2006 (Larkin et al. 2010). The barley rotation resulted in the lowest levels of wilt among the rotations, but wilt symptoms were severe in all plots and rotations. Productivity (yield), in general, also declined over time in the 2-year rotations. However, most of these effects were not evident until the third or fourth rotation cycle. Thus, although some rotation crops significantly reduced soilborne disease relative to other 2-year rotations, no 2-year rotation was effective in preventing long-term increases in diseases such as Verticillium wilt and common scab (Larkin et al. 2010).

With the 3-year rotations, initial analyses following the first full rotation cycle indicated that potato crops following canola, barley, or sweet corn provided the lowest levels of Rhizoctonia disease and best tuber quality, whereas potato crops following clover or soybean resulted in disease problems (Larkin and Honeycutt 2006). Crops associated with increased disease levels, such as green bean and soybean, did not adversely affect the potato crop as long as the crop did not immediately precede potato (e.g. soybean-barley, green bean-sweet corn). With analyses of subsequent potato crops in 2003 and 2004 (second rotation cycle), effects due to cropping sequence (the specific sequence of rotation crops) became more evident, in addition to the effects of the crop preceding potato (Larkin et al., unpublished). For example, by the second cycle, rotations containing soybean and green bean (Sb-Ba, Sc-Sb, Gb-Sc) did not significantly reduce scurf severity, whereas all rotations that included canola (C-Sc, Sc-C, and Sb-C) significantly reduced scurf (35–50% reduction), even when canola did not precede potato (Fig. 2.2a) Thus, by the second full rotation cycle, it appeared that the entire cropping sequence and not just the crop preceding potato was influential in shaping the soilborne disease characteristics. In 2004, the Ba-Cl, C-Sc, and Sb-Ba also significantly reduced severity of common scab relative to continuous potato and some of the other rotations (Fig. 2.2b). As with the 2-year rotation studies, each rotation resulted in distinctly different effects on soil microbial community characteristics (as determined by SU and FAME profile analyses) and were correlated to some degree with potato disease and yield measurements (Larkin and Honeycutt 2006). In these studies, a full year of clover preceding potato was associated with high levels of Rhizoctonia disease. However, in other studies, the 3-year barley-clover rotation emerged as a desirable rotation with low levels of disease, particularly Rhizoctonia diseases (Peters et al. 2003, 2004).

2.3 Cover Crop Effects on Soilborne Diseases

A cover crop is defined as a crop grown primarily to cover the soil in order to protect it from soil erosion and nutrient losses between periods of crop production (Sarrantonio and Gallandt 2003). Benefits and uses of cover crops, including reduction of water runoff and soil erosion, addition of organic matter, improved soil structure and tilth, addition and recycling of nitrogen, greater soil productivity, and weed, pest, and disease control, have been documented and summarized in numerous reviews, general references, and practical guides (Fageria et al. 2005; Hartwig and Ammon 2002; Magdoff and van Es 2000; Sustainable Agriculture Network 1998; Sarrantonio and Gallandt 2003; Snapp et al. 2005; Thorup-Kristensen et al. 2003). There are numerous references to reductions of plant diseases with cover crops;

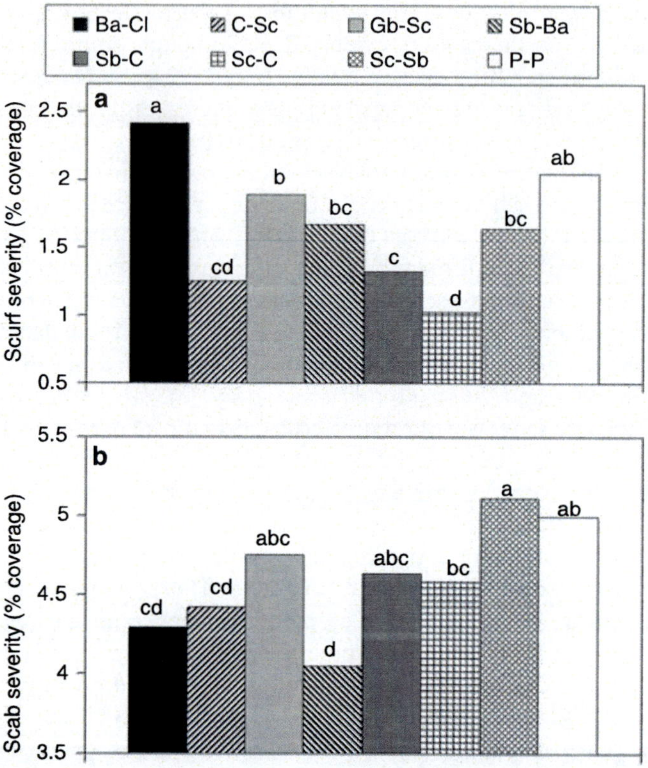

Fig. 2.2 Effects of different 3-year rotations on disease severity of potato tubers (% surface coverage) for (**a**) black scurf, and (**b**) common scab diseases after two rotation cycles (6 years). Crop rotations (crop 1 and crop 2) were: *Ba-Cl* = barley/clover-clover, *C-Sc* = canola-sweet corn, *Gb-Sc* = green bean-sweet corn, *Sb-Ba* = soybean-barley/clover, *Sb-C* = soybean-canola, *Sc-C* = sweet corn-canola, *Sc-Sb* = sweet corn-soybean, and *P-P* = consecutive potato (control), with potato the third year crop in all rotations. Bars topped by the same letter are not significantly different according to Fisher's protected LSD test ($P < 0.05$)

however, very few studies dealing strictly with cover crops (and not specifically green manures) have been related to potato pathogens or diseases.

In our field studies with different 2-year rotations (previous section), each rotation crop was also evaluated with and without the addition of a fall cover crop of winter rye, planted soon after harvest of the rotation and/or potato crop (Larkin et al. 2010). In the case of the barley/clover rotation, since this rotation already included a cover crop, a different cover crop (ryegrass) was substituted for the cover crop comparison. These cover/no cover comparisons were conducted from 2003 to 2006 within the long-term rotation plots established in 1997 and 1998. In these studies, modest, but significant benefits were attributed to the addition of the winter rye cover crop across all rotations. Although effects varied somewhat depending on the rotation used, overall, addition of the cover crop reduced severity of black scurf and common scab tuber diseases by 12.5% and 7.6%, and increased tuber yield by 3.6%, when averaged over all rotations for three potato cropping seasons (Larkin et al. 2010). Black scurf severity was significantly reduced in all rotations except rapeseed (where disease was already low), with reductions due to cover crop ranging from 10% to 19%. Common scab was significantly reduced (by 8–21%) with cover crop addition in the barley, canola, rapeseed, and potato rotations. Changing the cover crop underseeded with barley from red clover to ryegrass also resulted in significant reductions in black scurf and common scab (18.6% and 19.3%, respectively), supporting previous studies indicating that red clover can cause scurf and scab problems on potato crops (Brewer 2003; Larkin and Honeycutt 2006). Thus, ryegrass may be a preferable cover crop for underseeding with barley.

Although reductions in soilborne disease associated with the use of cover crops were generally modest (5–20%), the combined effects of using a cover crop in conjunction with an effective rotation crop were quite substantial and greater than using either approach separately. For example, combining the winter rye cover crop with a canola or rapeseed rotation reduced black scurf and common scab by 25–41% relative to a continuous potato rotation, and 21–37% relative to the barley/clover standard rotation (Larkin et al. 2010).

2.4 Green Manure Crop Effects on Soilborne Diseases

Green manuring refers specifically to the incorporation of fresh plant material for the purpose of soil enrichment (Pieters 1927). Thus, green manure crops are grown solely to be incorporated into soil as organic matter while still fresh and green. Green manuring generally results in larger organic matter inputs than traditional crop rotations or cover crops, producing improvements in soil fertility and structure (Abdallahi and N'Dayegamiye 2000; Grandy et al. 2002; Little et al. 2004; MacRae and Mehuys 1985; Thorup-Kristensen et al. 2003), as well as significant changes in soil microbial community characteristics (Collins et al. 2006; Stark et al. 2007). Green manures result in increased microbial biomass and activity, but also change microbial communities in ways that are distinctly different from other types of organic matter amendments, such as manure or sawdust (Elfstrand et al. 2007).

Green manures of different kinds have long been associated with reductions in potato diseases, particularly common scab and Verticillium wilt, although results have been variable. Early work indicated that green manures of rye and cowpea could reduce common scab (Weinhold et al. 1964). In long-term field trials in California, soybean grown as a green manure prevented the build-up of common scab over a period of 13 years, maintaining minimal disease levels compared to a rapid increase in common scab with pea or barley green manures and in the potato control (Weinhold et al. 1964). However, the soybean green manure was not effective in reducing common scab in a field with an already established high population of pathogen and high disease pressure.

Two years of green manure treatments using sudangrass, winter pea, rapeseed, rye, oats, or corn resulted in significant reductions in the incidence of Verticillium wilt (30–80% reduction) in the subsequent potato crop, with sudangrass providing substantially greater reductions in disease and significantly greater yield than most of the other crops (Davis et al. 1996). The sudangrass, rapeseed, and oat green manure treatments also reduced wilt in a second successive potato crop the following year. In these trials, disease reduction was not necessarily associated with reduction of pathogen inoculum, but was associated with reductions in root infections, increased microbial activity, and specific changes in microbial populations (Davis et al. 1996). Multiple follow-up studies with sudangrass green manures in subsequent years also showed reduced Verticillium wilt disease and improved tuber quality and yield, and that the benefits of sudangrass treatments extended beyond disease suppression and were correlated with increases in *Fusarium* spp. populations, microbial activity, and soil fertility factors (Davis et al. 2004).

Single season applications of green manures of sudangrass, winter pea, and broccoli at different biomass rates reduced inoculum density of *V. dahliae* and severity of wilt, but resulted only in marginal (nonsignificant) potato yield increases (Ochiai et al. 2007). Single-season green manure treatments of buckwheat or canola have resulted in significantly less Verticillium wilt and marginally less common scab as well as increased potato yield relative to fallow control plots (Wiggins and Kinkel 2005a). In these experiments, green manure treatments were also associated with an increase in the density and pathogen-inhibitory activity of indigenous Streptomycetes toward multiple soilborne potato pathogens (*S. scabies, V. dahliae, F. oxysporum,* and *R. solani*) (Wiggins and Kinkel 2005a, b).

Green manures of certain crops are also associated with biofumigation, which refers to the breakdown of plant metabolites in soil to produce volatile compounds that can reduce populations of weeds, nematodes, and plant pathogens (Matthiessen and Kirkegaard 2006; Sarwar et al. 1998). Crops in the *Brassicaceae* family produce glucosinolates that break down to produce isothiocyanates, whereas Sudangrass produces cyanogenic glucosides that break down to produce hydrogen cyanide, that are toxic to many soil organisms. Use of these plants as green manure crops has been shown to reduce pathogens or diseases (Brown and Morra 1997; Kirkegaard et al. 1996; Muelchen et al. 1990; Olivier et al. 1999; Smolinska and Horbowicz 1999), nematodes (Buskov et al. 2002; Mohjtahedi et al. 1993), and weeds (Boydston and Hang 1995; Brown and Morra 1995), and to improve soil characteristics and crop yield (McGuire 2003). Although biofumigation is the presumed mechanism of action for these crops, further research has indicated that additional mechanisms, including specific changes in soil microbial communities not related to levels of glucosinolate or other toxic metabolites, are also important in the reduction of soilborne diseases by *Brassica* crops, particularly for the control of *Rhizoctonia* (Cohen et al. 2005; Davis et al. 1996, 2004; Larkin and Griffin 2007; Mazzola et al. 2001).

Because of the potential for biofumigation effects, interest and use of Brassica crops as rotation and green manure crops has been increasing in recent years, in various fruit, vegetable, and field crops (Mazzola and Mullinix 2005; Subbarao et al. 1999; Zasada et al. 2003), as well as in potato cropping systems (Gies 2004;

McGuire 2003). In commercial applications of Brassica green manures in the Pacific Northwest, white mustard (*Sinapis alba*) and oriental mustard (*Brassica juncea*) green manures resulted in comparable tuber yield, quality, and disease control as fumigation with Metam sodium, improved certain soil properties, such as infiltration rate, and also provided an economic benefit (McGuire 2003). In our own research with Brassica green manures, a variety of Brassica crops (canola, rapeseed, Indian mustard, yellow mustard, turnip, and oilseed radish) and barley green manure treatments reduced soil inoculum levels of R. solani (20–56% reduction), and Indian mustard, rapeseed, and radish also reduced subsequent seedling diseases in greenhouse tests by 40–83% (Larkin and Griffin 2007). In on-farm field trials, Indian mustard, rapeseed, canola, and ryegrass grown as green manure rotation crops reduced powdery scab in the subsequent potato crop by 15–40%, and canola and rapeseed reduced black scurf by 70–80% relative to a standard oats rotation. At another field site, an Indian mustard green manure reduced common scab by 25%, and rapeseed, yellow mustard, and ryegrass also reduced black scurf relative to a standard rotation (Larkin and Griffin 2007). Disease reductions were not always associated with higher glucosinolate-producing crops and were also observed with non-Brassica green manures (barley and ryegrass), indicating other mechanisms and interactions were important, particularly for control of R. solani. Overall, Indian mustard (high-glucosinolate, *B. juncea*) was most effective for reducing powdery scab and common scab diseases, but rapeseed and canola (relatively low glucosinolate-producing, *B. napus*) were most effective in reducing *Rhizoctonia* diseases (Larkin and Griffin 2007). In recent trials in a field severely infested with Verticillium wilt, a single green manure crop of a mustard blend ('Caliente 119', a mixture of *Sinapis alba* and *Brassica juncea*) or sorghum-sudangrass hybrid resulted in a 20–25% reduction in Verticillium wilt in the subsequent potato crop compared to a standard barley rotation (Larkin et al. 2011a). The mustard blend also reduced other soilborne diseases and increased tuber yield relative to the barley control. However, by the second rotation cycle (second potato crop), wilt was high in all treatments, and was not effectively reduced by the green manures, indicating that a 2-year green manure-potato rotation was not sustainable for disease management (Larkin et al. 2011a). Our research with Brassica green manures is continuing with evaluations of the best rotation crops for disease control, enhancing the extent of disease reduction, determining the best methods of application and implementation, and the economic consequences of these rotations.

2.5 Cropping System Approaches and Irrigation Effects on Soilborne Diseases: Case Study

Beginning in 2004, large-scale field trials were established in Maine as part of a project designed to identify and establish the relative contribution of factors that limit cropping system sustainability in the Northeast. In this ongoing study, the effects of different cropping systems designed with specific management goals were evaluated on all aspects of potato crop production, as an attempt to better understand how these various factors and interactions may be involved in constraining productivity. For this study, 3-year cropping systems designed to specifically address management issues of soil conservation, soil improvement, and disease suppression were established and their subsequent effects on crop production were determined and compared with standards representing a typical potato cropping system and a non-rotation control under both irrigated and non-irrigated conditions. In this component of the overall research project, these cropping systems were evaluated for their effects and interactions on soilborne diseases and soil microbial community characteristics. Data were collected in the potato cropping year of each system following the complete rotation cycle. To date, results have been compiled for 5 full years following the first complete rotation cycle (2006–2010). Results regarding soilborne diseases and soil microbiology for the first 3 data years has been published thus far (Larkin et al. 2011b). Some cumulative results for the first five seasons are presented here. Subsequent reports will provide results from our interdisciplinary evaluation of these cropping systems on soil physical and chemical properties, plant nutrition, plant growth, tuber yield, economic viability, and other properties.

2.5.1 Cropping Systems

In this study, the cropping systems consisted of five different systems designed to address specific management goals of soil conservation, soil improvement, and disease suppression, as well as a system representing a typical standard rotation currently used in the Northeast US, and a non-rotation control of continuous potato (*Solanum tuberosum* L.). The standard or 'status quo' rotation (SQ) consisted of a 2-year rotation of barley (*Hordeum vulgare* L.) underseeded with red clover (*Trifolium pretense* L.) as a cover crop, followed by potato the following year. This system includes regular spring and fall tillage each year. The soil conserving system (SC) consisted of a 3-year rotation of barley underseeded with the forage grass timothy (*Phleum pratense* L), and then the timothy would overwinter and be allowed to continue undisturbed for a full year (2nd year), and then followed by potato in the third year. In this system tillage was also substantially reduced, with no tillage except immediately prior to planting potato and as needed for normal maintenance and harvest of the potato crop. In addition, straw mulch was applied after potato harvest to further conserve soil resources. The soil improving system (SI) consisted of the same basic rotation as SC (3-year, barley/timothy-timothy-potato, limited tillage, straw mulch), but with yearly additions of compost, to provide C and organic matter to improve soil quality. The disease-suppressive system (DS) was designed to make use of multiple strategies for suppressing soilborne diseases, including the use of disease-suppressive rotation crops, a longer rotation period, crop diversity, green manures, and fall cover crops. The DS system consisted of a 3-year rotation with the disease-suppressive *Brassica* 'Caliente 119' Mustard Blend

(blend of oriental and white mustard seeds, *Brassica juncea* L. and *Sinapis alba* L.) grown as a green manure, followed by a fall cover crop of rapeseed (*Brassica napus* L.'Dwarf Essex') in the first year. In the 2nd year, a disease-suppressive Sorghum-Sudangrass hybrid (*Sorghum bicolor* x *S. bicolor* var. *sudanense* L.) was grown as a green manure, followed by a fall cover crop of winter rye (*Secale cereale* L.), with potato in the 3rd year. Continuous potato (PP) was the non-rotation control consisting of a potato crop planted in the same plots each year. All cropping systems were evaluated under both irrigated and rainfed management. Full experimental details and methodologies are provided elsewhere (Larkin et al. 2011b).

2.5.2 *Effects on Soilborne Disease*

The primary potato diseases observed throughout the study were Rhizoctonia stem and stolon canker on the potato plants, and black scurf and common scab on the tubers. Rhizoctonia diseases generally occurred at low to moderate severity levels (0.3–1.5% surface coverage) and common scab occurred at relatively high severity levels (3.0–9.0% surface coverage). Overall, all cropping systems reduced Rhizoctonia stem and stolon canker relative to continuous potato (PP) with individual yearly reductions of 10–50% and overall average reductions (all years together) of 20–30%, whereas irrigation had no significant effect on stem and stolon canker (Larkin et al. 2011b).

Black scurf severity levels and cropping system effects varied somewhat from year to year, but average levels combined from multiple years were representative of the levels observed. Averaged over all 5 years, DS averaged significantly lower scurf severity than all other cropping systems under both irrigated and non-irrigated conditions (with reductions of 27–58%), and irrigated plots averaged 20% higher scurf severity than non-irrigated plots (Fig. 2.3a). Cropping system effects on common scab also varied somewhat from year to year, but overall consistent results were observed Overall, DS reduced scab severity better than all other cropping systems (25–40%), and irrigation tended to increase scab levels in each cropping system, averaging a 21% increase over all systems (Fig. 2.3b).

2.5.3 *Yield Effects*

In all years, under non-irrigated conditions, both total and marketable yield tended to be highest in the SI system, with yield increases of up to 40–60% higher than other cropping systems. DS also increased both total and marketable yield in all years relative to PP, with increases of 13–35%. With irrigation, tuber yields for all systems (except SI) increased by 20–40%, generally bringing them comparable to

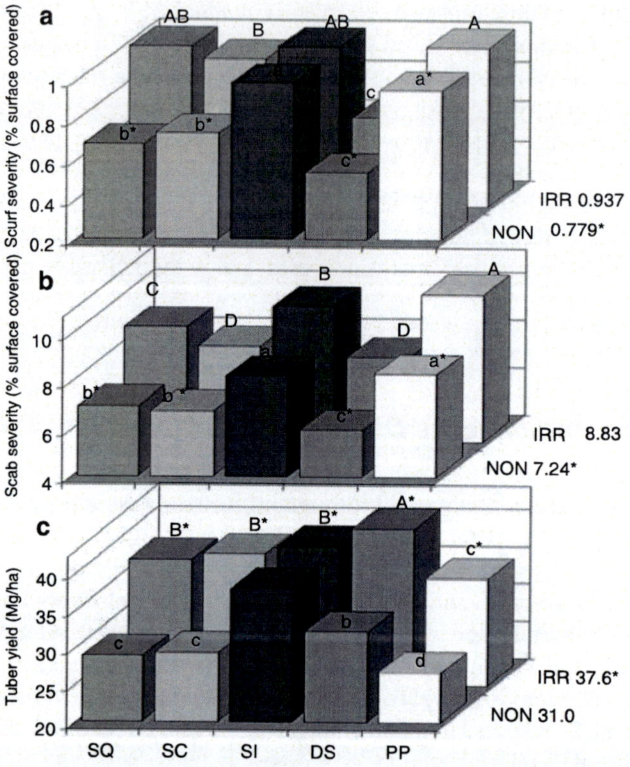

Fig. 2.3 Effect of different cropping system approaches with and without irrigation on (**a**) severity of black scurf and (**b**) common scab tuber diseases, and (**c**) total tuber yield (combined data from 5 years of data, 2006–2010). *SQ*=status quo system, standard 2-year rotation (barley/clover – potato); *SC*=soil-conserving system, 3-year, limited tillage, (barley/timothy – timothy – potato); *SI*=soil-improving system, 3-year, system same as *SC* but with yearly compost amendments; *DS*=disease-suppressive, 3-year (mustard blend green manure/rapeseed cover crop – sudangrass green manure/rye cover crop – potato); *PP*=continuous potato, nonrotation control: *IRR*=irrigated; *NON*=non-irrigated (rainfed). Bars within irrigation regime topped by the same letter are not significantly different from each other based on ANOVA and Fisher's protected LSD test ($P<0.05$) Bars topped by an asterisk represent values that are significantly different than their corresponding value for that cropping system in the other irrigation regime ($P<0.05$)

yield levels of SI without irrigation. Under irrigated conditions, DS produced the highest numerical yields, averaging slightly higher than SI in all years, and all cropping systems yielded better than PP (by 11–24%). Averaged over all 5 years, SI increased yield by 41% under rainfed conditions, and irrigation increased total yield by 18–29% in different cropping systems and 21% over all cropping systems (Fig. 2.3c). The combined effects of improved rotation and irrigation were best illustrated with the irrigated DS system, which resulted in average yields 53% higher than in non-irrigated PP, and 42% higher than the non-irrigated standard rotation (SQ) over the 5-year period.

2.5.4 Soil Microbial Communities

Soil microbial community data from soil samples collected from all plots in the spring of each year prior to planting potatoes indicated consistent and highly significant effects due to cropping system for most parameters in all years, whereas irrigation had little effect on soil microbial characteristics at this time of sampling. Combined data averaged over 3 crop years illustrated the overall effects (Larkin et al. 2011b).

Overall average populations of culturable bacteria and fungi were higher in SI soils than all other cropping systems, with bacterial populations in DS soils next highest, and lowest microbial populations observed in PP soils. General microbial activity, as estimated by average substrate utilization across numerous carbon sources (AWCD), indicated highest activity in the SI and SQ soils, and lowest activity in the PP soils. Substrate diversity and richness tended to be higher in DS soils (Larkin et al. 2011b). Analysis of the soil fatty acid data indicated that soil FAME profiles were distinctly different for each of the cropping systems, and that overall characteristics for each system were fairly consistent from year to year. Graphical depiction of the first two canonical variates (CV 1 and CV2) from canonical variates analysis from combined data for all 3 years illustrated that soil from each cropping system had microbial characteristics that were very distinct from each other system (Fig. 2.4). Irrigation also had some effects on soil microbial characteristics but these effects were minor in comparison to cropping system effects, as indicated by the greater separation among the cropping systems and closer proximity of the irrigated and not irrigated values for each cropping system. Interestingly, the SI and DS systems showed the greatest effects on soil microbiology (evident by the greater separation from the SQ and PP systems), but each showed very different effects (SI showing lower values for CV1 and DS lower values for CV2).

Analysis of FAME parameters and characteristics also indicated significant differences among the cropping systems regarding FAME structural classes and indicator biomarkers. Overall, DS soil had the highest proportion of the FAME biomarker for fungi, among cropping systems, as well as a high proportion of biomarkers for actinomycetes, and a high ratio of Gram-positive:Gram-negative FAMEs. SI soil had the highest proportion of the mycorrhizae biomarker among cropping systems, whereas DS and PP had the lowest level of the mycorrhizae biomarker. PP soil also had the lowest levels of the fungi marker and the fungi:bacteria ratio (Larkin et al. 2011b). All these differences indicate that each cropping system affected microbial communities in distinct and different ways.

2.5.5 Cropping System Approaches Summary and Conclusions

Each of the designated cropping systems accomplished their respective management goals. SI, through addition of organic matter as composted dairy manure, had the greatest effects on soil properties, and improved soil quality through numerous

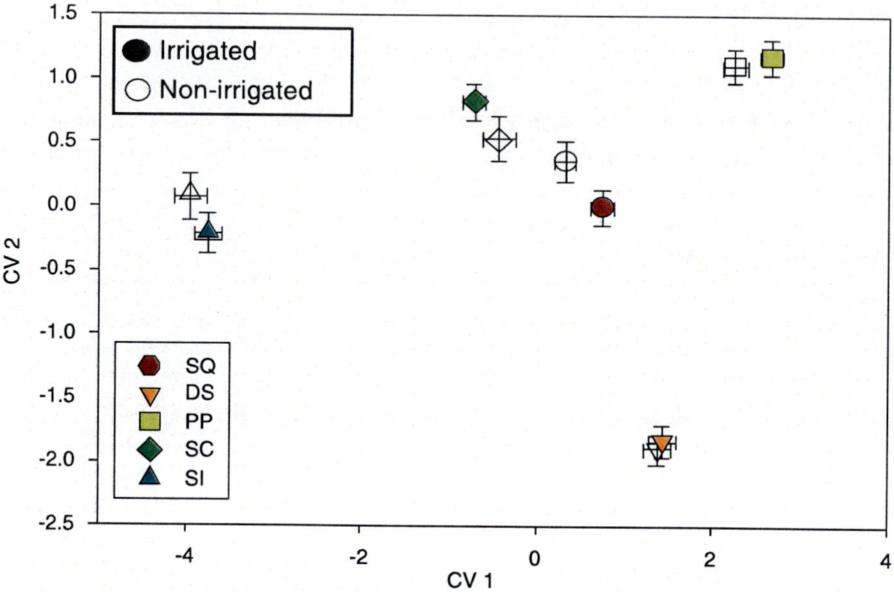

Fig. 2.4 Effect of different cropping system approaches with and without irrigation on soil microbial community characteristics, as represented by canonical variates (CV) 1 and 2 from CV analysis of soil fatty acid methyl-ester profiles (combined data from three field seasons, 2006–2008). *SQ*=status quo system, standard 2-year rotation (barley/clover – potato); *SC*=soil-conserving system, 3-year, limited tillage, (barley/timothy – timothy – potato); *SI*=soil-improving system, 3-year, system same as SC but with yearly compost amendments; *DS*=disease-suppressive, 3-year (mustard blend green manure/rapeseed cover crop – sudangrass green manure/rye cover crop – potato); *PP*=continuous potato, nonrotation control: filled shapes=irrigated; open shapes=non-irrigated

measured parameters, and also affected plant growth characteristics. SC improved soil conservation (relative to the standard, 2-year, SQ rotation) by limiting tillage, reducing erosion, and lengthening the rotation period. However, in this study, these changes in the SC system did not result in significant effects on soil properties, diseases, yield, or most soil microbial assessments relative to the SQ rotation. Several other studies have also observed that these types of changes (limited tillage, additional year of rotation) often do not result in significant changes within the first few years of implementation (Carter and Sanderson 2001; Carter et al. 2005; Griffin et al. 2009), but would be expected to show significant reductions in soilborne diseases over longer periods (Carter and Sanderson 2001; Peters et al. 2003, 2004). The DS system, consisting of *Brassica* and Sudangrass green manure crops, fall cover crops, and high crop diversity, resulted in the greatest reductions in stem and stolon canker, black scurf, and common scab relative to the other rotations, under both irrigated and non-irrigated conditions, thus accomplishing the disease-suppressive objective. DS also produced significant shifts in soil microbial community characteristics different from all other rotations. Thus, the strategies used for the disease suppression system, the use of disease-suppressive biofumigation rotation crops, cover crops, and crop diversity, successfully reduced disease relative to other rotations. Irrigation improved yield substantially in most cropping systems (all but SI) throughout, even in the wetter years. However, irrigation also tended to increase disease problems for both black scurf and common scab.

Overall, soil water, soil quality, and soilborne diseases were all important factors involved in constraining productivity, and systems addressing these constraints enhanced productivity. However, for future long-term productivity, it will likely be necessary to balance the need for low disease levels with sustainable cropping practices to optimize yields. Thus, a cropping system incorporating the DS system approaches (disease-suppressive rotations, cover crops, and crop diversity), with irrigation and some of the SC/SI system soil quality inputs (limited tillage, organic amendments), and a profitable third year crop, may provide potential for enhanced sustainable production and disease management.

2.6 Overall Conclusions

Crop rotations in all forms, whether as harvested crops, cover crops, or green manures, can have significant effects on soilborne diseases. And it is clear that the effects of crop rotations extend far beyond that of serving as a simple break in the host-pathogen cycle, with extensive influences on soil microbial community dynamics and characteristics, and potential for inhibiting, suppressing, and inactivating pathogen growth, survival, and disease development. All aspects of the rotations (rotation crop, crop genotype, rotation length, crop sequence, crop management, etc.) can significantly affect soil microorganisms and have potential effects on the development of soilborne diseases. Different uses of crop rotations (as full-season, cover, or green manure crops) also will result in different effects on soil microorganisms, and potentially different resultant effects on soilborne diseases. Overall, green manure crops may have the greatest potential for management of soilborne diseases, with their large additions of organic matter, specific effects on soil microbial communities, and potential for direct antagonism of pathogens through toxic breakdown compounds, but significant reductions in soilborne diseases can be observed with all types of rotations. In many cases, crop rotations are effective at reducing multiple diseases with a single cropping practice, although other cases exist where control of one disease results in increases in some other diseases. In general, disease control from crop rotation ranges from moderate to substantial, but will almost certainly never result in complete control of a pathogen or pathogens. Thus, crop rotations are best implemented as an important component of an integrated disease management program and not as the only control means for soilborne diseases (Lazarovits 2010). Combining crop rotations with other cultural, biological, or chemical approaches can substantially increase disease control and help achieve greater sustainability. The use of crop rotations as a crucial component in the active management of soil microbial communities to develop disease suppression and greater crop productivity is a viable, worthwhile, and achievable goal. However, much additional research is needed to determine the specific effects and interactions among different crop rotations and soil microorganisms, the roles and effects of the changes in soil microbial communities on soilborne diseases, and establishment and maintenance of stable disease suppression through management of soil physical, chemical, and biological attributes.

Acknowledgements We thank the many people who have worked on or contributed to this research over the years, from summer workers and student aides, filed crews, technicians, colleagues, and miscellaneous expertise. In particular, we thank L. Matthiesen for her thorough and capable technical assistance on all aspects of this work; D. Cowperthwaite and D. Torrey for managing and maintaining the field sites; P. Pinette, G. Trusty, E. Champaco, B. LeGasse, and E. Mallory for additional technical support; and all the others without whom these large-scale, long-term studies could never be completed. We also thank the Maine Potato Board and USDA Potato Research Program for providing additional funding for portions of this research.

References

Abdallahi MM, N'Dayegamiye A (2000) Effects of green manures on soil physical and biological properties and on wheat yields and N uptake. Can J Soil Sci 80:81–89

Ball BC, Bingham I, Rees RM, Watson CA (2005) The role of crop rotations in determining soil structure and crop growth conditions. Can J Plant Sci 85:557–577

Boydston RA, Hang HA (1995) Rapeseed (*Brassica napus)* green manure crop suppresses weeds in potato (*Solanum tuberosum*). Weed Technol 9:669–675

Brewer MT (2003) Effects of biological control and a ryegrass rotation on rhizoctonia disease of potato. Thesis (M.S.) in Plant, Soil and Environmental Sciences, University of Maine, Orono, ME

Bridge J (1996) Nematode management in sustainable agriculture. Annu Rev Phytopathol 34:201–225

Brown PD, Morra MJ (1995) Glucosinolate-containing plant tissues as bioherbicides. J Agric Food Chem 43:3070–3074

Brown PD, Morra MJ (1997) Control of soilborne plant pests using glucosinolate-containing plants. Adv Agron 61:167–231

Buskov S, Serra B, Rosa E, Sorense H, Sorensen JC (2002) Effects of intact glucosinolates and products produced from glucosinolates in myrosinase-catalyzed hydrolysis on the potato cyst nematode (*Globodera rostichiensis*). J Agric Food Chem 50:690–695

Carter MR, Sanderson JB (2001) Influence of conservation tillage and rotation length on potato productivity, tuber disease and soil quality parameters on a fine sandy loam in eastern Canada. Soil Till Res 63:1–13

Carter MR, Holmstrom D, Sanderson JB, Ivany JA, DeHaan R (2005) Comparison of conservation with conventional tillage for potato production in Atlantic Canada: crop productivity, soil physical properties and weed control. Can J Soil Sci 85:453–461

Cohen MF, Mazzola M, Yamasaki H (2005) *Brassica napus* seed meal soil amendment modifies microbial community structure, nitric oxide production and incidence of *Rhizoctonia* root rot. Soil Biol Biochem 37:1215–1227

Collins HP, Alva AK, Boydston RA, Cochran RL, Hamm PB, McGuire A, Riga E (2006) Soil microbial, fungal, and nematode responses to soil fumigation and cover crops under potato production. Biol Fertil Soils 42:247–257

Cook RJ (1986) Interrelationships of plant health and the sustainability of agriculture, with special reference to plant diseases. Am J Altern Agric 1:19–24

Cook RJ (2000) Advances in plant health management in the twentieth century. Annu Rev Phytopathol 38:95–116

Davis JR, Huisman OC, Westermann DT, Hafez SL, Everson DO, Sorensen LH, Schneider AT (1996) Effects of green manures on verticillium wilt of potato. Phytopathology 86:444–453

Davis JR, Huisman OC, Westermann DT, Everson DO, Schneider A, Sorensen LH (2004) Some unique benefits with sudangrass for improved U.S.#1 yields and size of Russet Burbank potato. Am J Potato Res 81:403–413

Elfstrand S, Hedlund K, Martensson A (2007) Soil enzyme activities, microbial community composition and function after 47 years of continuous green manuring. Appl Soil Ecol 35:610–621

Fageria NK, Baligar VC, Bailey BA (2005) Role of cover crops in improving soil and row crop productivity. Commun Soil Sci Plant Anal 36:2733–2757

Gies D (2004) Commercial use of mustards for green manure and biofumigation in the United States. Agroindustria 3:403–405

Grandy AS, Porter GA, Erich MS (2002) Organic amendment and rotation crop effects on the recovery of soil organic matter and aggregation in potato cropping systems. Soil Sci Soc Am J 66:1311–1319

Griffin TS, Larkin RP, Honeycutt CW (2009) Delayed tillage and cover crop effects in potato systems. Am J Potato Res 86:79–87

Halloran JM, Griffin TS, Honeycutt CW (2005) An economic analysis of potential rotation crops for Maine potato cropping systems. Am J Potato Res 82:155–162

Hartwig NL, Ammon HU (2002) Cover crops and living mulches. Weed Sci 50:688–699

Hide GA, Read PJ (1991) Effects of rotation length, fungicide treatment of seed tubers and nematicide on diseases and the quality of potato tubers. Ann Appl Biol 119:77–87

Hoekstra O (1989) Results of twenty-four years of crop rotation research at 'De Schreef' experimental site. In: Vos J et al (eds) Effects of crop rotation on potato production in the temperate zones. Kluwer Academic Publishers, Dordrecht, pp 37–43

Honeycutt CW, Clapham WM, Leach SS (1996) Crop rotation and N fertilization effects on growth, yield, and disease incidence in potato. Am Potato J 73:45–61

Karlen DL, Eash NS, Unger PW (1992) Soil and crop management effects on soil quality indicators. Am J Altern Agric 7:48–55

Karlen DL, Hurley EG, Andrews SS, Cambardella CA, Meek DW, Duffy MD, Mallarino AP (2006) Crop rotation effects on soil quality at three northern corn/soybean belt locations. Agron J 98:484–495

Kirkegaard JA, Wong PTW, Desmarchelier JM (1996) In vitro suppression of fungal root pathogens of cereals by Brassica tissues. Plant Pathol 45:593–603

Krupinsky JM, Bailey KL, McMullen MM, Gossen BD, Turkington TK (2002) Managing plant disease risk in diversified cropping systems. Agron J 94:198–209

Larkin RP (2003) Characterization of soil microbial communities under different potato cropping systems by microbial population dynamics, substrate utilization, and fatty acid profiles. Soil Biol Biochem 35:1451–1466

Larkin RP, Griffin TS (2007) Control of soilborne diseases of potato using *Brassica* green manures. Crop Prot 26:1067–1077

Larkin RP, Honeycutt CW (2006) Effects of different 3-year cropping systems on soil microbial communities and Rhizoctonia diseases of potato. Phytopathology 96:68–79

Larkin RP, Griffin TS, Honeycutt CW (2010) Rotation and cover crop effects on soilborne potato diseases, tuber yield, and soil microbial communities. Plant Dis 94:1491–1502

Larkin RP, Honeycutt CW, Olanya OM (2011a) Management of Verticillium wilt of potato with disease-suppressive green manures and as affected by previous cropping history. Plant Dis 95:568–576

Larkin RP, Honeycutt CW, Griffin TS, Olanya OM, Halloran JM, He Z (2011b) Effects of different potato cropping system approaches and water management on soilborne diseases and soil microbial communities. Phytopathology 101:58–67

Lazarovits G (2010) Managing soilborne diseases of potatoes using ecologically based approaches. Am J Potato Res 87:401–411

Little SA, Hocking PJ, Greene RSB (2004) A preliminary study of the role of cover crops in improving soil fertility and yield for potato production. Commun Soil Sci Plant Anal 35:471–494

MacRae RJ, Mehuys GR (1985) The effect of green manuring on the physical properties of temperate-area soils. Adv Soil Sci 3:71–94

Magdoff F (2000) Concepts, components, and strategies of soil health in agroecosystems. J Nematol 33:169–172

Magdoff F, van Es H (2000) Building soils for better crops, 2nd edn, Cover crops and crop rotations. Sustainable Agriculture Publications, University of Vermont, Burlington, pp 87–108

Matthiessen JN, Kirkegaard JA (2006) Biofumigation and enhanced biodegradation: opportunity and challenge in soilborne pest and disease management. Crit Rev Plant Sci 25:235–265

Mazzola M, Mullinix K (2005) Comparative field efficacy of management strategies containing *Brassica napus* seed meal or green manure for the control of apple replant disease. Plant Dis 89:1207–1213

Mazzola M, Granatstein DM, Elfving DC, Mullinix K (2001) Suppression of specific apple root pathogens by *Brassica napus* seed meal amendment regardless of glucosinolate content. Phytopathology 91:673–679

McGuire AN (2003) Mustard green manures replace fumigant and improve infiltration in potato cropping system. Crop Manage. (Online) doi:10.1094/CM-2003-0822-01-RS

Mohjtahedi H, Santo GS, Wilson JH (1993) Managing Meloidogyne chitwoodii on potato with rapeseed as green manure. Plant Dis 77:42–46

Muelchen AM, Rand RA, Parke JL (1990) Evaluation of crucifer green manures for controlling Aphanomyces root rot of peas. Plant Dis 74:651–654

Ochiai N, Crowe FJ, Dick RP, Powelson ML (2007) Effects of green manure type and amendment rate on Verticillium wilt severity and yield of russet burbank potato. Plant Dis 91:400–406

Olivier C, Vaughn SF, Mizubuti EG, Loria R (1999) Variation in allyl isothiocyanate production within *Brassica* species and correlation with fungicidal activity. J Chem Ecol 25:2687–2701

Pankhurst CE, Doube BM, Gupta VVSR (1997) Biological indicators of soil health. CAB International, Oxon

Peters RD, Sturz AV, Carter MR, Sanderson JB (2003) Developing disease-suppressive soils through crop rotation and tillage management practices. Soil Tillage Res 72:181–192

Peters RD, Sturz AV, Carter MR, Sanderson JB (2004) Influence of crop rotation and conservation tillage practices on the severity of soil-borne potato diseases in temperate humid agriculture. Can J Soil Sci 84:397–402

Pieters JA (1927) Green manuring, principles and practices. Wiley, New York

Sarrantonio M, Gallandt E (2003) The role of cover crops in North American cropping systems. J Crop Prod 8:53–74

Sarwar M, Kirkegaard JA, Wong PTW, Desmarchelier JM (1998) Biofumigation potential of Brassicas. III. In vitro toxicity of isothiocyanates to soil-borne fungal pathogens. Plant Soil 210:103–112

Scholte K (1987) The effect of crop rotation and granular nematicides on the incidence of *Rhizoctonia solani* in potato. Potato Res 30:187–199

Smolinska U, Horbowicz M (1999) Fungicidal activity of volatiles from selected cruciferous plants against resting propagules of soil-borne fungal pathogens. J Phytopathol 147:119–124

Snapp SS, Swinton SM, Labarta R, Mutch D, Black JR, Leep R, Nyiraneza J, O'Neil K (2005) Evaluating cover crops for benefits, costs, and performance within cropping system niches. Agronomy J 97:322–332

Specht LP, Leach SS (1987) Effects of crop rotation on Rhizoctonia disease of white potato. Plant Dis 71:433–437

Stark C, Condron LM, Stweart A, Di HJ, O'Callaghan M (2007) Influence of organic and mineral amendments on microbial soil properties and properties. Appl Soil Ecol 35:79–93

Subbarao KV, Hubbard JC, Koike ST (1999) Evaluation of broccoli residue incorporation into field soil for verticillium wilt control in cauliflower. Plant Dis 83:124–129

Sustainable Agriculture Network (1998) Managing cover crops profitably, 2nd edn. Sustainable Agriculture Network, Beltsville

Sumner DR (1982) Crop rotation and plant productivity. In: Rechcigl M (ed) CRC handbook of agricultural productivity. CRC Press, Boca Raton, pp 273–313

Thorup-Kristensen K, Magid J, Jensen LS (2003) Catch crops and green manures as biological tools in nitrogen management in temperate zones. Adv Agron 79:227–302

Thurston HD (1990) Plant disease management practices of traditional farmers. Plant Dis 74:96–101

Weinhold AR, Oswald JW, Bowman T, Bishop J, Wright D (1964) Influence of green manures and crop rotation on common scab of potato. Am Potato J 41:265–273

Wiggins BE, Kinkel LL (2005a) Green manures and crop sequences influence potato diseases and pathogen inhibitory activity of indigenous streptomycetes. Phytopathology 95:178–185

Wiggins BE, Kinkel LL (2005b) Green manures and crop sequences influence alfalfa root rot and pathogen inhibitory activity among soil-borne streptomycetes. Plant Soil 268:271–283

Zasada IA, Ferris H, Elmore CL, Roncoroni JA, MacDonald JD, Bolkan LR, Yakabe LE (2003) Field applications of brassicaceous amendments for control of soilborne pests and pathogens. Plant Health Prog. (Online) doi:10.1094/PHP-2003-1120-01-RS

Chapter 3
Early and Late Blight Potential on Russet Burbank Potato as Affected by Microclimate, Cropping Systems and Irrigation Management in Northeastern United States

O. Modesto Olanya, C. Wayne Honeycutt, Zhongqi He, Robert P. Larkin, John M. Halloran, and Jonathan M. Frantz

Abstract Soil and irrigation management have been used to optimize crop production; however, their effects on microclimate, development, and control of potato diseases have not been adequately quantified. The effects of soil, crop, and water management on development of potato early blight and late blight were quantified in a potato cropping systems experiment from 2006 to 2008. Microclimate, (soil temp, air temp, relative humidity, soil water content and leaf wetness) was not significantly impacted by cropping systems, and varied within seasons and across years. Irrigation management had little impact on microclimate, suggesting that treatment induced effects were not significant. Early blight incidence, severity, and lesion numbers were, however, impacted by management systems and years. Disease incidence was significantly ($P < 0.05$) greater in Continuous Potato (PP) than Disease Suppressive (DS), Soil Conserving (SC), Soil Improving (SI) and Status Quo (SQ) systems. Due to fungicide applications, no late blight was recorded in field plots, however; the potential for late blight development, based on theoretical

O.M. Olanya (✉)
USDA-ARS, ERRC, Wyndmoor, PA 19308, USA
e-mail: Modesto.Olanya@ars.usda.gov

C.W. Honeycutt
USDA, Natural Resources Conservation Service, Washington, DC 20250, USA
e-mail: Wayne.Honeycutt@wdc.usda.gov

Z. He
USDA-ARS, SRRC, New Orleans, LA 70122, USA
e-mail: zhongqi.he@ars.usda.gov

R.P. Larkin • J.M. Halloran
USDA-ARS, New England Plant, Soil, and Water Laboratory, Orono, ME 04469, USA
e-mail: bob.larkin@ars.usda.gov; John.halloran@ars.usda.gov

J.M. Frantz
USDA-ARS, Application Technology Research Unit, Toledo, OH 43606, USA
e-mail: jonathan.frantz@ars.usda.gov

Z. He et al. (eds.), *Sustainable Potato Production: Global Case Studies*,
DOI 10.1007/978-94-007-4104-1_3,

late blight indices such as hours of RH > 90%, predicted area under disease progress curves (AUDPC), severity values, and blight units had similar values among cropping systems and water management. Microclimatic variables were not significantly correlated to early blight or late blight potential, perhaps due to the small-scale size of experimental plots and influence of the surrounding environment, or lack of significant treatment effects. In addition to the positive attributes often associated with potato cropping systems (increased crop growth, yield, soil health, economic returns, production sustainability), this research demonstrated improved management of potato early blight with cropping systems, but no effect on late blight, a potentially explosive foliar disease. Nevertheless, cropping systems and irrigation management provide useful tools for the enhanced sustainability of potato production systems.

3.1 Introduction

Early blight caused by *Alternaria Solani* Sorauer and late blight incited by *Phytophthora infestans* (mont) de Bary are two of the most important potato diseases in North America and the world (Christ 1990; Erwin and Ribeiro 1995). The economic damage caused by early blight disease is much less severe compared to late blight, and is often associated with premature defoliation of potato plants, which often leads to tuber yield reduction. The dry rot of potato tubers has also been reported as a type of damage incited by early blight (Nnodu et al. 1982). Although yield losses attributed to early blight are known to vary depending on cultivars, the stage of potato crop maturity, management practices, location and cropping year or season, yield losses in the United States have been reported in the range of 20–30% (Christ and Maczuga 1989). Although increases in early blight disease have been reported, recent studies on yield loss in North-eastern United States have not been conducted.

Late blight is an economically important disease of potato in both temperate and tropical regions of the world. In North America, periodic epidemics have been documented, resulting in tremendous losses of potato crop (Guenthner et al. 2001). The primary damage is incited by pathogen infection of foliage and tubers, resulting in complete destruction of foliage and death of plants. Tuber rot in the field and tuber or seed decay in storage may occur as a result of co-infection by late blight and other fungal or bacterial pathogens.

The development of early and late blight diseases is affected by atmospheric and soil variables. Air temperature, relative humidity, wetness duration or moisture is required for development of early blight on potatoes (Adams and Stevenson 1990). Similarly, the presence of moisture, dew or high relative humidity on potato canopy is favorable for conidia germination of the early blight pathogen. Air temperature in the range of 18–28°C can increase infection of potato plants by *A. solani* as well as pathogen sporulation (Vloutoglou and Kalogerakis 2000). It is also worth noting that infection of potato plants or other hosts by *A. solani* can occur under dry and wet conditions (Rotem et al. 1978). Splashing rainfall and wind also play a critical role in dispersal of the pathogen conidia. Similar to the above, *P. infestans* has many of the same environmental requirements for its development. Air temperature, relative humidity, rainfall (soil moisture), and leaf wetness impact various developmental stages of the pathogen. Air temperature impacts the germination of sporangia (Mizubuti and Fry 1998), pathogen infection of potato and lesion development as well as sporulation and inoculum survival (Minogue and Fry 1981; Sato 1994). Relative humidity and moisture are also important environmental considerations for dispersal of *P. infestans* sporangia and its germination, since sporangia can easily be desiccated when moisture is limited.

Due to the fact that the epidemiology of early and late blight development are substantially influenced by edaphic factors mentioned above, it is crucial to assess how modification of atmospheric and soil conditions in potato field plots by irrigation management may impact development of those diseases. It is also well known that the potato crop is very sensitive to water stress and plant available water is required for optimum yield (Starr et al. 2008). It has been documented that irrigation or water requirements often differ among locations, soil types and crop management practices. It has also been shown that water management may also impact the potato crop and subsequently alter potato physiology, thereby making the crop more resistant to some diseases (Rotem and Palti 1969). For example, optimum water supply may increase potato vigor, growth and development, while excess water adversely impacts yield by enhancing development of early and late blight, perhaps due to favorable environmental conditions in the field and less vigorous plants. However, the quantitative data to ascertain the relationships of water management to microclimatic parameters and effects on disease are not well documented.

Cropping systems or crop management practices may affect microclimate and potato disease development in diverse agricultural situations (Olanya et al. 2007, 2009, 2010a, b; Starr et al. 2008). For example, it has been shown that potato foliar diseases may be selectively enhanced by crop and soil management techniques, while soil-borne diseases decreased by the same practices (Olanya et al. 2006). Various rotation schemes and patterns have also been shown to be effective for the management of certain soil-borne potato diseases (Larkin et al. 2011; Olanya et al. 2010a). Cropping systems and alternative management options can potentially influence early blight and late blight diseases. Cropping systems approaches include soil amendments, crop rotations, soil fertility, and irrigation applications. In this research, five different potato crop management systems designed to address specific management goals of soil conservation, soil improvement and disease suppression, as well as a system representing a typical standard rotation for the northeastern US and a non-rotation control, were established as previously described in Chap. 2. Thus, the systems established included: Status Quo (SQ), Soil Conserving (SC), Soil Improving (SI), Disease Suppressive (DS), and Continuous Potato (PP), under both rainfed and irrigated management, and these were evaluated for their potential impacts on early and late blight diseases. In this chapter, the effects of cropping systems and irrigation management on microclimate, early and late blight potential will be summarized. Additional details have been previously published (Olanya et al. 2009). For additional information regarding the experimental set-up, treatments, and design, please refer to Chap. 2 or Larkin et al. (2011).

3.2 Microclimate in Potato Fields

3.2.1 Microclimate

Sensors were deployed within potato canopies from May to September and air temperature, soil temperature, relative humidity, and leaf wetness were recorded. Soil water mark sensors were deployed at 15 cm below the soil surface. The results of microclimate measurements indicated that average ambient temperature (°C), soil temperature (°C) and relative humidity (%) were similar among cropping systems and between years, implying that crop management had little impact on relative humidity, air or soil temperatures in 2006 and 2007 (Table 3.1). This implies that potato systems management had little effect on the above microclimate parameters recorded in this experiment. Crop density has been shown to affect microclimate due to limitation on moisture evaporation in dense canopy (Rotem and Palti 1969); however, management effects on canopy temperature were not detected. Research results from other studies have also documented that the air temperature and relative humidity effects on potato canopy were influenced by irrigation types (Olanya et al. 2007) and the ambient temperature and humidity in the neighboring fields rather than treatment induced effects (Rotem and Palti 1969).

Average seasonal air temperatures did not vary significantly between irrigated and rainfed treatments, suggesting limited impact of water application on air temperature in our experiment. In previous research, it was noted that when ambient temperatures in adjacent fields were 10–20°C, sprinkling irrigation lowered temperatures by only 1°C. However, when ambient temperatures were 29–32°C, differences in temperature effects were 8–9°C (Rotem and Palti 1969). Slight differences

Table 3.1 Effects of cropping systems on atmospheric and soil variables quantified from a Russet Burbank potato canopy in field experiments at Presque Isle in 2006 and 2007

	2006					2007				
Variable	PP	SQ	SC	SI	DS	PP	SQ	SC	SI	DS
Soil Temperature (°C)	17a*	18a	18.a	17a	17a	18a	18a	18a	18a	18a
Soil Water (kPa)	65b	63b	71.a	74a	63a	56a	63a	58a	56a	67a
RH (%)	81a	81a	82a	79a	81a	75a	73a	76a	75a	75a
Air Temperature (°C)	17a	17a	17a	17a	17a	18a	18a	18a	18a	18a

Cropping system treatments were: Continuous potato (*PP*), Status Quo (*SQ*) consisting of potato-barley rotation, disease suppressive (*DS*), soil conserving (*SC*), and soil improving (*SI*). Data was averaged across rainfed and irrigation treatments in each year

RH relative humidity, *KPa* KiloPascal

*Values with the same letter indicate that there are no significant differences in environmental variables among cropping systems treatments in each year

in hours of RH > 90% were observed between irrigated and rainfed treatments (data not shown) suggesting that water application may result in limited changes in cumulative hours of RH in excess of 90% relative humidity in potato canopy. However, differences in cumulative RH > 90% hours were also observed between months and years, perhaps reflecting climatic variations between years. This is in contrast with findings of Rotem and Palti (1969) who noted that crop densities affect RH and other microclimate conditions. Waggoner (1965) observed that fluctuations in air temperature and relative humidities in crop canopies are not totally unexpected, and therefore discrepancy in micro-climatic variables can be encountered in the same or different set of experimental conditions depending on canopy density, wind velocity and duration, radiation, and many other factors. Although we did not record leaf temperature in this experiment, it is possible that differences between irrigated and rainfed and perhaps soil management treatments may occur.

Soil water content varied slightly among potato cropping systems and irrigation treatments when data were averaged across months within a year. This implies that crop management strategies may influence soil water content, perhaps through retention of soil moisture. Although treatments of SC (limited tillage plus forage grass rotation), SI (addition of compost amendments plus 3-year forage rotation) were expected to influence soil moisture retention in field plots through the addition of organic matter, these differences were not always consistent among system management, and showed minimum variation between irrigated and rainfed treatments in all cropping systems (Table 3.2). It is possible that the duration of time above a threshold dryness may explain some of the differences in management systems. Previous research showed that spatial and temporal variation in soil water content on a Russet Burbank potato field could be attributed to spatial variation in soil characteristics (Starr 2005). In our experiments, we noted variation in soil water content among months (data not shown). From June to September, soil water content values in rainfed treatments ranged from −34.3 to −107.6 Kilo Pascals (KPa) when data were averaged across cropping systems. Differences in soil water content / moisture attributed to irrigation on a Russet Burbank potato have been previously reported (Starr et al. 2008).

Soil water content, however, varied significantly prior to and after irrigation events. Based on the data in 2008 cropping season, soil water content increased substantially after irrigation events in all cropping systems, when data were averaged 24-h prior to irrigation event and compared to 24-h post irrigation event (Fig. 3.1). This indicates that irrigation treatment were effective in increasing soil water content. In our experiment, the maximum duration of irrigation event was 2 h, with amount of 2 cm of water per application event.

The accumulation of leaf wetness (hours) was numerically higher in irrigated than rainfed potatoes (Fig. 3.2). This indicates that water application can result in greater leaf wetness duration on potato canopy in irrigated than rainfed treatment. A diurnal pattern of leaf wetness was observed in both irrigated and rainfed treatments, suggesting that day and night cooling patterns influenced leaf wetness in addition to the water application effects. Sinusoidal pattern of leaf wetness duration have been previously recorded on Russet Burbank potato canopy irrespective of the irrigation treatments (Starr et al. 2008; Olanya et al. 2007). Similarly, it has been recorded that the majority of wetting events in a potato canopy may be attributed to

Table 3.2 Effects of cropping systems and irrigation management on soil water content collected from a Russet Burbank potato cropping system experiment at Presque Isle, ME in 2006 and 2007

	Irrigation	2006					2007				
Soil water	treatment	PP	SQ	SC	SI	DS	PP	SQ	SC	SI	DS
Mean KPa	Rn	61a*	75a	87a	99a	75a	64a	72a	72a	54a	75a
	Irri	69a	51a	54b	50b	52a	48a	53a	45a	57a	58a
Min KPa	Rn	5a	9a	24a	22a	19a	8a	8a	8a	11a	10a
	Irri	4a	8a	6a	7a	7a	8a	5a	4a	6a	5a
Max KPa	Rn	124a	140a	145a	160a	136a	138a	153a	153a	108a	159a
	Irri	69b	51b	54b	49b	52b	124a	125a	116a	129a	124a

Cropping system treatments were: Continuous potato (*PP*), Status Quo (*SQ*) consisting of potato-barley rotation, disease suppressive (*DS*), soil conserving (*SC*), and soil improving (*SI*)

Rn Rainfed, *Irri* Irrigated treatment, *KPa* KiloPascal, Min and Max refers to minimum and maximum values

*Values with the same letter indicate that there were no significant differences in soil water content among cropping systems treatments in each year

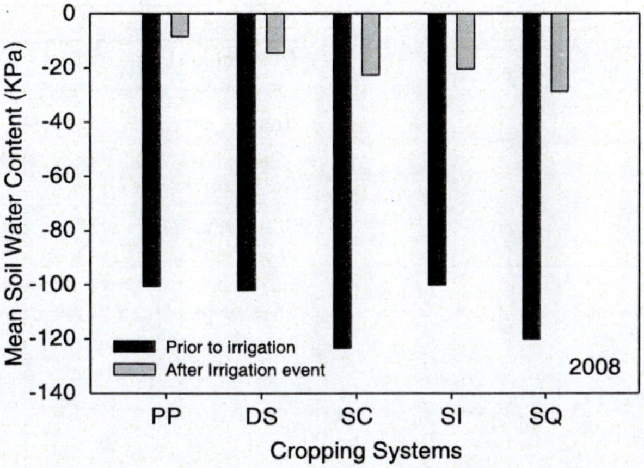

Fig. 3.1 Soil water content (*KPa* KiloPascal) recorded in a Rust Burbank potato cultivar recorded 24-h prior to and after irrigation event in 2008

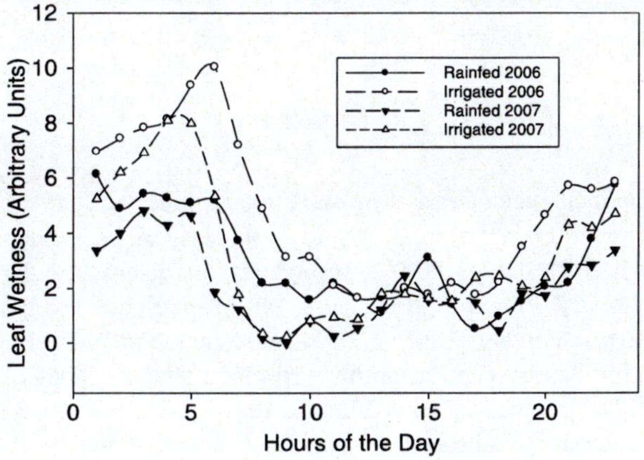

Fig. 3.2 Cumulative leaf wetness in irrigated and rainfed field plots in 2006 and 2007 cropping years, at Presque Isle, Maine

night cooling pattern of leaves below the dew temperature of the surrounding air (Wilson et al. 1999). Therefore, our findings are similar to those of other research results. The effects of cropping systems on leaf wetness duration were not analyzed since wetness sensors were not deployed in all cropping systems treatments. A similar pattern was also shown in cropping systems treatments in which similar diurnal effects of leaf wetness was recorded in all systems treatments (Fig. 3.2).

3.3 Early Blight Disease Levels

Experimental plots were assessed for early blight incidence, disease severity, and lesion numbers using a visual evaluation of disease levels on potato foliage. In determination of early blight incidence, the number of plants with symptoms

Table 3.3 Effects of cropping systems on early blight incidence

Cropping systems	2006	2007	2008
Continuous Potato (PP)	63.6a*	43.1a	42.6a
Disease Suppressive (DS)	54.5ab	18.3b	12.7b
Status Quo (SQ)	50.2b	11.6b	11.5bc
Soil Conserving (SC)	38.2bc	12.5b	16.6b
Soil Improving (SI)	30.9c	14.7b	7.8c

*Different letters within a column indicate significant differences in disease levels among cropping systems. Data refers to disease incidence averaged across irrigation treatments. Disease incidence (%) was computed as number of plants diseased/total plants assessed × 100

of early blight disease was expressed as percentage of total plants evaluated while early blight severity was expressed as percentage of foliage area with disease.

3.3.1 *Early Blight Incidence*

The incidence of early blight (%) was significantly ($P<0.05$) impacted by cropping systems (Olanya et al. 2009). In 2006, early blight incidence ranged from 31% (SI) to 64% (PP). During 2007 cropping season, disease incidence ranged from 12% (SQ) to 43% (PP) and in 2008, early blight incidence levels were 8% to 43% in SI and PP, respectively (Table 3.3). Disease incidence was significantly greater ($P<0.05$) in PP than in other cropping systems in all the 3 years of the experiment.

The progress of early blight disease varied across all crop management systems. No disease was detected during the first sampling period, but a progressively greater amount of disease was detected in subsequent sampling times in all treatments (Olanya et al. 2009). Variation in disease progress was detected among the 3 assessment years (2006 to 2008).

3.3.2 *Pathogen Lesions*

Similarly, the pathogen lesion numbers also varied among cropping systems and years during the 3 years of the experiment. The average number of *A. solani* lesion numbers ranged from 13.5 (SI) to 26.3 (PP) in 2007. Mean lesion numbers were in the range of 6.7 (SC) to 36.1 (PP), and 2.4 (SI) to 14.4 (PP) in 2007 and 2008, respectively (Fig. 3.3). These observations suggest that cropping systems significantly ($P<0.05$) impacted the number of pathogen lesions on potato foliage, since the greatest amount of lesions were consistently detected in PP compared to DS, SC, SI, and SQ rotations.

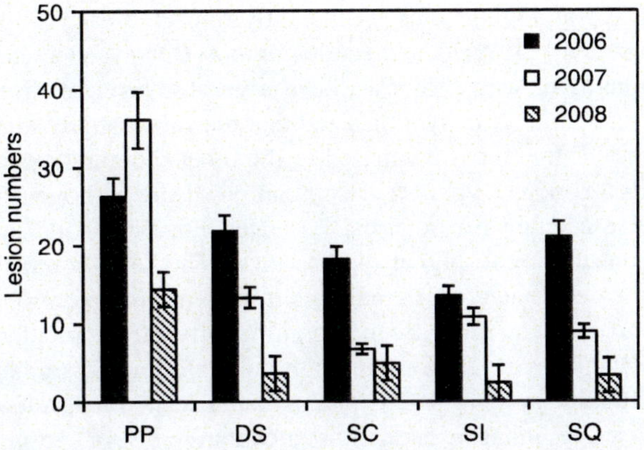

Fig. 3.3 Mean *Alternaria solani* lesion numbers on a Russet Burbank potato foliage in a potato cropping systems experiment, Presque Isle, Maine. *PP* continuous potato, *DS* disease suppressive, *SC* soil conserving, *SI* soil improving, *SQ* status quo

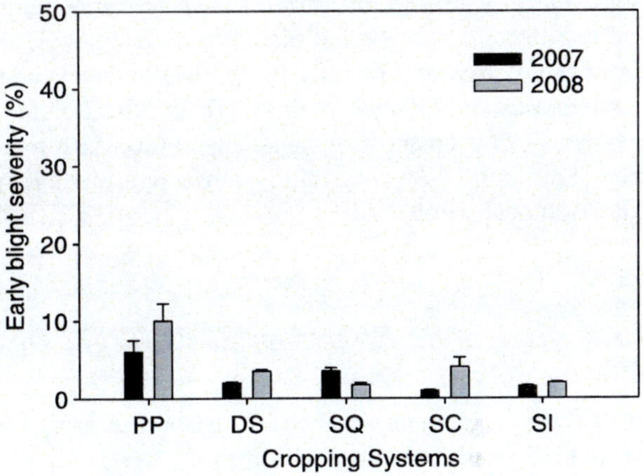

Fig. 3.4 Early blight severity in potato cropping systems experiment at Presque Isle in 2007 and 2008. Disease was assessed as percentage of leaf area with early blight on 10 plants/plot. *PP* continuous potato, *DS* disease suppressive, *SQ* status quo, *SC* soil conserving, and *SI* soil Improving

3.3.3 Early Blight Severity

Unlike early blight incidence and pathogen lesions, very low levels of early blight severity were detected in the cropping systems experiments. In 2007 and 2008, the highest level of disease severity was detected in PP treatment in both years; however, numerically low levels of early blight severity were recorded in SC and SI treatments (Fig. 3.4). Differences in early blight severity among cropping systems were statistically significant in 2007 and 2008, with PP treatment resulting in significantly greater early blight severity than other treatments (Fig. 3.4). No data on early blight severity was recorded in 2006. The low severity may also be explained by the application of fungicides such as chlorothalonil and mancozeb in this experiment.

Cropping systems significantly impacted early blight disease based on disease incidence, severity and lesion numbers (Olanya et al. 2009). Continuous potato (PP) treatment was shown to have a greater level of disease compared to the other treatments. It is hypothesized that the substantially higher amount of host tissue in the PP treatment compared to the other cropping systems may have been a major factor in survival and dissemination of the pathogen from year to year. This may also account for potentially greater inoculum build-up in the PP system, thereby contributing to higher disease level. This is in contrast to previous studies, which showed that crop management systems (conventional, biological and reduced input treatments) did not significantly affect incidence of early blight disease (Olanya et al. 2006). Although there are limited studies that address the impact of cropping systems and irrigation management on development of early and late blight diseases, it has been demonstrated that *A. solani* can survive as a saprophyte during off-season period (Pscheidt 1985). The variation in early blight disease levels may be accounted for by the possible differences in inoculum levels that may be present from year to year. In these studies, *A. solani* inoculum density was not quantified and therefore, cannot be correlated to disease levels. Similarly, we did not detect significant correlation of early blight disease levels and environmental variables in previous studies (Olanya et al. 2009). In other related studies, it has been shown that the use of crop and soil management (non-host rotation crop, soil amendments, crop sanitation) may contribute to effective management of early blight disease (Harrison et al. 1965).

3.3.4 Irrigation Effects on Early Blight Disease

Irrigation management resulted in variation in early blight disease levels in some, but not all years. Within each cropping system, pathogen lesion numbers were significantly impacted by water management treatment in 2007. There was significantly greater number of pathogen lesions in PP than the rest of the cropping systems treatments irrespective of the irrigation treatment (Fig. 3.5). Pathogen lesion number was also comparatively greater in irrigated than non-irrigated treatment in 2007. Lesion development was comparatively lower in 2006 and in 2008 cropping years.

When the irrigation effects on disease development were averaged across cropping systems and compared between the two irrigation management regimes, mean lesion numbers differed significantly ($P < 0.05$) in 2007 and averaged 22 for irrigated and 9.6 for rainfed treatments. In 2006 average lesion numbers were not significantly different between irrigation treatments (19.6 and 20.5 for irrigated and rainfed treatments, respectively). Similarly, during 2008, lesion numbers per plant averaged 6.7 for irrigated and 4.6 for rainfed water management. With regard to disease incidence, irrigation management significantly ($P < 0.05$) impacted incidence levels in 2008 in which disease incidences were 21.9 and 14.2 for irrigated and rainfed treatments, respectively. The variation in irrigation effects on

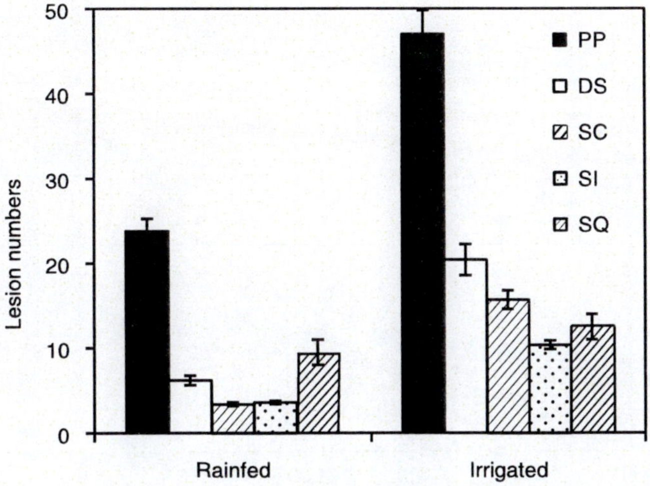

Fig. 3.5 Effects of cropping system and irrigation management on mean number of lesions of ***A. solani* on potato foliage in 2007**. *PP* continuous potato, *DS* disease suppressive, *SC* soil conserving, *SI* soil improving and *SQ* status quo

disease levels may be explained by the splash dispersal of conidia of *A. solani* in irrigated treatment allowing for increase in disease incidence as opposed to rainfed treatment.

In previous studies, it has been shown that supplemental irrigation in potato production systems can lead to increases in some foliar and soil-borne diseases (Olanya et al. 2006, 2009), while crop rotation and soil amendments with or without supplemental water application resulted in lower disease levels. This was attributed to the presence of additional moisture for pathogen infection and disease development and dispersal. The amount of wetness duration on foliage of crop plants is an important epidemiological factor since infection and pathogen development requires free moisture (Ingold 1978; Rotem et al. 1978; Lacey 1986). In this study, the effects of irrigation were varied implying that disease development were enhanced by certain irrigation treatments and not influenced in others. In general, the frequency, duration, amount and methods of water application can affect pathogen survival and disease development in potato production systems.

3.4 Late Blight Potential as Affected by Potato Cropping Systems and Irrigation Management

Late blight is a significant constraint to potato production. The disease has a foliar and tuber blight component and is capable of destroying potato foliage within 2 weeks under favorable environmental conditions (Fry 2007, 2008). The impact of cropping system on late blight potential has not been extensively investigated

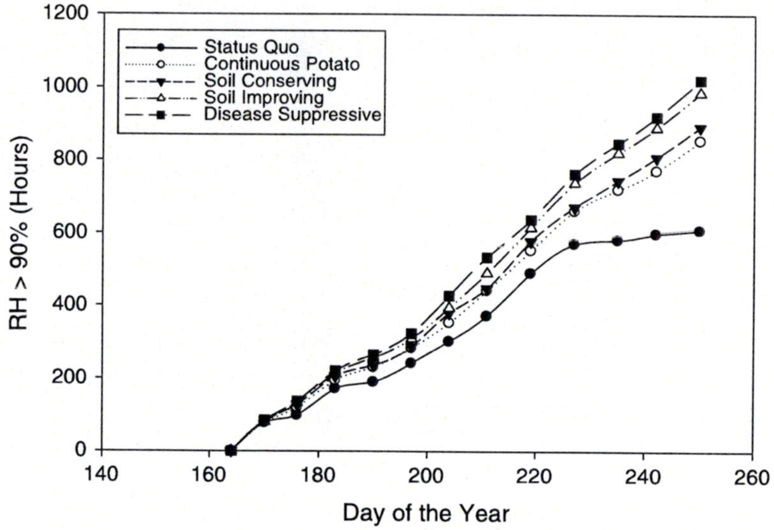

Fig. 3.6 Effects of cropping system on total cumulative hours of Relative Humidity (*RH*) % indicative of late blight potential

(Olanya et al. 2006). The hourly environmental outputs in the same potato canopies within the same crop management systems already described were utilized to compute severity indices (hours of RH>90%, blight units, severity values, risk values, simulated area under disease progress curves) which would indicate the potential for late blight development. Due to the devastating potential for late blight disease, *P. infestans* was effectively controlled by mancozeb and chlorothalonil fungicide applications in these experiments to enable evaluation of systems effects across a broad range of disciplines. Therefore, protectant and contact fungicides such as chlorothanonil (Syngenta Crop Protection, Greensboro, NC) and mancozeb (Dupont Agricultural Products, Wilmington, DE) were applied to potato foliage at recommended rates during crop development period, based on late blight forecasting (Krause et al. 1975) in 2006, 2007 and 2008.

3.4.1 Cumulative Hours of RH>90%

Potato late blight is heavily favored by environmental factors such as high humidity, wetness duration and low temperature (Melhus 1915; Harrison 1992). When temperatures are conducive (range of 7–22°C), relative humidity (RH) in excess of 90% favor late blight development (Krause et al. 1975). Therefore, assuming the presence of pathogen inoculum, the cumulative hours of RH in excess of 90% is strongly related to late blight development. We used cumulative hours of RH as an indicator of late blight potential, which was recorded with a data logger.

In general, over 90% RH values were recorded in all treatments. Relative humidity values in excess of 90% increased steadily during the cropping season from day of year (DOY) 164 to 250 (Fig. 3.6). Crop management had similar effects on hours of RH>90 accumulation indicating that favorable conditions for blight development occurred in all treatments. Total RH>90% ranged from 600 in SQ to 994 in DS treatments. After DOY 226, which was near crop maturity, cumulative severity values increased less rapidly in SQ compared to SI, SC, PP and DS treatments. Although crop management practices such as addition of compost and manure (SI) or rotation of potato with barley / timothy (SQ) may enhance potato growth and canopy architecture, these effects did not impact relative humidity across treatments. In contrast to this finding, other research has documented that crop densities can influence microclimatic conditions in the canopy of some crops (Rotem and Palti 1969). However, the researchers documented that the RH of the surrounding environment had greater influence in determining the RH of field plots. The cumulative hours of RH>90% were similar among cropping system and irrigation treatments in 2006, 2007 and 2008, suggesting that late blight potential was not significantly impacted by management.

3.4.2 Simulated Area Under Disease Progress Curves (AUDPC)

Due to the effective control of late blight in cropping systems, no disease was detected in these experiments. In the absence of late blight occurrence, simulation of foliar blight development provides an important tool to assess potential effects of cropping systems and irrigation management on late blight disease. The epidemiology of late blight has been extensively investigated and it has been shown that environmental factors such as temperature, relative humidity and rainfall (moisture) are crucial for disease development (Bruhn and Fry 1981; Raposo et al. 1993). Therefore, we used a late blight simulation model previously described (Andrade-Piedra et al. 2005) in computing predicted disease values. Crop emergence, cultivar reaction category (susceptible, moderately resistant and resistant), pathogen factors such as latent period (LP), infection efficiency (IE) and lesion growth rate (LGR) parameters were also specified for the model. Microclimatic data (air temperature, RH, rainfall) from sensors deployed in each plot were averaged for each hour of the day over the 2 growing years. Assuming the presence of inoculum, the hourly data was used as input variables for the simulation runs. In the simulation runs (Olanya et al. 2007), predicted disease severity and area under disease progress curves (AUDPC) were generated as output variables for the simulation. The simulation runs were computed by Macro SAS program (SAS Institute Inc., Cary, NC).

Based on simulations, cropping systems did not significantly impact potential disease levels (predicted AUDPC – % disease day) in 2006 and 2007 experiments. During the 2006 experiment, simulated AUDPC ranged from 2,420 (SI) to 2,551 (DS), but were non-significant, implying that cropping systems treatment had no effect on potential for blight development (Table 3.4). In 2007, predicted AUDPC values were numerically less than in 2006, and values were in the range of 1,773

Table 3.4 Effect of cropping systems on simulated area under disease progress curves (AUDPC) on a Russet Burbank potato cultivar, computed from environmental variables for treatments at Presque Isle, ME during 2006 and 2007 cropping years

	AUDPC (% disease days)	
Cropping systems	2006	2007
Continuous Potato (PP)	2526a	1773a
Disease Suppressive (DS)	2551a	1822a
Soil Conserving (SC)	2470a	1913a
Soil Improving (SI)	2420a	1933a
Status Quo (SQ)	2520a	1816a
LSD (0.05)	430	312

(PP) to 1,933 (SI), but treatment differences were also non-significant. Similarly, irrigation effects were also not statistically significant (data not shown). The lower values for blight prediction in 2007 compared to 2006 suggests that microclimatic variables used in predicting late blight values were less favorable in 2007 compared to 2006. Predicted AUDPC were not computed for 2008 cropping year. The lack of significant differences in predicted blight values suggest that microclimatic variables did not vary significantly in the cropping systems experiments to impact potential disease levels. The plot sizes in these experiments are 15 × 6 m (length × width) and therefore were greatly influenced by the microclimate of the surrounding environment. Even though treatments were replicated five times, crop management systems did not influence microclimate, and subsequently late blight potential.

It is hypothesized that if late blight were allowed to developed in these sort of experiments, disease would probably develop with relatively no differences among treatments. In previous studies on potential for late blight development in irrigated potato, no significant differences were recorded in field plots with similar dimensions (Olanya et al. 2007). It was also shown that the microclimate of the surrounding field had much greater impact on field plots than the crop induced effects. The non-significant differences in water management effects on predicted AUDPC indicate that the duration and amount of water application in our experiments did not sufficiently influence potential for disease development. It is possible that under intense water application, predicted disease values may be different. In contrast, it has been documented that irrigation may enhance or increase late blight, particularly if the amount and duration is intense enough to influence soil moisture and air/leaf temperature (Rotem and Palti 1969; Johnson et al. 2003).

3.4.3 *Late Blight Severity Values and Blight Units*

Late blight severity value (SV), computed by the BLITECAST Model (Krause et al. 1975), is one of the most widely used indices for evaluation of late blight potential and prediction across a wide range of environments. Based on hours of

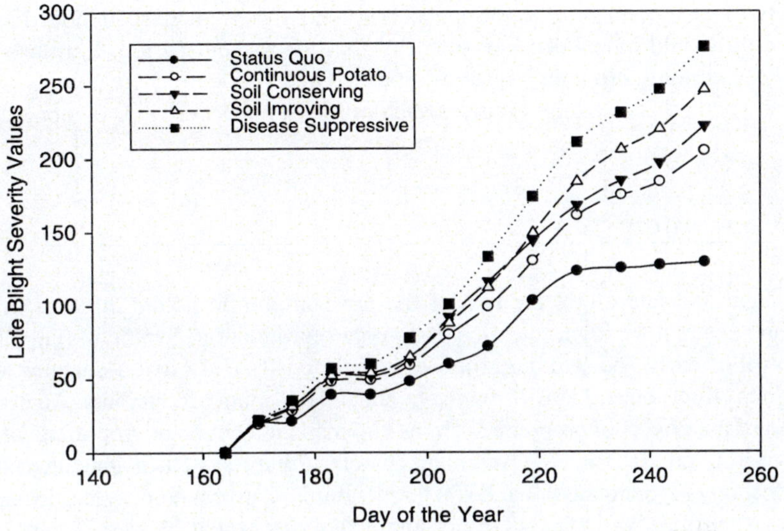

Fig. 3.7 Potential for late blight development in potato cropping systems based on computed severity values derived from measurements of air temperature and relative humidity in field plots at Presque Isle, 2008

RH≥90% and three temperature ranges of 7.2–11.6°C, 11.7–15°C and 15.1–26.7°C, severity values were computed for different treatments. No differences in severity value accumulation were observed among cropping systems and irrigation management in 2006 and 2007 (data not shown). In 2008, the total SV accumulations were also similar among crop management systems (Fig. 3.7). At day of year 254, SV ranged from 111 in SQ to 270 in the DS treatments, in which risk values did not differ significantly among SQ, PP, SI and SC systems. The threshold of 18 SV, indicative of imminent outbreak of late blight (assuming the presence of pathogen inoculum) was achieved on June 18 in 2008 for all cropping system treatments, indicating that system management did not impact the onset of late blight in any of our treatments.

The SV accumulation increased linearly in all treatments except in SQ, where the trend was not linear. Late blight severity values have often been used to predict the onset of late blight epidemics and monitoring environmental conditions for improved fungicide applications, rather than compare treatment effects on potential disease development in a given set of experiments. In this study, the lack of significant differences in late blight potential is not totally unexpected. Microclimate (temperature and relative humidity) did not differ significantly among treatments (Olanya et al. 2009), and therefore differences in late blight potential was not detected in our study. If field plot sizes were larger than our experimental plots or irrigation applications sufficiently intense and for longer hours, it may have been possible to detect treatment differences in blight potential. The variation of SV in the last assessment period observed in SQ treatment indicates that potential for late blight was less conducive in SQ than other treatments, however; this is inconsequential in that potato crop would have been harvested by that assessment period. Similarly, no differences among cropping systems and between irrigation treatments were detected when other late blight indices such as blight units were also computed.

3.5 Conclusion

Early blight and late blight are two of the most important potato diseases in the world and account for significant losses in potato production. Synthetic fungicides for control of early and late blight are extremely costly and often associated with deleterious effects on soil health, environment and optimum productivity. An understanding of the effects of cropping systems and management on microclimate of the potato crop is crucial for effective management. Potato system management had little effect on microclimate (air and soil temperature, relative humidity, leaf wetness) implying that these variables were not appreciably impacted by crop density and management. Although treatments of SC (limited tillage plus forage grass rotation), SI (addition of compost amendments plus forage rotation), and SQ (barley-potato rotation) were expected to influence soil moisture retention in field plots through the addition of organic matter, these differences were not always consistent among system management, but showed temporal variation in soil water content, particularly in response to irrigation treatment. Similarly, average seasonal averages of microclimate data did not differ much among cropping systems and irrigation management in potato fields in Maine.

The 3-year crop rotation and soil management systems impacted early blight development as shown in our data. Disease incidence and lesion numbers were significantly higher in PP than in the SC, SI, SQ or DS systems. The presence of non-host rotation crop such as timothy in SI and SC or barley in SQ systems could have reduced pathogen inoculum, thereby reducing early blight incidence in those systems. Disease levels differed among the 3 years of the experiment, implying possible variation in pathogen inoculum among years. Although unexpected, irrigation management had little effect on disease levels.

Based on several late blight severity indices that were computed from microclimate data in field plots, the potential for late blight development was not significantly impacted by cropping systems or irrigation management during the 3 years of the experiment. The cumulative hours of RH>90%, predicted AUDPC, late blight severity values, and blight units had similar values and trends among cropping systems and irrigation treatments. This implies that crop rotation and irrigation application did not influence microclimate to affect late blight potential or that microclimatic data were relatively uniform across treatments. This is not unexpected, considering that plot sizes were not large and that the microclimate of the surrounding field can outweigh field induced effects. Perhaps, under different treatment scenarios, late blight potential may be impacted. Other than the singularly positive attributes often associated with cropping systems (increased crop growth and yield, increased in soil health, better economic returns, production sustainability), this research demonstrated improved management of early blight disease with cropping systems, but no effect on late blight, a potentially explosive disease. Nevertheless, cropping systems and irrigation management provides very useful tools for enhanced sustainability of potato production.

References

Adams SS, Stevenson WR (1990) Water management, disease development and potato production. Am Potato J 67:3–11

Andrade-Piedra JL, Hijmans RJ, Juarez HS, Forbes GA, Shtienberg D, Fry WE (2005) Simulation of potato late blight in the Andes: validation of the LATEBLIGHT model. Phytopathology 95:1200–1208

Bruhn JA, Fry WE (1981) Analysis of potato late blight epidemiology by simulation modeling. Phytopathology 71:612–616

Christ BJ (1990) Influence of potato cultivars on the effectiveness of fungicide control of early blight. Am Potato J 67:419–425

Christ BJ, Maczuga SA (1989) The effect of fungicide schedules and inoculums levels on early blight severity and yield of potato. Plant Dis 73:695–698

Erwin DC, Ribeiro OK (1995) Phytophthora diseases worldwide. APS Press, St. Paul, 596pp

Fry WE (2007) The canon of potato science: late blight and early blight. Potato Res 50:243–245

Fry WE (2008) *Phytophthora infestans*: the plant (and R gene) destroyer. Mol Plant Pathol 9:385–402

Guenthner JF, Michael KC, Nolte P (2001) The economic impact of potato late blight on US growers. Potato Res 44:121–125

Harrison JG (1992) Effects of the aerial environment on late blight of potato foliage – a review. Plant Pathol 41:384–416

Harrison MD, Livingston CH, Oshima N (1965) Epidemiology of potato early blight in Colorado: initial infection, disease development and the influence of the environmental factors. Am Potato J 42:279–291

Ingold CT (1978) Water and spore liberation. In: Kozlowski TT (ed) Water deficits and plant growth; water and plant disease, vol 5. Academic, London, pp 119–140

Johnson DA, Martin M, Cummings TF (2003) Effect of chemical defoliation, irrigation water, distance from the pivot on late blight tuber rot in center-pivot irrigated potatoes in the Columbia basin. Plant Dis 87:977–982

Krause RA, Massie LB, Hyre RA (1975) BLIGHTCAST: a computerized forecast of potato late blight. Plant Dis Rep 29:95–98

Lacey J (1986) Water availability and fungal reproduction patterns of spore production, liberation and dispersal. In: Ayres PG, Boddy L (eds) Water, fungi and plants. Cambridge University Press, Cambridge, pp 65–86

Larkin RP, Honeycutt CW, Griffin TS, Olanya OM, Halloran JM, He Z (2011) Effects of different potato cropping system approaches and water management on soilborne diseases and soil microbial communities. Phytopathology 101:58–67

Melhus IE (1915) Germination and infection with fungus of the late blight of potato (*Phytophthora infestans*). Agric Exp Stn Univ WI Res Bull 37:1–64

Minogue KP, Fry WE (1981) Effects of temperature, relative humidity and rehydration rate on germination of dried sporangia of *Phytophthora infestans*. Phytopathology 71:1181–1184

Mizubuti ESG, Fry WE (1998) Temperature effects on the developmental stages of isolates from three clonal lineages of *Phytophthora infestans*. Phytopathology 88:837–843

Nnodu EC, Harrison MD, Parke RV (1982) The effects of temperature and relative humidity on wound healing and infection of potato tubers by *Alternaria solani*. Am Potato J 59:297–311
Olanya OM, Lambert DH, Porter GA (2006) Effects of pest and soil management systems on potato diseases. Am J Potato Res 83:397–408
Olanya OM, Starr GC, Honeycutt CW, Griffin TS, Lambert DH (2007) Microclimate and potential for late blight development in irrigated potato. Crop Prot 26:1412–1421
Olanya OM, Honeycutt CW, Larkin RP, Griffin TS, He Z, Halloran JH (2009) The effect of cropping systems and irrigation management on development of potato early blight. J Gen Plant Pathol 75:267–275
Olanya OM, Porter GA, Lambert DH, Larkin RP, Starr GC (2010a) The effect of Supplemental irrigation and soil management on potato tuber diseases. Plant Pathol 9:65–72
Olanya OM, Porter GA, Lambert DH (2010b) Supplemental irrigation and cultivar effects on potato tuber diseases. Aust J Crop Sci 4:29–36
Pscheidt JW (1985) Epidemiology and control of potato early blight, caused by *Alternaria solani*. Ph.D. dissertation, University of Wisconsin-Madison, WI, USA
Raposo R, Risks DS, Fry WE (1993) Evaluation of potato late blight forecasts modified to include weather forecasts: a simulation analysis. Phytopathology 83:103–108
Rotem J, Palti J (1969) Irrigation and plant diseases. Ann Rev Phytopathol 8:267–288
Rotem J, Cohen Y, Bashi E (1978) Host and environmental influences on sporulation in vivo. Ann Rev Phytopathol 16:83–101
Sato N (1994) Effect of sporulating temperature on the limit temperature in indirect germination of sporangia of *Phytophthora infestans*. Ann Phytopathol Soc Japan 60:60–65
Starr GC, Rowland D, Griffin TS, Olanya OM (2008) Soil water in relation to irrigation, water uptake and potato yield in a humid climate. Agric Water Manage 95:292–300
van der Walls JE, Korsen L, Aveling TAS (2001) A review of early blight of potato. Afr Plant Protect 70:91–102
Vloutoglou I, Kalogerakis SN (2000) Effects of inoculum concentration, wetness duration and plant age on development of early blight (*Alternaria solani*) and shedding of leaves in tomato plants. Plant Pathol 49:339–345
Waggoner PE (1965) Microclimate and plant diseases. Ann Rev Phytopathol 3:103–126
Wilson TB, Bland WL, Norman JM (1999) Measurement and simulation of dew accumulation and drying in a potato canopy. Agric Forest Meteorol 93:111–119

Chapter 4
Comparison of Soil Phosphorus Status and Organic Matter Composition in Potato Fields with Different Crop Rotation Systems

Zhongqi He, C. Wayne Honeycutt, O. Modesto Olanya, Robert P. Larkin, John M. Halloran, and Jonathan M. Frantz

Abstract Cropping management practices influence soil phosphorus (P) availability and soil organic matter (SOM) quality. This chapter summarizes the impact of cropping systems and water management on soil phosphorus status and organic matter characteristics after the first full cycle of the 3-year crop rotations. These data indicated that the 3-year crop rotations impacted more on labile P and organic matter fractions and relevant biochemical parameters (i. e. water extractable P and organic matter, mild modified Morgan soil test P, microbial biomass C and P, phosphatase and urease activities). However, these influences were not always consistent and statistically significant ($P=0.1$ or 0.05). Generally, irrigation had a greater influence on stable P and organic matter fractions than crop rotations. Continuous analysis of P and SOM from soils after the completion of the second rotation cycle of the 3-year crop rotations would provide more insights on the improvement of soil fertility and biochemical quality for potato production by crop rotations.

Z. He (✉) • R.P. Larkin • J.M. Halloran
USDA-ARS, New England Plant, Soil, and Water Laboratory, Orono, ME 04469, USA
e-mail: Zhongqi.He@ars.usda.gov; bob.larkin@ars.usda.gov; John.halloran@ars.usda.gov

C.W. Honeycutt
USDA, Natural Resources Conservation Service, Washington, DC 20250, USA
e-mail: Wayne.Honeycutt@wdc.usda.gov

O.M. Olanya
USDA-ARS, ERRC, Wyndmoor, PA 19038, USA
e-mail: modesto.olanya@ars.usda.gov

J.M. Frantz
USDA-ARS, Application Technology Research Unit, Toledo, OH 43606, USA
e-mail: jonathan.frantz@ars.usda.gov

Z. He et al. (eds.), *Sustainable Potato Production: Global Case Studies*,
DOI 10.1007/978-94-007-4104-1_4,

4.1 Introduction

Phosphorus (P) is an essential element for plant growth. Its input to cropland is necessary to maintain profitable production. Potato is a species with a great demand for soil P (Pursglove and Sanders 1981). For this reason, it was estimated that more P fertilizer is probably used in potato production than for any other crop in British agriculture (Cogliatti and Clarkson 1983). However, the high P demand is not due to a high content of P in potato plant, rather due to the low efficiency of acquiring soil P by the potato plants (Pursglove and Sanders 1981; Alvarez-Sanchez et al. 1999). By isotopic labeling of inorganic fertilizer P, Pursglove and Sanders (1981) found only 64–50% of P entering the potato organs were from fertilizer at 43 and 63 days after planting, and that only 4% of fertilizer P was recovered by potato plants. McArthur and Knowles (1993) characterized morphological and biochemical aspects of P deficiency in potato as affected by vesicular-arbuscular mycorrhizal fungi. They found that these fungi partially alleviated P deficiency stress, but did not completely compensate for inadequate abiotic P supply. Leggewie et al. (1997) cloned cDNAs of two potato phosphate transporters. They found that one phosphate transporter is expressed in roots, tubers, source leaves as well as floral organs, and another is mainly expressed in root organs when plants are deprived of P. It is no doubt that all these studies are useful to understand the mechanism, and improve the efficiency of P uptake by potatoes. However, the dramatic improvement of P up-take efficiency would be limited due to the intrinsic limitation of sparse rooting systems.

In order to increase P uptake, some plants can directly modify the rhizosphere in order to gain access to previously unavailable soil P pools through developing more extensive root systems, exuding organic acids and phosphatases, or through association of roots with mycorrhiza (Li et al. 1997; Vosatka and Gryndler 1999; Dechassa et al. 2003). Furthermore, it is reasonable to hypothesize that these P-efficient plants will also improve soil P availability for other plants planted after them as the organic acids, phosphatase activity, and biomass (natural organic matter) exudated by these plant roots and associated microorganism (fungi) remain in soil for a while. In other words, rotation plants could improve the P uptake by potato plants. However, this kind of information is not available at present time.

Cropping management practices influence soil organic matter (SOM) content and quality (Ohno et al. 2009, 2010). Parts of SOM can be sequentially extracted by water (water extractable organic matter, WEOM) and pyrophosphate (pyrophosphate extracted organic matter, PEOM) (Kaiser and Ellerbrock 2005; Kaiser et al. 2007). In functional characterization of SOM fractions from different cropping practices including potato-winter rye rotation management, Kaiser and Ellerbrock (2005) found that the composition of both WEOM and PEOM fractions depended on the type of crop as WEOM and PEOM from soils with maize and winter rye cropping differed in their C=O content. Furthermore, WEOM represents a less stable and younger SOM fraction whereas PEOM represents a higher proportion of total SOM and more carboxylic group than WEOM fraction. Thus, Kaiser et al. (2007) proposed that Fourier transform infrared (FT-IR) analysis of the C=O groups

Table 4.1 Total soil P (mg kg^{-1} of dry soil) in rainfed and irrigated potato fields with continuous potato (*PP*), *status quo* potato-barley (*SQ*), disease suppressive (*DS*), soil conserving (*SC*), and soil improving (*SI*) crop systems

	Rainfed		Irrigated	
	A	SE	A	SE
PP	1,671	111	1,628	121
SQ	1,672	89	1,664	83
DS	1,616	91	1,517	128
SC	1,758	36	1,614	86
SI	1,693	57	1,531	108
Means	1,682	23	1,592	29

Data are presented in average (*A*) of five field replicates with standard error (*SE*, n=5) (He et al. 2011c)

in the PEOM fraction is preferable over WEOM or unrestricted soil samples for identifying long-term effects of crop rotation and fertilization. However, they cautioned that for crop rotations with potatoes, possible short-term soil management-induced effects on composition of PEOM required further studies.

Potato yield in Maine U.S.A. has remained relatively constant for over 50 years. To identify and quantify constraints to potato productivity, we established five different potato crop management systems, designed to address specific management goals of soil conservation, soil improvement, and disease suppression, as well as a system representing a typical rotation and a non-rotation control, under both rainfed and irrigated management. These systems are named as Soil Conserving (SC), Soil Improving (SI), Disease Suppressive (DS), Status Quo (SQ), and Continuous Potato (PP), respectively. Each system is evaluated by our interdisciplinary team for plant growth and productivity, soil chemical-physical-biological properties, tuber diseases, soil borne diseases, foliar diseases, economics, and their interactions (see Chap. 2 for the overview of this case study). Based on published data (He et al. 2010, 2011a, b, c), this chapter summarizes the impact of cropping system and water management on soil phosphorus status and organic matter after the first full cycle of the 3-year crop rotations, and discusses the correlationships between potato yields and several relevant soil parameters.

4.2 Soil Phosphorus Status in Potato Fields

4.2.1 Total Soil Phosphorus Levels

The concentration of total soil P in the ten potato fields was around 1,600 mg kg^{-1} of soil (Table 4.1). Whereas some additional P from compost might have been added to the Soil Improving plots, the total soil P level was not changed. Indeed, all differences in total soil P among different cropping rotations were smaller than the standard errors of the five replicate plots with the same cropping rotation. In addition, there was no statistically significant difference between the rainfed soil and irrigated soil with the same cropping systems. However, the total soil P level in the rainfed soil was consistently higher than that in the corresponding irrigated soils. The means of the total soil P in the five rainfed and five irrigated cropping systems were 1,682 and 1,591 mg kg^{-1} of soil with comparable standard errors of 23 and 29, respectively. The difference of 91 mg kg^{-1} of soil suggested that 3-year irrigation lowered the surface (0–20 cm) soil P by an average of 5.4% in these fields (He et al. 2011c).

4.2.2 *Soil Test Phosphorus Levels*

Three types of soil tests for P in soil have been evaluated (He et al. 2011a). For Olsen P content, soils (1.0 g) were extracted by 25 mL of 0.5 M $NaHCO_3$ (pH 8.5) for 30 min. For ammonium oxalate extractable P (oxalate P), soils (2.0 g) were extracted by 200 mL of 0.2 M $(NH_4)C_2O_4$ (pH 3.0) for 4 h in the dark. For modified Morgan P (MMP), soils (4.0 g) were extracted by 20 mL of 0.62 M NH_4OH /1.25 M CH_3COOH (pH 4.8) for 15 min. Ammonium oxalate extracted P was the highest with a level of more than 1,400 mg P per kg of dry soil. Olsen P was modest in a range from 100 to 120 mg P per kg of dry soil. Modified Morgan P was lowest with a level not more than 10 mg P per kg of dry soil. Similar to the total soil P, neither Olsen P nor ammonium oxalate extractable P was significantly changed by the 3-year potato cropping rotation and irrigation. These observations suggest that these differences are small enough so that it may not be necessary to monitor P levels annually. Evaluation of these changes after one or more complete cycles of crop rotation would be appropriate (He et al. 2011a).

Unlike total soil P, however, the MMP levels were significantly impacted by crop rotation (He et al. 2011a). Figure 4.1 presents MMP_t concentrations measured by ICP (MMP_t) and measured by the colorimetric method (MMP_i). A previous study on soil MMP in the Northeast USA reported that the ICP method measured average 1.5 mg kg^{-1} more MMP than colorimetry (Heckman et al. 2006). In our samples, we found that MMP_t were twice that of MMP_i. Regression analysis of the 10 paired data revealed a linear relationship of $MMP_t = 1.42\ MMP_i + 4.64$ ($R^2 = 0.702$, $p < 0.01$). Although the high correlation existed in the two sets of MMP data, the significantly different MMP levels measured by the two methods should be a further research topic as the different values may lead to a different recommendation for P fertilizer application (Ohno et al. 2007).

The greatest concentration of MMP in the SI plot could be due to extra labile P applied from the compost. This observation suggested the compost amendment had a greater impact on mild soil test P than total soil P as the latter was not affected by the compost amendment (Table 4.1). Plant-available P (i. e. soil test MMP) has been previously observed to be higher in soils from compost- and animal manure-amended potato fields than in unamended potato soils (Erich et al. 2002). The other eight rainfed and irrigated fields had received the same amounts

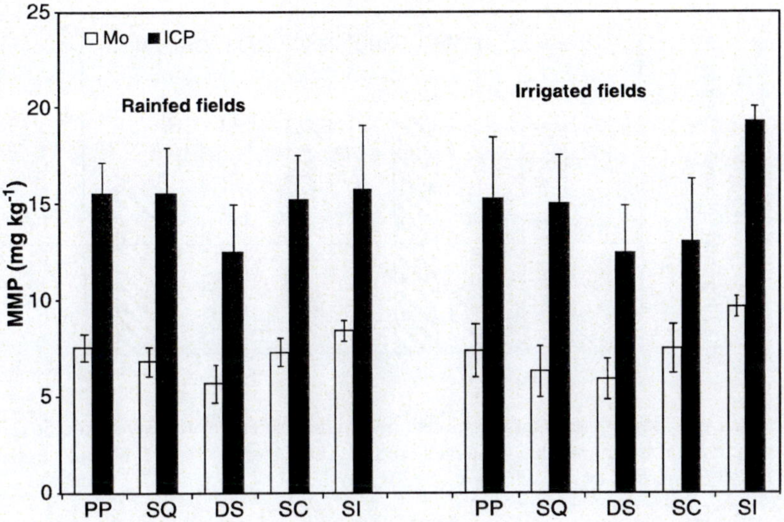

Fig. 4.1 Modified Morgan P (*MMP*) in potato fields after the end of the 3-year rotations. *MMP* was measured by both modified molybdate blue method (*Mo*) and ICP-AES. Rotation systems were (1) continuous potato (*PP*); (2) Status Quo (*SQ*); (3) disease suppressive (*DS*); (4) soil conserving (*SC*); and (5) soil improving (*SI*) (Data are presented in average with standard deviations (n=5) He et al. 2011a)

Table 4.2 Linear regressions between MMP (P_i or P_t) and relative potato yields (Y) in rainfed and irrigated field with same annual rates of P application (He et al. 2011a)

	Linear regression	R^2
Rainfed	$Y=-0.112P_i+1.849$	0.919*[a]
	$Y=-0.052P_t+1.841$	0.602
Irrigated	$Y=-0.158P_i+2.285$	0.774^
	$Y=-0.0022P_t+1.250$	0.0005
Both	$Y=-0.139P_i+2.099$	0.614*
	$Y=-0.040P_t+1.720$	0.178

[a] Symbol ^, and * represent statistical significance at $P=0.1$, and 0.05, respectively

of fertilizer P. Thus, MMP changes in these fields reflected the impacts of the crop rotation management. The change of the MMP levels in the eight rainfed and irrigated fields was in the reverse order of the yields of potato tuber harvested in the season prior to the soil sample collection. In other words, the decrease of the MMP levels seemed due to more P up take by potato. Therefore, we quantitatively compared the Spring-sampled MMP levels with the previous year's potato yields in these fields (Table 4.2). The linear but negative regressions indicated that MMP_i was more related to the potato yield than MMP_t. In other words, MMP_i is a better indicator to reflect P taken up by the potato plants in these soils although more research is needed to confirm it.

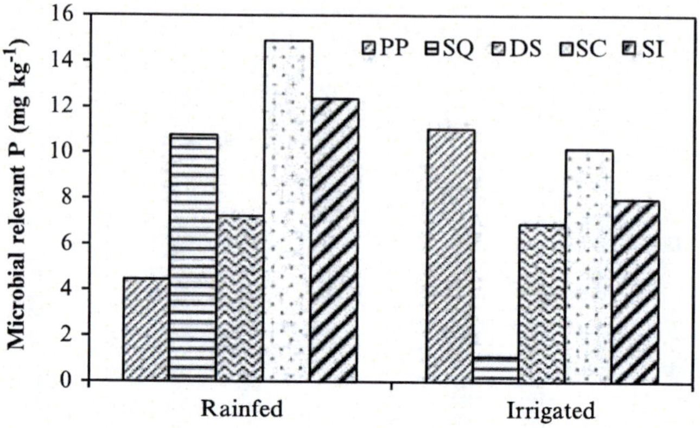

Fig. 4.2 Microbial biomass relevant P in potato fields after the end of the 3-year rotations. Soil samples were chloroform fumigated (The data are the differences in Olsen P between untreated and treated soil samples adapted from He et al. (2011a))

4.2.3 Microbial Biomass Phosphorus Levels

For soil microbial biomass P measurement, soil samples were treated by chloroform fumigation or microwave irradiation prior to extraction. When microwaved soils were used to measure microbial biomass P, the P contents in these samples were lower than that of the corresponding untreated soils. Therefore, this result demonstrated that the microwave method was not appropriate for microbial biomass P determination. On the other hand, fumigated soils showed greater P contents than their corresponding untreated soils, indicating that fumigation was more effective than microwaving in releasing microbial biomass P (He et al. 2011a). Based on all soils including those of rotation crops, fumigation increased extracted P by approximately 10.8 and 7.0 mg kg^{-1} soil under rainfed and irrigated management, respectively. As the fumigation-extraction method may underestimate microbial P due to the incomplete release of microbial P during fumigation and adsorption of released microbial P onto the mineral soil surface, a correction factor ranging from 0.33 to 0.57 has been suggested (McLaughlin et al. 1986). Thus, it was estimated that the rainfed soil contained microbial biomass P in a range from 18.9 to 32.7 mg kg^{-1}, and the irrigated soil contained microbial biomass P in a range from 12.3 to 21.2 mg kg^{-1} (He et al. 2011a). Similar levels have been reported by Wu et al. (2007).

Figure 4.2 presents the microbial biomass relevant P in the five potato fields. Compared to the PP system, crop rotation increased microbial P in the rainfed fields. However, crop rotation negatively impacted microbial levels in the irrigated fields. Further comparison of these two sets of data revealed that irrigation increased microbial biomass relevant P by 6.58 mg P kg^{-1} in the Continuous Potato soils. However, irrigation decreased microbial biomass relevant P in other four field soils by 9.65, 0.33, 4.71 and 4.37 mg P kg^{-1}, respectively.

4.2.4 *Sequentially Extracted Phosphorus Fractions*

Sequential fractionation is a common method used in evaluating the impacts of soil management practices on P distribution in soil and other environmental samples (Toth et al. 2011). The extracted P fractions typically comprised H_2O-P, 0.5 M $NaHCO_3$-P_i and -P_o, 0.1 M NaOH-P_i and -P_o, 1 M HCl-P_i, and residual-P fractions. Negassa and Leinweber (2009) review more than 100 articles published in soil science and related journals in the past 25 year related to the subject. To improve on our understanding on the mechanisms of change of P lability in these different cropping systems and irrigation, soil P in the 10 potato fields were sequentially extracted and separately quantified (He et al. 2011c). Phosphorus in these soils distributed in the five pools separated by the sequential fractionation in the order H_2O-P < residual P < $NaHCO_3$-P ≈ HCl-P < NaOH-P. Whereas the relative concentrations of P_o in the sequentially extracted H_2O, $NaHCO_3$, and NaOH fractions of the soils with the potato production was comparable to those in previous studies (He et al. 2004a; Waldrip-Dail et al. 2009), substantial enzymatically-hydrolyzable P_o amounts were not observed in the three fractions (He et al. 2011c). These data supported that, as inorganic fertilizer was applied to these fields, these management practices mainly impacted the distribution of inorganic P fractions, with little significant changes of organic P fractions.

With the sequential fractionation procedure, only the most labile H_2O extractable P pool was statistically impacted by the cropping systems as the highest water extractable P found in the PP and the SI cropping systems (He et al. 2011c). This observation implies that the labile portion of soil P, either water extractable P or MMP_i, was changed by short-term (3-year) cropping managements, thus impacting the plant uptake and runoff potential. However, this change would not reflect on more stable and recalcitrant soil P pools that accounted for the majority of soil P.

Interestingly, irrigation impacted P distribution in these soils with a pattern different from that of crop rotations as irrigation impacted P fractions in the order of NaOH > HCl > residual > $NaHCO_3$ > H_2O fractions. As NaOH-P was the most abundant pool in most soils, it is believed that it is the primary sink of soil P in fields and laboratory incubation with animal manure (He et al. 2004b, 2006b; Zheng et al. 2004). In these potato soils, irrigation seemed to mobilize and redistribute this portion of NaOH-P. The average reduced amount in the NaOH fractions by irrigation was 199 mg P_i kg^{-1} soil. The mobilized P portion was partly converted to P_o in the same NaOH fractions with the majority transferred to more stable or recalcitrant HCl and residual fractions (Table 4.3). The combined increase of P_i in the HCl fractions and P_o in the NaOH fractions was 116 mg kg^{-1} soil, accounting for 58% of P_i decrease in the NaOH fractions. In other words, another 42% of P_i in the NaOH fractions reduced by irrigation might have been transferred to $NaHCO_3$ and H_2O fractions. As P in these two fraction was more labile, this portion of P from NaOH-P_i did not accumulate in the two fractions, but was rather lost by plant uptake, runoff, or leaching from the irrigated soils. A complementary pattern of changes of P in H_2O, $NaHCO_3$, and NaOH fractions observed in laboratory incubation

Table 4.3 Impacts of irrigation on soil P fraction distribution

	Inorganic P		Organic P		Total P	
	R	I	R	I	R	I
H_2O	7.1	7.4[a]	3.9	3.4[ns]	11.0	10.9[ns]
$NaHCO_3$	176	153**	86	83[ns]	262	236*
NaOH	771	572***	303	350[ns]	1061	922***
HCl	175	235***	ND[b]	ND	175	235***
Residual	NA[c]	NA	NA	NA	170	189**

Data are presented in the means of P concentrations (mg kg^{-1} of dry soil) in the five cropping systems under rainfed (*R*) or irrigation (*I*) (He et al. 2011c)

[a]Symbols ns, *, ** and *** are for not statistically significant, and statistically significant at $P=0.05$, 0.01 and 0.001, respectively

[b]*ND* not detected

[c]*NA* not applicable

experiments could support this hypothesis (He et al. 2004b, 2006b; Waldrip-Dail et al. 2009). Another possibility was that the downward movement of P in the irrigated surface soils was faster than that in the rainfed surface soils. This downward movement has been observed with long term P fertilization (Wang et al. 2007; He et al. 2009c) although there are no reports on the effect of irrigation.

4.3 Soil Phosphatase and Urease Activities

Soil phosphatase activity is a significant mechanism for plant and microbial use of organic P in soil environments by converting organic P compounds into bioavailable inorganic phosphate (Yadav and Tarafdar 2003; Senwo et al. 2007). Similarly, urease activity could be an indicator of the potential rate of degradation of nitrogen compounds in soil (Banik et al. 2006) as it is an important enzyme in soil that mediates the conversion of organic N to inorganic N by hydrolysis of urea to ammonia. Thus, acid phosphatase (acPase), alkaline phosphatase (alPase), phosphodiesterase (diPase), and urease activities were measured to evaluate the impact of potato crop rotations on these soil enzymes (He et al. 2010).

4.3.1 Phosphatase Activities

In the rainfed PP plot, the three soil phosphatase activities were 0.691 mg p-NP g^{-1} dry soil for acPase, 0.116 mg p-NP kg^{-1} dry soil for alPase, and 0.122 mg p-NP kg^{-1} dry soil for diPase. All three enzyme activities increased in the soils with crop rotations (Fig. 4.3). The increase was relatively greater for alPase and diPase activities than for acPase activity although the specific activity of the latter was 3–5 fold greater than alPase and diPase. For all three enzymes, the smallest increase was observed in soil with the SC rotation, and the greatest increase was in soil with the SI rotation. Where the rotation crops and planting sequences were exactly the same

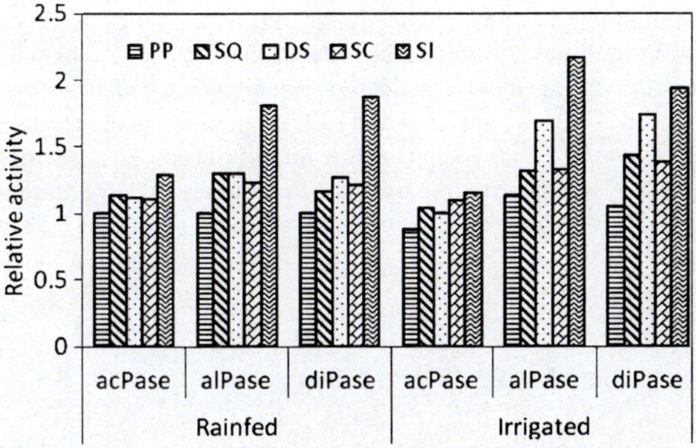

Fig. 4.3 Effect of cropping systems on soil phosphatase activities under rainfed and irrigation managements. The relative activity is based on 0.691 mg p-NP kg^{-1} dry soil h^{-1} for acid phosphatase (*acPase*), 0.116 mg p-NP kg^{-1} dry soil h^{-1} for alkaline phosphatase (*alPase*), and 0.122 mg p-NP kg^{-1} dry soil h^{-1} for phosphodiesterase (*diPase*) (Data are adapted from He et al. (2010)

for the SC and SI rotations, the only difference was addition of compost in each crop phase of the SI rotation, indicating the increased phosphatase activities were directly attributable to compost addition. Similar patterns of crop rotation impacts on soil phosphatase activities were observed in soils collected from the irrigated fields. Activity of alPase were 14% greater and activity of diPase was 31% greater in the DS systems, compared to the Continuous Potato system.

Although crop rotation increased phosphatase activities in soils under both rainfed and irrigated management, irrigation impacted phosphatase activities in different ways (Fig. 4.3). Soil acPase activity in all five cropping systems decreased with irrigation. The least negative impact was observed in the soil with the SC system (2% decrease) whereas a 9–12% decrease was observed in soils of the other four cropping systems. Irrigation had positive impacts on soil alPase activity although these increases were only substantial in the soils in POT (14%), DS (31%), and SI (15%) systems. Irrigation increased soil diPase activity by 18% in the BP rotation, but less than 1% in the SI rotation.

4.3.2 *Urease Activities*

Soil urease activity could be measured under either buffered (pH 9.0) or non-buffered conditions (Banik et al. 2006). The activity in the Continuous Potato system ranged from 88 mg NH_3-N kg^{-1} dry soil 20 h^{-1} under non-buffered conditions to 110 mg NH_3-N kg^{-1} 20 h^{-1} under buffered conditions (He et al. 2010). Cropping system influenced soil urease activity, with the most dramatic difference between the PP and SI systems. A higher soil urease activity in PP was observed with irrigation than with rainfed management, ranging from 100 mg NH_3-N kg^{-1} 20 h^{-1} under non-buffered conditions to 138 mg NH_3-N kg^{-1} 20 h^{-1} under buffered conditions. Antonious (2003) reported soil urease activity in the rhizosphere of potato plants under different soil management practices in Kentucky ranged from 42 to 92 mg NH_3-N kg^{-1} dry soil 24 h^{-1}. Kandeler et al. (1999) reported soil urease activity to be lower in soils following potato compared to rotation crops of sugar beet, winter wheat, spring barley, and alfalfa. It is possible that increased soil urease activity in the SQ, DS and SC systems reflected greater secretion of urease by the rotation plants than potato.

4.4 Soil Organic Matter Composition

4.4.1 *Microbial Biomass C Measurement*

Soil microbial biomass C was determined based on the chloroform fumigation extraction method described by Horwath and Paul (1994). Soil microbial biomass C under rainfed management increased from 69 mg kg^{-1} in the PP system to 155 mg kg^{-1} in the SI system (Fig. 4.4). Irrigation seemed to increase soil microbial activity compared to rainfed management in PP (Fig. 4.4). However, when analyzed across all cropping systems, soil microbial biomass C did not differ among irrigated cropping systems (He et al. 2010).

Under rainfed management, microbial biomass C was highly correlated with phosphatase and urease activities (Table 4.4a). These enzyme activities were also significantly correlated ($P \leq 0.05$), indicating the possible microbial origins of these enzymes. The poor but certain correlation ($P = 0.1$) with potato yields implies that soil microbial activity made some contributions to improving soil fertility. In contrast to microbial biomass C, the data of microbial biomass P was not related ($P \leq 0.1$) to

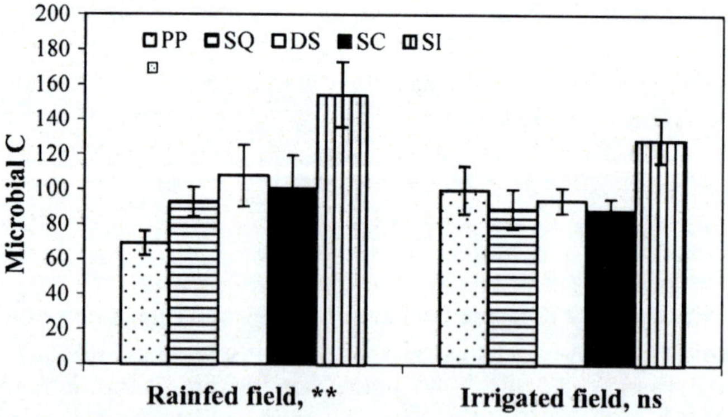

Fig. 4.4 Soil microbial biomass C levels under rainfed and irrigated management. Data represent the average of five field replicates with standard error bars (He et al. 2010). The symbol ** or ns indicates statistical significance at $P = 0.01$ or no significance at $P = 0.05$, respectively

Table 4.4 Correlation coefficients among soil biochemical parameters and potato yield

	Microbial C	acPase	alPase	diPase	Urease-u	Microbial P
(A). Rainfed management						
Microbial C	1					
acPase[a]	0.960**	1				
alPase[b]	0.980**	0.989*	1			
diPase[c]	0.984**	0.945*	0.981**	1		
Urease-u[d]	0.969**	0.917*	0.966**	0.992**	1	
Microbial P	0.536	0.592	0.518	0.466	0.356	1
Yield	0.852^	0.855^	0.870^	0.831^	0.848^	0.145
(B). Irrigated management						
Microbial C	1					
acPase[a]	0.400	1				
alPase[b]	0.801	0.689	1			
diPase[c]	0.580	0.753	0.946*	1		
Urease-u[d]	0.147	0.821^	0.487	0.562	1	
Microbial P	0.216	−0.236	−0.005	−0.262	0.196	1
Yield	0.467	0.036	0.651	0.661	−0.203	−0.368

Data are adapted from He et al. (2010, 2011a)

Symbol ^, *, and ** represents statistical significance at $P=0.1$, 0.05 and 0.01, respectively

[a] acPase, acid phosphatase

[b] alPase, alkaline phosphatase

[c] diPase, phosphodiesterase

[d] Urease-u, urease activity measured under non-buffered conditions

any of these parameters in Table 4.4a. It seems that microbial biomass relevant P is not a good indicator of soil microbial activities. This is probably partly due to the interference of microbial P measurement by the high back ground Olsen P in the soils (He et al. 2011a). A recent report (Zhao et al. 2008) suggested that the part of Olsen P, if>60 mg kg^{-1}, should be removed by either anion resin or 0.5 M sodium bicarbonate ($NaHCO_3$) solution (pH 8.5) prior to fumigation. Adoption of the pretreatment could improve the accuracy and statistical significance in future evaluation of the long-term impacts of the potato crop rotation and irrigation on the microbial biomass P levels. While strong correlations were found under rainfed management, much weaker correlations were observed for irrigated management (Table 4.4b). Reasons for this observation cannot be elucidated with the current data. However, these data imply that water management had a dramatic influence on soil enzymes and microbial biomass C and P in this study.

4.4.2 *Elemental Analysis of Water and Pyrophosphate Extracted Soil Organic Matter*

Water extractable organic matter (WEOM) fraction is the most labile and mobile fraction of soil organic matter (SOM) (Gregorich et al. 2003; Ohno et al. 2009). Sodium pyrophosphate extractable organic matter (PEOM) fraction is a relative

Table 4.5 Selected elemental contents in water extracts of rainfed and irrigated soils from the continuous potato (*PP*), status quo (*SQ*), and disease suppressive (*DS*), soil conserving (*SC*), and soil improving (*SI*) systems

	C	N	P	Na	K	Ca	Mg	S	Al	Fe
	mg kg^{-1} dry soil									
	Rainfed									
PP	130.9	28.3	3.37	9.37	28.7	33.8	16.8	18.3	4.00	4.07
SQ	114.4	42.1	2.73	2.87	30.1	35.0	19.1	16.9	4.53	5.73
DS	114.2	40.7	3.00	3.03	34.1	34.1	18.3	17.0	2.07	2.70
SC	89.7	35.6	2.43	5.80	35.4	34.3	17.7	17.2	2.72	3.40
SI	139.0	41.6	2.60	6.93	51.7	64.8	26.5	28.4	5.93	7.20
SI-d[a]	147.1	4.6	1.20	0.80	0.57	8.1	2.0	14.4	4.30	5.9
	Irrigated									
PP	112.9	24.7	3.20	14.17	15.7	25.3	12.8	16.8	3.00	4.00
SQ	120.1	40.9	3.30	10.37	27.8	36.1	18.8	17.6	6.57	8.83
DS	186.2	49.2	3.93	16.77	58.6	51.0	27.6	26.8	2.17	2.47
SC	133.4	58.0	3.93	16.77	58.6	51.0	27.6	26.8	2.17	2.47
SI	174.6	42.4	2.60	8.33	43.4	55.0	26.1	23.8	3.20	4.27
SI-d	153.0	9.4	1.50	0.90	0.93	10.2	3.0	20.3	6.70	9.07

Data are adapted from He et al. (2011b)

[a] SI-d, dialyzed against water in cellulose dialysis tubing (Molecular weight cutoff: 12,000) for four times (12 h each) prior to elemental analysis

labile SOM pool (Ellerbrock et al. 2005; Kaiser et al. 2007). To investigate the impact of management practices on SOM compositions, WEOM and PEOM from the 10 potato field soils were sequentially extracted (He et al. 2011b).

The elemental contents of water extracts of the 10 soil samples are listed in Table 4.5. The major elements in these soil water extracts were C, N, K, Ca, Mg, and S, with the concentrations in mg L^{-1} of extracts or tens mg kg^{-1} of dry soil. P, Na, Al, and Fe were in the middle levels of tenths mg L^{-1} of extracts or mg kg^{-1} of dry soil. Although the levels of these elements varied among the five systems, clear differences among the treatments, except for Na ion, was not observed. Irrigation seemed to consistently increase the water extractable Na levels of all five production systems. The C levels in the 10 water extracts were comparable to those in water extracts of four types of Italian soils (Provenzano et al. 2010) and a calcareous soil in the French Mediterranean region (Hassouna et al. 2010). There are no data reported on the other elemental contents of WEOM, either in un-treated water extracts (Ohno et al. 2009; Provenzano et al. 2010) or purified fractions (Kaiser et al. 2007; Hassouna et al. 2010). Dialysis did not change the C content of rainfed and irrigated SI extracts. However, the contents of N, Na, K, and Mg were reduced greatly by an order of magnitude, indicating that these elements were mainly in the inorganic forms, and not associated with the WEOM. The contents of other measured elements were partially reduced or not changed at all. This observation indicated that parts of these elements were associated with WEOM, either directly trapped in the complicated structures of organic matter or complexing through organo-mineral bridging bonding.

Table 4.6 Selected elemental contents in pyrophosphate-extracted organic matter of rainfed and irrigated soils from the continuous potato (*PP*), status quo (*SQ*), disease suppressive (*DS*), soil conserving (*SC*), and soil improving (*SI*) systems

	C	N	P	Na[a]	K	Ca	Mg	S	Al	Fe
	% of dry matter		mg kg^{-1} of dry matter							
	Rainfed									
PP	24.4	2.13	62.4	1159	14.2	16.1	7.0	73.5	178	288
SQ	24.5	2.14	79.5	1179	11.1	15.0	6.0	61.9	152	325
DS	30.0	2.70	63.2	1149	14.4	12.9	6.6	72.5	161	237
SC	39.5	3.67	66.1	1118	12.2	12.0	5.7	67.5	136	207
SI	41.9	3.35	51.3	1119	8.4	17.9	4.9	80.7	65	107
	Irrigated									
PP	32.6	2.86	45.6	1165	20.3	16.1	5.5	61.2	100	202
SQ	23.8	1.97	68.0	1167	9.2	17.4	5.5	62.2	123	292
DS	30.6	2.54	56.7	1152	10.6	16.4	5.6	71.2	98	156
SC	22.6	1.82	53.7	1115	8.9	13.2	4.9	65.5	86	133
SI	31.0	2.17	39.8	1122	8.1	18.0	4.6	72.8	56	86

Data are adapted from He et al. (2011b)
[a]Na was from 0.05 M NaOH used to dissolve the organic matter solid

The elemental contents of the 10 PEOM samples are listed in Table 4.6. The contents of C and N were in the ranges from 22.6% to 41.9% and from 1.8% to 3.7% of dry matter, respectively. Under rainfed conditions, soil conservation and improving practices seemed to increase the C content of PEOM. Although these organic matter samples were extracted by pyrophosphate, P in these samples were very low (<80 mg kg^{-1} of dry matter). The P level was indeed lower than P content in some humic fractions in which case P level may be up to 2 g kg^{-1} of dry matter (He et al. 2006a, 2009b). The P concentration in a peat humic acid extracted by alkaline pyrophosphate solution (0.1 M NaOH+0.1 M $Na_4P_2O_7$) was as high as 84.7 g kg^{-1} of dry matter, but the humic acid from the same peat source extracted by 0.1 M alone contained less than 100 mg kg^{-1} of dry matter (Francioso et al. 1998). The difference suggests incorporation of pyrophosphate in the peat humic acid during alkalinized pyrophosphate extraction (Francioso et al. 1998). The much lower P concentration in PEOM samples from the potato soils implied that the pyrophosphate incorporation into PEOM did not occur apparently due to the neutral pyrophosphate extraction conditions.

The Na concentration reported in Table 4.5 was apparently from 0.05 M NaOH (i.e. 1.15 g Na L^{-1} of organic matter solution) that was used to dissolve the samples. The contents of Al and Fe were the highest with the levels > hundreds mg kg^{-1} of dry matter in most samples. Compared to the limited literature on element contents of purified OM, the contents of Ca, Mg, Al, and Fe measured in soil mobile humic acid and recalcitrant Ca humate fractions extracted by 0.1 M NaOH (He et al. 2009b) were generally higher than those in these PEOM samples. The contents of Ca, Mg, Fe, Cu, Mn, and Zn measured in peat humic acid extracted by alkaline pyrophosphate solution (Francioso et al. 1998) were also higher than the corresponding contents of these PEOM samples. This observation implied the neutral pyrophosphate solution

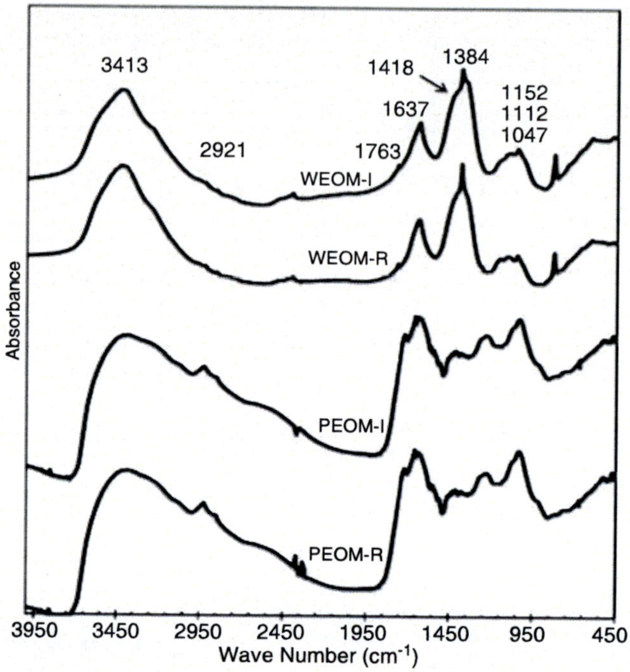

Fig. 4.5 FT-IR spectra of water extracted organic matter (*WEOM*) and pyrophosphate extracted organic matter from rainfed (*R*) and irrigated (*I*) soils with Soil Improving rotation (Data are adapted from He et al. (2011b))

might be less capable of extracting metal-organic matter complexes than alkaline solution. Whereas the impacts of cropping managements on most elements were not observed, the contents of Al and Fe were lower in SC and SI PEOM samples than in the other three soil management systems. In addition, irrigation seemed to lower the two metal contents in PEOM. Francioso et al. (1998) found that Fe was much more concentrated in the high molecular weight peat humic fractions. Thus, these management practices might have increased the low molecular weight PEOM fractions and/or decreased the high molecular weight PEOM fractions (He et al. 2011b).

4.4.3 *FT-IR Spectroscopic Analysis of Water and Pyrophosphate Extracted Soil Organic Matter*

The typical Fourier transform infrared (FT-IR) spectra of WEOM and PEOM are shown in Fig. 4.5. Except for the sharp band at 1,384 cm^{-1} (inorganic nitrate), these features of these soil WEOM samples were typical to those of plant- and animal manure-derived WEOM samples (He et al. 2009a), but different from those of

Table 4.7 Relative FT-IR band heights of soil water (*WEOM*)- and pyrophosphate (*PEOM*)-extractable organic matter samples from soils from the continuous potato (*PP*), status quo (*SQ*), disease suppressive (*DS*), soil conserving (*SC*), and soil improving (*SI*) systems

	Band A/Band C[a]		Band B/Band C	
	WEOM	PEOM	WEOM	PEOM
	Rainfed			
PP	0.23	0.28	1.40	4.28
SQ	0.23	0.30	1.17	4.31
DS	0.12	0.31	1.79	4.32
SC	0.10	0.32	3.39	4.18
SI	0.11	0.30	2.69	4.51
	Irrigated			
PP	0.26	0.28	0.91	4.58
SQ	0.02	0.30	1.84	4.57
DS	0.16	0.30	1.33	4.45
SC	0.05	0.31	2.31	4.20
SI	0.02	0.30	2.13	4.61

Data are adapted from He et al. (2011b)

[a]Two bands in 3,020–2,800 cm^{-1} were grouped Band A. Two bands in 1,740–1,600 cm^{-1} were grouped Band B. A band at 1,100–1,030 cm^{-1} was named Band C

humic substances (Stevenson and Goh 1971; He et al. 2006a). Compared to those of WEOM, FT-IR spectra of PEOM did not show the sharp peak at 1,384 cm^{-1} as soluble nitrate compounds should be associated with the purified PEOM. PEOM samples showed more apparent band intensities at 2,921, 1,715, 1,231, and 1,028 cm^{-1}. Stevenson and Goh (1971) assigned infrared spectra of humic acids and related substances to three types. Type I spectra show equally strong bands at 1,720 cm^{-1} and 1,600 cm^{-1} with no discernible absorption being evident at 1,640 cm^{-1}. Type II spectra show a very strong 1,720 cm^{-1} band, a shoulder at 1,650 cm^{-1} and the absence of a 1,600 cm^{-1} band. Type III spectra are similar to Type I with additional strong bands between 2,900 and 2,840 cm^{-1}. Therefore, the spectra of the PEOM samples were similar to those of Type III, but with less band intensity at 1,715 cm^{-1}. It is reasonable that the spectra of PEOM, not WEOM, were closely similar to that of humic acid as PEOM should be a more humified SOM fraction than the most labile WEOM pool (He et al. 2011b).

Semi-quantitative FT-IR analysis showed management practices impacted the relative abundance of both aliphatic (2,921–2,853 cm^{-1}) and aromatic groups (1,635 cm^{-1}) (Table 4.7). In the five rainfed samples, the relative heights of band A and B were in a complementary mode. That is, the band A/band C ratio was the highest in the PP and SQ WEOM samples, but the band B/band C ratio was the highest in the SC and SI WEOM samples. In addition, the ratio of band B to band C is also the highest with the two soil conserving systems. These data suggested that aromatic groups were increased in the WEOM fractions of soil conserving and improving fields. Irrigation reduced the relative intensities of both band A and B in PP, DS, SC and SI samples, but the intensity pattern was similar to that of rainfed samples. This observation implied that irrigation might have lowered both aliphatic and aromatic fractions in the WEOM pool. Compared to its rainfed counterpart, irrigated SQ WEOM showed a much lower band A/band c ratio, but increased the band B/band C ratio. Thus, irrigation seemed enriching aromatic fraction in the SQ WEOM, which may need to be further investigated and confirmed.

The relative intensities of both band A and B relative to the internal reference band C were higher in the spectra of PEOM samples than WEOM samples, indicating more stable aliphatic, aromatic, and carboxyl groups in stable PEOM than in labile WEOM fractions of soil organic matter (He et al. 2011b). However, the values of band A/band C and band B/band C were basically the same among the 10 PEOM samples. These data indicated that neither 3-year crop rotation nor irrigation affected the structural composition of PEOM, although long-term (20 year) tillage management can significantly change the chemical characteristics of SOM fractions (Ding et al. 2002). Kaiser et al. (2007) characterized WEOM and PEOM extracted from soils of two (sandy and clayey) of the oldest agronomic long-term filed experiments in Germany. These rotations include a 2-year winter rye-potato, a 4-year sugar beet-summer barley-potato-winter-wheat, and an 8-year sugar beet-summer barley-potatoes-winter wheat-alfalfa-alfalfa-potato-winter wheat. Whereas their work suggested that analysis of the stable PEOM is preferable for the long-term (about 40 years or longer) impacts of potato crop rotation on SOM, the work by He et al. (2011b) complementarily demonstrated that WEOM is more labile to a short–term effect of crop rotations.

4.5 Conclusion

Appropriate evaluation of soil P availability is a prerequisite for best P management in potato production systems. A 3-year crop rotation and irrigation treatment did not significantly change soil total P in the sandy loam potato fields in Maine, USA. However, the modified Morgan P level differed between soils samples with different rotation management. Correlation analysis revealed that the impact of crop rotation on the modified Morgan inorganic P was negatively correlated to the previous year's potato yield for fields receiving the same amount of P fertilizer, indicating that the modified Morgan P was a better soil test P for evaluating the P nutrient requirement for potato grown in these fields. Sequential fractionation of P in these soil samples revealed that the distribution of P in different labile pools had changed under these crop management practices. Crop rotation mainly increased water extractable inorganic P. Irrigation had a greater impact as it caused stable P pools in NaOH and HCl fractions inter-changed.

Compared to continuous potato production, crop rotation practices consistently increased microbial biomass C and P, soil phosphatase and urease activities in rainfed soils, and somewhat correlated to the potato yield ($P=0.1$), suggesting the improvement of soil biochemical quality by crop rotations. FT-IR analysis revealed that the impacts of rotation and irrigation practices reflected on the changes of aliphatic groups and aromatic compounds of WEOM samples. The 3-year crop rotation and irrigation changed the relative abundances of these functional groups only in WEOM, not in PEOM. Continuous analysis of P and SOM from soils after the completion of the second round of the 3-year crop rotations would provide more insights on the improvement of soil fertility and biochemical quality for potato production by crop rotations.

References

Alvarez-Sanchez E, Etchevers JD, Ortiz J, Nunez R, Volke V, Tijerina L, Martinez A (1999) Biomass production and phosphorus accumulation of potato as affected by phosphorus nutrition. J Plant Nutr 22:205–217

Antonious GF (2003) Impact of soil management and two botanical insecticides on urease and invertase activity. J Environ Sci Health B B38:479–488

Banik P, Ghosal PK, Sasmal TK, Bhattacharya S, Sarkar BK, Bachi DK (2006) Effect of organic and inorganic nutrients for soil quality conservation and yield of rainfed low land rice in subtropical plateau region. J Agron Crop Sci 192:331–343

Cogliatti HH, Clarkson DT (1983) Physiological changes in, and phosphate uptake by potato plants during development of, and recovery from phosphate deficiency. Physiol Plant 58:287–294

Dechassa N, Schenk MK, Claassen N, Steingrobe B (2003) Phosphorus efficiency of cabbage (*Brassica oleraceae* L. var. *capitata*), carrot (*Daucus carota* L.), and potato (*Solanum tuberosum* L.). Plant Soil 250:215–224

Ding G, Novak JM, Amarasiriwardena D, Hunt PG, Xing B (2002) Soil organic matter characteristics as affected by tillage management. Soil Sci Soc Am J 66:421–429

Ellerbrock RH, Gerke HH, Bachmann J, Goebel M-O (2005) Composition of organic matter fractions for explaining wettability of three forest soils. Soil Sci Soc Am J 69:57–66

Erich MS, Fitzgerald CB, Porter GA (2002) The effect of organic amendments on phosphorus chemistry in a potato cropping systems. Agric Ecosyst Environ 88:79–88

Francioso O, Ciavatta C, Tuganoli V, Sanchez-Cortes S, Gessa C (1998) Spectroscopic characterization of pyrophosphate incorporation during extraction of peat humic acids. Soil Sci Soc Am J 62:181–187

Gregorich EC, Beare MH, Stocklas U, St-Georges P (2003) Biodegradability of soluble organic matter in maize cropped soils. Geoderma 113:237–252

Hassouna M, Massiani C, Dudal Y, Pech N, Theraulaz F (2010) Changes in water extractable organic matter (WEOM) in a calcareous soil under field conditions with time and soil depth. Geoderma 155:75–85

He Z, Griffin TS, Honeycutt CW (2004a) Enzymatic hydrolysis of organic phosphorus in swine manure and soil. J Environ Qual 33:367–372

He Z, Griffin TS, Honeycutt CW (2004b) Evaluation of soil phosphorus transformations by sequential fractionation and phosphatase hydrolysis. Soil Sci 169:515–527

He Z, Ohno T, Cade-Menun BJ, Erich MS, Honeycutt CW (2006a) Spectral and chemical characterization of phosphates associated with humic substances. Soil Sci Soc Am J 70:1741–1751

He Z, Senwo ZN, Mankolo RN, Honeycutt CW (2006b) Phosphorus fractions in poultry litter characterized by sequential fractionation coupled with phosphatase hydrolysis. J Food Agric Environ 4(1):304–312

He Z, Mao J, Honeycutt CW, Ohno T, Hunt JF, Cade-Menun BJ (2009a) Characterization of plant-derived water extractable organic matter by multiple spectroscopic techniques. Biol Fertil Soils 45:609–616

He Z, Olk DC, Honeycutt CW, Fortuna A (2009b) Enzymatically- and ultraviolet-labile phosphorus in humic acid fractions from rice soils. Soil Sci 174:81–87

He Z, Tazisong IA, Senwo ZN, Honeycutt CW, Zhang D (2009c) Nitrogen and phosphorus accumulation in pasture soil from repeated poultry litter application. Commun Soil Sci Plant Anal 40:587–599

He Z, Honeycutt CW, Griffin TS, Larkin RP, Olanya M, Halloran JH (2010) Increases of soil phosphatase and urease activities in potato fields by cropping rotation practices. J Food Agric Environ 8(2):1112–1117

He Z, Honeycutt CW, Olanya M, Larkin RP, Halloran JH (2011a) Soil test phosphorus and microbial biomass phosphorus in potato fields. J Food Agric Environ 9(1):540–545

He Z, Honeycutt CW, Zhang H (2011b) Elemental and Fourier transform infrared spectroscopic analysis of water and pyrophosphate extracted soil organic matter. Soil Sci 176:183–189

He Z, Zhang H, Zhang M (2011c) Irrigation-induced changes in phosphorus fractions of Caribou sandy loam soil under different potato cropping systems. Soil Sci 176(12):676–683

Heckman JR et al (2006) Soil test calibration for predicting corn response to phosphorus in the northeast USA. Agron J 98:280–288

Horwath WR, Paul EA (1994) Microbial biomass. In: Weaver RW (ed) Methods of soil analysis, part 2. Microbiological and biochemical properties. Soil Science Society of America, Madison, pp 753–773

Kaiser M, Ellerbrock RH (2005) Functional characterization of soil organic matter fractions different in solubility originating from a long-term field experiment. Geoderma 127:196–206

Kaiser M, Ellerbrock RH, Gerke HH (2007) Long-term effects of crop rotation and fertilization on soil organic matter composition. Eur J Soil Sci 58:1460–1470

Kandeler E, Stemmer M, Klimanek E-M (1999) Response of soil microbial biomass, urease and xylanase within particle size fractions to long-term soil management. Soil Biol Biochem 31:261–273

Leggewie G, Willmitzer L, Riesmeier JW (1997) Two cDNA from potato are able to complement a phosphate uptake-deficient yeast mutant: identification of phosphate transporters from higher plants. Plant Cell 9:381–392

Li M, Osaki M, Rao IM, Tadano T (1997) Secretion of phytase from the roots of several plant species under phosphorus-deficient conditions. Plant Soil 195:161–169

McArthur DAJ, Knowles NR (1993) Influence of vesicular-arbuscular mycorrhizal fungi on the response of potato to phosphorus deficiency. Plant Physiol 101:147–160

McLaughlin MJ, Alston AM, Martin JK (1986) Measurement of phosphorus in the soil microbial biomass: a modified procedure for field soils. Soil Biol Biochem 18:437–443

Negassa W, Leinweber P (2009) How does the Hedley sequential phosphorus fractionation reflect impacts of land use and management on soil phosphorus: a review. J Plant Nutr Soil Sci 172:305–325

Ohno T, Hoskins BR, Erich MS (2007) Soil organic matter effects on plant available and water soluble phosphorus. Biol Fertil Soils 43:683–690

Ohno T, He Z, Tazisong IA, Senwo ZN (2009) Influence of tillage, cropping, and nitrogen source on the chemical characteristics of humic acid, fulvic acid, and water-soluble soil organic matter fractions of a long-term cropping system study. Soil Sci 174:652–660

Ohno T, He Z, Sleighter RL, Honeycutt CW, Hatcher PG (2010) Ultrahigh resolution mass spectrometry and indicator species analysis to identify marker components of soil- and plant biomass-derived organic matter fractions. Environ Sci Technol 44:8594–8600

Provenzano MR, Caricasole P, Brunetti G, Senesi N (2010) Dissolved organic matter extracted with water and a saline solution from different soil profiles. Soil Sci 175:255–262

Pursglove JD, Sanders FE (1981) The growth and phosphorus economy of the early potato (*Solanum tuberosum*). Commun Soil Sci Plant Anal 12:1105–1121

Senwo ZN, Ranatunga TD, Tazisong IA, Taylor RW, He Z (2007) Phosphatase activity of Ultisols and relationship to soil fertility indices. J Food Agric Environ 5(1):262–266

Stevenson FJ, Goh KM (1971) Infrared spectra of humic acids and related substances. Geochim Cosmochim Acta 35:471–483

Toth JD, Dou Z, He Z (2011) Solubility of manure phosphorus characterized by selective and sequential extractions. In: He Z (ed) Environmental chemistry of animal manure. Nova Science Publishers, Inc., Hauppauge, pp 227–251

Vosatka M, Gryndler M (1999) Treatment with culture fractions from Pseudomonas putida modifies the development of Glomus fistulosum mycorrhiza and the response of potato and maize plants to inoculation. Appl Soil Ecol 11:245–251

Waldrip-Dail H, He Z, Erich MS, Honeycutt CW (2009) Soil phosphorus dynamics in response to poultry manure amendment. Soil Sci 174:195–201

Wang J, Lester DW, Guppy CN, Lockwood PV, Tang C (2007) Changes in phosphorus fractions at various soil depths following long-term P fertilizer application on a black Vertosol from southeastern Queensland. Aust J Soil Res 45:524–532

Wu J et al (2007) Dynamics in microbial immobilization and transformations of phosphorus in highly weathered subtropical soil following organic amendments. Plant Soil 290:333–342

Yadav RS, Tarafdar JC (2003) Phytase and phosphatase producing fungi in arid and semi-arid soils and their efficiency in hydrolyzing different organic P compounds. Soil Biol Biochem 35:745–751

Zhao XR, Li GT, Lin QM (2008) Interference of soil-extractable phosphorus in measuring soil microbial biomass phosphorus. Commun Soil Sci Plant Anal 39:1367–1374

Zheng Z, MacLean AJ, Sanderson JB, Lafond J (2004) Soil phosphorus dynamics after ten annual applications of mineral fertilizers and liquid dairy manure: fractionation and path analyses. Soil Sci 169:449–456

Chapter 5
Comparing Modelled Productivity to Historical Data in New England Potato Production Systems

Jonathan M. Frantz, Robert P. Larkin, Georgette Trusty, C. Wayne Honeycutt, Zhongqi He, O. Modesto Olanya, and John M. Halloran

Abstract Potato yields in Northern Maine have remained fairly constant for the last 70 years. Many long-term projects have sought to identify the limitations to potato yield, but identifying limiting factors is difficult without first identifying the upper limits of potato production. A simple, light-driven mechanistic model is validated with specific case studies, and then, potential yield limitations to potato production in this region are identified based on analysis of the model. It was found that meteorologically-limited productivity peaks at about 55 Mg ha^{-1}, which is about 80% higher than historical averages. Most yield increases in those specific case studies examined were due to enhancement of radiation capture, which was achievable either by improved water management or disease suppression. Strategies for sustained yield improvements should continue to improve on radiation capture, either by increasing the peak radiation capture potential, prolonging the radiation capture duration, or by shifting radiation capture to coincide with available light. This model is useful to set realistic productivity goals for this region, can be easily adapted to other regions, and indicates strategies for potato yield improvement.

5.1 Introduction

Northern Maine, the primary potato production area in the New England region, has a climate with average temperatures ideally suited for potato production (16.8–18.6°C average temperature from June to August; C.I.A 2011), ample rainfall

J.M. Frantz (✉) • R.P. Larkin • G. Trusty • Z. He • O.M. Olanya • J.M. Halloran
USDA-ARS, New England Plant, Soil, and Water Laboratory, Orono, ME 04469, USA
e-mail: jonathan.frantz@ars.usda.gov

C.W. Honeycutt
USDA, Natural Resources Conservation Service, 14th and Independence Ave., SW Rm. 5006-S, Washington, DC 20250, USA

Z. He et al. (eds.), *Sustainable Potato Production: Global Case Studies*,
DOI 10.1007/978-94-007-4104-1_5,

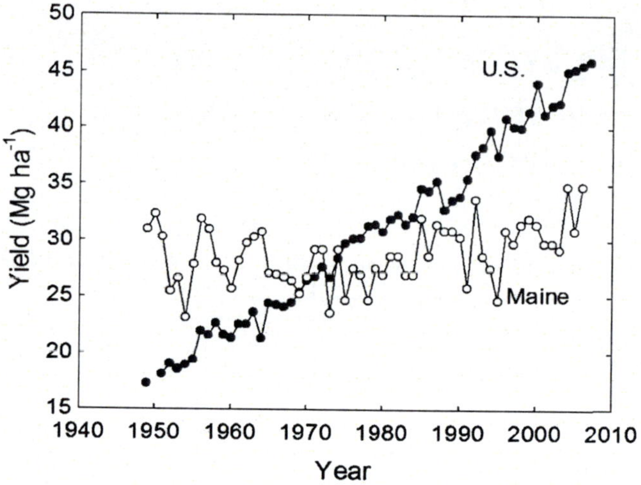

Fig. 5.1 Average potato yield from 1949 to 2008 in the US and Maine. Data are adapted from USDA National Agricultural Statistics Services (2011a, b)

(930 mm year^{-1} average), and a growing season of between 100 and 110 days. These characteristics allowed Maine to have a strong potato industry, relative to other US states, since 1870 (USDA National Agricultural Statistics Services 2011b). Today, Maine remains in the top ten producing states in the U.S. (USDA National Agricultural Statistics Services 2011a), in spite of yield rates that have remained fairly constant for the last 70 years (Fig. 5.1). The lack of improvement in potato yields per area over the last seven decades even suggests that some upper limit may have already been reached. How much can possibly be produced in this region – or any region?

There are many environmental factors that could be limiting potato production. Cool air and soil temperatures early in the season may delay shoot emergence. Whereas total rainfall amounts are often adequate over the course of a growing season, uneven distribution of rainfall can lead to flood or drought stress. Long periods of humidity, rain events, and cool temperatures can exacerbate foliar disease pressure including late blight (Olanya et al. 2007, 2010). On average, there are over 160 days per year with measurable precipitation (at least 2.5 mm), which decreases potential sunlight (C.I.A 2011). Long-term potato culture with historically little crop rotations has depleted the soil organic matter, which altered water and nutrient holding dynamics (Carter and Sanderson 2001; Griffin and Porter 2004). Late spring freeze events or even snowfall can delay field planting shortening the growing season.

In the simplest terms, potatoes must intercept sunlight in order to photosynthesize, and transfer that fixed carbon efficiently to the tubers. Stress in the early canopy development can negatively influence leaf expansion and leaf emergence, or can impact photosynthesis once the leaves have emerged or the canopy has formed. Weed competition and insect and disease pressure can impact both the plant's ability to gather light and partition resources efficiently to expanding tubers, potentially reducing yield. Finally, stress later in the crop production cycle can negatively influence tuber initiation and expansion, or shift resources from tubers to other plant parts to, for example, explore the soil for water and nutrients.

Several long-term projects have sought to identify the limitations to potato yield and improve management practices in ways that boost potato yield. But it is difficult to identify the limitations to potato yield without first identifying the upper limits of potato production (Bugbee and Salisbury 1988).

This chapter seeks to describe the theoretical limits of potato productivity in this region through a simple, light-driven model. The goal in the initial modelling is to illustrate where the upper limit to potato productivity might be if conditions were ideal. This approach was successfully demonstrated in the Estonian region for potato production (Kadaja and Tooming 2004). Then, from the ideal production levels, we work backwards to predict the yields of several specific case studies in an effort to identify some additional yield limitations to potato production in this region. Case studies of both state-wide production and smaller-scale field plots are compared to the model to see how further improvements might be made. This assessment would help identify parameters that should be measured more often in the field.

5.2 Field Site and Model Conceptual Description

A detailed description of the research site and management can be found in Larkin et al. (2011). Briefly, the research plots are located in Presque Isle, Maine on Caribou-type soil. There are five crop management systems: SQ is status quo and represents a 2-year barley-potato rotation; PP is a potato monoculture (no rotation); DS is disease suppressive, which includes a 3-year rotation of mustard green manure, Sudangrass green manure, and winter rye-potato; SC is soil conserving, which is a 3-year rotation of no-till barley interseeded with timothy, timothy sod, and potato; and SI is soil improving, which builds upon the SC treatment by adding compost in each phase of the rotation. Each treatment is grown under rainfed (unirrigated) and irrigated conditions. Irrigation treatments are applied based on the average of 20 tensiometers located throughout the research site. 17.5 mm of water is applied at each irrigation for those treatments. Fertilization is based on pre-plant soil tests each season. Planting times at the research site range from May 24 to June 1, and tend to be a week to 10 days later than commercial fields in the same area. Unirrigated plots are replicated six times and irrigated plots are replicated five times.

There are many models for potato production that can reliably replicate field data in a variety of production systems (e.g. POMOD, Kadaja and Tooming 2004; SPUDSIM, Fleisher et al. 2010; SUBSTOR-potato, Ritchie et al. 1995; SIMPOTATO, Hodges et al. 1992). Their detail and complexity have allowed for comprehensive analyses of mechanisms of stress and steady-state production, in part due to their ability to accept site, cultivar, and management-specific parameters.

As early as three decades ago, Monteith (1977) proposed a simple, light-driven model that could estimate yield potential in an area given few inputs. This model has been modified and adapted since its development and includes an "Energy Cascade" model useful for controlled environment studies (Volk et al. 1995). Conceptually, the energy from light "flows" through various efficiencies such as radiation capture efficiency, photosynthetic efficiency, respiration efficiency, and partitioning efficiency until a yield per area is calculated. Mathematically, yield is calculated in this model as:

$$\text{Yield}\,(\text{g yield}/\text{d}/\text{m}^2) = \text{PPF}\times\text{RCE}\times\text{CQY}\times\text{CUE}\times\text{HI}\times\text{CF}$$

Where PPF is photosynthetic photon flux (moles photons available/d/m^2), RCE is radiation capture efficiency (moles photons absorbed/moles photons available), CQY is canopy quantum yield (moles C fixed/moles photons absorbed), CUE is carbon use efficiency (moles C remaining in plant/moles C fixed), HI is harvest index (moles C harvested/moles C remaining in plant), and CF is a conversion factor that takes into account C content of tubers and percent dry weight (g yield/mole C harvested). The model is run in 1-day time steps, and the sum of the daily values is the total yield. These parameters can either be measured directly or calculated from direct measurements. Additionally, the model lends itself to expansion for more mechanistic analysis, as sub-models addressing each parameter are developed, feeding into the main model's equation.

5.3 Description of Model Parameters

Idealized PPF was determined based on location latitude from the plant growth model PlantMod (version 4.0.7, IMJ Software). Idealized PPF is the PPF with no cloud cover, so the more cloud-bearing weather events a site has, the more this term would over-estimate actual PPF (Table 5.1).

Typical PPF was obtained from a 30-year historical weather dataset (Marion and Urban 1995). This dataset is not average values for a period of time, but typical or representative values for that date and time at a given location, in this case, Caribou, ME approximately 10 km north of the field site. An example set of data may utilize data from Feb. 1987 followed by Mar. 1982 because these months were deemed most representative of the weather at that location (Marion and Urban 1995). The value of this, as opposed to using averages, is that day-to-day light fluctuations are simulated. A single day's simulation, therefore, may not accurately predict that day every year, but simulations over longer periods of time would likely predict that site with reasonable precision yet show typical variability from month to month or week to week that can be expected at that location. Actual PPF was calculated based on radiation measurements from a weather station located adjacent to the research plots in Presque Isle, Maine. A pyranometer (model 200X, LI-COR, Lincoln, NE) measured total W m^{-2} every minute and the average of 10 min was stored. It was empirically determined that approximately 48.3% of the measured radiation from

Table 5.1 Model parameters and the values utilized for Idealized, Typical, and Actual scenarios

	Modeling scenario		
	Idealized	Typical	Actual
PPF (mol supplied m^{-2} day^{-1})	Light data obtained from PlantMod 4.0.7 assuming no clouds during the production season	Light data from 30-year database describing typical radiation at the field site	Measured light data at the field site for each year
RCE (mol captured/ mol supplied)	Highest possible from measured canopies, pooled over 4 years	Highest possible from measured canopies, pooled over 4 years	Measured from each year and each treatment
CQY (mol C fixed/ mol captured)	0.03	0.03	0.03
CUE (mol C in plant/ mol C fixed)	0.6	0.6	0.6
HI (mol C in harvestable yield/mol C in plant)	0.8	0.8	0.8
CF (g fresh weight/mol C in harvestable yield)	(30 g potato/mol C harvested in tubers)/0.8 water content	(30 g potato/mol C harvested in tubers)/0.8 water content	(30 g potato/mol C harvested in tubers)/0.8 water content

PPF photosynthetic photon flux, *RCE* radiation capture efficiency, *CQY* canopy quantum yield, *CUE* carbon use efficiency, *HI* harvest index, *CF* conversion factor

that pyranometer was in the 400–700 nm range, so radiation measurements were converted to PPF using $2.07 \times \mathrm{W\ m^{-2}} = \mu\mathrm{mol\ m^{-2}\ s^{-1}}$ (1/0.483 = 2.07; this is a site and/or sensor-specific conversion but is expected to be a reasonable estimate of other sites; see also discussion in Nobel 1991). These instantaneous values were converted to the proper units by multiplying by $3600\ \mathrm{s\,h^{-1}}/1{,}000{,}000\ \mu\mathrm{mol\ mol^{-1}}$. Together, the Idealized, Typical, and Actual model simulations are based on these PPF datasets to provide an optimized, realistic, and actual view of PPF at a given site. The names of these model simulations refer only to the PPF, not necessarily to the selection of other model terms (Table 5.1).

RCE can be estimated on wide spatial scales non-destructively by photographic image analysis (Klassen et al. 2003) or spectroscopic analysis that correlates canopy N-content with light capture (Major et al. 2003). In our case, RCE was determined from weekly light measurements made throughout the production seasons over 4 years of field experiments. The measurements were made with a line quantum sensor (SunScan, Delta-T Devices, Cambridge, UK) that accounts for intercepted and reflected radiation based on canopy characteristics and light measurements above and below the potato canopy. Four years of weekly RCE measurements were pooled and for the Idealized and Typical modelling scenarios (Table 5.1), the pooled values were the highest RCE values for a given day after emergence, which developed a maximum RCE throughout the growing season. The peak RCE would occur if there were no canopy disturbances that occurred due to intermittent drought, herbivory, wind or other disturbances. RCE in individual years were used to evaluate more real-world production scenarios in the Actual modelling scenarios that would include management-specific canopy disturbances.

CQY is a more difficult parameter to measure directly in the field, and only a few measurements have ever been made on whole potato plants or communities (Timlin et al. 2006; Fleisher et al. 2006). Fortunately, CQY varies in predictable ways for C_3 plants and reasonably accurate estimates can be made based on a few known environmental conditions (Long 1991). Photosynthetic efficiency varies with temperature and CO_2 concentration. The peak efficiencies reported for C_3 plants vary from a ratio of 1 CO_2 fixed for every 8 photons of light (quantum efficiency of 0.125 moles CO_2 fixed per mole photons absorbed; Thornley and Johnson 1990) to 1 CO_2 fixed for every 12 photons (quantum efficiency of 0.083 moles CO_2 fixed per mole photons absorbed; Lal and Edwards 1995). In a model for a typical C3 plant, Harley and Tenhunen (1991) assumed a value of 0.06 mol mol^{-1}, whereas Björkman (1981) report a similar value in C_3 plants over a range of temperatures from 20°C to 30°C. High light, such as that found at the tops of canopies, and high temperatures decrease quantum yield in ambient CO_2 concentrations (about 400 μmol mol^{-1} CO_2). In a canopy-scale study of potato photosynthesis, Fleisher et al. (2006) report peak CQY values of above 0.1 mol mol^{-1} in low light and temperatures. In higher light and temperatures, values were measured in the 0.013–0.038 mol mol^{-1} range. We therefore assumed a CQY value of 0.03 mol mol^{-1} for this model, which closely matches those values found by Fleisher et al. (2006) in similar light and temperature environments. Small changes in the value of this term (for example, 0.03–0.04) lead to large changes (in this example, 25% greater) in the overall predictions, and it is for this reason that this term is not modified among simulations since it is based on lab measurements of potato.

CUE is a calculated term that describes the amount of carbon incorporated into the plants divided by the total amount of carbon fixed in photosynthesis. Essentially, it is a term describing how well plants can incorporate the carbon fixed during the day into biomass gain. It requires accurate measurements of canopy net photosynthesis, and night respiration. Because day-time respiration cannot be measured directly, day-time respiration is estimated as some percentage of night-time respiration corrected for changes in day-night temperature. Many estimates of this parameter have been made on a variety of plants grown in different environments. Laboratory measurements tend to be slightly higher than field measurements, but the overall range of CUE reported is 0.50–0.65 in steady-state or actively growing conditions (Gifford 1994, 1995; Frantz et al. 2004). Lower values have been measured for seedlings or in low growth rates due to stressful environments (van Iersel 2003b). Overall, this term is less well studied compared to other parameters within this model. We used a value of 0.60 for this model to reflect healthy, field-grown plants (Gifford 1995).

HI is measured by harvesting the plant at the end of the experiments, weighing tubers separately from the shoot and non-tuberous roots. If, at the end of an experiment the HI is a value of 0.8 (80% of the mass is in the tubers), it can be assumed for simplicity for each day that 80% of retained carbon is partitioned into the tubers.

This would not be the case if yield was to be predicted throughout the growing season as many other conditions can influence HI such as temperature, stage in growth, genetics, N supply, water, light, etc. Potato HI has been reported to be between 0.2 and 0.8, with lower values found in higher temperature environments (Fleisher et al. 2006; Tibbitts et al. 1994). We used a value of 0.8 for this model to reflect HI typically found in lab-based studies performed at the temperatures commonly encountered during field production in this region.

Finally, a conversion factor (CF) is used to convert moles of carbon in potatoes to grams dry weight yield. Since potatoes are predominantly starch and other carbohydrates (Kolbe and Stephen-Beckman 1997), a conversion factor of 30 g/mol harvested potato tubers is used. Potatoes are assumed to be 80% moisture, based on previous reports (Tibbitts et al. 1994; Kolbe and Stephen-Beckman 1997).

5.4 Overall Model Performance

Overall, the model over-predicted yield in five distinct management systems by, on average, 1.4 Mg ha^{-1} (Fig. 5.2). However, the model adequately described the year-to-year variation. This is not in itself surprising since the primary drivers for the

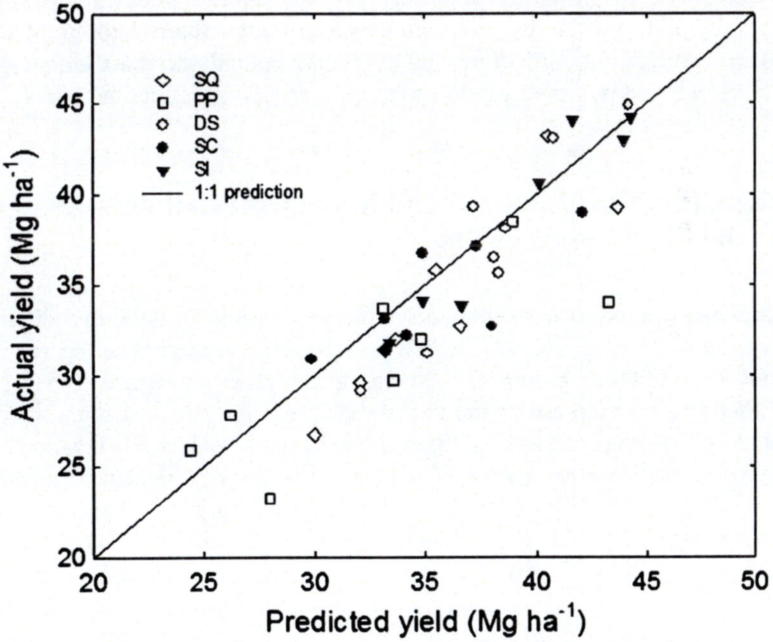

Fig. 5.2 Comparison of predicted (modelled) and actual (measured) yield in the five potato management systems. In general, the model over-estimated the yield by an average of 1.4 Mg ha^{-1}, perhaps due to differences in harvest index, carbon use efficiency, and/or canopy quantum yield

model, light and light capture, were measured directly in each management system. However, it does speak to the reasonableness of the other, estimated parameters such as CQY, HI, and CUE. It has been observed that models using a RCE approach tend to over-estimate the yield perhaps due to the differences in quantum yield among different leaf layers (Fleisher et al. 2010). This effect should have been minimized since estimates of CQY were made not from biochemical or leaf-scale models, but on whole-canopy measurements. The over-prediction likely resulted from small errors in overestimating CQY, CUE, and/or HI. Each parameter was selected based on published reports of plant canopies and, where possible, potato plant communities. CQY value of 0.03 is half the value used in other plant community models (Harley and Tenhunen 1991) and is less than the peak CQY reported in potato canopy studies previously (Fleisher et al. 2006). CUE and HI values of 0.6 and 0.8, respectively, were selected as the highest reported values of potato plants or plants in general in the temperature, CO_2, and light ranges encountered in these field plots. However, many of those reports were based on controlled environment conditions rather than in the field. It is feasible that both CUE and HI would decline slightly in field conditions given less-than-ideal conditions (Nemali and van Iersel 2004; van Iersel 2003a) from water, temperature, insect, or disease stress. These stresses occurred to varying degrees in the different management environments, likely contributing to the discrepancies in predicted versus actual yields in the different treatments. The sporadic nature of stress events make parameterizing a model based on controlled environment studies challenging with potatoes. For example, a single, short-term drought event during tuber bulking can inhibit future bulking of those potatoes and result in initiation of new tubers. These not only decrease potato grade but lower overall yields, which would show up as decreased HI in this model.

5.5 Results of the Measured and Modelled Productivities at the Field Experiments

In the idealized (i.e. no clouds) simulation, the upper limit for potato productivity was predicted to be 85.0 Mg ha^{-1}, which is about 70% greater than top reported yields in the area (Table 5.2) and 180% greater than historical averages (Fig. 5.1). A yield of 85.0 Mg ha^{-1} represents the upper yield limit for potato if there were no clouds and RCE was maximized for the entire growing season at this location.

Utilizing typical weather patterns from the 30-year historical record, which incorporates cloudy conditions typically found in a potato growing season in this area, the maximum potential yield is 55.5 Mg ha^{-1} or 10% greater than maximum measured yields and 80% greater than historical averages (Table 5.2). In other words, best-case yields utilizing realistic weather from Northern Maine represent significant gains above historical averages and have nearly been achieved in small areas on commercial farms in some years. Utilizing actual weather recorded during 2006–2009 and starting with planting dates from those years, the best case yields (maximum RCE) would have been 52.7, 55.8, 49.2, and 58.3 Mg ha^{-1}in a rainfed,

Table 5.2 Potential and measured potato yields in northern Maine

	Potential	Measured
	Mg ha^{-1}	
Maximum	85.0	50.5
Typical climate	55.5	30.9
2006 climate	52.7	34.3
2007 climate	55.8	33.1
2008 climate	49.2	30.3
2009 climate	58.3	30.9

For all potential values, peak radiation capture measured throughout growing seasons over a 4-year period was used. For potential maximum yield, theoretical light based on latitude and no cloud cover was used, while typical climate used historical averages for climate including sunlight to predict upper limit of yield. Maximum measured yield has been observed on smaller areas of commercial farms but has been difficult to reach consistently on large-scale production

SQ management system most similar to industry standards in this region. These yields compare favourably with the yields predicted from the "typical" weather dataset, indicating that as a reference, the typical weather provides realistic conditions for simulation over the course of a potato growing season at this location. The simulated yields compare to reported state-wide averages of 34.3, 33.1, 30.3, and 30.9 Mg ha^{-1} in 2006–2009 respectively (Table 5.2). The differences in yields for the different years are likely due to differences in radiation capture from year to year compared to the optimal RCE that should be possible if no stress is experienced.

If actual radiation capture from those years are used, our predicted yields were 35.5±3.6 Mg ha^{-1}, 33.2±4.6 Mg ha^{-1}, 32.1±1.9 Mg ha^{-1}, and 30.0±0.9 Mg ha^{-1}. Actual yield was 35.8, 31.4, 29.6, and 26.7 Mg ha^{-1} (SQ treatment; Fig. 5.3). Altering management can significantly influence yields of potatoes. Continuous potato production results in lower yield, with less measured radiation capture in this treatment (Figs. 5.3 and 5.4). Managing the soil system to increase resiliency against diseases improves yield above the barley-potato rotation (DS treatment, Fig. 5.3). Reduction of foliar diseases would result in more radiation capture, increasing yield, while reduced root diseases could limit water stress and wilting or direct loss of potatoes. SC and SI treatments also boost yield above the barley-potato rotation, likely due to improvements in water availability and, therefore, reduction in water stress during periods of drought (Fig. 5.3; Porter et al. 1999).

Adding irrigation tends to boost yield, especially in years with less rain or infrequent rain events (Fig. 5.5). Much of the variability in yields among the management strategies could be attributed to radiation capture differences (Fig. 5.4). The largest difference between irrigated and rainfed plants is the radiation capture throughout the season; with irrigation, leaves intercept more light. This could be due to more leaves, larger leaves, or properly oriented leaves (turgid versus wilted) to collect the available light. Peak radiation capture measured in traditional barley-potato plots in unirrigated areas was 81%±6%, while the peak capture in

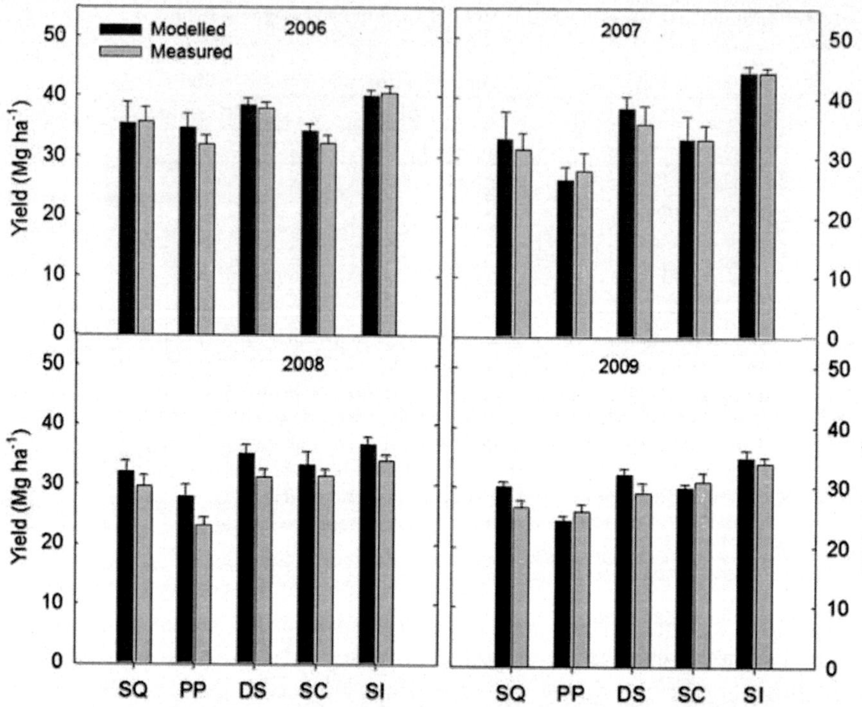

Fig. 5.3 Measured and modelled yields in rainfed field plots managed in different ways for 2006–2009. SQ is status quo barley-potato rotation, PP is continuous potato-potato production, DS is disease suppressive, SC is soil conserving, and SI is soil improving. Mean yield values are shown with standard errors (n = 6)

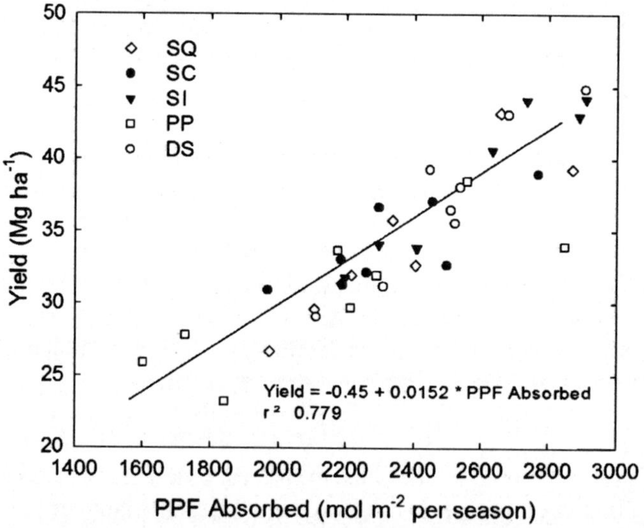

Fig. 5.4 The relationship between PPF (photosynthetic photon flux) absorption and yield of potatoes from 2006 to 2009. SQ is status quo barley-potato rotation, PP is continuous potato-potato production, DS is disease suppressive, SC is soil conserving, and SI is soil improving. Data were collected by taking weekly radiation capture measurements and calculating the fraction of available light absorbed throughout each growing season

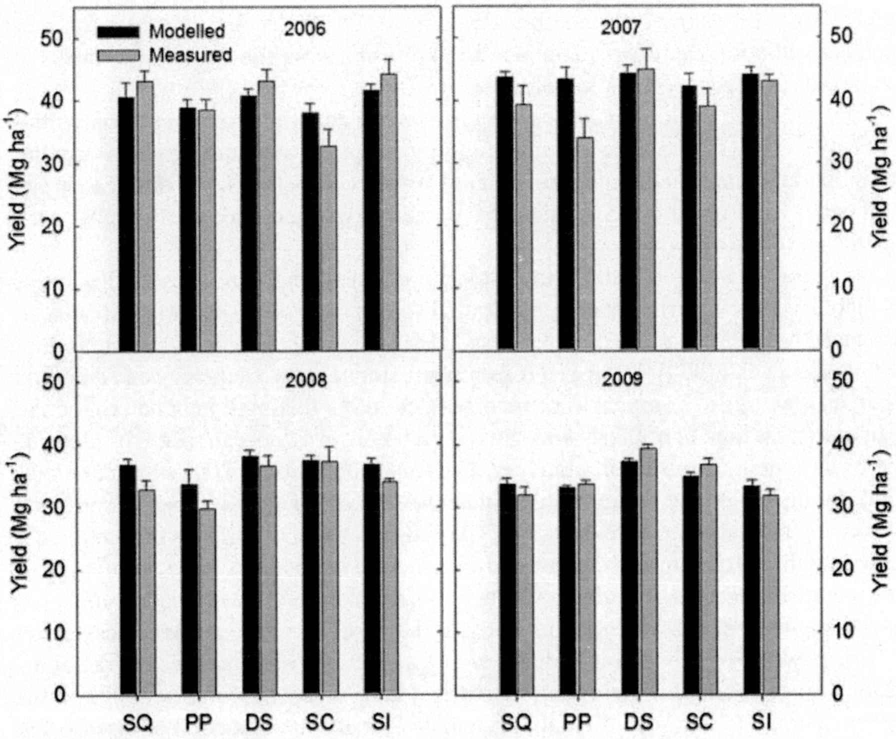

Fig. 5.5 Measured and modelled yields in irrigated field plots managed in different ways for 2006–2009. SQ is status quo barley-potato rotation, PP is constant potato-potato production, DS is disease suppressive, SC is soil conserving, and SI is soil improving. Mean yield values are shown with standard errors (n=5)

irrigated plots was 94%±3%. This resulted in predicted yield of 40.5±2.5 Mg ha^{-1} due to improved radiation capture. Measured yield in those plots was actually 43.2 Mg ha^{-1}, indicating that the increase in yield was due in part, but not exclusively, to improved radiation capture of the canopy.

5.6 What Does the Model Tell Us About Improving Yields Based on the PPF Component of the Model?

Comparing theoretical peak yields to actual yield or predicted yield using real-time weather information suggests some strategies for improving yields. The fact that yield predictions were so close to measured yield when measured PPF and RCE values were used to simulate 2006–2009 seasons strongly indicates the assumed values of CQY, HI, and CUE were reasonable. If values for HI and CUE used in the model are within 0–5% of actual HI and CUE in the field, then the most likely target for improved production is enhancing radiation capture. This can be done by (1) improving the peak radiation capture potential, (2) improving the duration that the peak radiation capture is maintained, (3) by enhancing the light that reaches the canopy, or (4) by a combination of these.

Increasing the peak radiation capture is done by ensuring proper spacing within and between rows to balance maximum resource (light, water, and nutrients) capture for individual plants and minimizing plant-to-plant competition. Recommendations for spacing and resource application have been developed and, for the most part, optimized for this production area.

Irrigation and/or soil management techniques that improve water and nutrient availability have clearly been demonstrated to improve yield in large part due to enhanced radiation capture over the season (Figs. 5.3 and 5.5), which is consistent with Porter et al. (1999). Improved recommendations for the frequency and amounts of green manures or compost to be incorporated into a field will help determine the economic cost/benefit of these strategies so that more growers can take advantage of the clear benefits from these practices. Currently, only about 20% of commercial fields in this region are equipped to irrigate their fields on a regular basis. The costs of meeting regulatory requirements for capturing and using water have proven to be a large inhibition to more widespread use of irrigation in potato fields. A combination of organically-derived soil amendments and irrigation holds the most promise for effectively maximizing radiation capture for the duration of the growing season.

Insect, weed, and disease avoidance would also effectively improve the radiation capture since those stresses can decrease leaf area or create conditions that limit light reaching the canopy. Soil management that targets the reduction of disease has also been shown to minimize crop losses (Larkin and Griffin 2007), and improve yields in part due to enhanced radiation capture (Figs. 5.2 and 5.4). It is important to note that changes in soil or water management can shift the disease risk from one threat to another (Olanya et al. 2010).

It is often said that potato yield in the New England area is limited due to a short growing season. Comparing available light with radiation capture of the potato canopy (Fig. 5.6), it can clearly be seen that the weather-limiting problem has as much to do with crop timing (occurrence during the year) as it does the length of the growing season. Predicted potential PPF ranges from 65 mol m^{-2} day^{-1} to 68 mol day^{-1} from emergence to the Summer Solstice. After that time, PPF steadily declines until harvest, when predicted PPF reaches about 40 mol m^{-2} day^{-1} in early October. At the Summer Solstice, the plants have just emerged, so only about 5% to 10% has been measured to be absorbed, with the variation due to differences in planting date. When the plant has produced a canopy to intercept light, the summer solstice has already been reached. In other words, the peak light environment occurs when there is no canopy to intercept the light, and when the canopy is present, potential light is declining. This makes late-season cloudy days or rain events especially damaging because there are fewer days with less light to compensate for the lost light. Selecting varieties that can emerge in cooler soil conditions or perhaps pre-treating the seeds with a longer dormancy breaking period after cold storage are, theoretically, approaches to improve early potato stand establishment. This model predicts a 10% and 17% increase in yield if planting could occur 2 and 4 weeks earlier in the

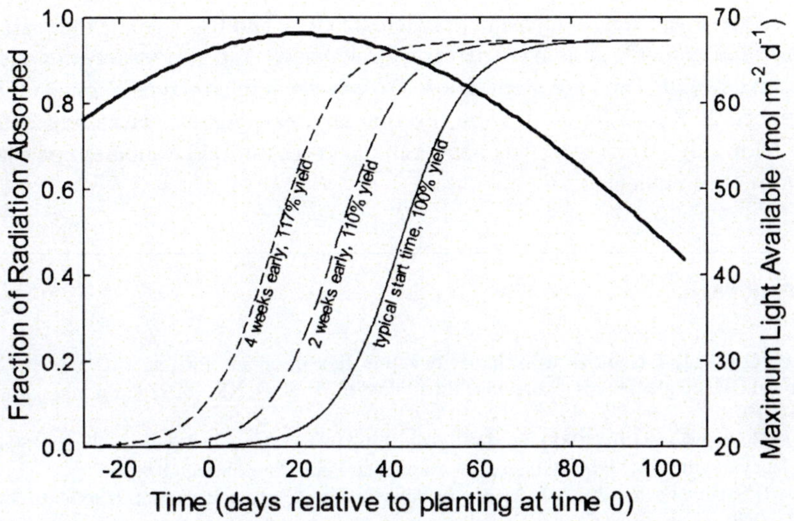

Fig. 5.6 Available light (thick line, right axis) and optimized radiation capture (thin line, right axis) throughout a potato growing season. In a typical year, only 5% to 10% of peak light at the summer solstice is captured due to a later planting time or delayed emergence. Planting 2 (long-dashed line) or 4 weeks earlier (short-dashed line) better captures more light so that yield can be improved by up to 17% over the same growing season length

season, respectively, with no additional days of growth. In other words, without changing the length of the season but when the season occurs, it is possible that significant yield improvements could be made. However, getting into wet fields early in the planting season remains a significant engineering hurdle to overcome in order to plant earlier and avoid soil compaction. Late season frost is still an issue as well with earlier emerging varieties.

Yield was strongly correlated with radiation capture (Fig. 5.4) and some management strategies improved radiation capture. For example, irrigation consistently increased yields due to higher peak radiation capture; avoidance of drought led to fewer wilting or senescing leaves that would intercept little light. Higher disease pressure in continuous potato cultivation was a likely cause of decreased radiation capture (Fig. 5.4), while the disease suppressive system had less loss of canopy from disease and corresponding increases in yield. Finally, the soil improving system, with additions of compost as a cornerstone to the management strategy, had consistent radiation capture between irrigated and unirrigated treatments. This suggests that the compost helped maintain moisture availability throughout each season, thereby avoiding wilting or periodic water stress encountered in the other, rainfed systems.

There is a general lack of measurements of CQY, CUE, and to a lesser extent, HI made on potato plant communities in the field. It deserves to be stated again that in controlled environment studies, HI is known to be influenced profoundly by seasonal temperatures, management, and genetics (Tibbitts et al. 1994; Timlin et al. 2006), which are not addressed in this model. Larger, whole-plant gas exchange measurements are necessary to accurately measure or calculate CQY and CUE. Since there are no commercially available systems to perform these measurements, a variety of systems have been developed and are often tailored to fit a specific crop type (e.g. van Iersel and Bugbee 2000; Miller et al. 1996; Poni et al. 1997; Whiting and Lang 2001). To our knowledge, such work has not taken place yet for field evaluation of whole communities of potatoes.

References

Björkman O (1981) Responses to different quantum flux densities. In: Lange OL, Nobel PS, Osmond CB, Ziegler H (eds) Encyclopedia of plant physiology, NS, vol 12A. Springer, Berlin, pp 57–107

Bugbee BG, Salisbury FB (1988) Exploring the limits of crop productivity I. Photosynthetic efficiency of wheat in high irradiance environments. Plant Phys 88:869–878

Carter MR, Sanderson JB (2001) Influence of conservation tillage and rotation length on potato productivity, tuber disease and soil quality parameters on a fine sandy loam in eastern Canada. Soil Till Res 63:1–13

C.I.A. (2011) The World Factbook. https://www.cia.gov/library/publications/the-world-factbook/

Fleisher DH, Timlin DJ, Reddy VR (2006) Temperature influence on potato leaf and branch distribution and on canopy photosynthetic rate. Agric J 98:1442–1452

Fleisher DH, Timlin DJ, Yang Y, Reddy VR (2010) Simulation of potato gas-exchange rates using SPUDSIM. Agric Forensic Met 150:432–442

Frantz JM, Cometti NN, Bugbee B (2004) Night temperature has a minimal effect on the growth and respiration and rapidly growing plants. Ann Bot 94:155–166

Gifford RM (1994) The global carbon cycle: a viewpoint on the missing sink. Aust J Plant Phys 21:1–15

Gifford RM (1995) Whole plant respiration and photosynthesis of wheat under increased CO_2 concentration and temperature: long-term vs. short-term distinctions for modeling. Global Change Biol 1:385–396

Griffin TS, Porter GA (2004) Altering soil carbon and nitrogen stocks in intensively tilled two-year rotations. Biol Fertil Soils 39:366–374

Harley PC, Tenhunen JD (1991) Modeling the photosynthetic response of C3 leaves to environmental factors. In: Boote KJ, Loomis RS (eds) Modeling crop photosynthesis: from biochemistry to canopy. CSSA Special Publication 19, pp 17–39

Hodges T, Johnson SL, Johnson BS (1992) A modular structure for crop simulation models: implemented in the SIMPOTATO model. Agric J 84:911–915

Kadaja J, Tooming H (2004) Potato production model based on principle of maximum plant productivity. Agric Forensic Met 127:17–33

Klassen SP, Ritchie G, Frantz JM, Pinnock D, Bugbee B (2003) Real time imaging of ground cover: relationships with radiation capture, canopy photosynthesis, and relative growth rate. Spec Pub Crop Sci Soc Am 66:3–14

Kolbe H, Stephen-Beckman S (1997) Development, growth, and chemical composition of the potato crop (*solanum tuberosum* L.). II. Tuber and whole crop. Potato Res 40:135–153

Lal A, Edwards GE (1995) Maximum quantum yields of O_2 evolution in C4 plants under high CO_2. Plant Cell Physiol 36:1311–1317

Larkin RP, Griffin TS (2007) Control soilborne diseases of potato using *brassica* green manures. Crop Protect 26:1067–1077

Larkin RP, Honeycutt CW, Olanya OM, Halloran JM, He Z (2011) Impacts of crop rotation and irrigation on soilborne diseases and soil microbial communities. In: He Z, Larkin RP, Honeycutt CW (eds) Sustainable potato production: global case studies. Springer, Dordrecht

Long SP (1991) Modification of the response of photosynthetic productivity to rising temperature by atmospheric CO_2 concentrations: has its importance been underestimated? Plant Cell Environ 14:729–739

Major DJ, Baumeister R, Toure A, Zhao S (2003) Methods of measuring and characterizing the effects of stresses on leaf and canopy signatures. Spec Pub Crop Sci Soc Am 66:81–93

Marion W, Urban K (1995) User's manual for TMY2s: typical meteorological years. Natl Renewable Energ Lab Spec Publ 463–7668, Golden, CO

Miller DP, Howell GD, Flore JA (1996) A whole-plant, gas-exchange system for measuring net photosynthesis of potted woody plants. Hort Sci 31:944–946

Monteith JL (1977) Climate and the efficiency of crop production in Britain. Philos Trans R Soc Lond B 281:277–294

Nemali KS, van Iersel MW (2004) Light effects on wax begonia: photosynthesis, growth respiration, maintenance respiration, and carbon use efficiency. J Am Soc Hort Sci 129:416–424

Nobel PS (1991) Physicochemical and environmental plant physiology. Academic Press, San Diego, pp 200–202

Olanya OM, Starr GC, Honeycutt CW, Griffin TS, Lambert DH (2007) Microclimate and potential for late blight development in irrigated potato. Crop Protect 26:1412–1421

Olanya OM, Porter GA, Lambert DH, Larkin RP, Starr GC (2010) The effects of supplemental irrigation and soil management on potato tuber diseases. Plant Path J 9:65–72

Poni S, Magnanini E, Rebucci B (1997) An automated chamber system for measurements of whole-vine gas exchange. Hort Sci 32:64–67

Porter GA, Opena GB, Bradbury WB, McBurnie JC, Sisson JA (1999) Soil management and supplemental irrigation effects on potato: I. Soil properties, tuber yield, and quality. Agric J 91:416–425

Ritchie JT, Griffin TS, Johnson BS (1995) SUBSTOR: functional model of potato growth, development and yield. In: Kabat P (ed) Modelling and parameterization of the soil-plant-atmosphere system. Wageningen Pers, Wageningen, The Netherlands, pp 401–435

Thornley JHM, Johnson IR (1990) Plant and crop modeling. Clarendon, Oxford

Tibbitts TW, Cao W, Wheeler RM (1994) Growth of potatoes for CELSS. NASA Contractor Report 177646, NASA, CA

Timlin D, Rahman SML, Baker J, Reddy VR, Fleisher D, Quebedeaux B (2006) Whole plant photosynthesis, development, and carbon partitioning in potato as a function of temperature. Agric J 98:1195–1203

USDA National Agricultural Statistics Services (2011a) Crop production 2010 summary. U.S. Dept. Ag, Washington, DC

USDA National Agricultural Statistics Services (2011b) Maine potatoes 2010 crop acreage, yield, size, and grade. U.S. Dept. Ag. New England Agricultural Statistics, Concord, NH

van Iersel MW (2003a) Carbon exchange rates of four bedding plant species as affected by short-term temperature changes. J Am Soc Hort Sci 128:100–106

van Iersel MW (2003b) Carbon use efficiency depends on growth respiration, maintenance respiration, and relative growth rate. A case study with lettuce. Plant Cell Environ 26:1441–1449

van Iersel MW, Bugbee B (2000) A multiple chamber, semicontinuous, crop carbon dioxide exchange system: design, calibration, and data interpretation. J Am Soc Hort Sci 125:86–92

Volk T, Wheeler R, Bugbee B (1995) An approach to crop modeling with the energy cascade. Life Support Biosph Sci 1:119–127

Whiting MD, Lang GA (2001) Canopy architecture and cuvette flow patterns influence whole canopy net CO_2 exchange and temperature in sweet cherry. Hort Sci 36:691–698

Part III
Linking Irrigated Potato Cropping Systems to Sustainable Agriculture in the West USA

Chapter 6
Potato Tuber Yield, Tuber Size Distribution, and Quality as Impacted by Preceding Green Manure Cover Crops

Samuel Y.C. Essah

Abstract Sustainable potato production systems are needed to maintain higher tuber yield and quality. Studies were conducted from 2005 to 2008 on commercial potato farms to evaluate the effect of preceding green manure cover crops on potato tuber yield, tuber size distribution, and quality. Sorghum sudan (*Sorghum sudanensis* var super sweet), Sorghum sudan (*Sorghum sudanensis* var sordan 79), mustard (*Brassica spp.*), and canola (*Brassica napus*) were planted prior to the 2006 and 2007 potato crop. Another treatment included sordan 79 planted but the above ground biomass removed for hay before incorporating the stubble and roots into the soil. A wet fallow plot where no cover crops were planted was included as a control. Additional green manure cover crops planted prior to the 2008 potato crop included barley (*Hordeum vulgare* L.), barley plus applied compost, sunflower (*Helianthus annus*), peas (*Pisum sativum*), and annual rye grass (*Lolium spp.*). Results from these studies suggest that green manure cover crops can increase potato tuber yield by increasing tuber size, and can improve tuber quality by reducing tuber external defects such as knobs, growth cracks, and misshapes. The positive impact of sorghum sudan suggest that it is possible to harvest the above ground biomass for hay and still obtain high tuber yield and quality when the remaining stubble and roots are plowed into the soil prior to planting potatoes.

6.1 Introduction

Green manure cover crops are grown in rotation with cash crops primarily for their biomass and ground cover to reduce erosion and to incorporate organic matter. The green manure crop may or may not be harvested by grazing or haying.

S.Y.C. Essah (✉)
Department of Horticulture and Landscape Architecture, Colorado State University, San Luis Valley Research Center, 0249 East Road 9 North, Center, CO 81125, USA
e-mail: sessah@lamar.colostate.edu

Z. He et al. (eds.), *Sustainable Potato Production: Global Case Studies*,
DOI 10.1007/978-94-007-4104-1_6,

Green manure cover crops may also be incorporated into the soil while still actively growing (as a green manure) primarily for soil improvement or pathogen suppression. In the western United States and other parts of the world where ground water for irrigated agriculture is limiting, green manure cover crops may be planted in rotation with cash crops to reduce the consumptive use of the ground water. Regardless of the primary purpose, green manure cover crops should be evaluated for their complimentary or antagonistic effects on the following cash crop.

6.1.1 Effect of Preceding Green Manure Cover Crops on Potato Tuber Yield and Quality Due to Disease Suppression

There are a lot of soil pathogens that impact potato yield and quality, but the incorporation of green manure cover crops preceding a potato crop can control the pathogens and result in increased tuber yield and quality (Davis et al. 2010). Green manure cover crops that have been identified to provide disease suppression include barley (*Hordeum vulgare* L.), corn (*Zea mays*), rape (*Brassica rapa*), oats (*Avena sativa* L.), rye (*Secale cereal* L.), sudan grass (*Sorghum sudanensi* L.), and wheat (*Triticum spp.*), with sudan grass showing the highest potato yield response for marketable size tubers (Davis et al. 1994b).

Davis and his colleagues (2010) working in Idaho and California reported that once a suppressive effect has been established, a green manure treatment for a single season was sufficient to either maintain or to re-establish the control of verticillium wilt (VW) of potato. In their study, VW was controlled in the first year of Russet Burbank potato cropping following 2–3 successive years of green manure treatment. With the disease controlled, tuber yields were increased in the first year of potato cropping, but were reduced during the second consecutive year of potato cropping. When a green manure cover crop, such as, Austrian winter pea (*Pisum spp.*), sudan grass, Dwarf Essex rape, Bridger rape, oats, rye, or sweet corn was planted for a single season as a rotation, VW was again controlled and yields of Russet Burbank were increased to higher levels or to levels that were equivalent to initially observed yields following either 2 or 3 successive seasons of green manure treatment. Davis and his associates also demonstrated the significant effects of green manure crops on soil microbial activity, which were inversely related to the incidence of VW. For example, when sudan grass was planted as a green manure crop preceding potatoes, it had a lasting effect on *Fusarium avenaceum* populations with quantitative changes that extended 5 years beyond the time of soil incorporation with sudan grass (Davis et al. 2010). Davis et al. (2010) did demonstrate that the use of sweet corn as a green manure for 2 or 3 seasons suppressed VW by 60–70% and increased potato yields. In addition, corn green manures increased populations of several soil borne fungi. In their study, corn varieties differed for effectiveness as green manure, which could be accounted for by differences of biomass. Results of such studies demonstrate the importance of green manures and soil ecology to the management of potato.

The level of organic matter in the soil has been shown to be a key factor (Davis et al. 2001) related to soil borne pathogens, thereby providing a partial explanation for the positive effect of green manures. A series of green manures evaluated by Davis et al. (1994a, 1996, 1997, 2005), found no evidence of biofumigation. They repeatedly showed a suppression and control of VW without reducing populations of the pathogen. They demonstrated disease suppression when soil borne levels of *V. dahliae* were increased. Davis et al. (2008) hypothesized that the disease suppression was based on biological control. They demonstrated that *V. dahliae* readily grows on organic substrate in the laboratory, but in the field it acts as an obligate parasite and selectively attacks only the apical feeder roots, which in turn may be targeted by *Fusarium* spp. (Huisman 1982). Since green manures can increase populations of *Fusarium* spp. in the soil, they can increase the potential for biological control. Davis et al. (2010) observed that barley green manures selectively increase populations of *Fusarium Solani* in soil. Davis et al. (1999) observed yield increase of potato with sudangrass preceding the potato crop as a green manure. This occurred even though soil borne disease inocula had increased by more than six fold. This phenomena suggest a sustainable potato cropping system with green manure as part of the rotation. Davis and his group have stated that this phenomena has never been observed with fumigation treatments such as picfume or metam sodium.

Mustards (*Brassica spp*) are commonly grown as green manures in Washington and Idaho, with the belief that they act as biofumigants (Sherwood 2007). Davis et al. (2010) stated that this believe is based on the fact that mustards are high in methyl isothiocyanates – the active ingredient of metam sodium. Since metam sodium is the most commonly used fumigant against soil borne pathogens, it is often believed that this plant constituent will be similarly effective. Davis et al. (1996) observed a suppression of VW by using two cultivars of rape seed. They concluded that while the rape seed cultivars could suppress VW, the suppression was not equivalent to the effects of either sudan grass or to sweet corn – two cultivars which produce no methyl isothiocyanate, but caused increased suppression of the pathogen. In another study where Davis et al. (1996) compared three spp. of Brassica for disease suppression (*B. hirta, B. juncea, and B. napus*) with a non-green manure control, they found no evidence for a significant reduction of soil pathogens populations; again showing no evidence for biofumigation.

6.1.2 *Impact of Green Manure Cover Crops on Soil Nutrients and Potato Tuber Yield and Quality*

The intensification of potato production in recent decades has resulted in generally shorter potato rotations, with decreased use of legume and non-legume rotational crops, leading to depletion of soil organic matter (Stark and Porter 2005). Cropping systems that return small amounts of residues to soil can reduce soil C and N status over relatively short time periods (Zielke and Christenson 1986). Therefore, the use of cropping systems that increase organic residues from green manure crops can increase the sustainability of intensive potato production (Stark and Porter 2005). Such green manure crops are used primarily as soil amendments and as nutrient sources for subsequent crops (Cherr et al. 2006).

The slow release of nitrogen (N) from decomposing green manure residues are better timed with plant uptake, thereby increasing N uptake efficiency and crop yield while reducing N leaching losses (Wivstad 1997; Bath 2000). Green manures may drive long-term increases of soil organic matter and microbial biomass, further improving nutrient retention and N uptake efficiency (Augustine et al. 1999). Most green manure crops when incorporated into thc soil help enhance input of organic matter to the soil. Forage grasses are generally efficient in the uptake of soil mineral N, and the decomposition of the incorporated green manure crop releases mineral nitrogen to the succeeding potato crop. The quantity of N mineralized following incorporation of forage grass is increased with increased maturity of the forage crop, due to accumulation of a greater quantity of organic N overtime (Whitehead et al. 1990). The rate of N fertilization of a preceding green manure crop can determine the quantity of N mineralized following incorporation of a cover crop due to the C/N ratios of the plant residues (Whitehead et al. 1990). Incorporation of low C/N ratio residues generally results in net N mineralization, whereas high C/N ratio residues result in net immobilization (Kumar and Goh 2000). The amount of N mineralized from a crop residue varies not only with the C/N ratio but also the composition of the residue (Thorup-Kristensen et al. 2003). Management of green manure crop residues may also influence N mineralization. The N mineralization from a preceding leguminous white clover (*Trifolium repense* L.) or pea (*Pisum sativum* L.) crop was greater when the residue was incorporated by plowing than when it was surface mulched, whereas the reverse was true for non-leguminous ryegrass and wheat crops due to net immobilization following plowing (Kumar and Goh 2002). Early fall plow-down of a preceding red clover crop reduced soil nitrate content in the following spring compared with late-fall or spring plow-down (Sanderson et al. 1999). This finding was attributed to increased fall mineralization and subsequent nitrate leaching during the winter period with early-fall plow-down. Studies by Simard and N'dayegamiye (1993) reported that after plowing in grassland, net mineralization of N in sandy soils can reach 55–80 kg N ha^{-1}.

Studies by Essah et al. (2010) in Colorado, have indicated that harvesting the above ground biomass of sorghum sudan (*Sorghum bicolor* L.) for hay, and incorporating the stalks and roots into the soil have increased tuber yield and improved tuber size of Rio Grande Russet and Russet Norkotah potato.

Following green manure incorporation, cover crops may supply 20–55% of the recovered N to the subsequent crop (Sims and Slinkard 1991; Malpassi et al. 2000), and subsequent fertilizer requirements can be reduced or eliminated (Griffin et al. 2000). Growing green manure cover crops as part of a crop rotation system can help build soil fertility, and are particularly useful when grown before crops which need a lot of N, such as potato. Growing green manure cover crops can be important tools to reduced N losses and increase N supply for the succeeding crops (Stute and Posner 1995). The incorporated green manure crops affect not only the amount of organic N available for the main crops but also the depth distribution of the available

N in the soil. Nitrogen supplied to soil from legume cover crops will likely reduce or possibly replace the need for fertilizer N. In addition to the N added to the soil, green manure legumes may increase organic matter, lower soil bulk density, increase soil microbial biomass, increase infiltration of water (McGuire et al. 1998), reduce crop pathogens (Honeycutt et al. 1996), help to control weeds, and prevent soil erosion. Cropping systems in which large amounts of organic residues from green manure crops are applied to the soil have the potential for improved soil nutrient availability, tilth, water-holding capacity, and aeration (Bullock 1992; Honeycutt et al. 1995; Macrae and Mehays 1985; Smith et al. 1987). Extensive use of crop residues can be a primary factor contributing to the sustainability of a crop production system due to the increased use of renewable on-farm resources and improvements in soil quality (Stark and Porter 2005).

Legumes as green manure crops have received considerable attention as an important component of sustainable cropping systems because they can supply biologically fixed N to subsequent crops. The potential benefits of growing legumes prior to potatoes include, contribution of biologically fixed N to the cropping system, improved yield and quality, improved soil physical properties, suppression of soil-borne potato diseases, and N contributions to subsequent crops (Griffin and Hesterman 1991). Total N content of legume crops grown in potato rotations can be as high as 240 kg N ha^{-1} (Griffin and Hesterman 1991), most of which is released during the first year after incorporation (Fox and Piekielek 1988).

Several green manure cover crop treatments were tested on their effect on subsequent potato crops by Griffin and Hesterman (1991). They observed that 238 kg N ha^{-1} was plowed down in sweet clover, the highest N yielding legume. In that study, the potato crops following legume, produced more dry matter than potatoes in non-legume crops, as well as higher N uptake, but gave no higher tuber yield. Indeterminate potato varieties are particularly sensitive to the seasonal pattern of N availability from plow down green manure crops (Westermann and Kleinkopf 1985). Nitrogen released too early from green manure crops can delay tuber bulking and promote excessive vine growth, resulting in reduced yield and an increased proportion of immature tubers (Ojala et al. 1990). On the other hand, N released too late in the growing season can reduce N use efficiency and increase the potential for nitrate leaching. Nitrogen release rates from legumes can vary widely depending on the crop species and variety grown as green manure (Sainju and Singh 1997), stage of crop growth at incorporation (Frankenberger and Abdelmagid 1985), and environmental factors such as soil temperature and moisture (Honeycutt 1994, 1999; Honeycutt and Potaro 1990). For a given soil, the N mineralization rate can be effectively related to temperature, although the relationship can be confounded by soil moisture effects on microbial activity (Honeycutt 1994, 1999).

Nitrogen release patterns from green manure crops can influence performance of succeeding potato crop. Tindall (1991) observed that when crop residues and spring regrowth were chisel-plowed incorporated just before potato planting, plant available N from barley/alfalfa and barley/red clover residues peaked about 35 days after planting potatoes and then gradually decreased during the growing season. Potatoes began emerging 10–15 days before available soil N had reached its peak. The timing of N release may be very different for fall-incorporated residues. Incorporating legumes in late summer when soil temperatures and microbial activity are usually high typically results in rapid residue decomposition and N mineralization. This increased N mineralization can result in increased soil NO_3-N concentration that would be susceptible to leaching during the winter and early spring. As a result, N contributions to the subsequent potato crop could be substantially reduced. By comparison, late fall legume incorporation would delay residue decomposition, thereby reducing the potential for NO_3-N leaching. Griffin and Hesterman (1991) found that late-season potato vine dry matter and N content were 61–100% and 75–145% higher, respectively, following legumes such as alfalfa, clover, birdsfoot trefoil, and hairy vetch, than following non-legume. However, total and/or marketable tuber yields were unaffected by rotation crop. They concluded that the differential response of potato vegetative and tuber growth following legumes indicate that legume N eventually become available, but probably not early enough to benefit tuber growth. Similarly, N uptake patterns were observed by Van Cingel (1992) suggesting that legume N is sometimes released too late in the season to provide yield benefits to potato. Plotkin (2000) demonstrated that the ratio of vine N to tuber N increases with increasing N rate and when potatoes follow legumes rather than cereal grains. The delayed timing of N release and greater partitioning to the vines can result in a lack of yield benefit, but perhaps more importantly may reduce the quality and maturity of the potato crop. These drawbacks may be prevalent when indeterminate varieties are grown in short season environments and would be less troublesome in regions with long growing seasons or when determinate varieties are grown.

There have been studies on the effect of nitrogen fertilizer inputs and nitrogen cycling from cover crops on potato tuber yields. For example, Neeteson (1988) reported higher potato yields at low N fertilizer rates following leguminous crops that have a lower carbon-to-nitrogen ratio and a higher nitrogen cycling (nitrogen mineralization) potential, such as red clover and alfalfa. Similarly, Neeteson (1988) reported lower yields were observed for potato following oats, which is a cover crop with higher carbon and nitrogen ratios and lower potential to mineralize nitrogen. Neeteson (1988) found that at optimal N fertilizer rates, potato tuber yields were slightly lower following legumes. Results from studies conducted by Sincik et al. (2008) indicated that potatoes following legume cover crops produced approximately 36–38% higher tuber yields compared to potatoes following winter wheat when zero N was applied. In other legume studies, Odland and Sheehan (1957) and Emmond and Ledingham (1972) reported higher potato yields following legumes than non-legumes crops, but Murphy et al. (1967) found no yield benefits following legumes. The author of this chapter suggest that these effects of legumes and/or non- legume cover crops on tuber yield responses could have been in part due to potential responses of potato varieties to the increased availability of nitrogen. For example, Essah and Delgado (2009) found that excessive application of N fertilizer reduced potato tuber yields and tuber quality and that this response was dependent on the type of potato varieties. In other words, in cases where the amount of nitrogen is increased to higher levels than needed, a negative effect could then be observed (Essah and Delgado 2009). Further, when nitrogen is applied in better synchronization with the nitrogen demands of a given potato variety (and the nitrogen that is cycled is accounted for when applying N), tuber yields could be increased (Essah and Delgado 2009). Since cover crops have the potential to affect the nitrogen balances of the following crop (Delgado 1998; Delgado et al. 2001, 2010), this could be one of the factors that could potentially contribute to effects on tuber yields and quality (Delgado et al. 2007; Essah and Delgado 2009), but it needs to be kept in mind that cover crops have effects on soil pathogens, weeds, water balances, and other factors that can also impact yields.

Mishra and Srivastava (2004) reported a 14.2% increase in potato yield when soybean preceded potatoes. Sincik et al. (2008) observed that potatoes following common vetch and faba bean cover crops produced 12.7% and 15.0% more tuber yield ha^{-1}, respectively, compared to potatoes following winter wheat plots. Sincik et al. (2008) concluded that green manure legume cover crops help to increase soil fertility and are particularly useful when grown before potato, and added that when green manure legumes are grown as cover crops, they can lower N fertilizer rate use. The number of growing seasons a field is cropped to perennial forage prior to potato would likely have a strong influence on the rotation benefits resulting from the use of legumes (Angers et al. 1999). However, several studies comparing potato yields following legume crops provide evidence that legumes do not provide potato yield benefits when fertilized at the optimum N rate relative to potatoes following grain crops (Griffin and Hesterman 1991; Neeteson 1988; Neeteson and Zwetsloot 1989; Porter et al. 1999; Plotkin 2000; Van Cingel 1992).

This chapter discusses research studies conducted in Colorado, Western United States, to evaluate the effect of different green manure cover crops as a management tool in potato cropping systems. The effect of using sorghum sudan as a cover crop with the tops harvested for hay was compared with the non-harvesting of sorghum sudan (incorporated as a green manure crop), and other green manure cover crops. These comparisons were made for tuber yield, tuber size distribution, and quality.

6.2 Case Study: Response of Irrigated Potato to Preceding Cover Crops in Colorado

6.2.1 Experimental Procedure

The present study was conducted at the San Luis Valley in south-central Colorado (latitude 37° 40′ N, longitude 106° 9′ W, and 2310 m altitude). Field studies were conducted under commercial grower operations from 2005 to 2008, for center-pivot irrigated potatoes using the traditional best management practices recommended by Colorado State University. Cover crops and potatoes were grown under center-pivot irrigation over a coarse-textured sandy soil with low soil organic matter (<1.5%). Each year, a randomized block design with five replicated plots, 10.5 m long by 3.6 m wide each, was established to plant the cover crops. The cover crop plots were established on a commercial field where the rest of the field was planted to sorghum sudan, except the randomized block area. Limited irrigation was applied each year to minimize the cost of irrigation. Three to five irrigation events were applied for a total irrigation of about 180 mm with the randomized block area receiving the same amount of irrigation as the commercial field.

Cover crops planted prior to the 2006 and 2007 Rio Grande Russet potato cultivar included: (1) sorghum-sudan (*Sorghum sudanensis* var. super sweet); (2) Sordan 79 (*Sorghum sudanensis* var. sordan 79); (3) Sordan 79 with the tops removed for hay (Sordan hay removed); (4) mustard (*Brassica spp.*); (5) and canola (*Brassica napus*). A wet fallow ground treatment was included as a control, where no cover crops were planted, but the same amount of irrigation was applied as the treatments with cover crops. Additional cover crops that were added to the study to precede the 2008 Russet Norkotah cultivar included: (1) barley (*Hordeum vulgare* L); (2) barley plus applied compost; (3) sunflower (*Helianthus annus*); (4) peas (*Pisum sativum*); (5) annual rye grass (*Lolium spp.*); and the previous cover crops and fallow system used for the previous years of 2006 and 2007. The barley crop was harvested for grain before the remaining straw was incorporated into the soil by plowing in the fall. After removing the above ground biomass of sordan 79 for hay, the remaining stubble and roots were incorporated into the soil by plowing in the fall. All other preceding cover crops were incorporated into the soil while still actively growing (as green manures) during the summer.

Potato plots were harvested for the 2006, 2007, and 2008 treatments. Each randomized block was set up so that each plot had four rows spaced at 0.85 m, with tubers planted at in-row seed spacing of 28–30 cm. Potatoes were planted between May 10th and 15th and harvested between September 15th and 20th by harvesting the two middle rows of each plot using an experimental plot potato digger at the farmers field, the week before commercial farming harvesting operations at the field site started. For each plot, total tuber yield was measured, and tubers were sorted into various size distribution groups based on weight (<114 g, >114 g, >284 g, 114–284 g and 284–454 g tubers). Additionally, tuber external defects were evaluated.

Statistical analysis was conducted using analysis of variance (ANOVA) (SAS Institute Inc. Cary, NC, Version 9.2). ANOVA was performed for total tuber yield, tuber size distribution groups, and tuber quality parameters. Differences among treatment means were compared using the least significant difference test (LSD) at the 0.05 level of probability.

6.2.2 Results and Discussion

6.2.2.1 Tuber Yield and Tuber Size Distribution

When averaged over 2006 and 2007, Rio Grande Russet tubers responded to cover crop treatments compared to a wet fallow system (Fig. 6.1). For total tuber yield, there were no significant differences among treatments (Fig. 6.1a). However, Rio

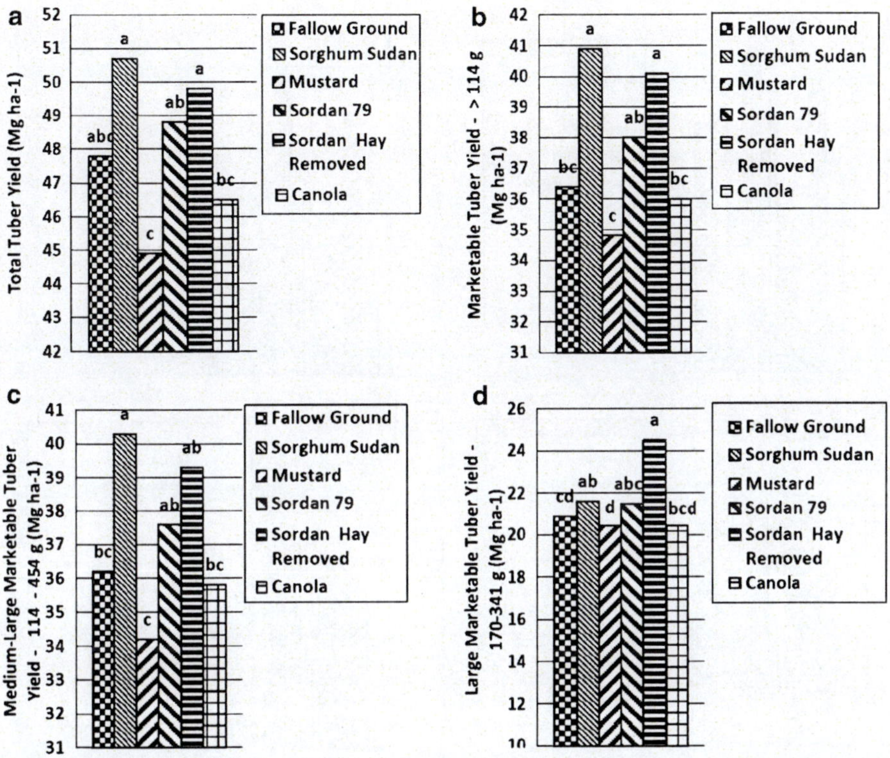

Fig. 6.1 Effect of preceding cover crop on (**a**) total tuber yield, (**b**) marketable tuber yield, (**c**) medium-large marketable tuber yield, and (**d**) large marketable tuber yield of Rio Grande Russet potato crop

Grande Russet marketable tuber yield increased and tuber quality (larger tubers) improved when potato followed sorghum sudan green manure, and the sorghum sudan variety sordan 79 with above-ground biomass harvested for hay (Fig. 6.1b, c, d). Larger tubers (tubers > 114 g and between 170–341 g) were produced in both sorghum sudan as green manure crop and sordan 79 with the above ground biomass harvested for hay compared to the wet fallow treatment.

When the quality of tubers (larger tubers) following sorghum sudan green manure and sordan 79 with the above ground biomass harvested for hay was compared to the quality of the tubers following green manure canola and mustard, it was found that both sorghum sudan varieties contributed to higher total and larger tubers than the canola and mustard green manure crops. Canola, mustard and wet fallow total yield and marketable size yields were not significantly different among themselves.

In summary, for the average of 2006 and 2007, sorghum sudan and the sorghum sudan variety sordan 79 green manure cover crops contributed to higher yields and better tuber quality (larger tubers) than potatoes following a wet fallow system.

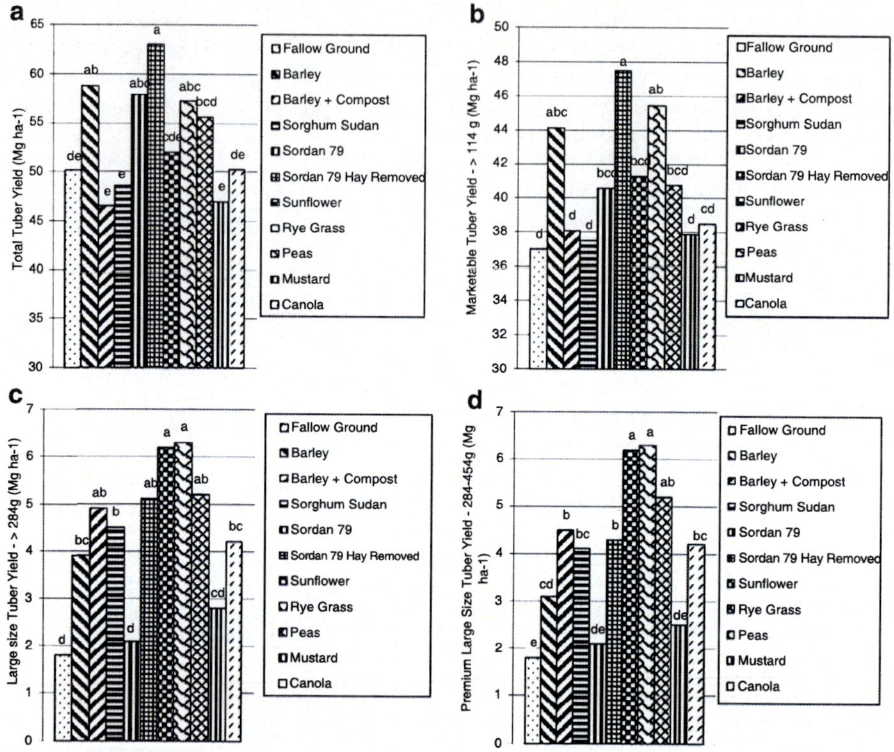

Fig. 6.2 Effect of preceding cover crops on (**a**) total tuber yield, (**b**) marketable tuber yield, (**c**) large size tuber yield, and (**d**) premium large size tuber yield of Russet Norkotah crop

In 2008, Russet Norkotah tuber production and quality responded to cover crop treatments compared to a wet fallow system (Fig. 6.2). In this year, both sorghum-sudan varieties incorporated as green manure had a positive impact on tuber production and/or quality. Sordan 79 increased tuber yield and quality compared to wet fallow. Total yield production of Russet Norkotah was higher following the sordan 79 with all aboveground biomass incorporated and/or harvested for hay treatments than following a wet fallow system (Fig. 6.2a). Marketable-size tuber yields (>114 g) and yields of better-quality tubers (>284 g and 284–454 g) following sordan 79 with all aboveground biomass harvested for hay were higher than when following wet fallow (Fig. 6.2b, c, d). Additionally, when Russet Norkotah followed the sorghum sudan super sweet variety, there was also an increase in the quantity of larger tubers (>284 g and 284–454 g) compared to when following a wet fallow system (Fig. 6.2c, d)

Russet Norkotah following a barley stubble incorporated or ryegrass green manure cover crop had higher total and marketable-size tuber yields than when following wet fallow. The barley stubble incorporated and ryegrass green manure crops also contributed to better tuber quality (tubers > 284 g and 284–454 g) than the wet fallow system. Russet Norkotah following a barley cover crop that received compost, or canola, sunflower, or pea green manure cover crops also had better-quality tubers (>284 g and 284–454 g) than when it followed a wet fallow system.

When Russet Norkotah followed a ryegrass or sunflower green manure crop, the quality of the tubers (>284 g and 284–454 g) improved compared to when it followed a barley stubble incorporated. Russet Norkotah following sordan 79 with all aboveground biomass harvested for hay, or the pea green manure crop, had increased tuber quality as far as the size of the tubers within the compartment 284–454 g, when compared to Russet Norkotah following barley stubble incorporated. The barley and compost applied treatments also increased the quality of the tubers when compared to barley alone (284–454 g); however, compost reduced the total and marketable tuber production when compared to barley alone.

In summary, both varieties of sorghum sudan, sunflower, canola, peas and ryegrass as green manures increased the quantity and/or quality of the tubers when compared to wet fallow. Ryegrass, sunflower, sordan 79 with all aboveground biomass harvested for hay, and peas, increased the quantity and/or quality of the tubers when compared to barley stubble incorporated. Manure application to the preceding barley cover crop increased tuber quality when compared to barley alone, but reduced total production. Mustard did not improve the quality of the tuber when compared to wet fallow or barley. Mustard as a green manure crop did not improve total tuber yields and marketable yields, but total and marketable tuber yields with mustard as green manure were lower than with barley stubble incorporated.

6.2.2.2 Tuber External Defects

For 2006, Rio Grande Russet responded to cover crop treatments with improved tuber quality, represented by a reduced percentage of tuber external defects ($P<0.05$, Table 6.1). In 2006, the treatment with the highest percentage of external defects was the wet fallow treatment (no cover crops); over 3% of the tubers produced following this treatment had external defects. Sorghum sudan super sweet variety, mustard, sorghum-sudan variety sordan 79 with all aboveground biomass harvested for hay, and canola all reduced the percentage of external defects such as cracks, knobs and misshapes, when compared to the wet fallow treatment. Most of the cover crop treatments reduced the external defects by 50%, bringing the percentage down to approximately 1.5% or lower. Mustard as green manure crop eliminated nearly all external defects, lowering the percentage to about 0.3%.

For 2007, the external defects data is not presented, since there were minimal external defects at the site (across all treatments<0.5%). Only the sorghum-sudan super sweet variety and the sorghum-sudan sordan 79 variety showed any external defects, which were less than 0.5%. All other treatments had zero external defects.

For 2008, Russet Norkotah also responded to cover crop treatments compared to a wet fallow system. In this year, the treatment with the highest rate of external defects was the wet fallow treatment, with close to 2% occurrence of defects (Table 6.1). On average, all of the cover crop treatments reduced the percentage of external defects by about 50%,

Table 6.1 Effects of preceding cover crop and/or treatment on tuber external defects of Rio Grande Russet (2006) and Russet Norkotah (2008)*

	2006	2008
Cover crop/Treatment	Tuber external defects (%)**	
Wet fallow	*3.1a*	*1.8a*
Barley	–	*0.9b*
Barley and compost applied	–	*0.9b*
Sunflower	–	*0.6bc*
Sordan 79	*2.3ab*	*0.2c*
Sordan 79 with hay removed	*1.7bc*	*0.8b*
Sorghum Sudan	*1.4c*	*0.9b*
Canola	*1.1cd*	*0.7bc*
Mustard	*0.3d*	*0.2c*
Peas	–	*1.0b*
Ryegrass	–	*0.9b*
LSD	*0.8*	*0.5*

*LSD compares means between cover crops and/or treatments. Values in the same column and with different letters are significantly different at $P < 0.05$
**Includes growth cracks, knobs, and misshapes

with only 1% (or less) of the tubers showing external defects. Sorghum-sudan variety sordan 79 and mustard as green manure resulted in a lower percentage of external defects than the barley with the stubble incorporated before potato.

6.2.3 *Summary*

The goal of this chapter is to present the effects of cover crop treatments on potato tuber production and quality. In the case study presented, field studies were conducted under commercial farm operations to assess the effects of different cover crop treatments on the yield and quality of the potato tubers that followed, as measured by tuber size (with larger tubers being considered of better quality) and appearance (e.g., reduced percentage of cracks, knobs, and misshapes). These studies were conducted from 2005 to 2008, and on average cover crop treatments provided a significant advantage over wet fallow, contributing to increased yields and better tuber quality (larger tubers with less external defects). Sorghum sudan as green manure cover crop showed an advantage in increasing tuber yields and/or quality, providing the farmer with additional income compared to a wet fallow system. Only for two of three studies is tuber external defects reported, since in one study the percentage of external defects for all treatments was less than 0.5%. For the two studies with measurable external defects above 2% for some treatments, the cover crop treatments were beneficial and reduced the percentage of tuber external defects. For these two studies, cover crop treatments reduced external defects on average by 50%, helping to minimize the potential for losses in profit. The percentage of external defects was higher for the tubers produced after wet fallow. It is important to note that the use of canola, mustard, and sordan 79 as green manure crops, reduced external defects in the tubers that followed, compared to tubers that followed the use of barley stubble incorporated. Mustard did not provide as great of an advantage over the wet fallow. The case study presented in this chapter shows that there are several other green manure cover crops in addition to sorghum-sudan that can provide tuber production and/or quality advantages. Data presented in this chapter suggest that the responses in potato tuber yields and quality to the preceding cover crop treatments could be due to a series of factors other than just nutrients.

When we look at the tuber responses as far as production (yields) and quality (size and defects), during 2006, positive effects were achieved when potato followed either sorghum-sudan variety. Even when the aboveground biomass for the variety sordan 79 was removed for hay, there was still a positive response in potato tuber production and quality. The results suggest that both the belowground material and the aboveground litter left behind after harvesting the cover crop for hay are playing an important role, potentially contributing to soil biological and biogeochemical factors that may be providing the mechanism for the potatoes' physiological response.

Additionally, several cover crop treatments such as the ryegrass and sordan 79 with all the aboveground biomass harvested for hay, also showed the potential to produce the same marketable-size yields with better tuber quality (larger tubers with less external defects) than the barley stubble incorporated treatment. Delgado et al. (2007) found a correlation between tuber yields and the nutrient content of the preceding cover crop. Davis et al. (2010) found that the preceding green manure crops affected soil biology, contributing to yield responses. The underlying mechanisms that are causing these physiological responses by the potato crop are unknown and are beyond the scope of this chapter. However, the responses of potato following a grain cover crop, a leguminous crop, or even a grain cover crop with compost, all of which are presented in this chapter, suggest that the mechanisms are very complex. Additional research in this area will be needed to model some of these physiological responses and to better understand the soil-plant interface in green manure cover crop–potato systems.

6.3 Conclusions

The purpose of the 2005–2008 studies was to evaluate the effects of certain types of cover crop treatments on tuber yield, as well as to obtain new, additional information on the effects of cover crop treatments on tuber size distribution and/or external defects. This unique set of studies has enormous implications for sustainability. The results suggest that although there is not yet a clear understanding of the mechanism (possible mechanisms could include but are not necessarily limited to soil biology impacts, suppression of diseases, biogeochemistry pathways and/or availability of macro and micro nutrients, and the physiological responses by the potato at the root, tuber and/or aboveground compartments, which can even influence the external appearance of a tuber), cover crops can potentially contribute to a sustainable system that can improve potato tuber yields and quality and minimize external tuber defects. Additionally, positive results from the sorghum sudan crop suggest that there is even potential to use cover crops for hay production while still keeping the sustainability and tuber quality benefits of these crops.

Again, although the mechanism behind these effects is not yet certain, it is proposed that cover crops may impact soil biology and biogeochemical processes that impact the soil-root-plant system, triggering these physiological responses by the potato that impact yields, tuber size and quality (external defects). There is a need for additional research on the soil-plant-microbiological interactions. This is a complex system; for example, Manter et al. (2010) reported that soil microbes inside the root were correlated with tuber yields. More information is needed to understand how we can manage systems to continue to achieve greater production and higher-quality tubers, and the potential economic benefits for farms that may result. It is clear from the 2005–2008 studies that the tested cover crop treatments are generating a positive, or in some instances, a negative effect on potato tuber yields and quality, and additional research is needed to understand the underlying mechanisms of these effects.

Understanding how these production and quality effects are accomplished is important not only for potential economic benefits for farms, but also for conservation and sustainability. Due to the interrelated factors of limited soil and water resources, continuing population growth, and global climate change, there is a need to apply conservation practices that will contribute to sustainable systems and help conserve soil and water quality. It is proposed that the use of the summer green manure cover crops–potato systems with limited irrigation (<200 mm), which was discussed in this chapter and was found to contribute to higher yields and tuber quality while conserving water resources, is among the potential practices that can be used to mitigate climate change and adapt to it effects. Additional studies will be needed to understand the mechanism causing these potato systems to respond positively to these cover crop treatments.

References

Angers DA, Edwards LM, Sanderson JB, Bissonette N (1999) Soil organic matter quality and aggregate stability under eight potato cropping sequences in a fine sandy loam of Prince Edward Island. Can J Soil Sci 79:411–417

Augustine EO, Ortal CI, Pascua SR Jr, Santa Cruz PC, Padre AT, Ventura WB, Obien SR, Ladha JK (1999) Role of indigo in improving the productivity of rainfed lowland rice-based cropping systems. Exp Agr 35:201–210

Bath B (2000) Matching the availability of N mineralized from crops with the N-demand of field vegetables. Dissertation, Swedish University of Agricultural Science

Bullock DG (1992) Crop rotation. Crit Rev Plant Sci 11:309–326

Cherr CM, Schlberg JMS, McSorley P (2006) Green manure approaches to crop production: a synthesis. Agron J 98:302–319

Davis JR, Stark JC, Sorensen LH, Schneider AT (1994a) Interactive effects of nitrogen and phosphorus and verticillium wilt of Russet Burbank potato. Am Potato J 71:467–486

Davis JR, Pavek JJ, Corsini DL, Sorensen LH, Schneider AT, Everson DO, Westermann DT, Huisman OC (1994b) Influence of continuous cropping of several potato clones on the epidemiology of verticillium wilt of potato. Phytopathology 84:207–214

Davis JR, Huisman OC, Westermann DT, Hafez SL, Everson DO, Sorensen LH, Schneider AT (1996) Effects of green manures on Verticillium wilt of potato. Phytopathology 86:444–453

Davis JR, Huisman OC, Westermann DT, Everson DO, Sorensen LH, Schneider AT (1997) Alternative approaches for control of Verticillium wilt of potato with sudangrass. In: Tjamos EC, Rowe RC, Heale JB, Fravel DR (eds) Advances in verticillium research and disease management. APS, St. Paul, pp 292–295

Davis JR, Huisman OC, Everson DO, Sorenson LH, Schneider AT (1999) Control of Verticillium wilt of the Russet Burbank potato with corn and barley (abstr.). Am J Potato Res 76:367

Davis JR, Huisman OC, Everson DO, Schneider AT (2001) Verticillium wilt of potato. A model of key factors related to disease severity and tuber yield in southern Idaho. Am J Potato Res 78:291–300

Davis JR, Huisman OC, Everson DO, Schneider AT, Sorensen LH (2005) Suppression of Verticillium wilt with wheat and improved yield and quality of the Russet Burbank potato (Abst.). Am J Potato Res 82:64

Davis JR, Huisman OC, Everson DO, Westermann DT, Schneider AT (2008) The mode of action of green manures for the suppression of Verticillium wilt of potato (abst). Am J Potato Res 85:7

Davis JR, Huisman OC, Everson DO, Nolte P, Sorensen LH, Schneider AT (2010) Ecological relationships of verticillium wilt suppression of potato by green manures. Am J Potato Res 87:315–326

Delgado JA (1998) Sequential NLEAP simulations to examine effect of early and late planted winter cover crops on nitrogen dynamics. J Soil Water Conserv 53:241–244

Delgado JA, Riggenbach RR, Sparks RT, Dillon MA, Kawanabe LM, Ristau RJ (2001) Evaluation of nitrate-nitrogen transport in a potato-barley rotation. Soil Sci Soc Am J 65:878–883

Delgado JA, Dillon MA, Sparks RT, Essah SYC (2007) A decade of advances in cover crops. J Soil Water Conserv 62:110A–117A

Delgado JA, Del Grosso SJ, Ogle SM (2010) ^{15}N Isotopic crop residue exchange studies suggest that IPCC methodologies to assess N_2O-N emissions should be reevaluated. Nutr Cycling Agroecosyst 86:383–390

Emmond GS, Ledingham RJ (1972) Effects of crop rotation on some soil-borne pathogens of potato. Can J Plant Sci 52:605–611

Essah SYC, Delgado JA (2009) Nitrogen management for maximizing tuber yield, quality and environmental conservation. In: Yanful EY (ed) Appropriate technologies for environmental protection in the developing world. Springer, The Netherlands

Essah SYC, Delgado JA, Dillon M (2010) Response of irrigated potato (Solanum tuberosum L.) to green manure cover crops. In: Annual Meetings Abstract. ASA/CSSA/SSSA. October 31–November 3. Long Beach, CA, USA, 120pp

Fox RH, Piekielek WP (1988) Fertilizer N equivalence of alfalfa, birdsfoot trefoil, and red clover for succeeding corn crops. J Prod Agric 1:313–317

Frankenberger WT Jr, Abdelmagid HM (1985) Kinetic parameters of nitrogen mineralization rates of leguminous crops incorporated into soil. Plant Soil 87:257–271

Griffin TS, Hesterman OB (1991) Potato response to legume and fertilizer nitrogen sources. Agron J 83:1004–1012

Griffin T, Liebman M, Jemison J Jr (2000) Cover crops for sweet corn production in a short-season environment. Agron J 92:144–151

Honeycutt CW (1994) Linking N mineralization and plant N demand with thermal units. Soil testing: prospects for improving nutrient recommendations. Soil Sci Soc Am Spec Publ 40, Madison, WI
Honeycutt CW (1999) Nitrogen mineralization form soil organic matter and crop residues: field validation of laboratory procedures. Soil Sci Soc Am J 63:134–141
Honeycutt CW, Potaro LJ (1990) Field evaluation of heat units for predicting crop residue carbon and nitrogen mineralization. Plant Soil 125:213–220
Honeycutt CW, Clapham WM, Leach SS (1995) Influence of crop rotations on selected chemical and physical properties in potato cropping systems. Am Potato J 72:721–735
Honeycutt CW, Clapham WM, Leach SS (1996) Crop rotation and nitrogen fertilization effects on growth, yield and disease incidence in potato. Am potato J 73:45–61
Huisman OC (1982) Interrelations of root growth dynamics to epidemiology of root-invading fungi. Ann Rev Phytopathol 20:303–327
Kumar K, Goh KM (2000) Crop residues and management practices: effects on soil quality, soil nitrogen dynamics, crop yield, and nitrogen recovery. Adv Agron 68:197–319
Kumar K, Goh KM (2002) Management practices of antecedent leguminous and non-leguminous crp residues in relation to winter wheat yields, nitrogen uptake, soil nitrogen mineralization and simple nitrogen balance. Eur J Agron 16:295–308
Macrae RJ, Mehays GR (1985) The effect of green manuring on physical properties of temperate area soils. Adv Soil Sci 3:71–93
Malpassi RN, Kaspar TC, Parkin TB, Cambardell CA, Nubel NA (2000) Oat and rye root decomposition effects on nitrogen mineralization. Soil Sci Soc Am J 64:208–215
Manter DK, Delgado JA, Holm DG, Stong RA (2010) Pyrosequencing reveals a highly diverse and cultivar-specific bacterial endophyte community in potato root. Microbial Ecol 60:157–166
McGuire AM, Bryant DC, Denikson RF (1998) Wheat yields, nitrogen uptake, and soil moisture following winter legume cover crop vs fallow. Agron J 90:404–410
Mishra KK, Srivastava JS (2004) Soil amendments to control common scab of potato. Potato Res 47:101–109
Murphy HJ, Carpenter PN, Goven MJ (1967) Potato fertilization-rotation studies on Aroostook farm permanent fertility plots 1951–1965. Maine Agr Exp Stn Bull 645
Neeteson JJ (1988) Effects of legumes on soil mineral nitrogen and response of potatoes to nitrogen fertilizer. In: Vos J et al (eds) Effects of crop rotation on potato production in the temperate zones. Kluwer Academic Publishers, Dordrecht
Neeteson JJ, Zwetsloot HJC (1989) An analysis of the response of sugarbeet and potatoes to fertilizer N and soil mineral N. Neth J Agric Sci 37:129–141
Odland TE, Sheehan JE (1957) The effect of redtop and red clover on yields following crops of potatoes. Am Potato J 34:282–284
Ojala JC, Stark JC, Kleinkopf GE (1990) Influence of irrigation and nitrogen management on potato yield and quality. Am Potato J 67:29–44
Plotkin J (2000) The effect of green manure rotation crops on soils and potato yield and quality. Dissertation, University of Maine
Porter GA, Opena GB, Bradbury WB, McBurnie JC, Sisson JA (1999) Soil management and supplemental irrigation effects on potato: 1. soil properties, tuber yield, and quality. Agron J 91:416–425
Sainju UM, Singh BP (1997) Winter cover crops for sustainable agricultural systems: influence on soil properties, water quality and crop yields. Hort Sci 32:21–28
Sanderson JB, MacLeod JA, Kimpinski J (1999) Glyphosate application and timing of tillage of red color affects potato response to N, soil N profile, and root and soil nematodes. Can J Soil Sci 79:65–72
Sherwood P (2007) Spread the news: mustard can drive away pests. Panna Magazine, Springer
Simard RR, N'dayegamiye A (1993) Nitrogen mineralization of meadow soils. Can J Soil Sci 73:27–38
Sims JR, Slinkard AE (1991) Development and evaluation of germplasm and cultivars of cover crops. In: Hargrove WL (ed) Cover crops for clean water. PRC International Conference, West Tennesse Exp Stn, Jackson. 9–11 April 1991. Soil and Water Conserv Soc, Ankeny, IA

Sincik M, Metin TZ, Tanju GA (2008) Responses of potato (*Solanum tuberosum* L) to green manure cover crops and nitrogen fertilization rates. Am J Potato Res 85:150–158
Smith MS, Frye WW, Varco JJ (1987) Legume winter cover crops. Adv Soil Sci 7:95–139
Stark JC, Porter GA (2005) Potato nutrient management in sustainable cropping systems. Am J Potato Res 82:329–338
Stute JK, Posner JL (1995) Synchrony between legume nitrogen release and corn demand in the upper Midwest. Agron J 87:1063–1069
Thorup-Kristensen K, Magid J, Jensen JS (2003) Catch crops and green manures as biological tools in nitrogen management in temperate zones. Adv Agron 79:227–302
Tindall TT (1991) Potato available nitrogen from barley/legume underseedings and reduced tillage/ridge tillage potatoes. Dissertation, University of Maine
Van Cingel EH (1992) Nitrogen equivalence of rotation crops and its effect on potato yield and tuber maturation. Dissertation, University of Maine
Westermann DT, Kleinkopf GE (1985) Nitrogen requirements of potatoes. Agron J 77:616–621
Whitehead DC, Bristow AW, Lockyer DR (1990) Organic matter and nitrogen in the unharvested fractions of grass swards in relation to the potential for nitrate leaching after ploughing. Plant Soil 123:39–49
Wivstad M (1997) Nitrogen mineralization and yield of spring wheat as influenced by a source of nitrogen in cropping systems, Agraria 34. SLU, Uppsala, Sweden
Zielke RC, Christenson DR (1986) Organic carbon and nitrogen changes in soil under selected cropping systems. Soil Sci Soc Am J 50:363–367

Chapter 7
Mustard Green Manure Use in Eastern Washington State

Andrew M. McGuire

Abstract In the irrigated region of Eastern Washington State, the use of mustard green manures in potato production has grown from 720 ha in 1999 to over 12,000 ha in 2010. Farmers are using the practice mainly before potatoes, and primarily for soil quality benefits. However, the practice has also shown potential for management of soilborne pests, including *Verticillium dahliae* and parasitic nematodes (*Meloidogyne chitwoodi* and *M. hapla*). Research has confirmed these pest related benefits in a short wheat-potato rotation but results have been inconsistent in other rotations after a single green manure crop. Because of the low organic matter soils in this region, farmers are finding the soil quality benefits, such as increased infiltration rates and resistance to wind erosion, to be of greater value than the pest control benefits. Ongoing research to determine the pest suppression mechanisms of the practice holds promise of greater use and benefits to potato production in this region.

7.1 Introduction

In the early 1990s, Dale Gies was looking for a solution. His problem was obtaining enough land each year to produce the amount of potatoes that could support him and his family. His irrigated farm, just South of Moses Lake, Washington (USA), was too small if he used the 3- or 4-year rotations that were the norm in this region. He could lease land, but its availability was uncertain and its quality variable. What he needed was something that would allow him to grow potatoes in a 2-year rotation on his own land without the risk of yield reductions due to soilborne pests or deteriorating soil quality.

A.M. McGuire (✉)
Center for Sustaining Agriculture and Natural Resources (CSANR),
Washington State University, Ephrata, WA 98823, USA
e-mail: amcguire@wsu.edu

Z. He et al. (eds.), *Sustainable Potato Production: Global Case Studies*,
DOI 10.1007/978-94-007-4104-1_7,

While searching for a solution, Gies happened to read about research in nearby Idaho where mustard was being tested for control of cyst nematodes in sugar beets. Intrigued, Gies obtained seed of one of the varieties being tested (*Sinapis alba* var. Martigena) and began growing the mustard as a green manure following his spring wheat and preceding potatoes in his desired 2-year rotation.

After 6 years of apparent success with his experiment, Gies invited Washington State University (WSU) to test his observations. Through on-farm trials, WSU confirmed that Gies' cropping system allowed him to both maintain his potato yields and improve his soil. In addition, he could eliminate his normal fumigation with metam sodium (sodium N-methyldithiocarbamate). Although a few other farmers were also trying mustard or other green manures in potato production at that time, it was these surprising, counter-intuitive, results, disseminated though a WSU Extension program, that got the attention of potato farmers in the Columbia Basin of Eastern Washington State.

7.2 Past Research

Green manuring is an old technology, used by farmers since at least the fifth century BC. Immigrants brought the practice to the USA from Northern Europe and its use reached a peak in the early 1900s. Since then, most farmers have replaced green manures with synthetic inorganic fertilizers. Recently, however, there is renewed interest in green manures for their pest related benefits, in addition to their long-known soil building benefits.

Even while Dale Gies was searching for a solution to his problem, researchers were looking at mustards and other green manure crops for their potential in potato production. Mustard and other *Brassica* crops have been shown to suppress nematodes (Mojtahedi et al. 1993; Riga and Collins 2004), weeds (Boydston and Hang 1995), and soilborne fungal pathogens (Olivier et al. 1999; Smolinska and Horbowicz 1999) but not wireworms (Horton 2008). *Verticillium dahliae* has been suppressed by green manures of sudangrass, wheat, and sweet corn and mustard (Davis et al. 1996, 2007; Ochiai et al. 2007; Larkin et al. 2011a). Research in other regions has shown that green manures preceding potatoes can suppress *Rhizoctonia* (Sexton et al. 2007), common scab and powdery scab (Larkin and Griffin 2007; Larkin et al. 2010, 2011b). One study found that non-mustard Brassica green manures increased the total fungal populations but reduced those of *Pythium* while a non-Brassica green manure resulted in increases in both total fungal and Pythium populations (Lazzeri and Manici 2001). Matthiessen and Kirkegaard (2006) have reviewed the use of *Brassica* green manures for management of soilborne pests and diseases.

Much of this research has been done in the Pacific Northwest of the United States. In this region, potato production is very important, but also very challenging due to soilborne diseases and nematodes.

7.3 Potato Production in the Columbia Basin

The Columbia Basin of Washington and Oregon lies just east of the Cascade mountain range. It is characterized by low precipitation (13–23 cm annually) and light soils (sands to silt loams) with low organic matter contents (0.3–1.6%). With irrigation (mainly center pivot systems with overhead sprinklers), these soils are able to produce high yields of many crops.

Washington State produces 23% of all U.S. potatoes with the Columbia Basin region producing 96% of the state's total. The Columbia Basin region also accounts for 77% of Oregon's potato crop. The climate and long growing season of the Columbia Basin result in the highest potato yields in the U.S. Most of the region's production comes from long-season potatoes grown for processing.

The combination of long season potatoes and a warm growing season make potato early dying (PED) and parasitic nematodes major threats. The fungus *Verticillium dahliae* is the most important component of the PED disease complex (Stevenson 2001) When *V. dahliae* is found with the root-lesion nematode (*Pratylenchus penetrans*) in soils, the two pests can interact synergistically causing increased yield losses in potatoes (Powelson 1985), up to 50% if not managed (Rowe et al. 1987).

Root-knot and stubby-root nematodes also cause economic damage to potatoes in the Pacific Northwest. Root-knot nematodes cause little or no yield loss but can cause entire potato crops to be rejected by potato processors because of the bumps and blemishes they cause when female nematodes infect the surface of tubers. These defects darken the tuber during processing making the product unattractive. The predominant species in the Pacific Northwest is the Columbia root-knot nematode (*Meloidogyne chitwoodi*). A pre-plant population of one *M. chitwoodi* nematode per 250 grams of soil is enough to result in rejection of the entire field of potatoes because of the reproductive capacity of this pest during the summer. The Northern root-knot nematode (*M. hapla*), while present in many Pacific Northwest potato fields, causes less severe tuber damage.

Stubby-root nematodes (*Paratrichodorus* spp.) do not themselves damage potatoes, but as a vector of Tobacco rattle virus, they can cause corky ring-spot. This disease causes internal necrosis in tuber that can result in the rejection of entire fields of tuber by processors.

To manage these pests, potato processors have required 2 or 3 years between potato crops to maintain yield and quality. Numerous rotation crops are grown including alfalfa, spring and winter wheat, grain corn, sweet corn, green peas, onions and dry edible beans. However, to get the control required, fumigation is also often needed. Columbia Basin farmers fumigate 90% of the potato fields each year. Most fields receive metam sodium (or metam potassium) to control PED, while those with nematodes receive 1,3 dichloropropene (1,3-D). About 30% of fields are fumigated with both products. The total cost of this fumigation for the region is estimated at $25 million each year.

Given these serious pests and the cost of managing them, farmers took notice when it was found that mustard green manures were apparently allowing Dale Gies to produce Norkotah potatoes (highly susceptible to Verticillium wilt) in a 2-year rotation, without fumigation.

7.4 Benefits of Mustard Green Manures

From 1999 to 2003, WSU conducted research on the Gies farm. Measurements showed that the green manures had improved the quality of the sandy loam and loamy sand soils (McGuire 2003). After three or four cycles (several fields were surveyed) of his 2-year rotation (spring wheat/mustard green manure – potato), organic matter levels were 50% above levels in adjacent fields not receiving green manures. Water infiltration rates in Gies fields were 2–4 times those of neighboring fields with the same soil type (Table 7.1) and the green-manured soils had more stable aggregates. They were also more resistant to wind erosion than adjacent fields not receiving green manures (McGuire, unpublished). Empirical evidence from other local farmers using mustard green manures point to other related benefits including improved stability of potato hills in sandy soils, increased harvest speed, cleaner tubers at harvest, and increased water-holding capacity of soils.

In addition to improved soil quality, research on the Gies farm revealed the mustard green manure's ability to maintain yields without metam sodium fumigation. Three fumigant replacement studies (McGuire 2003) were conducted in 2000 and 2001 comparing potato yields following mustard green manures with and without

Table 7.1 Average water infiltration rates (cm min^{-1}), after consecutive 2.5 cm applications of ponded water to soil as affected by mustard green manure (MGM) rotations and crop harvest

	Average infiltration rate*, cm min^{-1}	
Date and point in rotation**	1st application	2nd application
September 3, 1999		
After wheat harvest (MGM)	3.53a	1.22a
After wheat harvest	0.33b	0.46b
November 2, 2000		
After potato harvest (MGM)	0.51a	0.48a
After sugar beet harvest	0.99a	0.13b
March 7, 2001		
Potatoes/winter (MGM)	1.45a	.25a
Sugar beets/winter	0.15b	0.13b
March 5, 2002		
Potatoes/winter (MGM)	0.35a	0.23a
Fallow/winter	0.25a	0.13b

*Mean separation, within date and application by PLSD at 0.01 level
**The fields had no tillage operations after harvest of the crops in the first two sampling dates, and no spring tillage operations in the latter two dates

fumigation. Yields averaged 72.8 Mg ha^{-1} total, and 62.5 Mg ha^{-1} for U.S. No. 1 (>113 g) potatoes. Fumigation did not result in any significant yield or tuber quality benefits despite the presence of damaging levels of *V. dahliae*. Yields were higher than the regional average. These results confirmed the observations of Gies, who in 2004 stopped applying metam sodium to his potato fields.

Although Gies does not have root-knot nematodes in his fields, the levels of other parasitic nematodes have decreased under his rotation. Moreover, despite the increased frequency of potatoes in his rotation, there appears to be no buildup of *Verticillium dahliae*.

However, in similar trials conducted on other farms, the results were inconsistent. In a 2001 replicated trial, with long-season potatoes grown after a mustard green manure, yields with metam sodium fumigation were not different from those without fumigation (McGuire, unpublished). Specific gravities of the tubers grown after mustard were higher than after fallow. Other fields (unreplicated plots) on the same farm, however, produced lower yields where the fumigant had been omitted. Similarly, other farmers in the region have observed inconsistent results when replacing metam sodium with mustard. There are several possible reasons for these inconsistent results:

1. There is a cumulative effect of multiple green manure crops, which may also be affected by the frequency of the practice. The soil in the Gies trials had received three green manure crops over 6 years, but the soils in the other trials had received only one green manure crop.
2. There were differences in the way the green manures were managed such that the incorporated biomass was different in quantity or quality.
3. Management of the potato crops differed.

Where farmers use the green manures to replace fumigation, there are economic benefits as well. In 2009, farmers could save an estimated $269 USD ha^{-1} by making this switch, assuming no significant effects on potato yield or quality. This does not take into account additional benefits due to improved soil quality, possible reductions in fumigant rates for nematode control, the potential for increasing potato frequency in rotation, and the possible value-added marketing based on more sustainable production practices.

7.5 Mechanisms of Pest Suppression

The soil building-benefits of green manures are the result of various mechanisms linked to the input of organic matter to the soil. The same is probably true for green manure's pest control-benefits. Davis et al. (2001) found organic matter to be the only manageable factor that predicted both wilt suppression (*V. dahliae*) and tuber yield in 100 surveyed potato fields. However, because of the interaction of organic matter, soil biology, pest biology and soil fertility, it is difficult to match specific mechanisms with specific pest control benefits.

Despite these challenges, recent research has offered several possible mechanisms of green manure mediated pest suppression. Most studies have presumed that only one mechanism is at work, while it is perhaps more likely that multiple mechanisms contribute to the observed effects.

One proposed pest control mechanism is due to specific chemicals produced by *Brassica* crops, most likely glucosinolates. The effect of these chemicals on soilborne pests, when incorporated as a green manure has been termed biofumigation (Kirkegaard and Sarwar 1998) because of the similarities between some breakdown products and synthetic fumigants. Biofumigation can be narrowly defined as "the use of isothiocyanate-generating Brassicas as biologically-active green manures to emulate the use of the synthetic pesticide metam sodium" (Matthiessen and Kirkegaard 2006). However, the term has come to be used for any pest suppression mechanism involving the chemicals released by an organic amendment. Many *Brassicaceae* plants contain glucosinolate compounds that release isothiocyanates upon their decay. These isothiocyanates have been shown in controlled laboratory experiments to kill or suppress growth of pathogens and pests (Brown and Morra 1997). The decay of organic amendments can also produce toxic gases, including NH_3 and HNO_2, and volatile fatty acids, particularly in low pH soils (Bailey and Lazarovits 2003). However, chemistry-based biofumigation effects have more often been assumed than confirmed, and other factors have been found to contribute to suppression where biofumigation was at first suspected (Cohen et al. 2005; Mazzola and Mullinix 2005).

Development of suppressive soils is another possibility. This can occur through either general or specific suppression (Baker and Cook 1974). General suppression of pathogens occurs through the increase of biological activity after application of organic amendments. Stimulation of overall soil microbial activity can reduce relative pathogen population or virulence through increased competition, parasitism, or inhibition. Darby et al. (2006) found that the effect is dependent on the quality and quantity of the amendment, and decreases over time without further amendment. Using paper mill residuals, Rotenberg et al. (2007) found that initiation of root rot suppression occurred between the second and third annual amendment. In this study, the quality of the amendment was also found to be important with particulate organic matter C:N ration being correlated to disease severity in the field. In addition, once suppression was established, it was no longer correlated with total microbial activity. With Verticillium wilt in potatoes, Davis et al. (2007) found that grass green manures gave better suppression than *Brassicas* and that *Fusarium* spp. were dominant among the competitors on the potato roots.

Specific suppression by particular antagonistic microorganisms has been suggested as another mechanism. Wiggins and Kinkel (2005) found that canola and buckwheat green manures increased pathogen inhibitory activity of indigenous Streptomycetes and reduced Verticillium wilt in potatoes. Cohen et al. (2005) also noted that disease suppression with *Brassica* seed meal amendments was associated with higher populations of Streptomycetes and nitric oxide-producing bacteria, but was not related to glucosinolate content. Larkin et al. (2010, 2011b) have also reported various changes in soil microbial community characteristics associated with *Brassica* green manures. Davis et al. (2004) reported increases in populations of specific fungal species associated with suppression of *V. dahliae* in potatoes by green manures.

Aside from any direct effect on the pest organisms, disease control may also be due to or enhanced by systemic resistance in the plant. Cohen et al. (2005) found that low-glucosinolate *B. napus* reduced apple root infection by *Rhizoctonia solani* not by reducing growth of the fungal pathogen but by potentially stimulating plant defense responses.

Systematic approaches have been proposed to test for biofumigation (Matthiessen and Kirkegaard 2006) but have not been conducted for mustard green manures in potatoes. Similar strategies are needed to test for other potential mechanisms.

The control of nematode by green manures is often very effective. Proposed mechanisms for the nematode reductions are biofumigation, non-host status of the green manure, and the green manure as a trap crop (Widmer and Abawi 2002). The control of cyst nematode by mustards in sugar beets (Koch and Gray 1997) and of *M. hapla* with arugula (Melakeberhan et al. 2006) are examples of the latter. Sudangrass, as a non-host cover crop, can effectively reduce populations of *M. hapla*, but further reductions have been documented when the sudangrass is incorporated as a green manure, probably due to nematicidal properties of the released decomposition products (Widmer and Abawi 2002). Biofumigation is probably the mechanism with *Brassica* green manures, which can reduce the impact of *M. chitwoodi* on potatoes in the Pacific Northwest of the USA by 50–80% (Riga et al. 2003).

7.6 Management of Mustard Green Manures

Although the effective mechanisms have not been identified, WSU has made recommendations for mustard green manure management assuming that several potential mechanisms are important. When possible, these recommendations are made to maximize the effects of these presumed mechanisms. The selection of green manure species and variety exemplifies this strategy.

These priorities have guided selection of green manure species and variety:

1. Crop rotation. Priority was given to crops not currently being grown in the region.
2. Biomass production. We have assumed that incorporating large amounts of biomass into our low organic matter soils leads to many of the benefits we see, especially those that affect the soil's physical properties. Mustards (*S. alba* or *Brassica juncea*) produce well in the Columbia Basin and have advantages over small grains (see #3 below)
3. Parasitic nematode management. Our minimum requirement has been that the species/variety selected is not a good host for the parasitic nematodes in the region (*M. hapla, M. chitwoodi*) and if possible, that incorporation of the green manures would reduce nematode populations.
4. Biofumigation. We have assumed that the chemical properties of the selected species/variety are important for control of nematodes and/or soilborne diseases. In searching for high-glucosinolate varieties, we have worked closely with the Italian Research Institute for Industrial Crops (ISCI). ISCI has selected several mustard varieties for their biomass production and glucosinolate content. Most recently, a variety of arugula (*Eruca sativa*) was selected and is being tested for nematode suppression.

Given these desired characteristics, mustards, both *Brassica juncea* and *Sinapis alba*, perform well. In addition, they are small-seeded, which decreases seeding rates and cost. They fit in current rotations, have not increased pest pressure, and the seed can be produced locally. Even if biofumigation were not a factor, mustards would still be favored because of their combination of high biomass production, frost resistance, competitiveness with weeds, and poor host status with problem nematodes (when grown as short-cycle crops). In addition, they do not have any weedy characteristics such as hard or dormant seed.

Currently, farmers in the region are planting several blends of ISCI-licensed *B. juncea* and *S. alba* varieties. The blends, collectively called "Caliente," are being marketed worldwide. In addition to the Caliente blends, Pacific Gold (*B. juncea*) and Ida Gold (*S. alba*) varieties from the University of Idaho breeding program are being grown and a small amount of Martigena (*S. alba*) is still being used. WSU conducts mustard variety trials to help farmers choose varieties.

Most farmers in the Columbia Basin plant the green manures after wheat harvest but they can also follow early sweet corn or green peas. When following wheat, the mustard is typically planted in August. If a suitable drill is available, direct seeding the mustard through the wheat residue works well. Leaving the residues on the soil surface has the following advantages:

1. It saves on tillage costs, as the residue will be tilled under with the mustard anyway.
2. It saves on fertilizer costs, as the incorporated crop residue will tie up nitrogen requiring more fertilizer for the same mustard production.
3. It incorporates fewer wheat seeds resulting in less volunteer wheat to compete with the mustard.
4. After the mustard canopy closes, the wheat residue on the ground will begin to break down. Mushrooms, the fruiting bodies of basidiomycetes are often seen under the mustard on direct seeded fields (Fig. 7.1). These beneficial fungi are known to improve soil quality (Caesar-TonThat and Cochran 2000).
5. When the residues are left aboveground to be incorporated with the mustard, they may buffer the N release from mustard decomposition and reduce the potential for leaching of nitrate over the winter (Starovoytov et al. 2010)

However, because suitable drills are often not available, other planting methods are also used. The seed can be flown on before wheat harvest. After harvest, the field is then packed to increase seed-to-soil contact. If the straw is removed or incorporated, various broadcast methods can be successful. One of the most popular is to blow the seed on with dry fertilizer followed by a pass with a light harrow to incorporate the seed.

Biomass is generally maximized if the plant can be kept vegetative for as long as possible. This requires optimum soil fertility and moisture levels so that stress does not cause the plants to flower early. For a 75-day growing season, approximately 134 kg of available nitrogen ha^{-1} is needed. Although research has not confirmed the need, sulfur is often added with the nitrogen and thought to be necessary for highest yields. Phosphorus and potassium are not normally applied. Soil moisture is maintained by sprinkler irrigation. Because volunteer wheat is a host for parasitic nematodes and because it competes well with mustard, it is killed with one of several selective herbicides available.

Fig. 7.1 Mushrooms in wheat residues under a mustard crop canopy

Fig. 7.2 A mustard green manure crop being chopped and incorporated

By the end of September, the mustard is in full bloom and in October, farmers incorporate the crop into the soil. This is usually done by first mowing the crop with a flail chopper and then disking twice (Fig. 7.2). The timing of incorporation depends on farmers' fumigation needs, mustard planting dates, and irrigation water availability.

Incorporation occurs from early October to mid-November with the peak being the last 2 weeks of October. Quick incorporation of the fresh, green biomass into a moist soil is thought to maximize any biofumigation effects (Morra and Kirkegaard 2002). The fields are left undisturbed through the winter until field preparation for potatoes starts the following spring (February–April).

7.7 Extent of Use

WSU Extension has made farmers in the region aware of the benefits and management of mustard green manures through a series of field days, presentations, and publications. Word-of-mouth has also been important in disseminating this information. As awareness has grown, so too has use of the practice. By 2002, Columbia Basin farmers were growing mustard green manures on over 8,000 ha annually, about 14% of the area used for potato production. This increased to over 12,000 ha in 2010 (Fig. 7.3). Most of this use is before potatoes but also before onions and green peas.

Other regions of the Western U.S. are also using mustard green manures. Seed sales indicate that annually, farmers in Idaho, California, and Arizona are using the practice on an additional 12,000 ha of farmland.

In 2010, Washington state potato growers were surveyed on their use and views of metam-sodium fumigation and mustard green manures. The survey was conducted by the Washington State University Social and Economic Sciences Research Center with input from the author. The survey consisted of 31 questions, seven pages, and

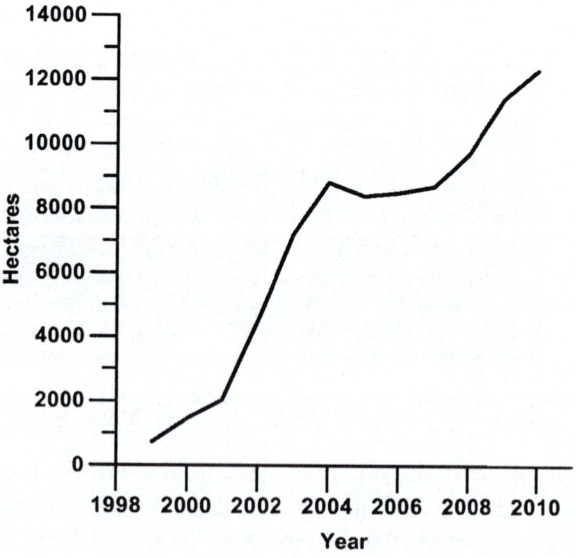

Fig. 7.3 Area under mustard green manures in the Columbia Basin

was available in printed or online format. Of the 235 surveys sent out, 70 were completed for a response rate of 33.3%. The main findings of the survey were:

- Those who had larger farms were more likely to have tried green manures than those with small farms
- The soil-building rather than pest-control benefits were the most important to farmers' decision to try green manures
- Better soil tilth, increased soil organic matter, reduced wind erosion, and improved water infiltration were the top rated benefits of using green manures before potatoes.
- Only 17% of respondents reported using green manures to replace metam-based fumigation. Of those, 64% reported results equal to those obtained with fumigation.
- 26% of respondents plan to use green manures to replace metam-based fumigation in the future. However, this would increase to 71% if additional scientific data was available showing that green manures could successfully replace metam-based fumigation.

Although the survey showed that farmers are motivated to use mustard green manures for their soil and pest control benefits, increased regulation of fumigants and market pressures are also driving farmers towards alternatives to fumigation. In 2008, the U.S. Environmental Protection Agency decided to re-register metam sodium, but with increased regulation. New regulations will require larger buffer zones, increased posting requirements, and worker protection and applicator training, all of which will increase the cost and decrease the ease-of-use of fumigation. Also in 2008, the McDonalds Corporation, the largest buyer of potatoes in the U.S., agreed to promote pesticide use reduction within its American supply chain. Responding to this, some potato processors have targeted fumigants for reduction because, by weight, fumigants are the highest use pesticide in potato production.

7.8 Conclusion

The demand for food, and thus the need for quality soils, will only increase. Although improved synthetic fertilizers and pesticides will continue to be important, they, by themselves, do not build soil quality. It will be through green manuring and other practices that increase or conserve soil organic matter that we will maintain and build our soils. However, our knowledge and level of management of these practices must increase.

There are still many questions regarding mustard green manures use in potato production. Does this work only because local soil organic matter levels are so low? How much annual biomass is enough? Are the effects cumulative? If so, would a shorter rotation work better than a longer one? Is there an optimum balance of the need for diversity in the rotation with the need for organic matter additions to the soil? Why does fresh biomass seem to be more effective than dried crop residues?

To begin to answer some of these questions, a joint project between Washington State University and Oregon State University was initiated in 2010. Over the next four years, it will attempt to determine the relative importance of green manure biomass quantity and quality to the benefits in potato production. It will also determine the ability of mustard green manures to replace metam sodium fumigation under a variety of conditions.

If we continue to improve this old technology by applying our growing knowledge of soil ecology, plant pathology, plant breeding, biochemistry, horticulture, and agronomy there are many possibilities:

- Crops bred for green manure use.
- Rotation of different green manure crops.
- Prescription green manure blends.
- Genetically modified green manure crops.

Any or all of these could be the future of green manuring if we choose to pursue them.

In 1927, Pieters wrote in *Green Manuring: Principles and Practice*, "Much is known of what goes on in the soil when organic matter is added, but much still remains to be learned." While we have added much to our cumulative knowledge since then, the same could be said today. The soil still has secrets, but with our increasing knowledge of both the soil and plants, green manuring could again become a common practice.

References

Bailey KL, Lazarovits G (2003) Suppressing soil-borne diseases with residue management and organic amendments. Soil Till Res 72(2):169–180

Baker EF, Cook RJ (1974) Biological control of plant pathogens. WH Freeman & Co, San Francisco

Boydston RA, Hang A (1995) Rapeseed (*Brassica napus*) green manure crop suppresses weeds in potato (*Solanum tuberosum*). Weed Tech 9(4):669–675

Brown PD, Morra MJ (1997) Control of soil-borne plant pests using glucosinolate-containing plants. Adv Agron 61:167–231

Caesar-TonThat TC, Cochran VL (2000) Soil aggregate stabilization by a saprophytic lignin-decomposing *basidiomycete* fungus I. Microbiological aspects. Biol Fertil Soils 32(5):374–380

Cohen MF, Yamasaki H, Mazzola M (2005) *Brassica napus* seed meal soil amendment modifies microbial community structure, nitric oxide production and incidence of *Rhizoctonia* root rot. Soil Biol Biochem 37(7):1215–1227

Darby HM, Stone AG, Dick RP (2006) Compost and manure mediated impacts on soilborne pathogens and soil quality. Soil Sci Soc Am J 70(2):347–358

Davis JR, Huisman OC, Westermann DT, Hafez SL, Everson DO, Sorensen LH, Schneider AT (1996) Effects of green manures on Verticillium wilt of potato. Phytopathology 86(5):444–453

Davis JR, Huisman OC, Everson DO, Schneider AT (2001) Verticillium wilt of potato: a model of key factors related to disease severity and tuber yield in southeastern Idaho. Am J Potato Res 78(4):291–300

Davis JR, Huisman OC, Westermann DT, Everson DO, Schneider A, Sorensen LH (2004) Some unique benefits with sundangrass for improved US #1 yields and size of Russet Burbank potato. Am J Potato Res 81(6):403–413

Davis JR, Huisman OC, Everson DO, Westerman DT, Schneider AT (2007) The mode of action of green manures for the suppression of Verticillium wilt of potato. Proc Annu Meet Potato Assoc Am, Idaho Falls, Idaho

Horton DR (2008) No effects of an autumn mustard crop on wireworm densities or damage to potatoes in the following growing season. Potato Prog 8:1–4

Kirkegaard JA, Sarwar M (1998) Biofumigation potential of brassicas. I. Variation in glucosinolate profiles of diverse field-grown brassicas. Plant Soil 201(1):71–89

Koch DW, Gray FA (1997) Nematode-resistant oil radish for control of *Heterodera schachtii*. 1. Sugar beet-barley rotations. J Sugar Beet Res 34:31–43

Larkin RP, Griffin TS (2007) Control of soilborne potato diseases using Brassica green manures. Crop Prot 26:1067–1077

Larkin RP, Griffin TS, Honeycutt CW (2010) Rotation and cover crop effects on soilborne potato diseases, tuber yield, and soil microbial communities. Plant Dis 94:1491–1502

Larkin RP, Honeycutt CW, Olanya OM (2011a) Management of Verticillium wilt of potato with disease-suppressive green manures and as affected by previous cropping history. Plant Dis 95:568–576

Larkin RP, Honeycutt CW, Griffin TS, Olanya OM, Halloran JM, He Z (2011b) Effects of different potato cropping system approaches and water management on soilborne diseases and soil microbial communities. Phytopathology 101:58–67

Lazzeri L, Manici LM (2001) Allelopathic effect of glucosinolate-containing plant green manure on *Pythium sp.* and total fungal population in soil. HortScience 36(7):1283–1289

Matthiessen JN, Kirkegaard JA (2006) Biofumigation and enhanced biodegradation: opportunity and challenge in soilborne pest and disease management. Crit Rev Plant Sci 25:235–265

Mazzola M, Mullinix K (2005) Comparative field efficacy of management strategies containing *Brassica napus* seed meal or green manure for the control of apple replant disease. Plant Dis 89(11):1207–1213

McGuire AM (2003) Mustard green manures replace fumigant and improve infiltration in potato cropping system. Crop Manag. doi:10.1094/CM-2003-0822-01-RS

Melakeberhan H, Xu A, Kravchenko A, Mennan S, Riga E (2006) Potential use of arugula (*Eruca sativa* L.) as a trap crop for *Meloidogyne hapla*. Nematology 8(5):793–799

Mojtahedi H, Santo GS, Wilson JH, Hang AN (1993) Managing *Meloidogyne chitwoodi* on potato with rapeseed as green manure. Plant Dis 77(1):42–46

Morra MJ, Kirkegaard JA (2002) Isothiocyanate release from soil-incorporated Brassica tissues. Soil Biol Biochem 34(11):1683–1690

Ochiai N, Powelson ML, Dick RP, Crowe FJ (2007) Effects of green manure type and amendment rate on Verticillium wilt severity and yield of Russet Burbank potato. Plant Dis 91(4):400–406

Olivier C, Vaughn SF, Mizubuti ESG, Loria R (1999) Variation in allyl isothiocyanate production within Brassica species and correlation with fungicidal activity. J Chem Ecol 25(12):2687–2701

Pieters AJ (1927) Green manuring. John Wiley & Sons, New York

Powelson ML (1985) Potato early dying disease in the Pacific Northwest caused by *Erwinia carotovora* pv. carotovora and *Erwinia carotovora* pv. atroseptica. Am Pot J 62(4):73–176

Riga E, Collins H (2004) Green manure effects on *Meloidogyne chitwoodi* and *Paratrichodorus allius*, economically important nematodes of potatoes in the Pacific Northwest of the USA. Agroindustria 3:321–322

Riga E, Mojtahedi H, Ingham RE, McGuire AM (2003) Green manure amendments and management of root knot nematodes on potato in the Pacific Northwest of USA. Nematol Monogr Perspect 2:151–158

Rotenberg D, Jimenez-Wells A, Chapman EJ, Whitfield AE, Goodman RM, Cooperband LR (2007) Soil properties associated with organic matter-mediated suppression of bean root rot in field soil amended with fresh and composted paper mill residuals. Soil Biol Biochem 39:2936–2948

Rowe RC, Davis JR, Powelson ML, Rouse DI (1987) Potato early dying: causal agents and management strategies. Plant Dis 71(6):482–489

Sexton P, Plant A, Johnson SB (2007) Effect of a mustard green manure on potato yield and disease incidence in a rainfed environment. Crop Manag. doi:101094/CM-2007-0122-02-RS

Smolinska U, Horbowicz M (1999) Fungicidal activity of volatiles from selected cruciferous plants against resting propagules of soil-borne fungal pathogens. Phytopathologische Zeitschrift 147(2):19–124

Starovoytov A, Gallagher RS, Jacobsen KL, Kaye JP, Bradley B (2010) Management of small grain residues to retain legume-derived nitrogen in corn cropping systems. Agron J 102(3):895. doi:10.2134/agronj2009.0402

Stevenson WR (2001) Compendium of potato diseases. American Phytopathological Society, St. Paul, Minnesota

Widmer TL, Abawi GS (2002) Relationship between levels of cyanide in Sudangrass hybrids incorporated into soil and suppression of Meloidogyne hapla. J Nematol 34(1):16–22

Wiggins B, Kinkel LL (2005) Green manures and crop sequences influence potato diseases and pathogen inhibitory activity of indigenous *Streptomycetes*. Phytopathology 95(2):178–185

Chapter 8
Yields of Potato and Alternative Crops Impacted by Humic Product Application

Mir M. Seyedbagheri, Zhongqi He, and Daniel C. Olk

Abstract Humic substances (HS – humic acid, fulvic acid, and humin) are a family of organic molecules made up of long carbon chains and numerous active functional groups, such as phenols and other aromatics. Humic substances play dynamic roles in soil physical, chemical, and biological functions essential to soil health and plant growth. This chapter reviews field trials conducted in the Western and Midwestern USA on the effects of application of commercial humic products on yields of potato and several other crops. Examination of these studies reveals that potato growth is more responsive to P fertilization and minimal soil fertility, but less responsive to N fertilization. Whereas some observations were not always consistent, the different soil properties and qualities of humic products from different supplies might have attributed to the inconsistencies. Thus, it is recommended that commercially available humic products be tested locally to determine benefits on potato and other crop production. Research on the impact of long-term humic application on potato production is especially needed, as little such information is currently available in the scientific literature for U.S. potato producing regions.

8.1 Introduction

Organic matter undergoes a biological degradation process termed "humification" by a community of soil macro- and microorganisms, where it is broken down and recycled for use as energy and substrate for cellular metabolism (Chen and Aviad 1990).

M.M. Seyedbagheri (✉)
Elmore Center, University of Idaho, 535 E. Jackson, Mountain Home, ID 83647, USA
e-mail: mirs@uidaho.edu

Z. He (✉)
USDA-ARS, Southern Regional Research Center, Robert E Lee Blvd, New Orleans, LA 70124, USA
e-mail: zhongqi.he@ars.usda.gov

D.C. Olk
USDA-ARS, National Laboratory for Agriculture and the Environment, Ames, IA 50011, USA

Z. He et al. (eds.), *Sustainable Potato Production: Global Case Studies*,
DOI 10.1007/978-94-007-4104-1_8,

Humic substances (HS) are the products of the humification process of plant and animal residues in various stages of decomposition and are found in soil and geologic deposits, including peat, lignite, and leonardite.

As the major constituents of soil organic matter (SOM), HS can be divided into three major fractions, humic acid (HA), fulvic acid (FA) and humin, based on their solubility in acid and alkali (Tan 2003). HS are recognized as the most chemically active compounds in soils, with cation and anion exchange capacities that far exceed those of clays (Stevenson 1982). Their properties of chelation, mineralization, buffer effect, clay-mineral organic interaction, and cation and ion exchange capacities profoundly influence soil physical, chemical and biological functions essential to soil health and plant growth (Stevenson 1982). Humic acid from lignite is a ready source for carbon and N for both plants and the microbial community. Ubiquitous in the environment, HS are an integral part of all ecosystems, and play an important role in the global cycling of nutrients and carbon (MacCarthy et al. 1990).

Because of their ability to aid in the formation of soil aggregates, HA can increase soil water holding capacity, reduce crusting, and improve tilth. HS support the biological activities of soil macro- and microorganisms, and serve as an adsorption and retention complex for inorganic plant nutrients. The positive effect of HS on plant growth is documented under laboratory and greenhouse conditions (Visser 1986; Chen and Aviad 1990). Whereas Chen and Aviad (1990) reported positive results from lab and greenhouse studies, they concluded that humic products cannot have any effect on crop growth in field conditions at the low application rates recommended by industry. Several authors (Nardi et al. 2002; Tan 2003) have demonstrated that the addition of HS in appropriate concentrations can stimulate root growth and enhance efficiency of the root system. Other claimed benefits of HS include increased N uptake by plants, which serves to increase soil N utilization efficiency and can enhance the uptake of K, Ca, Mg, and P, and improve availability of nutrient and trace mineral uptake to plants.

Over the last half century, several research groups have conducted applied field trials in the western U.S. to evaluate the impact of applications of different commercial humic products on yield and quality of potato and alternative crops. This chapter reviews some of these research findings. Summarization of these findings suggests that nutrient and health statuses of soil are critical factors for achieving positive results of humic product application on yield of potato and other crops.

8.2 Effects of Humic Product Application on Potato Yield

8.2.1 Effects of Humic Product Addition with Variable N Fertilization

Early studies of humic product effects were focused on whether the addition of humic products to a fertilizer might be beneficial for potato production (Kunkel and Holstad 1968; Lorenz et al. 1974). The humic product was from "Aqua Humus", a leonardite ore from which the insoluble fractions were removed. It contained ~60% humic and fulvic acid derivatives. It was a dark brown to black hydrophilic colloid with a very high base exchange capacity. In addition to the inherent organic constituents, the "Aqua Humus" was sometimes enriched with inorganic N, P, and K.

Kunkel and Holstad (1968) tested the effects of humic product application on the yield and quality of Russet Burbank potatoes grown in neutral (pH 7) Columbia soils. The experiments were conducted in 1963 and 1964 on a coarse silt loam with a moderate to slow water infiltration rate and an organic matter content of ~1%. Kunkel and Holstad (1968) mixed the humic product at 200 lb acre^{-1} (i. e. 224 kg ha^{-1}) with acid (pH 5.4, 16-16-16) and base (pH 8, solid 15-15-15, liquid 12-12-12) NPK fertilizers. These mixtures were applied in bands at planting time at different rates to provide 80–500 lb N acre^{-1} (i. e. 90–560 kg N ha^{-1}). In addition, humic product was also applied at 100 and 300 lb acre^{-1} with each solid base fertilizer rate to further test the impact of product application. These experiments show that potato yields continued to increase, but at a decreasing rate from the lowest to the highest rate of fertilizer applied. Addition of humic product to the acid fertilizer and solid base fertilizer did not change either total yield or yield of No. 1 grade potatoes. For example, the average yield for 16 plots with humic product addition was 59.8 Mg ha^{-1} whereas the average yield of those receiving the same fertilizer without humic product was 59.2 Mg ha^{-1}. When the dry base fertilizers were used at the equivalent rate, the yields were roughly 7% lower than those for acid fertilizer. On the other hand, the addition of humic product to the base liquid fertilizer significantly increased the potato yield at all five fertilizer rates tested. When humic product was applied, potato yield was roughly 21% higher than in plots that received only liquid base fertilizer. The authors (Kunkel and Holstad 1968) hypothesized that the high base exchange capacity of the organic colloidal humate might have reduced a salt effect early in the growing season, and concluded that the interaction between humate and forms of fertilizer was highly significant.

Kunkel and Holstad (1968) also found that, in some cases, humic product mixed with fertilizer increased the levels of several elements (e g. N, P, K, Mg, and Mo) in the petiole. Under conditions where these nutrients are marginal, humic product addition could increase potato production; however, they proposed that similar results could be achieved with adequate fertilization without humate addition.

Lorenz et al. (1974) conducted similar field experiments for furrow-irrigated White Rose or Kennebecs potatoes grown in fine sandy loam soils in three counties of California. These soils were light to medium textured, alkaline calcareous, with pH values about 7.6. They used the same "Aqua Humus" product as Kunkel and Holstad (1968), and evaluated the addition of the humic product to N-containing fertilizer in five experiments. Fertilizer with or without humic product was applied in bands (3 inches to each side and 2 inches below the seed) at time of planting. Comparisons were made with $(NH_4)_2SO_4$, fertilizer 16-20-0, and urea at two rates (Fig. 8.1). The results show that the addition of humate had no significant effect on yield-either positive or negative. Therefore, they suggested that this humic product did not improve fertilizer N uptake efficiency.

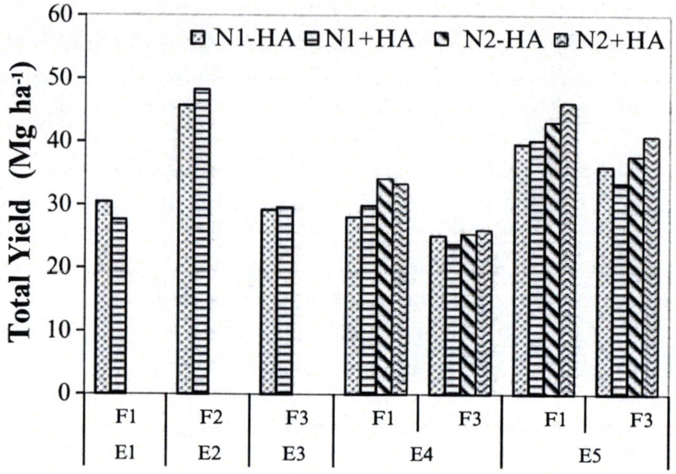

Fig. 8.1 Effects of "Aqua Humus" humate product on potato yield tested in California in 1960s. Refer to Lorenz et al. (1974) for details of the five experiments E1 to E5. F1, $(NH_4)_2SO_4$; F2, fertilizer 16-20-0; F3, urea. Fertilizer was applied at 134 (N1) and 269 (N2) kg N ha^{-1} with (+HA) or without (–HA) humic product

8.2.2 *Effects of Humate Addition on P Availability*

Precipitation of Ca phosphates negatively affects plant availability of P fertilizer applied to calcareous soils. Delgado et al. (2002) investigated availability improvement of applied P fertilizer from soils by a commercial liquid mixture of humic and fulvic acids (Solfer húmicos, Valencia, Spain). In this study, the mixture of HA and FA was applied to calcareous soils, with different levels of salinity and Na^+ saturation, which were fertilized with 200 and 2000 mg P kg^{-1} as $NH_4H_2PO_4$. Recovery was measured as the ratio of Olsen P-to-applied P after 30, 60 and 150 days. Their laboratory work (Delgado et al. 2002) indicated that application of the HA-FA mixture increased the amount of applied P that was recovered as Olsen P in all the soils, with the exception in one soil with the highest Na saturation. This observation implies the potential of humic product application in increasing crop production in soils where the low levels of available P are a limiting factor for crop production.

In a 3-year (2000–2002) study at the University of Idaho, Hopkins and Stark (2003) evaluated the effect of three rates of P (0, 60, and 120 lbs P2O5 acre^{-1}, i. e. 0, 29.4 and 58.7 kg ha^{-1}) applied in the mark-out band with and without humic product at a 10:1 v/v ratio. The field site was located at the University of Idaho Aberdeen R&E Center. The soil was a Declo sandy loam, calcareous (4–9% free lime), with pH ranging from 8.0 to 8.2. With medium soil test P levels (15–19 ppm) and low organic matter levels (1.1–1.3%), the properties of the soils used in this study were typical of potato producing regions in Idaho. The humic product was Quantum H (Horizon Ag). Seed pieces of Russet Burbank potatoes were planted

Table 8.1 Effect of P fertilizer with and without humic product on potato yield, specific gravity, petiole phosphorus and gross return. Combined (average) 3 years' data adapted from Hopkins and Stark (2003)

P_2O_5 (kg·ha^{-1})	Humic product (L·ha^{-1})	Total tuber yield (Mg ha^{-1})	Yield US No 1 tubers (Mg ha^{-1})	Yield tubers >283 g (Mg ha^{-1})	Specific gravity (g cm^{-3})	Petiole P (% dwt)	Gross return (US$ ha^{-1})
0	0	44.23	25.26	16.39	1.077	0.24	4,523
67.36	0	48.39	29.19	19.87	1.079	0.29	5,110
67.36	14	49.85	31.32	20.88	1.080	0.31	5,390
134.7	0	49.17	29.3	20.10	1.079	0.30	5,187
134.7	28	50.07	31.21	21.67	1.079	0.32	5,402
LSD (1%)		5.39	3.71	2.58	0.003	0.03	

with 12-inch spacing, and the 10-34-0 fertilizer, with and without added humic product, was applied in the mark-out band three inches to the side of the seed piece.

With the results from this experiment, Hopkins and Stark (2003) demonstrated that addition of humic product to the fertilizer band tended to increase total yield at both the high and the low P levels (Table 8.1). Similarly, U.S. No.1 yields generally increased as P was added at both rates, with a tendency for further yield increases occurring when the humic product was included in the fertilizer band. The primary effect of P and humic product treatment on U.S. No. 1 tuber yields was an increase in tuber size. In particular, yields of U.S. No. 1 tubers greater than 10 ounces increased in 2001 and 2002 with the application of P in combination with the humic product. Addition of P with the product also increased specific gravity in one of the 3 years of the study, compared with the untreated control, but the combined data suggested the general effects on specific gravity were negligible (Table 8.1). Addition of the humic product resulted in further increases (an average of 0.03%) in petiole P concentrations to levels greater than the marginal range in all 3 years at both rates. The authors proposed that the increases in petiole P were at least partially responsible for the increased tuber yield and size observed in their study.

Hopkins and Stark (2003) calculated that addition of humic product would increase gross revenue an average of \$248 ha^{-1} due to the tuber yield and quality increases. As average costs of humic product application were approximately \$25-50 ha^{-1}, application of the product to calcareous, low organic matter soil shows potential as a profitable management tool. However, Hopkins and Stark (2003) cautioned that growers wishing to apply humic acid amendments should work with reputable companies that can provide a consistent material with documented, non-biased data showing their product to work under local growing conditions. Furthermore, it is reasonable to assume that the effect of a humic product applied at relatively low rates is more effective if applied in a concentrated band. Although positive results were found for banded application of humic product with P fertilizer on potatoes grown on low organic matter, calcareous soil, the potential benefits of humic products with other soil types, crops, and fertilizer/amendment placements should be evaluated before expecting satisfactory results (Hopkins and Ellsworth 2005).

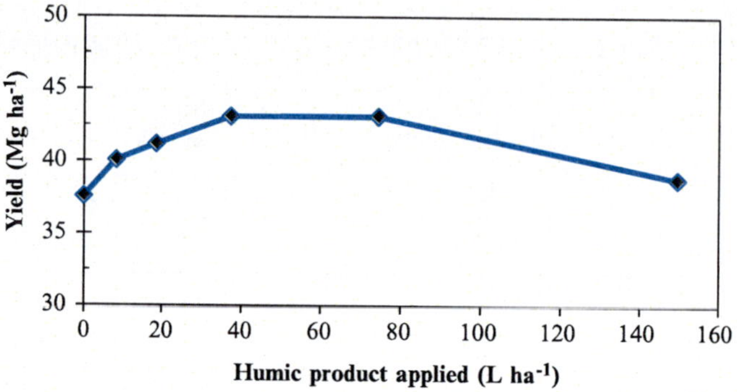

Fig. 8.2 Average potato yield affected by humate application at 3 sites in the Saylor Creek, ID experiment (Adapted from Seyedbagheri 2010)

8.2.3 *Effects of Humate Application Under Minimal Soil Fertility Conditions on Potato Yield and Quality*

In fields with minimal soil fertility, according to the University of Idaho Fertility Guide, Seyedbagheri (2010) conducted experiments at Saylor Creek and Mountain Home, ID, to evaluate the effects of different rates of humic product application on potato yield and quality. Climatic conditions were similar in the two areas, as both sites are semi-arid, with an annual rainfall of 152.4–203.2 mm. The soil in these fields was calcareous (5–7% free lime), pH was 8.0–8.2, and organic matter content was 0.9–1.0%. In these experiments, Russet Burbank seed pieces were planted by hand, spaced 25.4 cm apart. Each individual plot was 3.65-m wide and 7.6-m long and included four rows. The humic product used at the Saylor Creek fields had 6% HA by weight and was from Bio-Tech Company. At Mountain Home, granular humate (Agri-Plus) and liquid HA (Quantum-H) were applied. Liquid humic products were side-dressed, and granular humic product was top-dressed.

Figure 8.2 summarizes the effects of product application rates on potato yield at three farmers' fields at Saylor Creek, ID. These data are the average yields of three experimental fields. Evaluation of stand and vigor showed that plots treated with humic product rated very high (8 out of 10) in comparison with control plots (5 out of 10) (Seyedbagheri 2010). The Russet Burbank tuber yield increased from 37.6 to 43.1 T ha^{-1} (i.e. Mg ha^{-1}) from the control to product application at the rate of 37 L ha^{-1}. Yield declined when the application rate applied exceeded 75 L ha^{-1}. The non linear relationship implies the mechanism of humic product impact is quite complicated.

Moreover, this observation is in contrast with the early observation reviewed in section 8.2.1. The difference is, in this experiment, humate product was applied under minimal soil fertility. Moreover, the positive impacts of HA application observed in this study are consistent with studies under controlled conditions on HA application

Table 8.2 Effect of humic products on tuber yield (Mg ha^{-1}) in field trials conducted at Mountain Home, ID (Seyedbagheri 2010)

	Tuber size (g)					
Treatments	0–113.4	113.4–220.8	226.8–340.2	>340.2	Culls	Total
#1: Control	10.2 a*	17.9 a	8.2 a	5.0 a	3.4 a	44.6 a
#2: Granular humate only (Agri-Plus)	10.7 a	16.7 a	8.4 a	5.5 a	2.8 a	45.1 a
#3: Granular humate (Agri-Plus) + 46.5 L ha^{-1} Liquid humic acid (Quantum-H)	11.5 a	16.7 a	7.0 a	5.0 a	4.5 a	44.7 a
#4: Granular humate (Agri-Plus) + 93.0 L ha^{-1} Liquid humic acid (Quantum-H)	11.0 a	15.3 a	6.4 a	4.2 a	4.6 a	41.5 a

*Means followed the same letter in the same column are not significantly different at the 0.05 level (Neuman-Keul test)

and plant growth (Chen 1986; Chen and Aviad 1990; MacCarthy et al. 1990). In this study the product performed better in poor soil with high Ca (3500 to 5000 ppm, i.e. 5–7% free lime) than in more fertile soil (data not shown). The humic product used in this study seemed to enhance fertilizer use efficiency by increasing P, K, Zn and Fe uptake by the plants (Delgado et al. 2002). On the other hand, the HA could have had effects directly on the plant, not on soil nutrients. More research is needed to elucidate the mechanism of the humic product's role.

In the Mountain Home experiment, the potatoes were harvested and graded by weight (Table 8.2). Data in the table show that there was no significant statistical difference in potato tuber yield between control and humic product treatments. However, it is important to note that in the year following this study, the grower planted small grains in the same field. The area in the experimental plot that had been treated with the humic product showed a major yield difference (data not shown), which indicates that long-term field trials are also needed for evaluating the effects of humic products on plant growth.

8.3 Effects of Humic Product Application on Yields of Alternative Crops

Crop rotation is a sustainable cropping management practice for potato production (refer to Chaps. 2, 3, 4, 5, 6, and 7). Therefore, in this section, we present some data on several other crops for general information. Similar to the potato studies, the effects of humic product application on the yields of other crops are not always consistent. In Montana, Jones et al. (2007) conducted a greenhouse study to determine the effects of a low, commercially recommended rate of HA on P, Fe, and Zn

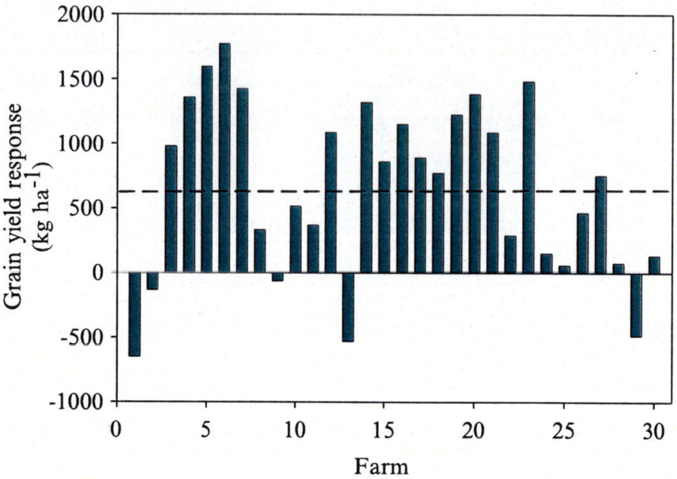

Fig. 8.3 Mean response of maize grain yield to application of a liquid humic product in 30 Iowan farms. Values were based on eight pairs of harvested plants and presume a planting density of 74,000 plants ha^{-1}

availability and spring wheat yields, in both a calcareous soil and a noncalcareous soil. Their greenhouse results suggest that low commercial HA rates (~1.7 kg humate ha^{-1}) may be insufficient to enhance spring wheat growth, as no significant differences were found in nutrient uptake, shoot biomass, or grain yield between humate and control treatments. On the other hand, Belgium scientists Verlinden et al. (2009) show that application of a liquid mixture of HA and FA (Commercial name Humifirst) resulted in consistent increases in crop yield and nutrient uptake. These crops included grass, maize, and spinach, in addition to potato. The observed effects were largest for the potato field, followed by the grasslands and were smallest for the maize fields.

A liquid humic product (Innovative Crop Solutions) is currently being evaluated in dryland maize production in Iowa. Test strips of the product were established on maize farms in 2 years. For 30 farms in the first year (2009), eight representative maize plants were hand-sampled from each test strip and another eight plants were sampled from adjacent, unamended maize. A numeric increase in grain weight was observed in 25 of the 30 farms (Fig. 8.3). If each farm is considered a replicate, this yield increase was highly significant ($P<0.01$). Presuming a planting density of 74,000 plants ha^{-1} on all farms, the mean grain yield increase with product application was 630 kg ha^{-1} (dashed line). On nearly 100 other farms, combine grain yield increased with product application in about 70% of the cases, and the mean increase was about 440 kg ha^{-1}, which was also highly significant (data not shown). Comparable results were obtained in 2010. These grain yield increases provide a several-fold return on the application cost of the product: only 3.5 L ha^{-1} was applied at a cost of about $ 22 ha^{-1}, and the product can be applied as part of routine pesticide applications.

8.4 Conclusion

Relevant applied field trial experiments in the past half century reviewed in this chapter showed inconsistent effects of humic products on the yields of potato and other crops. Whereas some studies showed no significant yield response, findings in others showed some yield increase. Crop response might have been affected by the application rate of humic products and the soil nutrient status. In this limited set of experiments, humic product efficacy seems more responsive to P fertilization than N fertilization, as the humic product could release phosphate bound to Ca. Yet it is difficult to generalize across studies, for the efficacy of humic products would seem to depend on a large number of factors, including solar radiation, weather damage, soil type, crop, yield level, and absence or presence of other yield constraints (disease, pests, weeds, water stress). The absence of any industry-wide standards for producing humic products could also contribute to differences in findings from multiple research groups who used different commercial humic products. None of these potential factors has been systemically evaluated. Currently, the humic product effects were evaluated in the application year with a maximum of 3 continuous years. Research is needed to evaluate the potential of long-term product application for improving soil fertility and quality.

Disclosure Mention of trade names or commercial products in this publication is solely for the purpose of providing specific information and does not imply recommendation or endorsement by the U.S. Department of Agriculture. USDA is an equal opportunity provider and employer.

References

Chen Y (1986) Organic matter reactions involving micronutrients in soil and their effect on plants. In: Piccolo A (ed) Humic substances in terrestrial ecosystems. Elsevier, Oxford, pp 507–529

Chen Y, Aviad T (1990) Effects of humic substances on plant growth. In: MacCarthy P, Clapp CE, Malcolm RL, Bloom PR (eds) Humic substances in soil and crop sciences: selected readings. ASA and SSSA, Madison, Wisconsin, pp 161–186

Delgado A, Madrid A, Kassem S, Andreu L, del Campillo MC (2002) Phosphorus fertilizer recovery from calcareous soils amended with humic and fulvic acids. Plant Soil 245:277–286

Hopkins BG, Ellsworth JW (2005) Phosphorus availability with alkaline/calcareous soil. In: Stevens WB (ed) Proceedings of Western Nutrient Management Conference 6:88–93

Hopkins BG, Stark JC (2003) Humic acid effects on potato response to phosphorus. In: Robertson LD et al (eds) Proceedings of Winter Commodity Schools 35:87–92

Jones CA, Jacobsen JS, Mugaas A (2007) Effect of low-rate commercial humic acid on phosphorus availability, micronutrient uptake, and spring wheat yield. Comm Soil Sci Plant Anal 38:921–933

Kunkel R, Holstad N (1968) Effects of adding humates to the fertilizer on the yield and quality of Russet Burbank potatoes. Am Potato J 45:449–457

Lorenz OA, Weir BL, Bishop JC (1974) Effect of sources of nitrogen on yield and nitrogen absorption of potatoes. Am Potato J 51:56–65

MacCarthy P, Malcolm RL, Clapp CE, Bloom PR (1990) An introduction to soil humic substances. In: MacCarthy P, Malcolm R., Clapp CE, Bloom PR (eds) Humic substances in crop and soil sciences: selected readings. Soil Science Society of America, Madison, WI, USA, pp 1–12

Nardi S, Pizzeghello D, Muscolo A, Vianello A (2002) Physiological effects of humic substances on higher plants. Soil Biol Biochem 34:1527–1536

Seyedbagheri MM (2010) Influence of humic products on soil health and potato production. Potato Res 53:341–349

Stevenson FJ (1982) Humus chemistry: genesis, composition, reactions. Wiley Inter-Science, New York

Tan KH (2003) Humic matter in soil and the environment. Marcel Dekker, New York

Verlinden G, Pycke B, Mertens J, Debersaques F, Verheyen K, Baert G, Bries J, Haesaert G (2009) Application of humic substances results in consistent increases in crop yield and nutrient uptake. J Plant Nutri 32:1407–1426

Visser SA (1986) Effects of humic substances on plant growth. In: Burns RG, Dell'Agnola G, Nardi S, Savoini G, Schnitzer M, Sequi P, Vaughan D, Visse SA (eds) Humic substances: effects on soil and plants, Reda, Rome, Italy, pp 80–135

Chapter 9
Late Blight Epidemics in the Columbia Basin

Dennis A. Johnson, Philip B. Hamm, Jeffrey S. Miller, and Lyndon D. Porter

Abstract Late blight was not originally expected to be a serious threat to potato in the semi arid environment of the Columbia Basin. However, the disease has occurred every year at various severities since 1990. Migration of *Phytophthora infestans* into the Columbia Basin in the 1990's is well documented. The US-1 strain predominated in 1992 and several unique isolates were discovered in 1993, which were likely the result of genetic recombination. The recombinants were ephemeral and were not found in 1994. The US-8 stain was first observed in 1994 and came to predominate in 1995 and in subsequent years. Epidemics in the Columbia Basin have been traced to infected seed tubers, refuse tubers and volunteers. Late blight spreads in fields by foci with foci enlarging in size, producing daughter foci, and coalescing. The process continued as favored by the environment. Sporangia of *P. infestans* have the capability of surviving in water for extended periods of time after detachment from sporangiophores on potato tissue. Late blight has been successfully forecasted and managed regionally in the Columbia Basin. Early season rain is an effective indicator of late blight outbreaks because moisture is important for the build-up of inoculum in fields during the early stage of epidemics. Early in epidemics, moisture promotes transmission of *P. infestans* from infected seed tubers to emerged shoots in fields. Method of fungicide application affects fungicide

D.A. Johnson (✉)
Department of Plant Pathology, Washington State University, Pullman, WA 99164-6430, USA
e-mail: dajohn@cahnrs.wsu.edu

P.B. Hamm
Department of Botany & Plant Pathology, Oregon State University, Hermiston Agricultural Research & Extension Center, Hermiston, OR 978383, USA

J.S. Miller
Miller Research, Rupert, ID, USA

L.D. Porter
Vegetable and Forage Crops Research Unit, USDA-ARS, Prosser, WA, USA

Z. He et al. (eds.), *Sustainable Potato Production: Global Case Studies*, DOI 10.1007/978-94-007-4104-1_9, © Springer Science+Business Media B.V. 2012

distribution and cost. The alternate use of air application and chemigation provides good protection at a reduced cost compared to only air or ground application methods. Application of phosphorous acid to tubers after harvest and prior to storage can result in a reduction in post-harvest infection by *P. infestans*.

9.1 Introduction

The Columbia Basin of south-central Washington and north-central Oregon is a major potato-growing region in North America. Over 65,000 ha of potatoes are grown in the region annually, with mean tuber yields exceeding 74 t/ha in 2010. The region is isolated and nearly completely bordered by mountains; the potato production area extends for approximately 180 km, from Umatilla and Morrow counties in north-central Oregon to Grant and Adams counties in south-central Washington. Potatoes are mainly planted in March through April and harvested from August through October. The environment is semiarid and the potato crop is irrigated mostly by center-pivot systems. The first center-pivot systems were introduced about 1956 and after 1973 quickly replaced surface irrigation by gravity flow (Easton 1982). An increase in seasonal occurrence of late blight occurred with the increased use of center pivot irrigation in the region (Johnson et al. 2003a; Easton 1982); an increase in the severity of late blight has similarly been associated with sprinkler irrigation in the arid environment of Israel (Rotem et al. 1970).

Phytophthora infestans, the oomycete that causes potato late blight, is dependent on a wet, humid environment with mild temperatures for sporulation and infection. Sporangia are sensitive to drying (Minogue 1981) and are disseminated most effectively from field to field during cloudy and rainy periods (Hirst and Stedman 1960; Sunseri et al. 2002). Late blight was not originally expected to be a serious threat to potato in the semi arid environment of the Columbia Basin. The disease was first identified in this region during the 1947 growing season when weather was unusually cool, cloudy, and wet (Easton 1982). It was next reported 27 years later in 1974 and was observed in fields 7 of 16 years between 1974 and 1989. The frequency of disease occurrence since then has greatly increased and late blight has been present in the Columbia Basin every year since 1990, with the most severe outbreaks occurring in 1993, 1995, 1997, 1998, 2004 and 2010. The monetary cost of managing late blight is high and approached $30 million for the Columbia Basin in 1995 (Johnson et al. 1997). The cost was $22.3 million in 1998 and included $19.8 million for fungicides and application, $1.1 million for canopy desiccation, and $1.4 million in losses due to tuber rot in storage (Johnson et al. 1998).

9.2 Migration

Phytophthora infestans is well known for global migrations from its center of genetic diversity in the Toluca Valley of Mexico to the United States and Europe in the 1840's. One of the earliest recorded outbreaks of late blight on cultivated

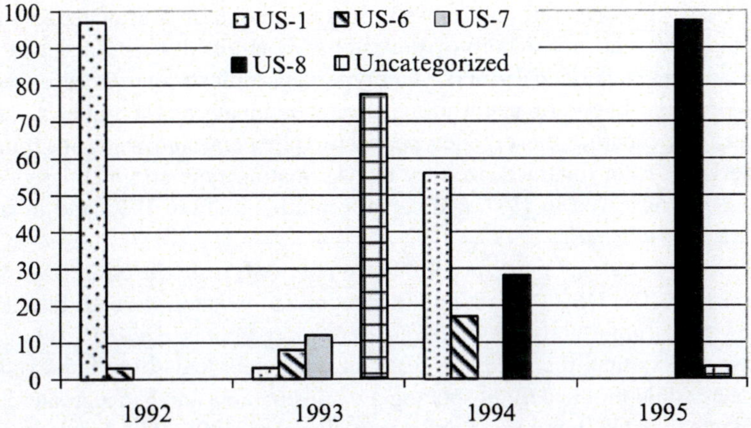

Fig. 9.1 Distribution of *P. infestans* strains in the Columbia Basin of Washington and Oregon from 1992 to 1995 (Adapted from Miller et al. 1997)

potatoes occurred in the United States in 1843 (Stevens 1933). Affected areas included all of New England and states as far west as Illinois and Wisconsin by 1845. In June of 1845, late blight was observed on potatoes in Belgium (Bourke 1964, 1991). The pathogen radiated in all directions that summer and reached Ireland by September (Bourke 1964). Late blight devastated Ireland resulting in the Irish potato famine from 1845 through 1852. A million people died and at least 1.5 million immigrated as a result (Large 1940).

Additional pathogen migrations occurred in the 1970's and 1990's (Fry et al. 1993). Evidence for these migrations was shown by an increased diversity in *P. infestans* populations (Goodwin et al. 1994a). Prior to these migrations some evidence pointed to a single clonal lineage or strain as being responsible for epidemics outside the Toluca Valley of Mexico (Goodwin et al. 1994b). Subsequent genetic analysis from herbarium specimens contradicts this view (Ristaino et al. 2001). During the initial population evaluation work with *P. infestans,* strains were identified as multilocus genotypes based on the following characteristics: (1) compatibility or mating type, (2) metalaxyl sensitivity, (3) glucose-6-phosphate (*Gpi*) and peptidase (*Pep*) allozyme genotypes, and (4) restriction fragment length polymorphism (RFLP) genotype (Goodwin et al. 1995). Strains were named consecutively based on the country in which they were first identified (e.g. US-1 was the first strain identified in the United States, CA-2 was the second strain identified in Canada, and PO-4 was the fourth strain identified in Poland).

The migration into the Columbia Basin is well documented and serves as an example for what happened in many areas of the world in the 1990's. The US-1 strain (characterized by the A1 mating type and metalaxyl sensitivity) predominated in 1992 (Fig. 9.1). This was the strain hypothesized to have been distributed globally outside Mexico prior to the migration in the 1970's (Goodwin et al. 1994a). The US-6 strain (characterized by the A1 mating type and metalaxyl resistance) was first observed that year, however. The frequency of the US-6 strain increased in

1993 and 1994, but was not found in 1995. Several unique isolates were discovered in 1993 and these did not fit any of the published strain descriptions, and were placed in an "uncategorized" group. Genotype data indicate that these uncategorized isolates were likely the result of genetic recombination (Gavino et al. 2000). The isolates representing these genotypes were ephemeral and were not found in 1994. The US-8 stain (characterized by the A2 mating type and metalaxyl resistance) was first observed in 1994 and came to predominate in 1995 and in subsequent years (*unpublished*).

Resistance to metalaxyl may have influenced the shift from the older US-1 strain to newer strains. The US-6, -7, -8, and uncategorized isolates were all resistant to metalaxyl. More importantly, newer strains are more aggressive and have greater competitive fitness than the older US-1 strain. Detached leaf studies showed that US-8 isolates could produce more sporangia per lesion area and had a greater lesion expansion rate than US-1 and US-6 isolates (Miller et al. 1998). US-8 isolates infect young shoots from infected seed more readily than other isolates (Marshall and Stevenson 1996). Additionally, when US-1 and US-8 isolates were released into a small field plot with equal frequency, US-8 isolates were recovered with greater frequency (96%) than US-1 isolates (1%) (Miller and Johnson 2000). The remaining 3% were non-parental phenotypes. From 1995 through 2009, the US-8 strain has been the predominant isolate recovered from potato fields in the Columbia Basin (*unpublished data*).

9.3 Origin of Epidemics

Oospores of *P. infestans* are not currently known to be a factor in overwintering in the Columbia Basin and a continuum of viable host tissue is essential for overwintering and transmission of the pathogen to new plant tissues in this region. The pathogen survives in infected tubers which may act as an inoculum source the following season (Melhus 1915; Van der Zaag 1956). However, infected tubers frequently rot during winter months and cease being a threat as an inoculum source when decomposed. Contemporary strains of *P. infestans* are highly aggressive and rapidly rot tubers, limiting the availability of viable host tissue (Kadish and Cohen 1992; Kirk et al. 2001). Epidemics of late blight characteristically start from low levels of initial inoculum originally arising from infected seed tubers, volunteer potato plants developing from infected tubers in the field, or from infected tuber refuse (Kadish and Cohen 1992; Johnson et al. 2003b; Zwankhuizen et al. 1998). The relative importance of the three types of late-blight-infected tubers varies as inoculum sources and depends in part on microclimates, local conditions, and the extent of infection the previous fall (Davidse et al. 1989; Gigot et al. 2009; Zwankhuizen et al. 1998).

Epidemics in the Columbia Basin have been traced to infected seed tubers, refuse tubers and volunteers (Johnson et al. 2003b). Observations in commercial fields and surrounding areas early in the course of epidemics revealed that volunteers are especially likely to pose a threat when potato plants in the field were infected the previous season, moreover, infected volunteers have been found in a field 2 years after an infected potato crop. Cull piles formed in late winter or early spring from tubers taken from storages can be a serious threat because infected tubers in storage are protected from external environmental variations in temperature and moisture, which may increase rot. Infected tubers may survive in cold storage at temperatures used to store seed tubers with little to no rot or symptom development (Johnson and Cummings 2009). Latent infections in seed tubers are a particular threat in generating new epidemics.

Transmission of *P. infestans* from infected tubers to plant tissues the next season may occur during seed-tuber handling, cutting and planting (Lambert et al. 1998) or in the field (Hirst and Stedman 1960; Melhus 1915). For secondary infection to occur during seed tuber handling and cutting, the pathogen must survive in intact tubers during the winter, sporulate, be dispersed, and infect additional tubers or foliage. Temperature and humidity within piles of cut seed tubers often favor sporulation, and sporangia have been observed on infected seed pieces within 19 h of cutting (Porter et al. 2001). Sporangia are readily transmitted by direct contact from infected tubers or seed pieces to noninfected seed pieces (Dowley and O'Sullivan 1991). Tubers infected prior to planting may be more likely to produce viable shoots than those infected in the field near harvest because of less time for rot to develop before shoot emergence. Under experimental conditions, transmission occurred from tubers to shoots when tubers were inoculated in the spring before planting, but not when tubers were inoculated in the fall (Gigot et al. 2009). Infection during seed-tuber cutting and handling increases the threat of late-blight outbreaks on foliage in the field. Fungicide seed piece treatments potentially reduce transmission from infected seed tubers (Inglis et al. 1999; Powelson et al. 2002).

The exact pathway by which *P. infestans* progresses from planted, infected seed tubers to plant foliage has been disputed (Andrivon 1995a, b; Boyd 1980; Melhus 1915). De Bary originally proposed that the pathogen spread by mycelial growth within infected seed tubers and advanced contiguously or followed growing shoots to produce lesions and sporangia on above-ground stems (De Bary 1876). However, the validity of De Bary's work on vegetation of the pathogen was questioned when not duplicated by other researchers as described by Melhus (1915). A moist environment plays an important role in the expression of transmission of the pathogen from infected seed pieces to shoots (Johnson 2010) and many of the studies contradicting De Bary's observations were done in relatively dry seasons or environments. Additionally, continuous lesions are not always observed on the below-ground stem between the infected seed piece and the resultant lesion on the above-ground stem (Johnson 2010) and mislead some of the previous researchers. No necrotic tissue or only slight streaking of reddish brown discolored tissue may be observed on below-ground stems when *P. infestans* is transmitted by mycelia growth within internal tissues (Fig. 9.2). De Bary's proposed pathway has been validated (Melhus 1915; Van der Zaag 1956) and was recently confirmed when *P. infestans* was detected in asymptomatic shoots emerging from infected tubers with the aid of the polymerase chain reaction (Appel et al. 2001; Hussain et al. 2007) and when sporangia and lesions developed on asymptomatic shoots placed in a moist environment (Johnson 2010).

Fig. 9.2 Streaking of reddish brown tissue on the below-ground stem where *Phytophthora infestans* moved internally in the below-ground stem from an infected seed piece to near the soil line and then formed a symptomatic lesion during a moist period

Emergence of infected shoots from infected seed pieces is often low and infected seed tubers frequently result in a reduced stand due to tuber rot and pre-emergence blighting of shoots (Boyd 1980; Melhus 1915). For example, over five consecutive years only 21 of 3260 (0.64%) infected seed tubers planted produced infected above ground shoots capable of sporulating (Hirst and Stedman 1960). In experiments in Oregon and Washington (Partipilo et al. 2000), transmission from artificially infected seed pieces to emerging shoots was 1.9–3.8% of the inoculated seed piece depending on the cultivar. In western Washington, tuber-to-sprout transmission was as high as 25% on plants held at 60–90% relative humidity in the greenhouse and transmission was greater with a US-8 than US-11 isolate (Gigot et al. 2009). Transmission is promoted in a moist environment (Johnson 2010). In a field experiment, transmission to emerged shoots did not occur from inoculated seed pieces until shortly after a rainy period following row closure. Eighty inoculated cut seed pieces were planted and late blight lesions developed almost simultaneously above the soil level on two separate main stems (Fig. 9.3). One of the two lesions occurred about 8 cm above the soil level. Sporangia formed with the developing lesions.

The transmission rate from an infected seed tuber to foliage does not need to be very high for a late blight epidemic to develop given the explosive polycyclic capabilities of *P. infestans* and the large amount of potato seed tubers planted in major

Fig. 9.3 Late blight lesion on an above-ground stem arisen from an infected seed piece planted in the field

production regions (Hirst and Stedman 1960; Vanderplank 1963; Van der Zaag 1956). For example, in the Columbia Basin in 2010 the number of seed pieces that was planted was over 2.625 x 10^9 with a total weight of 156,300 metric tons. Only a few infected tubers are needed to initiate an epidemic (Hirst and Stedman 1960; Vanderplank 1963; Van der Zagg 1956) and a few infected shoots arising from infected seed pieces in a commercial field is below the perception threshold and will not likely be noticed during the early stages of epidemics. Inoculum originating from infected seed tubers in commercial fields can be especially devastating because of the potential earliness of the initial inoculum, the rapidity with which it can be produced, and the proximity of inoculum to the crop. In addition, the moist conditions favoring emergence of infected shoots also favor sporulation and repeated infections in the field (Harrison 1992; Johnson 2010). As many as 300,000 sporangia can be produced from a single lesion demonstrating the explosive reproductive capabilities of the pathogen (Fry 2008). In addition, weather conditions in the Columbia Basin are usually the least stable in May and June, with greater likelihood of rain events, further encouraging the development of primary inoculum, spore movement and new infection.

9.4 Spatial Patterns of Epidemics

Spatial variation of late blight-infected foliage is distinctive in the pattern of disease foci (concentrated area of diseased foliage) in commercial fields. Initial foci are generally circular to asymmetrical and clearly defined, and daughter foci are associated with, but separated from parent foci. Daughter foci are often scattered over the field. Late blight is also generally more severe in particular locations of a potato circle. The disease frequently occurs earlier and is often more severe near the center of the pivot where more irrigation water is delivered and the time of leaf wetness is greater than elsewhere in the circle (Johnson et al. 2003a). Disease foci also frequently develop in areas of the field where surface water tends to accumulate such as in field depressions, in places where pivots or sprinkler systems overlap and along wheel lines where the soil is compacted resulting in long term water reservoirs. These areas favor late blight due to higher humidity and morc frcc water on plant surfaces and are the areas most severely infected in fields when the epidemic is not severe throughout the field.

Spatial variation of late blight in potato fields is aggregated and dynamic as the disease spreads within fields during epidemics. An aggregated rather than random or regular pattern is expected for late blight because after introduction of the disease into a field, new infections are more likely to occur near a previously infected plant (Miller and Johnson 2000). The pattern depends on the stage of late blight epidemic and the nature of primary infection. In epidemics in the Columbia Basin, late blight-infected plants were aggregated during epidemics in commercial fields. Aggregation increased as disease incidence increased in the early and mid-phases of epidemics in fields. A field where initial inoculum likely originated from infected seed tubers exhibited less initial aggregation than the other fields, perhaps due to the source of primary inoculum. In all fields examined, disease foci were sparse and scattered at low diseases incidences and became larger and more clumped as disease incidence increased. Disease foci were quite large at high incidence of disease. Consequently, late blight was observed to spread by foci with foci enlarging in size, producing daughter foci, and coalescing. The process continued as favored by the environment. Disease aggregation was found to decrease in some cardinal directions but continued to increase in other directions as disease incidence increased to the highest levels toward the end of epidemics (Johnson et al. 2003b). In contrast, disease aggregation was theoretically expected to rapidly decrease as disease incidence increased to relatively high levels (Ristanio et al. 2001). Few reports with quantitative data have been published concerning the degree of disease aggregation over the course of actual epidemics and quantitative data from the Columbia Basin is useful in evaluating existing theory on the spatial spread of disease.

9.5 Survival of Spores

Survival of zoospores and sporangia of *P. infestans* is an important variable of late blight epidemics, especially when overwintering oospores of *P. infestans* have not been found in the Columbia Basin. Many factors impact the survival of sporangia and zoospores including solar radiation (Mitzubuti et al. 2000; Rotem and Aust 1991; Rotem et al. 1985; Sunseri et al. 2002), temperature (Crosier 1933; Larance and Martin 1954; Martin 1949; Mizubuti and Fry 1998; Sato 1994), moisture

(Crosier 1933; Glendinning et al. 1963; Warren and Colhoun 1975), soil chemistry (Andrivon 1994a, b, 1995a, b; Ann 1994; Boguslavskaya and Filippov 1977; Hill et al. 1998), soil microorganisms (Kostrowicka 1959; Lacey 1965; Zan 1962) and spore physiology (Blackwell and Waterhouse 1930; De Weille 1964; McAlphine 1910; Rosenbaum 1917; Rotem and Aust 1991).

Survival of sporangia of *P. infestans* in air, water and soil has been observed in natural and controlled environments. Sporangia of *P. infestans* under ambient conditions during cloudy weather rarely survive for more than 3–4 h and in direct sunlight survival is rare beyond 1 h (Glendinning et al. 1963; Mitzubuti et al. 2000; Sunseri et al. 2002). Spores of isolates of *P. infestans* collected from the Columbia Basin survived in surface water between 14 and 21 days under ambient conditions (Porter and Johnson 2003). Therefore, spores of *P. infestans* have the capability of surviving in water for extended periods of time after detachment from sporangiophores from sporulating potato tissue. Overhead center-pivot irrigation on potato in the Columbia Basin makes a rotation every 18–24 h during hot weather. Spores surviving in surface water in the wheel tracks during a 3-week period could be dispersed approximately 56 times (504 h/18 h x 2 wheels per tower) by the wheels, providing opportunities for surviving spores to be possibly deposited on the top and under surfaces of adjacent potato plant tissue where infection can take place. Incidence of late blight tuber rot is often high in wet areas of fields and is also likely influenced by the survival of spores in surface water (Johnson et al. 2003b) or the movement of zoospores in these areas. Irrigation water is sometimes reused and spores could be transported to neighboring fields through infested water.

Zoospores are capable of surviving 10 days, sporangia 42 days, and mycelia 28 days *in vitro* in non-sterile soil at 22 °C (Zan 1962). The maximum survival of sporangia *in vitro* in non-sterile soil was 70–80 days (King et al. 1968; Larance and Martin 1954; Rotem et al. 1985; Zan 1962). Survival of *P. infestans* propagules under natural environmental conditions in naturally-infested soil and in artificially-infested soil in pots was 21–32 days dependent on soil type and moisture level (Lacey 1965; Murphy 1922). New clonal lineages of *P. infestans* from the Columbia Basin that were metalaxyl resistant survived under natural environmental conditions in soil between 23 and 30 days dependent on the soil type and moisture level (Porter and Johnson 2007).

An important means of survival is the overwintering of *P. infestans* in infected tubers. The effect of tuber depth, soil type and soil moisture on potato tuber infection by sporangia/zoospore inoculum was assessed in greenhouse studies for four soil types (Quincy fine sand, Quincy loamy fine sand, Quincy medium sand and Shano silt loam) commonly found in the Columbia Basin potato growing region (Porter et al. 2005). The majority of all infections were to tubers found at the soil surface and infected tubers were rarely found at 5 cm or deeper for all soil types. A Shano silt loam was effective at preventing any infection of tubers at 2 cm below an intact soil surface. Likely, this protection is due to the small pore size of the silt loam which prevents sporangia or zoospores from moving through the soil (Porter et al. 2005). However, additional factors such as oxygen concentrations (Cook and Papendick 1972; Uppal 1926; Zan 1962), microorganisms (Cook and Papendick 1972;

Murphy 1922), negative geotaxic tendencies of zoospores (Cameron and Carlile 1977; Carlile 1983) and tendency of zoospores to encyst on impact with foreign objects (Carlile 1983) are all factors that could be limiting spore movement and infection of subterranean tubers. Sporangia of *P. infestans* do not readily wash through soil and less than 1% of sporangia are found deeper than 5 cm in sandy soil (Jensen 1887; Murphy and Mckay 1925). Increased moisture levels were not a factor increasing the depth of tuber infection in the Columbian Basin soils that were assessed (Porter et al. 2005).

Fungicides impact the ability of spores of *P. infestans* to survive and infect tubers in the soil. Viability of sporangia/zoospores in soil previously treated with fungicides was determined using buried healthy whole tubers and by assaying infested soil applied to freshly cut tuber disks (Porter et al. 2006). Mancozeb and metiram, both ethylene bisdithiocarbamates, were fungicidal to sporangia and zoospores of *P. infestans* and when applied to soil are capable of acting as a fungicide barrier preventing tuber infection, however this protection only lasted up to 5 days under natural conditions (Porter et al. 2006). Cyazofamid was the only non-EBDC fungicide identified with sporangicidal activity when applied to soil. The duration of this protection has not been determined. Fluazinam and fenamidone were not sporangicidal when applied to soil; however, these fungicides significantly reduced whole tuber infections when applied to soil following an infestation of the soil with sporangia/zoospores of *P. infestans*. These two fungicides may disrupt mechanisms that enable sporangia/zoospores to find and/or infect whole tubers.

9.6 Improved Control with Resistance

Specific resistance, resistance that is expressed when genotypes of the host react differentially to different genotypes of the pathogen (Johnson and Gilmore 1980), has not been durable in potato against late blight (Fry 2008; Swiezynski et al. 1996). Partial or field resistance, resistance that reduces the rate of disease development, has been recognized and quantified in potato infected with *P. infestans* (Guzman-N 1964; Hodgson 1961; Thurston et al. 1962) and appears to be general and a durable type of resistance (Colon et al. 1995; Inglis et al. 2007). However, the development of potato cultivars with partial resistance has not received emphasis because of the relative convenience of selecting for specific resistance (Fry 2008), the availability of affordable late blight fungicides and because partial resistance is not always easily identified and can be modified by the environment (Johnson and Gilmore 1980). Combining partial resistance with integrated control tactics that reduce initial inoculum and the rate of disease development will be the most economic and stable management strategy for potato late blight (Stevenson et al. 2007). In contrast, growing susceptible and especially very susceptible late blight cultivars will aggravate the late blight situation in a region due to an increased production of inoculum. Late blight is often found first in the Columbia Basin on the very susceptible and early maturing cultivar Russet Norkotah.

Foliage of the commonly grown potato cultivars in the Columbia Basin is susceptible to *P. infestans* (Inglis et al. 1996; Porter et al. 2004). Fortunately, tubers of Umatilla Russet, Alturas, Gem Russet and a few others are moderately resistant (Porter et al. 2001, 2004). Incidence and severity of infection are less in moderately resistant tubers and late blight is much easier to manage in storage for tubers of moderately resistant cultivars. During the same storage season, losses may not be encountered for tubers of moderately resistant cultivars whereas they can be high for tubers of the very susceptible cultivar, Ranger Russet (Johnson et al. 2000). Foliage and tubers of Defender have a high level of partial resistance to *P. infestans* and the number of fungicide application can be reduced 3–6 times for successful control when compared to Russet Burbank. The mean economic returns associated with Defender were $6,196/ha; whereas they were only $4,388/ha for Russet Burbank (Stevenson et al. 2007).

9.7 Late Blight Forecasting

Late blight management has been augmented in regions of North America, Mexico and Europe by scheduling fungicide applications using predictive disease models (Beaumont 1947; Grunwald et al. 2000; Krause et al. 1975; Wallin 1962). Models developed in rain-fed agricultural regions that are based on leaf wetness or relative humidity and temperature in individual fields such as BLITECAST have not effectively predicted late blight outbreaks in the semiarid Columbia Basin and southern Idaho (Easton 1982; Henderson et al. 2007).

Late blight is forecasted and managed regionally in the Columbia Basin. This is because sporangia of *P. infestans* can become airborne in turbulent air currents and be quickly and widely disseminated within the region during cloudy and wet weather (Aylor et al. 2001; Sunseri et al. 2002), and when disease favoring weather of mild temperatures and rainy conditions occur, they usually prevail over the entire region. Additionally, the microclimate in fields can be similar throughout sections of the region after row closure. Row closure is when foliage between rows just touches and, for the main cultivars grown including Russet Burbank, generally extends from the second week of June in the southern Columbia Basin to the end of June in the northern Basin. Late blight has not been observed before row closure in the region. However, once row closure has occurred, microclimate conditions generally are favorable for late blight development whenever a field is irrigated (Easton 1982). Late blight is extremely difficult to manage once it is established in an irrigated field. For example, in a field with inoculum originating from infected seed tubers, incidence of late blight increased from 0.2% to 70% over a 4-week period after row closure even with nine applications of efficacious fungicides (Johnson et al. 2003b).

Two sets of forecasting models are used to regionally forecast the probability of late blight occurrence for the Columbia Basin. The first set of models identifies the probability of late blight occurrence early in the growing season and the second gives the probability of disease occurrence in midseason (Johnson et al. 1996, 1998). The models were derived empirically by examining the relationship of late blight occurrence in the region with meteorological variables at four vicinities in the Basin over 27 years using logistic regression analysis. Separate logistic regression models were derived for the early and mid-seasons for each of the four vicinities. Indicator variables for the early season logistic models include the presence of an outbreak during the preceding year and number of rainy days in April and May. Indicator variables for the midseason models include the presence of an outbreak the preceding year and either number of rainy days in July and August or number of rainy days in April and May and number of rainy days in July and August depending on the vicinity (Johnson et al. 1998). Variation in number of rainy days in the spring occurs among the four vicinities and having more than one location or vicinity has been beneficial in making reliable disease forecasts (Johnson et al. 1998).

The models had high sensitivity (percentage of years with late blight outbreaks classified correctly) and specificity (percentage of years without outbreaks classified correctly) when validated. As with predictive models used for most diseases, a high sensitivity is desired for the models used in the Basin. All years with late blight outbreaks and 96% of the total of 27 years of data used to develop the logistic models were correctly classified using data from at least one of the four locations (Johnson et al. 1998). All 13 years since models were first developed in 1997 through 2010 have been correctly classified at each of the four vicinities.

The probability of late blight occurrence for a given season can be calculated on 1 June when the number of rainy days in April and May is known. However, the probabilities can be estimated in early May from the actual number of rainy days in April and a 30-day rain forecast for May. This is done and a late blight forecast is usually given in early May. Early May is several weeks before row closure and sufficient time for growers to implement late blight management tactics. The advanced knowledge is beneficial because fungicides used for late blight are mostly protective, and to achieve the maximum effect, the first application must be made before the pathogen is introduced into the crop (Hirst and Stedman 1960). In addition, sufficient time and applications are also needed for fungicides to be adequately distributed throughout the plant canopy after the initial application, especially if the application is made by air.

The probabilities of late blight occurrence from the logistic models are then coupled with weather forecasts for occurrence of rainy days for 1 to 15-day and 1 to 30-day periods for the four vicinities (Johnson et al. 1998). The rain forecasts are obtained regularly throughout the growing season from a private weather forecasting group (Fox Weather, LLC, Fortuna, CA) and the short term (1 to 15-day) and long term (1 to 30-day) rain forecasts are derived independently. The probability of a late blight outbreak, weather forecasts, and crop canopy development are used to calculate a risk index and to determine initiation and intervals between recommended fungicide applications. Growers also use the late blight forecast to determine intensity of field monitoring for late blight. Late blight epidemiologists in Oregon and Washington provide oversight of the regional forecasting system. Fields are monitored for late blight throughout the growing season and the presence of any late blight in any fields is considered in scheduling recommended fungicide frequency and irrigation applications. Recommendations are available to growers via phone recordings, e-mails and a Website. Ambient temperatures are generally favorable for late blight development after row closure in the Basin and are not generally considered in scheduling disease management tactics.

A noted benefit of the late blight forecasting system used in the Columbia Basin is that growers have become more aware of environmental conditions that favor late blight outbreaks. Growers characteristically become lax in applying disease management tactics after several consecutive seasons with little or no disease. The regional late blight forecasting models have been beneficial in alerting growers when disease threats are high. Another benefit is that fungicide distributors have an idea of how much fungicide may be needed in the region for a particular year. For example, early spring in 1997 late blight fungicides were not available in the Pacific Northwest because of decisions by fungicide manufacturers to ship them elsewhere. The late blight forecast for the Columbia Basin predicted a severe epidemic in the region and this was used to convince manufacturers to ship fungicides to the Pacific Northwest. A severe epidemic did occur that year and fungicides were available to reduce the effects of the disease.

Early season rain is an effective indicator of late blight outbreaks because moisture is important for the build-up of inoculum in fields during the early stage of epidemics. Early in epidemics, moisture promotes transmission of *P. infestans* from infected seed tubers to emerged shoots in fields. Transmission from seed tubers to shoots bearing sporangia can occur within 24 h during rainy weather (Johnson 2010). Secondary infections will proceed almost immediately if a favorable environment with moisture continues. Moisture is also essential for effective dissemination of sporangia to additional fields. Additionally, solar irradiance is associated with incidence of late blight epidemics in the Columbia Basin (Johnson et al. 2009), but solar irradiance has not been incorporated in the late blight forecasting models.

9.8 Improved Fungicide Application for Late Blight Management

Effective management of late blight with protectant fungicides requires the distribution of an effective fungicide at an effective concentration throughout the canopy and field. Season long fungicide protection is expensive because of the need of repeated fungicide applications. Method of fungicide application affects fungicide distribution and cost. Fungicides are applied by air, ground and chemigation. There are many kinds of aircraft, as there are many types of ground applicators; whereas, chemigation, the adding of fungicide to the water stream and using the irrigation equipment for dispersal, generally means the use of a center-pivot irrigation system. In the Columbia Basin, the specific type of aircraft or ground applicator used is not as important as a consistent and properly developed and applied, fungicide program. That includes the use of late blight prediction models (Johnson et al. 1996, 1998).

Besides the general method used for application, other factors must be considered, such as how much water per hectare is used, frequency of application, time required to make an application, whether to add a spreader/sticker to the mixture, and fungicide rates. Cost of season long control can be significant, so choosing the appropriate application program is important (Johnson et al. 2000). Other considerations, dependent on method, include wind, humidity, droplet size, pressure, potential for skips, and nozzles. Each application method can provide adequate protection from late blight, but each has advantages and disadvantages. A discussion follows comparing application methodologies when using protectant fungicides to ultimately help in improving late blight control.

Applying fungicides by air allows for treatment of many hectares in a relatively short period of time. However, the cost of hiring a plane is moderately expensive, must be scheduled in advance, and should not be used under windy conditions and near natural or man-made obstructions such as trees, power lines and buildings. Care must be taken to accurately balance distance to canopy with pressure and nozzle orifice to ensure that droplet sizes are large enough to reduce evaporation during low humidity, so droplets containing fungicide reach the potato canopy (Jacobsen 1986), and are also of correct size to prevent off-site drift. Application skips used to be an issue prior to the use of GPS systems. The amount of water used per hectare (28–94 L) is not particularly important (Geary et al. 2004) if environmental and droplet size are appropriately considered, though as the amount of water used per hectare increases, so does cost and time needed to complete an application. The deposition pattern in the potato canopy immediately following application and the days that follow is extremely important. Prior to row closure, application of fungicides by air is evenly distributed in the potato canopy (Geary et al. 2004). Once between row closure occurs, the dynamics of fungicide coverage in the canopy change and fungicides do not readily reach the lower canopy. When using air application, the greater the amount of water used per hectare the more droplets of fungicide reach the canopy if droplet size is constant, but with larger water volumes the concentration of fungicide in each droplet is reduced. Regardless of the type of air craft, fungicide amounts are applied at the highest levels in the upper canopy, less in the middle canopy and very little in the lower canopy of the potato plants (Hamm and Clough 1999). Essentially the upper leaves "catch" the fungicide so less fungicide reaches the mid and lower levels of the canopy. Therefore at the day of application and for a time beyond, the lower levels of the canopy have less fungicide. In the Columbia Basin, frequent crop watering is essential due to light sandy soils. The application of water re-suspends the fungicide in individual spots and redistributes the material across the leaves and down the canopy. After a 7 day period, the amount in the upper canopy (on leaves present at the last application) is low while the amount present in the lower canopy is now higher from redistribution (Geary et al. 2004; Hamm and Clough 1999). Hence the reason and time for the next application to recharge and apply to new foliage in the upper canopy, which also serves to maintain fungicide levels in the mid to lower canopy through the next week when irrigation water is applied. The loss of fungicide through the canopy due to redistribution is a reason why full label rates of protectant fungicides are needed and why spreader/stickers do not aid in late blight protection (Geary et al. 2004).

In contrast to air application, chemigation is a slow method to apply fungicides. Often approximately 9 h is required to apply fungicides to a 50.6 hectare field whereas air may only take an hour. Chemigation can be accomplished over many fields simultaneously, since field equipment already present is being utilized, but each field still requires 6–12 h. Chemigation can be used under more windy conditions and costs are reduced since required equipment is already present in the field. Droplet and pressure balances are not as relevant, and neither is humidity. Application skips are eliminated as long as sprinklers, the pump injecting the fungicide into the water stream and the center pivot system continues to operate throughout the application.

The major difference between chemigation, air and ground application is the amount of water used during application. Amounts of water used are generally 28–47 L/ha for air application, 94–187 L/ha for ground application, and under the best reduced water situations, approximately 0.25 centimeters/ha or 25,245 L/ha for chemigation. Compared to air, the use of this large amount of water during chemigation creates a much different application pattern in and through the canopy, redistribution need, and time requirement to provide adequate late blight control throughout the canopy (Hamm and Clough 1999). Immediately following application, nearly equal amounts of fungicide are present throughout the canopy. Leaves and stems are adequately protected. However, in contrast to air application where relatively little fungicide initially reached the lower canopy, during chemigation large volumes of water pass through the canopy, moving the fungicide through the canopy. While fungicide levels are nearly equal in all canopy levels (upper, middle, lower) the day of application, the total amount of fungicide present is much less than with air application. With each application of irrigation water, fungicide is again redistributed, but given that less is present there is a greater chance of residue levels falling below the concentration needed to control late blight before the standard recommendation of the next application a week later. This is particularly true in the Columbia Basin where frequent watering occurs, though this is less likely where heavier soil and other factors allow less frequent applications of irrigation water. The large use of water is why full label rates of protectant fungicides should always be used and why the use of spreader/stickers is not justified. Trials using a boom attached to a center pivot irrigation system that applied approximately 675 L/ha of water applied significantly more fungicide to the canopy compared to chemigation with normal water amounts (Geary et al. 1999). While using an attached boom was not directly compared to air or ground application, the benefit of higher residues has been shown to provide better late blight control (Geary et al. 1999). Costs of the attached boom are initially high, but long term would reduce application costs by allowing grower application while also allowing grower controlled scheduling. Regardless of chemigation method used, this method can be an effective way to control later blight if done correctly.

Applying fungicides using a ground applicator may be the most common application method for late blight management in many potato producing areas in North America, but is the least used method in the Columbia Basin. Several factors that contribute to this are the cost of equipment/application, the time required to apply and the damage that occurs to the crop. Purchase costs are high for a ground applicator, regardless of the model, and so is the cost of hiring a ground applicator. While not as slow as chemigation for a given field (only a single field can be treated at a time in contrast to multiple fields by chemigation), ground application still requires a significant amount of time and water (94 – 187 L/ha is standard). Reduced yield occurs, either due to the wheel tracks causing soil compaction through the field or from planting skips specifically established for movement of ground equipment. Ground application is not as impacted by wind or evaporation compared to either air or chemigation, particularly if the appropriate balance between pressure and nozzles and boom distance to the canopy is used. Skips are easily prevented with careful observation of nozzles to confirm that they are properly working and with the use of GPS units.

A ground applicator is the most effective method for applying fungicide in a potato canopy. Nearly three times the amount of product can be found in the canopy the day of application compared to air, and many times more than chemigation. More material is found in the upper canopy, and reduced levels in the middle and lower canopy, but still much higher than that found in any canopy location compared to air or chemigation. Redistribution downward occurs as in air application, but through a 7 day period the fungicide levels are always higher at each canopy location compared to air. The level of fungicide in the lower canopy the day of the first application provides good protection but still requires at least one watering cycle to redistribute the fungicide from the individual fungicide spots, particularly in the mid to lower canopy levels, to provide complete coverage throughout the canopy. Given that irrigation water redistributes the fungicide, the use of spreader/stickers is not justified. However, a case could be made that reduced rates of fungicide could be used given the substantially higher levels that result from this application method. Seven day application schedules are still recommended due to the large amounts of new unprotected growth from the last application.

Long term in-season needs for fungicides equate to substantial application costs in the Columbia Basin given the long growing season and high late blight risk (Johnson et al. 2000). Given that, alternative methods to apply fungicides have been suggested that reduce costs while using the distinctiveness and effectiveness of air and chemigation methods. While air is expensive, the amount of fungicide applied to the canopy is high. Chemigation is the least expensive application method but also leaves the least amount of fungicide in the canopy. Trials have shown that beginning a 7-day application program using air, followed by chemigation, and continuing that alternation of methods effectively controls late blight with reduced costs in the Columbia Basin (Hamm et al. 2006).

In summary, late blight can be successfully managed by applying fungicides with each of the application methods, but each method has advantages and disadvantages. If late blight protection is needed immediately, then any method may work, given the time constraints needed for each to complete application. Chemigation is the least expensive but fungicide levels may fall below threshold levels within 7 days, depending on frequency and amounts of irrigation water applied. Ground application delivers the most fungicide to and throughout the canopy but is slow, expensive and reduces yield because of soil compaction in the wheel tracks. Air application requires at least one watering to ensure redistribution of fungicide droplets throughout the canopy after the initial fungicide application. The alternate use of air and chemigation provides good protection at a reduced cost compared to only air or ground application methods. Careful consideration and use of fungicide application methods as part of an integrated disease management program will help ensure season-long protection from late blight at the least cost.

9.9 Post-harvest Fungicides for Tuber Blight Control

Typically post-harvest products are applied as a low-pressure, low-volume spray as potatoes are being conveyed into storage. Some post-harvest disinfestants are applied through the humidification system during the storage season. Post-harvest fungicides are specific to a particular organism or class of organisms whereas disinfestants are general biocides with a wide spectrum against both bacteria and fungi. Additional products such as inorganic and organic salts, aromatic oils, bacteria, and other biological products have also been evaluated for potential post-harvest disease suppression properties.

The late blight pathogen can spread from tuber to tuber as a result of contact that occurs during tuber handling (Dowley and O'Sullivan 1991). This exposure may occur as tubers are lifted from the soil on the belt of commercial potato harvesters, as tubers collide on the harvester belt, as tubers are piled into trucks for transportation, or when tubers are delivered from trucks to storages, packing, or processing plants. Healthy tubers can become wounded during any of these phases, increasing tuber susceptibility to pathogen infection. Inoculum in soil adhering to tubers or from infected tubers may be present in the form of viable spores or mycelium and may be transferred to healthy tubers during these processes. Applications of post-harvest disinfestants and/or fungicides are aimed at reducing the viability of these potential inoculum sources on the surface of the healthy tubers prior to infection.

Post-harvest fungicides and disinfestants can be applied to potatoes as a low volume aqueous spray as the potatoes are conveyed into storage. The spray boom is generally located in an area where the potatoes may roll, such as a star-table or a drop from one conveyor to the next, to ensure adequate coverage of the tuber. Post-harvest applications may include one or multiple sets of spray nozzles. With all the post-harvest applied products, full coverage of the tuber is needed for optimal efficacy. The volume of product applied ranges from 0.25 gal to 1 gal/ton tubers. In general, 0.5 gal/ton is recommended to ensure adequate tuber coverage and to avoid excess water on the tuber surface and surrounding equipment.

The use of phosphite or salts of phosphorous acid was investigated as a post-harvest applied fungicide after research demonstrated that these products could control diseases on potatoes caused by Oomycetes (Johnson et al. 2004). Application of phosphorous acid to tubers after harvest and prior to storage can result in a reduction in post-harvest infection by *P. infestans*. In experiments where tubers were submersed in a suspension of *P. infestans* sporangia/zoospores, applications of phosphorous acid-based fungicides (= phosphonate, phosphites) significantly reduced tuber blight (Miller et al. 2006). Phosphorous acid fungicides are more effective for this purpose than general disinfestants such as hydrogen peroxide/peroxyacetic acid products (HPPA) or chlorine dioxide-based products.

Duration between the occurrence of inoculation and post-harvest treatment appears to impact the efficacy of the product applied. HPPA was effective in reducing late blight incidence when applied immediately after inoculation, but was not effective when treatment was made 1 h or more after inoculation (Miller et al. 2006). Phosphorous acid applications were effective in significantly reducing late blight up to 6 h after inoculation. Complete control of late blight was obtained with 12.8 fl oz/ton of phosphorous acid when applied in a larger scale trial (one ton of tubers stored for 77 days at 8.9 °C; *unpublished data*). As a result of this work, phosphorous acid is now being used more commonly for post-harvest control of late blight and pink rot.

References

Andrivon D (1994a) Dynamics of the survival and infectivity to potato tubers of sporangia of *Phytophthora infestans* in three different soils. Soil Biol Biochem 26:945–952

Andrivon D (1994b) Fate of *Phytophthora infestans* in a suppressive soil in relation to pH. Soil Biol Biochem 26:953–956

Andrivon D (1995a) Biology, ecology, and epidemiology of the potato late blight pathogen *Phytophthora infestans* in soil. Phytopathology 85:1053–1056

Andrivon D (1995b) Inhibition by aluminum of mycelial growth and of sporangial production and germination in *Phytophthora infestans*. Eur J Plant Pathol 101:527–533

Ann PJ (1994) Survey of soils suppressive to three species of *Phytophthora* in Taiwan. Soil Biol Biochem 26:1239–1248

Appel R, Adler N, Habermeyer J (2001) A method for the artificial inoculation of potato tubers with *Phytophthora infestans* and polymerase chain reaction assay of latently infected sprouts and stems. J Phytopathol 149:287–292

Aylor DE, Fry WE, Mayton H, Andrade-Piedra JL (2001) Quantifying the rate of release and escape of *Phytophthora infestans* sporangia from a potato canopy. Phytopathology 91:1189–1196

Beaumont A (1947) The dependence on the weather of the dates of outbreaks of potato late blight epidemics. Trans Br Mycol Soc 31:45–53

Blackwell EM, Waterhouse GM (1930) Spores and spore germination in the genus *Phytophthora*. Trans Brit Mycol Soc 15:294–310

Boguslavskaya NV, Filippov AV (1977) Survival rates of *Phytophthora infestans* (Mont) D. By. in different soils. Mikol Fitopatol 11:239–241

Bourke PMA (1964) Emergence of potato blight, 1843–46. Nature 203:805–808

Bourke A (1991) Potato blight in Europe in 1845: the scientific controversy. In: Lucas JA, Shattock RC, Shaw DC, Cooke LR (eds) Phytophthora. Cambridge University Press, Cambridge, UK

Boyd AEW (1980) Development of potato blight (*Phytophthora infestans*) after planting infected seed tubers. Ann Appl Biol 95:301–309

Cameron JN, Carlile MJ (1977) Negative geotaxis of zoospores of the fungus *Phytophthora*. J Gen Microbiol 98:59–602

Carlile MJ (1983) Motility, taxis, and tropism in *Phytophthora*. In: Erwin DC, Bartnicki-Garcia S, Tsao PH (eds) Phytophthora: its biology, taxonomy, ecology, and pathology. American Phytopathological Society, St. Paul, MN

Colon LT, Budding DJ, Keizer LCP, Pieters MMJ (1995) Components of resistance to late blight (*Phytophthora infestans*) in eight South American *Solanum* species. Eur J Plant Pathol 101:441–456

Cook RJ, Papendick RI (1972) Influence of water potential of soils and plants on root disease. Annu Rev Phytopathol 10:349–374

Crosier W (1933) Culture of *Phytophthora infestans*. Phytopathology 23:713–720

De Bary A (1876) Researches into the nature of the potato-fungus – *Phytophthora infestans*. J R Agr Soc 12:239–269

De Weille GA (1964) Forecasting crop infection by the potato blight fungus. K Ned Meteorologisch Inst Meded Verh 82:1–144

Dowley LJ, O'Sullivan E (1991) Sporulation of *Phytophthora infestans* (Mont.) De Bary on the surface of diseased potatoes and tuber to tuber spread of infection during handling. Potato Res 34:295–296

Easton GD (1982) Late blight of potatoes and prediction of epidemics in arid central Washington State. Plant Dis 66:452–455

Fry W (2008) *Phytophthora infestans*: the plant (and R gene) destroyer. Molec Plant Pathol 9:385–402

Fry WE, Goodwin SB, Dyer AT, Matuszak JM, Drenth A, Tooley PW, Sujkowski LS, Koh YJ, Cohen BA, Spielman LJ, Deahl KL, Inglis DA, Sandlan KP (1993) Historical and recent migrations of *Phytophthora infestans*: chronology, pathways, and implications. Plant Dis 77:653–661

Gavino PD, Smart CD, Sandrock RW, Miller JS, Hamm PB, Yun Lee T, Davis RM, Fry WE (2000) Implications of sexual reproduction for *Phytophthora infestans* in the United States: generation of an aggressive lineage. Plant Dis 84:731–735

Geary G, Johnson DA, Hamm PB, Cummings TF (1999) Fungicide application for late blight management: a boom attached to a center pivot irrigation system. Plant Dis 83:512–515

Geary G, Hamm PB, Johnson DA (2004) Deposition and redistribution of fungicides applied by air and chemigation for control of late blight in commercial potato fields. Am J Potato Res 81:305–315

Gigot JA, Gundersen B, Inglis DA (2009) Colonization and sporulation of *Phytophthora infestans* on potato tubers under northwestern Washington conditions. Am J Potato Res 86:1–14

Glendinning D, MacDonald JA, Grainger J (1963) Factors affecting the germination of sporangia in *Phytophthora infestans*. Trans Brit Mycol Soc 46:595–603

Goodwin SB, Cohen BA, Deahl KL, Fry WE (1994a) Migration from northern Mexico as the probable cause of recent genetic changes in populations of *Phytophthora infestans* in the United States and Canada. Phytopathology 84:553–558

Goodwin SB, Cohen BA, Fry WE (1994b) Panglobal distribution of a single clonal lineage of the Irish potato famine fungus. Proc Natl Acad Sci 91:11591–11595

Goodwin SB, Sujkowski LS, Dyer AT, Fry BA, Fry WE (1995) Direct detection of gene flow and probable sexual reproduction of *Phytophthora infestans* in northern North America. Phytopathology 85:473–479

Grunwald NJ, Rubio-Covarrubias O, Fry WE (2000) Potato late blight management in the Toluca Valley: forecasts and resistant cultivars. Plant Dis 84:410–416

Guzman-N J (1964) Nature of partial resistance of certain clones of three *Solanum* species to *Phytophthora infestans*. Phytopathology 54:1398–1404

Hamm PB, Clough GH (1999) Comparison of application methods on deposition and redistribution of chlorothalonil in a potato canopy and potential impact for control of late blight. Plant Dis 83:441–444

Hamm PB, Johnson DA, Cummings TF (2006) Comparison of deposition patterns in two programs for applying protectant fungicides to potato stems and leaves for the control of late blight. Am J Potato Res 83:473–484

Harrison JG (1992) Effects of the aerial environment on late blight of potato foliage – a review. Plant Pathol 41:384–416
Henderson D, Williams CJ, Miller JS (2007) Forecasting late blight in potato crops of southern Idaho using logistic regression analysis. Plant Dis 91:951–956
Hill AE, Grayson DE, Deacon JW (1998) Suppressed germination and early death of *Phytophthora infestans* sporangia caused by pectin, inorganic phosphate, ion chelators and calcium-modulating treatments. Eur J Plant Pathol 104:367–376
Hirst JM, Stedman OJ (1960) The epidemiology of *Phytophthora infestans* II. The source of inoculum. Ann Appl Biol 48:489–517
Hodgson WA (1961) Laboratory testing of the potato for partial resistance to *Phytophthora infestans*. Am Potato J 38:259–264
Hussain S, Lees AK, Duncan JM, Cooke DEL (2007) Development of a species-specific and sensitive detection assay for *Phytophthora infestans* and its application for monitoring of inoculum in tubers and soil. Plant Pathol 54:373–382
Inglis DA, Johnson DA, Legard DE, Fry WE, Hamm PB (1996) Relative resistances of potato clones in response to new and old populations of *Phytophthora infestans*. Plant Dis 80:575–578
Inglis DA, Powelson ML, Dorrance AE (1999) Effect of registered potato seed piece fungicides on tuber-borne *Phytophthora infestans*. Plant Dis 83:229–234
Inglis DA, Brown CR, Gundersen BG, Porter LD, Miller JS, Johnson DA, Lozoya-Saldana H, Haynes KG (2007) Assessment of *Solanum hougasii* in Washington and Mexico as a source of resistance to late blight. Am J Potato Res 84:217–228
Jacobsen BJ (1986) Methods of calibration and use of aircraft for applying fungicides. In: Hickey KD (ed) Methods for evaluating pesticides for control of plant pathogens. American Phytopathological Society, St. Paul, MN
Jensen JL (1887) Moyens de combattre et de detruire le *Peronospora* de la pomme de terre. Mem Soc Nat Agric Fr 131:31–156
Johnson DA (2010) Transmission of *Phytophthora infestans* from infected potato seed tubers to emerged shoots. Plant Dis 94:18–23
Johnson DA, Cummings TF (2009) Latent infection of potato seed tubers by *Phytophthora infestans* during long term cold storage. Plant Dis 93:940–946
Johnson DA, Gilmore EC (1980) Breeding for resistance to pathogens in wheat. In: Harris MK (ed) Biology and breeding for resistance. Texas A&M University, College Station, TX
Johnson DA, Alldredge JR, Vakoch DL (1996) Potato late blight forecasting models for the semiarid environment of south-central Washington. Phytopathology 86:103–106
Johnson DA, Cummings TF, Hamm PB, Rowe RC, Miller JS, Thornton RE, Pelter JQ, Sorensen EJ (1997) Potato late blight in the Colombia Basin: an economic analysis of the 1995 epidemic. Plant Dis 81:103–106
Johnson DA, Cummings TF, Hamm PB (1998) Expansion of potato late blight forecasting models for the Columbia Basin of Washington and Oregon. Plant Dis 82:642–645
Johnson DA, Cummings TF, Hamm PB (2000) Cost of fungicides used to manage potato late blight in the Columbia Basin: 1996 to 1998. Plant Dis 84:399–402
Johnson DA, Alldredge JR, Hamm PB, Frazier BE (2003a) Aerial photography used for spatial pattern analysis of late blight infection in irrigated potato circles. Phytopathology 93:805–812
Johnson DA, Martin M, Cummings TF (2003b) Effect of chemical defoliation, irrigation water, and distance from the pivot on late blight tuber rot in center-pivot irrigated potatoes in the Columbia Basin. Plant Dis 87:977–982
Johnson DA, Inglis DA, Miller JS (2004) Control of potato tuber rots caused by oomycetes with foliar applications of phosphorous acid. Plant Dis 88:1153–1159
Johnson DA, Cummings TF, Abi Ghanem R, Alldredge JR (2009) Association of solar irradiance and days of precipitation with incidence of potato late blight in the semiarid environment of the Columbia Basin. Plant Dis 93:272–280
Kadish D, Cohen Y (1992) Overseasoning of metalaxyl-sensitive and metalaxyl-resistant isolates of *Phytophthora infestans* in potato tubers. Phytopathology 82:887–889

King JE, Colhoun J, Butler RD (1968) Changes in the ultrastructure of sporangia of *Phytophthora infestans* associated with indirect germination and ageing. Trans Br Mycol Soc 51:269–281

Kirk WW, Niemira BA, Stein JM (2001) Influence of storage temperature on rate of potato tuber tissue infection caused by *Phytophthora infestans* (Mont.) de Bary estimated by digital image analysis. Potato Res 44:86–96

Kostrowicka M (1959) Interaction between *Phytophthora infestans* and *Rhizoctonia solani*. Proc 9th Int Bot Congr 2:201

Krause RA, Massie LB, Hyre RA (1975) BLITECAST: a computerized forecast of potato late blight (*Phytophthora infestans*). Plant Dis Report 59:95–98

Lacey J (1965) The infectivity of soils containing *Phytophthora infestans*. Ann Appl Biol 56:363–380

Lambert DH, Currier AI, Olanya MO (1998) Transmission of *Phytophthora infestans* in cut potato seed. Am J Potato Res 75:257–263

Larance RS, Martin WJ (1954) Comparison of isolates of *Phytophthora infestans* at different temperatures. Phytopathology 44:495

Large EC (1940) The advance of the fungi. Jonathan Cape, London

Marshall KD, Stevenson WR (1996) Transmission of *Phytophthora infestans* in cut potato seed. Am J Potato Res 73:370–371

Martin WJ (1949) Strains of *Phytophthora infestans* capable of surviving high temperature. (*Abstr.*) Phytopathol 39:14

McAlpine D (1910) Some points of practical importance in connection with the life-history stages of *Phytopthora infestans* (Mont.) de Bary. Ann Mycol 8:156–166

Melhus IE (1915) Hibernation of *Phytophthora infestans* in the Irish potato. J Agric Res 5:71–11

Miller JS, Johnson DA (2000) Competitive fitness of *Phytophthora infestans* isolates under semi-arid field conditions. Phytopathology 90:220–227

Miller JS, Hamm PB, Johnson DA (1997) Characterization of the *Phytophthora infestans* population in the Columbia Basin of Oregon and Washington from 1992 to 1995. Phytopathology 87:656–660

Miller JS, Johnson DA, Hamm PB (1998) Aggressiveness of isolates of *Phytophthora infestans* from the Columbia Basin of Washington and Oregon. Phytopathology 88:190–197

Miller JS, Olsen N, Woodell L, Porter LD, Clayson S (2006) Post-harvest applications of zoxamide and phosphite for control of potato tuber rots caused by Oomycetes at harvest. Am J Potato Res 83:269–278

Mitzubuti ESG, Aylor DE, Fry WE (2000) Survival of *Phytophthora infestans* sporangia exposed to solar radiation. Phytopathology 90:78–84

Mizubuti ESG, Fry WE (1998) Temperature effects on developmental stages of isolates from three clonal lineages of *Phytophthora infestans*. Phytopathology 88:837–843

Murphy PA (1922) The bionomics of the conidia of *Phytophthora infestans*. Sci Proc R Dublin Soc 16:442–466

Murphy PA, Mckay R (1925) Further experiments on the sources and development of blight infection in potato tubers. J Dep Lds Agric Dublin 25:10–21

Partipilo HM, Powelson ML, Inglis DA (2000) Seedborne *Phytophthora infestans*: rate of transmission and effect on stand in five potato cultivars. Am J Potato Res 77:415

Porter LD, Johnson DA (2003) Survival of *Phytophthora infestans* in surface water. Phytopathology 94:380–387

Porter LD, Johnson DA (2007) Survival of sporangia of new clonal lineages of *Phytophthora infestans* in soil under semiarid conditions. Plant Dis 91:835–841

Porter LD, Johnson DA, Cummings TF (2001) Development of *Phytophthora infestans* in potato tubers of nine clones in storage. Phytopathology 91:S188

Porter LD, Inglis DA, Johnson DA (2004) Identification and characterization of resistance to *Phytophthora infestans* in leaves, stems, flowers, and tubers of potato clones in the Pacific Northwest. Plant Dis 88:965–972

Porter LD, Dasgupta N, Johnson DA (2005) Effects of tuber depth and soil moisture on infection of potato tubers in soil by *Phytophthora infestans*. Plant Dis 89:146–152

Porter LD, Cummings TF, Johnson DA (2006) Effects of soil-applied late blight foliar fungicides on infection of potato tubers by *Phytophthora infestans*. Plant Dis 90:964–968

Powelson M, Ludy R, Partipilo H, Inglis D, Gundersen B, Derie M (2002) Seedborne late blight of potatoes. APSnet Plant Health Progress research article, http://www.planthealthprogress.org/current/management/poatolate/top.htm

Ristaino JB, Groves CT, Parra GR (2001) PCR amplification of the Irish potato famine pathogen from historic specimens. Nature 411:695–697

Rosenbaum J (1917) Studies of the genus *Phytophthora*. J Agric Res 8:233–276

Rotem J, Aust HJ (1991) The effect of ultraviolet and solar radiation and temperature on survival of fungal propagules. J Phytopathol 133:76–84

Rotem J, Palti J, Lomas J (1970) Effects of sprinkler irrigation at various times of the day on development of potato late blight. Phytopathology 60:839–843

Rotem J, Wooding B, Aylor DE (1985) The role of solar radiation, especially ultraviolet, in the mortality of fungal spores. Phytopathology 75:510–514

Sato N (1994) Effect of water temperature on direct germination of the sporangia of *Phytophthora infestans*. Ann Phytopathol Soc Jpn 60:162–166

Stevens NE (1933) The dark ages in plant pathology in America: 1830–1870. J Wash Acad Sci 23:435–446

Stevenson WR, James RV, Inglis DA, Johnson DA, Schotzko T, Thornton RE (2007) Fungicide spray programs for Defender, a new potato cultivar with resistance to late blight and early blight. Plant Dis 91:1327–1336

Sunseri MA, Johnson DA, Dasgupta N (2002) Survival of detached sporangia of *Phytophthora infestans* exposed to ambient, relatively dry atmospheric conditions. Am J Potato Res 79:443–450

Swiezynski KM, Domanski L, Sobkowiak S, Zaraycka H (1996) Reaction to *Phytophthora infestans* of potato genotypes with race-specific resistance. Potato Res 39:195–203

Thurston HD, Heidreick LE, Guzman J (1962) Partial resistance to *Phytophthora infestans (*Mont.) de Bary within the coleccion central Colombiana. Am Potato J 39:63–69

Uppal BN (1926) Relation of oxygen to spore germination in some species of the Peronosporales. Phytopathology 16:285–292

Van der Zaag DE (1956) Overwintering en epidemiologie van *Phytophtora infestans*, tevens einige nieuwe bestrijdingsmoelijkheden. Tijdschr Plantenziekten 62:89–156

Vanderplank JE (1963) Plant diseases; epidemics and control. Academic Press, New York, 349pp

Wallin JR (1962) Summary of recent progress in predicting late blight epidemics in United States and Canada. Am Potato J 39:306–312

Warren RC, Colhoun J (1975) Viability of sporangia of *Phytophthora infestans* in relation to drying. Trans Br Mycol Soc 64:73–78

Zan K (1962) Activity of *Phythophthora infestans* in soil in relation to tuber infection. Trans Br Mycol Soc 45:205–221

Zwankhuizen MJ, Govers F, Zakoks JC (1998) Development of potato late blight epidemics: disease foci, disease gradients, and infection sources. Phytopathology 88:754–763

Part IV
Improved Nitrogen Management in Rainfed Potato Production in Eastern Canada

Chapter 10
Nitrogen Fertilization Strategies in Relation to Potato Tuber Yield, Quality, and Crop N Recovery

Bernie J. Zebarth, Gilles Bélanger, Athyna N. Cambouris, and Noura Ziadi

Abstract In this chapter, we discuss the challenges to optimizing nitrogen (N) management in rain-fed potato production in eastern Canada, and evaluate a series of N fertilization strategies for their effects on tuber yield, size distribution and quality and on apparent recovery of fertilizer N by the potato crop. Selection of the optimal fertilizer N rate remains one of the most important decisions for growers. Optimal fertilizer N management is necessary to achieve economic goals associated with tuber yield and size, whereas over-fertilization greatly increases the risk of environmental losses of N and of reduced tuber quality. However, large variations in crop N demand and soil N supply among fields and among years, and also within fields, make selection of an optimal fertilizer N rate problematic. Improved predictions of crop demand and soil supply both in time and in space will be required to address this. Fertilizer N management can also be improved through appropriate timing of fertilizer application, fertilizer placement, and fertilizer formulation. Efficiency of N management can be improved through development of N management systems on a whole-field basis, or on a within-field basis using Site Specific Nutrient Management (SSNM), where soil-based tests are used to determine at-planting N management and plant-based or soil-based tests are used for in-season N management. In addition, use of controlled release fertilizer products can be beneficial in soils where the risks of leaching losses are high. In order to manage N efficiently and sustainably, it is important to consider N management as one component of an integrated cropping system.

B.J. Zebarth (✉)
Potato Research Centre, Agriculture and Agri-Food Canada,
PO Box 20280, 850 Lincoln Rd, Fredericton, NB, Canada E3B 4Z7
e-mail: Bernie.Zebarth@agr.gc.ca

G. Bélanger • N. Ziadi
Soils and Crops Research and Development Centre, Agriculture and Agri-Food Canada,
2560 Hochelaga Blvd, Quebec City, QC, Canada G1V 2J3

A.N. Cambouris
Soils and Crops Research and Development Centre, Agriculture and Agri-Food Canada,
979 de Bourgogne Ave, Quebec City, QC, Canada G1W 2L4

Z. He et al. (eds.), *Sustainable Potato Production: Global Case Studies*,
DOI 10.1007/978-94-007-4104-1_10,

10.1 Introduction

Potatoes are an economically important crop in eastern Canada (defined here as Quebec and Atlantic Canada). In 2009, approximately 145,000 ha of potatoes were harvested in Canada, of which about 50% was in eastern Canada (Statistics Canada 2010). A large proportion of this potato production is grown for processing (primarily French fry) and a smaller proportion for seed as well as for the table market.

Good nitrogen (N) management is a critical component of successful potato production (Zebarth and Rosen 2007). A sufficient N supply is important in achieving economically viable potato yields, and in meeting tuber size and quality targets in processing potato contracts. However, N is also easily lost from potato fields to water through nitrate leaching, and to the atmosphere as nitrous oxide, a greenhouse gas. Good N management is therefore required to meet both economic and environmental objectives in potato production.

In this chapter, we discuss the challenges to optimizing N management in rain-fed potato production in eastern Canada, and evaluate a series of N fertilization strategies for their effects on tuber yield, size distribution and quality and on apparent recovery of fertilizer N by the potato crop. Subsequent chapters address the use of soil- and plant-based test systems to improve fertilizer N recommendations (Chapter 11); N management in organic potato production systems (Chapter 12); and N losses to water (Chapter 13) and to the atmosphere (Chapter 14) in potato production systems in eastern Canada.

10.2 Challenges to Optimizing Fertilizer N Management

10.2.1 Climate, Soils and Cropping Systems

Potato production in eastern Canada occurs primarily in the provinces of Prince Edward Island (PEI), New Brunswick, and Quebec. The region has a cool temperate climate with a mean annual temperature ranging from 4 to 7°C, and humid soil moisture regimes with a mean annual precipitation ranging from 1050 to 1300 mm (Environment Canada 2011). The growing season is relatively short (approximately 120 days). Potential total tuber yield in New Brunswick is about 50 t ha^{-1} (Bélanger et al. 2000).

Soils on which potatoes are grown range from sandy to clayey soil texture, with loam and sandy loam soils most common. Potato fields range from flat to steeply sloping where the latter are commonly terraced to reduce soil erosion. Substantial within-field variation in soil texture, soil organic matter, soil drainage or topography occurs in many fields.

Cropping systems vary with potato (*Solanum tuberosum* L.)-barley (*Hordeum vulgare* L.)-red clover (*Trifolium pratense* L.) the most common crop rotation in PEI, potato-barley the most common rotation in New Brunswick, and potatoes commonly grown in rotation with corn (*Zea mays* L.) or cereal crops in Quebec. There has been greater diversity in crop rotations in recent years with increasing use of Italian ryegrass (*Lolium multiflorum* Lam.) and corn in potato rotations to address pest and disease issues, however this has not influenced the frequency of potatoes grown in the rotations.

Potato production in this region is primarily rain-fed. Use of irrigation is limited by lack of suitable water supplies, sloping fields, and heterogeneous soils. In addition, although tuber yield is frequently increased by irrigation, there is not an economic benefit from irrigation in all growing seasons (Bélanger et al. 2000a). There is, however, a trend towards increased irrigation due to significant crop losses when drought occurs in some years.

10.2.2 Challenges in Fertilizer N Management

Currently, general fertilizer N recommendations for potatoes in this region range from 125 to 200 kg N ha^{-1} (NBDAFA 2001; CRAAQ 2010). Fertilizer N is commonly banded all at planting in PEI and New Brunswick, where split application may limit tuber yield in some years (Porter and Sisson 1993; Zebarth et al. 2004a). In contrast, due to generally sandier soils used for potato production, fertilizer N is commonly applied as a split application in Quebec, with granular fertilizer N commonly applied at planting and 30 days after planting at first or final hilling. In comparison, rapid plant N uptake occurs during tuber initiation and set (from about 50 to 70 days after planting) and is reduced during tuber bulking (from about 70–90 days after planting) (Zebarth and Rosen 2007).

Optimizing N fertilization requires matching the supply of N to the crop N demand in space and time. Practical problems arise in making fertilizer N recommendations due to uncertainty in crop N demand and in soil N supply among and within fields and among years (Zebarth et al. 2009a).

Crop growth is regulated by the relative internal supplies of carbon and N (Lemaire and Millard 1999). Crop N demand and uptake under non-limiting N supply are primarily determined by crop growth (Gastal and Lemaire 2002) and there is commonly a close relationship between plant N uptake and plant dry matter accumulation (Vos 1997). Thus, crop N demand varies with factors that influence crop growth.

Crop growth and the resulting tuber yield vary with environmental conditions. This can be reflected in differences in the relationships between plant N accumulation and tuber yield (Fig. 10.1). Potato cultivars also vary in the relationship between growth and crop N uptake (Zebarth et al. 2004c), however, cultivar differences in tuber yield response to fertilizer N rate between Shepody and Russet Burbank, the two main potato cultivars used in the region, are often limited (Bélanger et al. 2000a). Tuber yield is sensitive to seasonal variation in temperature in this region which has a relatively short growing season. Cool, wet spring conditions can delay planting in some years, and thereby reduce yield (Fig. 10.1). Yield can also be

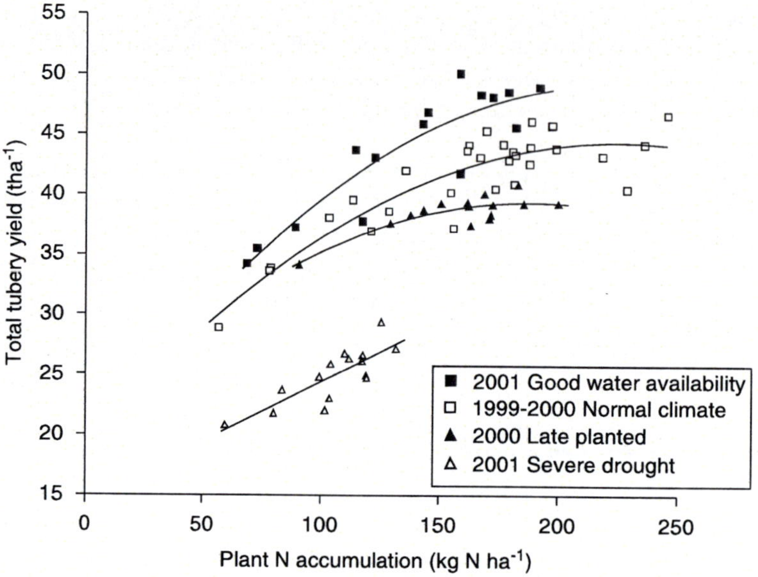

Fig. 10.1 The relationship between total tuber yield and N accumulation in the plant (tubers plus vines) prior to vine desiccation varies among years and environmental conditions (Adapted from Zebarth et al. 2004b, 2005b)

reduced in some years when drought occurs (Bélanger et al. 2000a), typically during tuber bulking, or because of disease or insect stress. As a result, the quantity of N which is required to obtain maximum tuber yield, and the amount of yield for a given quantity of N taken up, varies among sites and years (Fig. 10.1).

A study conducted at 12 site-years in New Brunswick illustrates the range in crop demand among fields and years. Maximum total tuber yield ranged from 24 to 51 t ha^{-1} in rain-fed production (Bélanger et al. 2000a). Where supplemental irrigation was applied to overcome drought stress, maximum total tuber yield at these sites still ranged from 32 to 50 t ha^{-1} (Bélanger et al. 2000a). This variation in yield among sites results in variable crop N demand among fields and years.

Soil N supply is also variable among and within fields and among years. High precipitation over the fall and winter period results in loss of most residual soil nitrate from the root zone, primarily as nitrate leaching (Zebarth et al. 2003a, 2009a). Soil N supply for a given growing season is therefore controlled primarily by soil N mineralization. Although soil mineral N measured in spring is often well correlated with growing season soil N supply (Sharifi et al. 2007), soil mineral N used alone is not an effective predictor of optimum fertilizer N rate for potatoes in this region (Bélanger et al. 2001a).

Soil N supply is controlled by the quantity and quality of soil mineralizable N in combination with environmental conditions during the growing season. Soil mineralizable N is influenced by history of organic amendment use (Sharifi et al. 2008a), crop rotation (particularly inclusion of legume and non-legume forage crops)

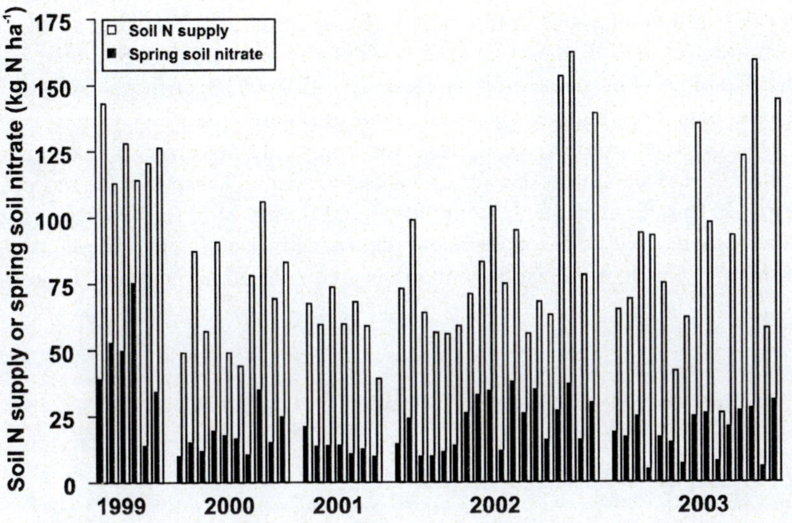

Fig. 10.2 Soil N supply (plant N accumulation in tubers plus vines at vine desiccation with no fertilizer N applied) varies among fields and years, and is commonly much higher than spring soil nitrate to 30 cm depth, indicating that most soil N supply is derived from soil N mineralization (Adapted from Zebarth et al. 2005a)

(Sharifi et al. 2009a), tillage (Sharifi et al. 2008b), soil properties and climatic zone (Dessureault-Rompré et al. 2010). In some cases, soil mineralizable N is influenced by management of the rotation crop (Sanderson et al. 1999; Zebarth et al. 2009b). Estimates of soil N supply in Atlantic Canada, based on plant N uptake in zero N fertilizer plots (Zebarth et al. 2005a), varied widely ranging from 26 to 162 kg N ha^{-1} (average 85 kg N ha^{-1}) (Fig. 10.2). In comparison, spring soil nitrate concentration to 30 cm depth ranged from 2 to 124 kg N ha^{-1} (average 22 kg N ha^{-1}) indicating that most of the soil N supply could be attributed to soil N mineralization in most years. The among field variation in soil N supply is also reflected in tuber yield response. For example, total tuber yield with no N applied under rain-fed production at 12 site-years in New Brunswick ranged from 19 to 47 t ha^{-1}, representing 61–93% of maximum yield (Bélanger et al. 2000a). It is important to note, however, that the environmental conditions which provide a higher crop growth typically also enhance soil N supply.

These uncertainties in crop N demand and soil N supply make fertilizer N management challenging. Under rain-fed production, foliar application of urea is the only practical option for fertilizer N application after hilling. Foliar applied urea is a relatively efficient way of applying N to the crop (Millard and Robinson 1990), but the quantity of N which can be applied is limited due to the potential for leaf damage. Consequently, most decisions with respect to fertilizer N management must be made early in the crop growing season when the options for use of plant-based measures of N status are limited (Zebarth et al. 2009a).

Soil tests based on residual nitrate, which are commonly used in many production areas (Greenwood 1986; Hergert 1987); are generally not effective in making fertilizer N recommendations in this region (Bélanger et al. 2001a) due to the loss of soil nitrate over the fall and winter period (Zebarth et al. 2009a). Soil tests based on mineralizable N show some promise (Sharifi et al. 2007; Sharifi et al. 2009b), but to date none have been adopted by commercial growers.

Here we discuss some of the options for improving fertilizer N management within the rain-fed potato production in eastern Canada. These include the potato crop response to rate, placement, formulation and timing of N fertilization, as well as options for site-specific N management. Options for use of soil- and plant-based tests to improve fertilizer N recommendations are explored in detail in Chapter 11.

10.3 Strategies to Improve Fertilizer N Management

10.3.1 Rate of N Fertilization

The rate of N fertilization is perhaps one of the most important decisions with respect to fertilizer N management, and will therefore be a focus of this chapter. General fertilizer N recommendations based on average response curves have been the main source of information for potato growers. In this section, the high variability in the response to N fertilization will be discussed along with its cause and consequences. It highlights the need for developing site-specific recommendations.

10.3.1.1 Tuber Yield and Size Response to N Rate

Potato tuber yield depends on the amount of intercepted radiation, the efficiency with which this intercepted radiation is used in the production of crop biomass (radiation use efficiency), and the harvest index (i.e. the proportion of the biomass partitioned to tubers). The amount of intercepted radiation depends on the leaf area index which is a function of the leaf area development and duration. Potato growth prior to emergence is controlled primarily by soil temperature (Yuan and Bland 2005) and the physiological maturity of the seed piece (Allen and Scott 1992); it is therefore not affected by N. Increased fertilizer N application increases leaf area index through increased size and number of leaves (Vos 1995). Increased fertilizer N application can also increase leaf longevity (Vos and Biemond 1992), thereby increasing leaf area duration. An adequate supply of N is required to achieve a canopy capable of intercepting most radiation, whereas excessive N can delay crop maturity, result in excessive vine growth, and increase the risk of foliar diseases (Allen and Scott 1992).

Nitrogen fertilization is known to affect the radiation use efficiency of several crop species but studies in potatoes have shown no effect of N deficiencies on radiation use efficiency (Vos and van der Putten 1998). The harvest index commonly decreases with increasing rates of N fertilization. For example, the harvest index averaged over six site-years in New Brunswick decreased from 80% with no N applied to 76% with 100 kg N ha^{-1} applied (Bélanger et al. 2001b).

Tuber yield is responsive to fertilizer N addition in almost all cases (Zebarth et al. 2009a). Fertilizer N application increases yield primarily though an increase in tuber mass (De la Morena et al. 1994). Tuber number per plant has been shown to increase, decrease or to be unaffected by N fertilization (Bélanger et al. 2002; De la Morena et al. 1994). Average fresh tuber weight increased with increasing N application rates with both Shepody and Russet Burbank in a 12 site-year study conducted in New Brunswick (Bélanger et al. 2002). Bulking rates, however, are not always affected by fertilizer N application. Fertilizer N application significantly increased bulking rates at two of six site-years in New Brunswick (Bélanger et al. 2001b) but in one case, a high N rate caused a decrease in tuber bulking rates. Although this has not often been quantified, over-fertilization can result in decreased tuber yield. In comparison, stem density is controlled primarily by cultivar and physiological age of the seed (Allen and Scott 1992).

The rate of fertilizer application required to achieve optimum yield varies with site, growing conditions, crop management, and incidence of disease and insects (Zebarth and Rosen 2007). Even for the same potato cultivar grown in the same year, yield response to N fertilization can vary widely among fields (Fig. 10.3a). This complex response reflects the fact that variation in soil N supply can often be as important as crop N demand in determining the optimal fertilizer N rate (Scharf et al. 2005; Lobell 2007). For example in Fig. 10.3a, sites S1 and S2 had different maximum marketable tuber yields but similar optimal fertilizer N rates whereas sites S1 and S3 had similar maximum marketable tuber yields but had much larger differences in optimal fertilizer N rates.

10.3.1.2 Tuber Quality Response to N Rate

Fertilizer N rate also has important effects on tuber quality. Tuber specific gravity, an important quality parameter for potato processing, is often unaffected by fertilizer N rate, or decreases with increasing fertilizer N rate, particularly when fertilizer N rate exceeds crop N requirement (Laboski and Kelling 2007). In eastern Canada, tuber specific gravity commonly decreases with increasing N rate across all rates applied (Bélanger et al. 2002; Zebarth et al. 2004a), although increasing from zero to a low fertilizer N rate may result in a small increase in tuber specific gravity in some cases (Zebarth et al. 2004a). This decrease in specific gravity with increasing N rates was shown to be greater for Shepody than for Russet Burbank (Bélanger et al. 2002).

Fertilizer N rate has inconsistent effects on chip or fry color. Chip color has been reported to be unaffected by fertilizer N rate (Silva et al. 1991; Long et al. 2004; McPharlin and Lancaster 2010), to improve when increasing N rate from zero to the optimal fertilizer N rate (Zebarth et al. 2004a) and to result in darker fry color with excessive N fertilization (Feibert et al. 1998). Increasing N rate from zero to the optimal fertilizer N rate was also reported to reduce after-cooking darkening (Wang-Pruski et al. 2007).

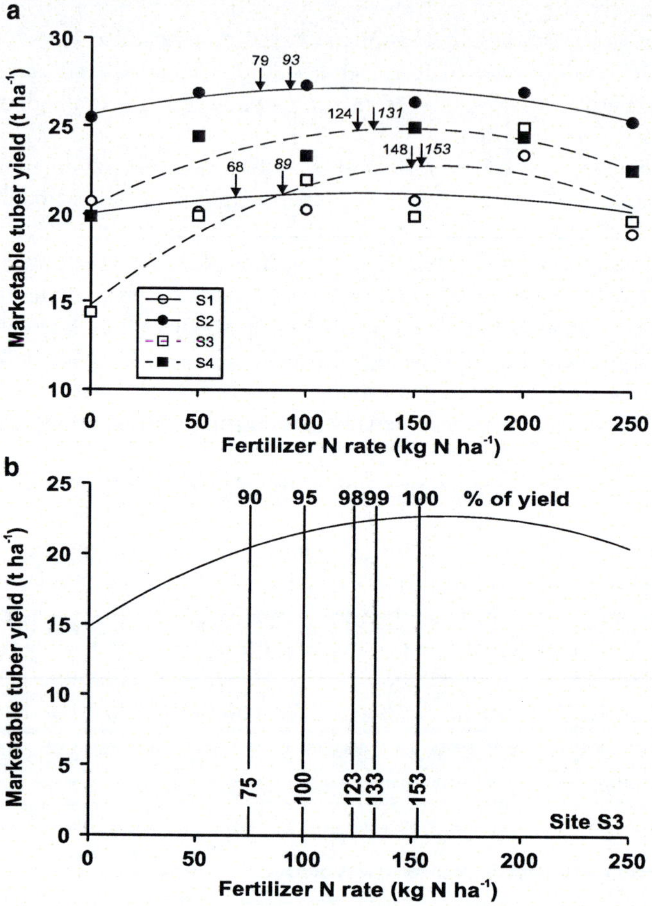

Fig. 10.3 (**a**) Tuber yield response varies among four sites planted to Shepody potatoes in 1995. For each site, the optimal fertilizer N rate at a cost:price ratio of 0.006 (italics) or 0.009 (normal) is indicated by arrows. (**b**) Using site S3 as an example, much of the fertilizer N applied results in a relatively small increase in tuber yield (Adapted from Bélanger et al. 2000a)

Increasing fertilizer N rate can also increase the incidence of internal tuber disorders such as hollow heart and internal brown spot, but these responses are inconsistent across sites (McPharlin and Lancaster 2010). Both insufficient and excess N fertility have been reported to increase the risk of sugar end disorder (Thompson et al. 2008).

Fertilizer N rate can also influence the human nutritional properties of tubers. Tuber nitrate concentration generally increases with increasing fertilizer N rate, with the highest tuber nitrate concentrations occurring when relative yield is at or close to 1.0 (Bélanger et al. 2002). In some cases, drought may be more important than N rate in contributing to high tuber nitrate concentrations (Zebarth et al. 2004a).

Most importantly, increasing fertilizer N rate has been reported to increase tuber concentrations of asparagine and reducing sugars, which are precursors to the production of acrylamide during frying (Gerendás et al. 2007; Lea et al. 2007).

10.3.1.3 Environmental Considerations of N Fertilization

Not all of the plant available N is utilized by the potato crop. Both the rate of fertilizer N application, and the relative efficiency of the potato crop in taking up the applied fertilizer, influence the potential for loss of N to the environment. Apparent recovery of fertilizer N in the potato plant is commonly 50–60% or less at commercial rates of fertilization (Vos 2009). Estimates of recovery of ^{15}N-labelled fertilizer and of apparent recovery in the whole plant range from 29–77% (as reviewed by Zebarth et al. 2009a). Apparent fertilizer N recovery typically decreases with increasing N rate, especially for above-optimal N rates (Vos 2009). Apparent recovery can also be reduced by factors that limit crop growth or N uptake such as delayed planting, drought, or incidence of diseases. Estimates of apparent N recovery in the whole potato plant in eastern Canada range from 29–70% on loamy soils in Quebec (Li et al. 2003) and 30 to 77% on loamy soils in New Brunswick (Zebarth and Milburn 2003; Zebarth et al. 2004b) and apparent N recovery in potato tubers from 21–62% on sandy soils in Quebec (Cambouris et al. 2008).

It is common for 70–85% of the N in the plant to be present in the tubers (Li et al. 2003; Zebarth et al. 2004b), with lower values occurring at high fertilizer N rates or in immature crops (Zebarth et al. 2009a). Nitrogen is mineralized rapidly from vegetable crop residues (Akkal-Corfini et al. 2010). Therefore, it is common for half or more of the applied N to remain in the field after tuber harvest, and for most of this N to be at risk of loss to the environment (Vos 2009).

Most nitrate is leached from the root zone over the autumn and winter period, therefore residual soil nitrate after potato production is commonly used as a measure of the risk of nitrate leaching loss (Zebarth et al. 2003a). On loamy soils in New Brunswick, Bélanger et al. (2003) found average residual soil nitrate to 90 cm depth for 12 site-years to range from 33 kg N ha^{-1} for non-fertilized plots to 160 kg N ha^{-1} in plots receiving 250 kg N ha^{-1}. Residual soil nitrate ranged from 46 to 99 kg N ha^{-1} at the optimal fertilizer N rate, and increased rapidly with increasing N application above the optimal rate (Fig. 10.4). In comparison, Zebarth et al. (2003a) measured residual soil nitrate to 30 cm depth of 3–250 kg N ha^{-1} in a survey of commercial potato fields. Residual soil nitrate generally increased with increasing fertilizer N rate and varied with potato cultivar with average values of 117, 56 and 43 kg N ha^{-1} to 30 cm depth for Russet Norkotah, Russet Burbank and Shepody, respectively. In a sandy soil in Quebec, Cambouris et al. (2008) reported residual soil nitrate to 70 cm depth to range from 53–114 kg N ha^{-1} for non-fertilized plots to a maximum value of 212 kg N ha^{-1} for a fertilizer N rate of 240 kg N ha^{-1}.

For N not recovered in tubers, there are two pathways of N loss to the environment which are of greatest concern: nitrate leaching to groundwater and emissions of nitrous oxide (N_2O), a greenhouse gas (Zebarth et al. 2009a). Nitrate leaching is

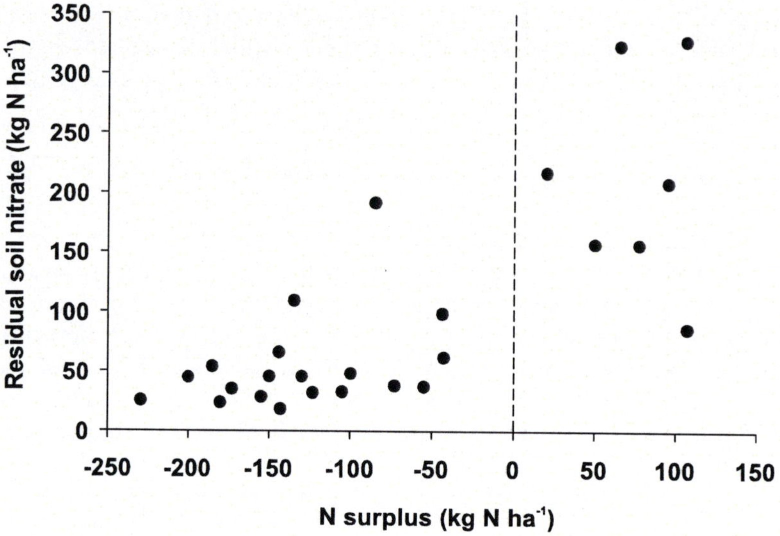

Fig. 10.4 Relationship between residual soil nitrate content to 30 cm depth and N surplus for seven site-years in New Brunswick. The N surplus is the fertilizer N rate minus the optimal fertilizer N rate (Adapted from Bélanger et al. 2003)

commonly the greatest pathway of loss in eastern Canada, with estimates of nitrate leaching losses from potato production ranging from 5–33 kg N ha^{-1} in loamy soils in New Brunswick (Milburn et al. 1990) and from 78–171 kg N ha^{-1} in sandy soils in Quebec (Gasser et al. 2002). Nitrate concentrations in leachate from potato fields commonly exceed the 10 mg NO_3-N L^{-1} drinking water guideline for nitrate (Milburn et al. 1990; Gasser et al. 2002; Vos and van der Putten 2004). In comparison, there are few estimates of denitrification or nitrous oxide emissions from potato production systems in eastern Canada. Burton et al. (2008) measured cumulative growing season emissions of N_2O ranging from 0.2–2.2 kg N ha^{-1} for potatoes grown in a loamy soil in New Brunswick, indicating that while this loss pathway can be of significant environmental importance, it has minimal impact from an agronomic standpoint.

10.3.1.4 Economic Considerations of N Fertilization

Fertilizer represents a significant cost to potato growers. For example, fertilizer was estimated to represent approximately 40% of direct input costs for potato production in PEI in 2007 (BDO Canada 2009). Thus, selection of the correct fertilizer N rate is of significant economic importance to growers. The economic risk associated with insufficient N fertilization, due to loss of tuber yield or size, is of far greater concern than the economic risk associated with excessive N fertilization, primarily due to low specific gravity. It is therefore common for growers to apply a sufficiently high fertilizer N to ensure that the crop N demand is met under most growing conditions.

The optimal fertilizer N rate for potatoes varies widely among experimental trials. For example, Neeteson and Wadman (1987) reported the optimal fertilizer N rate from 86 trials in The Netherlands to range from <50 kg N ha^{-1} to >350 kg N ha^{-1}. This optimal fertilizer N rate reflects the crop biological response, but is also influenced by economic considerations. The calculated optimal fertilizer N rate can also vary with the mathematical model used (Neeteson and Wadman 1987). Bélanger et al. (2000b) demonstrated that a quadratic model was suitable for potato experiments in New Brunswick.

This high variation in optimal N rate occurs even for the same potato cultivar grown in the same year. For example, the optimal fertilizer N rate for four field trials in New Brunswick where Shepody was grown ranged from 89–153 kg N ha^{-1} for a cost:price ratio of 0.006 and from 68–148 kg N ha^{-1} for a cost:price ratio of 0.009 (Fig. 10.3A).

The tuber yield response curves to fertilizer N rate in eastern Canada are commonly quite flat, and consequently there is a high degree of uncertainty associated with prediction of the optimal fertilizer N rate (Neeteson and Wadman 1987). In addition, the relatively flat yield response curve results in a limited increase in yield for a significant proportion of the fertilizer N applied. For example, Neeteson (1989) found that for trials in The Netherlands, a 25% reduction in recommended fertilizer rate resulted in a non-significant reduction in tuber yield. In Quebec, depending on the growing season, a yield reduction of 1.5% allowed a reduction in N rate ranging from 15–24% and from 15–22% for the total and the marketable tuber yield, respectively (Cambouris et al. 2007). Using the most responsive trial from Fig. 10.3A, fertilizer N rates of 75, 100, 123, 133 and 153 kg N ha^{-1} were predicted to result in 90%, 95%, 98%, 99% and 100% of the yield at the optimum fertilizer N rate (Fig. 10.3B).

For potatoes, the fertilizer N rate required to optimize net economic return is similar to, or in some cases even above, that required to achieve maximum biological tuber yield (Bélanger et al. 2000a; Bélanger et al. 2000b). For example, using sites S1–S4 from Fig. 10.3A under rain-fed production, the calculated economic optimum fertilizer N rate averaged 148 kg N ha^{-1} whereas the N rate predicted to achieve maximum biological tuber yield averaged 155 kg N ha^{-1}. This occurs because of the relatively low cost of N fertilizer compared with the value of potato tubers, and because tuber size increases in response to increasing fertilizer N over a wide range of fertilizer N rates (Zebarth and Rosen 2007). The optimal fertilizer N rate will also vary with changes in fertilizer N cost and tuber value, however this variation is commonly less than the among-field variation in optimal fertilizer N rate (Fig. 10.3A).

The choice of the optimal fertilizer N rate also has environmental implications. For most crops, residual soil nitrate or leaching potential begins to increase rapidly as the fertilizer N rate approaches that required to achieve maximum biological yield (Steenvoorden et al. 1986). Application of N above the optimal fertilizer N rate to potatoes can result in substantial increases in residual soil nitrate (Bélanger et al. 2003; Fig. 10.4) and consequently increase the potential for N losses to air and water.

10.3.2 Placement of Fertilizer N

Relatively few studies have examined the effects of fertilizer N placement in potato production. In Idaho under furrow irrigation, banded application of fertilizer N increased crop growth, tuber yield and plant N uptake compared with broadcast fertilizer N application (Westermann and Sojka 1996). Under rain-fed potato production in Germany, Maidl et al. (2002) found greater recovery of ^{15}N-labelled ammonium nitrate placed in the hill than applied as a broadcast. The benefit of fertilizer placement in this study was attributed primarily to a wet period between planting and emergence. When fertilizer was applied as a split application, there was little benefit of fertilizer N placement. On a fine-textured soil in Manitoba, banded application at planting increased petiole nitrate concentration, but not crop N uptake, compared with a pre-plant broadcast application (Zebarth et al. unpublished). Fertilizer placement may also change the shape of the yield response curve, because a small amount of fertilizer can be utilized more efficiently when placed than when broadcast (Harris 1992). Improperly placed fertilizer can damage germinating plants thereby reducing growth and tuber yield.

Most mineral fertilizer applied to potatoes in eastern Canada is banded at planting or surface broadcast just prior to hilling (commonly from emergence to about 50 days after planting) and incorporated by the hilling process. Pre-plant broadcast application of mineral fertilizer is avoided. Banding of fertilizer can increase efficiency of crop N uptake in several ways. First, banding places fertilizer N closer to the crop root system, which can enhance crop uptake, particularly early in the growing season. There is greater water infiltration in the furrow compared with the hill (Saffigna et al. 1976) and consequently banding fertilizer in the hill would also be expected to reduce the risk of nitrate leaching. Banding of fertilizer in the potato hill also delays nitrification due to very high concentrations of salts in the vicinity of the fertilizer band (Zebarth and Milburn 2003), maintaining more mineral N in ammonium form which is less susceptible to loss by leaching.

10.3.3 Timing of N Fertilization

Split fertilizer N application is a commonly used approach to improve crop fertilizer N utilization by improving the synchrony between N supply in soil and crop N demand (Zebarth and Rosen 2007). Several studies have reported split N application to increase recovery of fertilizer N in the potato crop compared with all fertilizer N applied at planting (Westermann et al. 1988; Vos 1999; Maidl et al 2002). Split N application is very effective in reducing nitrate leaching losses and increasing tuber yield and N uptake in irrigated production on sandy soils (Errebhi et al. 1998a). However in many studies, split application resulted in no effect or a modest increase in tuber yield and crop N uptake (Joern and Vitosh 1995; Vos 1999). This can be attributed to there being little or no benefit to split N application in situations where the risk of nitrate leaching is small (Harris 1992). In contrast, split N application

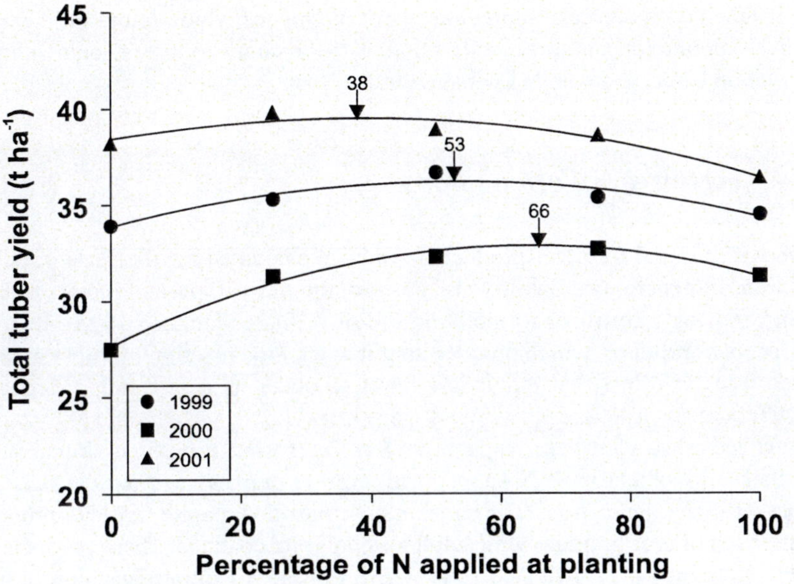

Fig. 10.5 Effect of proportion of N applied at planting on total tuber yield in 3 years on sandy soils in Quebec. The optimal timing ranged from 38% to 66% at planting with the remainder at final hilling (Adapted from Cambouris et al. 2007)

was reported to decrease crop N uptake and reduce tuber yield compared with all N applied at planting when dry soil conditions occur during the growing season (Porter and Sisson 1993; Zebarth et al. 2004a, b).

Some studies have identified a crop physiological response to N timing. High fertilizer N application early in the growing season can delay tuber initiation and bulking in indeterminate potato cultivars (Kleinkopf et al. 1981). However, other studies concluded that the timing of fertilizer N application does not have an effect on the time between tuber initiation and establishment of a high tuber bulking rate (Harris 1992). In some cases, late fertilizer N application may also reduce tuber specific gravity (Laboski and Kelling 2007) or increase second growth on tubers (Roberts et al. 1982). Split N application provides an additional practical advantage to growers with respect to flexibility; by applying a lower fertilizer rate at planting, a wider range of options are available with respect to in-season N management.

In eastern Canada under rain-fed production, timing of fertilizer N application on a medium-textured soil had little effect on crop N uptake or on tuber yield or quality parameters under normal rainfall conditions whereas split N application decreased tuber yield and crop N uptake under dry soil conditions (Zebarth and Milburn 2003; Zebarth et al. 2004a,b). In contrast, split N application under rain-fed production on sandy soils in Quebec increased tuber yield and N recovery in tubers compared with all fertilizer N applied at planting (Cambouris et al. 2007, 2008). For example, the proportion of N applied at planting to reach the maximum yield varied from 38% to 66% depending on the climatic conditions of the growing season (Fig. 10.5).

These results are in contrast to previous work in Quebec which found little benefit to split N application on seven soils ranging from sandy to sandy loam (Giroux 1982) and on a silty loam soil (Li et al. 1999).

10.3.4 Fertilizer N Formulations

A variety of mineral fertilizer products have been evaluated for their use in potato production. In general, availability and cost are the most important factors to consider in choosing a fertilizer formulation (Giroux 1982; MacLean 1983). In some cases, fertilizer products which produce an initial alkaline reaction in soil (e.g. urea) may result in yield loss, but this is less likely to occur on acidic soils (Meisinger et al. 1978; Giroux 1982).

Controlled release fertilizer products are fertilizer formulations that offer an alternative means to synchronize N supply with crop N demand without the need for multiple fertilizer applications. Currently, most controlled release fertilizer products are comprised of urea granules with sulphur or polymer coatings. These products are most effective on sandy, irrigated soils where the risk of nitrate leaching is high (Zebarth and Rosen 2007). Generally positive results are obtained with use of controlled release fertilizer products on potatoes in sandy soils (Zvomuya and Rosen 2001; Hutchinson et al. 2003), however negative responses can be obtained if the rate of fertilizer release is too slow to meet crop demand (Waddell et al. 1999). On sandy soils, the benefit may occur as increased tuber yield (Zvomuya and Rosen 2001; Ziadi et al. 2011), or as similar tuber yield with reduced N losses through nitrate leaching (Zvomuya et al. 2003; Wilson et al. 2009). These products also reduce or eliminate the additional cost associated with multiple fertilizer applications on sandy soils (Wilson et al. 2009). The primary limitation to use of controlled release fertilizer products has been increased cost relative to conventional mineral fertilizer N products (Simonne and Hutchinson 2005). Recently, new polymer-coated urea products have become available with better N release properties and at lower cost. This has resulted in increased use of these products in potato production.

In a recent study on sandy soils in Quebec, Ziadi et al. (2011) compared a controlled release fertilizer product (Environmentally Smart Nitrogen or ESN produced by Agrium Advanced Technologies, Calgary, AB; 44-0-0) with calcium ammonium nitrate in 3 years. The ESN increased marketable tuber yield by 12% compared with calcium ammonium nitrate (Table 10.1). The ESN also resulted in increased nitrate availability in soil during the growing season as measured using anion exchange membranes.

10.3.5 Site-Specific N Management

Crop N response can vary widely within fields (Shillito et al. 2009). Despite the recognition of significant spatial and temporal variation in soil N availability for crops within fields, the most common practice is still uniform applications of N.

Table 10.1 Effect of a controlled release N fertilizer (ESN) on marketable yield and yields of three potato size classes in a 3 year study in Québec

Treatment	Marketable tuber yield	Tuber size class[a] Jumbo	Medium	Small
	Mg ha^{-1}			
Unfertilized control	17.2	3.0	7.5	6.8
150 kg N ha^{-1} as Calcium ammonium nitrate (CAN)	26.0	7.4	10.9	7.3
150 kg N ha^{-1} as Environmentally Smart Nitrogen (ESN)	29.3	8.6	14.6	6.8
Contrasts (probability level)				
Control vs. others	<0.01	<0.001	<0.001	0.54
CAN vs. ESN	0.03	0.29	<0.01	0.34

Adapted from Ziadi et al. (2011)
[a]Jumbo > 227 g; medium < 277 g and > 5.1 cm long; small between 2.54 and 5.1 cm long

In some cases, this practice can result in under-fertilization and resulting yield loss in some parts of the field, and over-fertilization with implications for environmental N losses in others (Fiez et al. 1994; Kitchen et al. 1995; Vetsch et al. 1995). The goal of Site-Specific N management (SSNM) is to match N supply to crop N demand in space and time, and SSNM requires an understanding of the controls on within-field variation in crop N demand and soil N supply (Pan et al. 1997). Until recently, characterization of the spatial distribution of crop N demand and soil N supply was time consuming and costly which discouraged adoption by growers. Other limitations to adoption of SSNM include the absence of site-specific recommendations and lack of qualified services (Robert 2002).

Many strategies exist to characterize the spatial variability of soils and crops but the use of proximal or remote sensors is commonly most efficient. Yield monitors, soil apparent electrical conductivity instruments, instruments to map light reflectance from crop canopies, and airborne or satellite imagery can rapidly provide detailed information about soil and crop variability.

There are two main approaches to application of SSNM: (1) Variable Rate Application (VRA) and (2) use of Management Zones (MZ). The MZ approach, in which uniform management is applied to smaller more homogenous units, is generally more successful with soil-based than for the plant-based parameters because of the high heterogeneity over small distances of N variability in soil (Zebarth et al. 2009a). The VRA approach is generally more effective for plant-based parameters and can be applied using commercially available proximal sensors such as the Hydro N Sensor and the Greenseeker which measure crop N status using canopy light reflectance.

Few studies have been done on SSNM of potatoes in eastern Canada. In New Brunswick, Zebarth et al. (2003b) used the Hydro N Sensor to map N status in a potato field during two consecutive years. The Hydro N Sensor was generally effective in mapping spatial variability in crop N status, however application of a VRA approach was limited by uncertainty in what fertilizer application to assign to a given level of crop N status. In addition, an area with low apparent crop N status

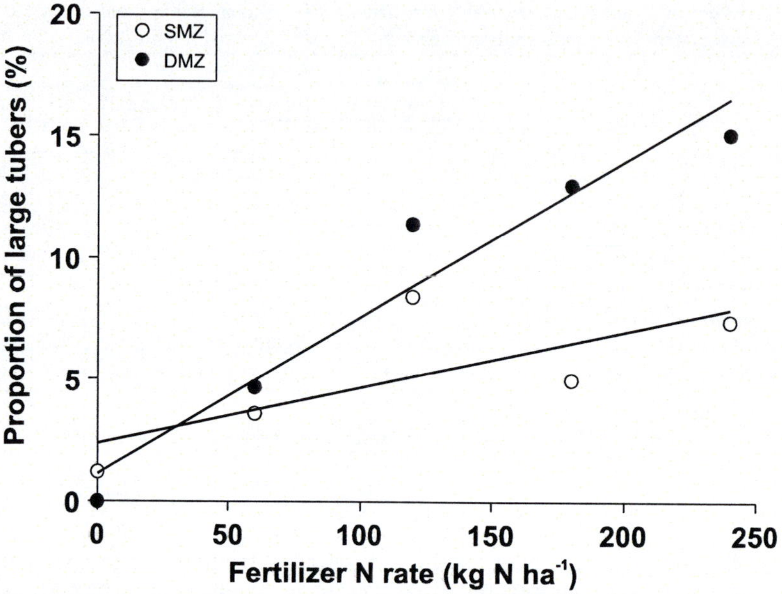

Fig. 10.6 Effect of fertilizer N rate on the proportion of large (88 mm ≤ diameter ≤ 112 mm) tubers from experimental trials in two management zones (SMZ vs DMZ) differing in soil water availability in Quebec (Adapted from Cambouris et al. 2007)

did not necessarily require higher fertilizer N application, rather it could reflect the presence of some other limitation to crop growth such as low stem density or excessive water.

The MZ approach was tested in Quebec using the Geonics EM38 sensor to map the spatial variability of soil in a 13-ha commercial potato field (Cambouris et al. 2006). Two management zones named SMZ and DMZ (to reflect shallow and deep soil depth over a clayey substratum, respectively) were delineated based on soil electromagnetic conductivity. The two MZ differed in tuber yield and quality due to differences in soil water holding capacity (Cambouris et al. 2006). In addition, response of tuber yield, size distribution and specific gravity to rate and time of N fertilization often differed between experiments located in the SMZ and DMZ zones (Cambouris et al. 2007). In some cases, these differences may be sufficiently large to justify different potato management practices (e.g., nutrient management, seedpiece spacing) to optimize potato production for the chip processing market. For example, the production of large tubers was more responsive to fertilizer N addition at the DMZ site compared with the SMZ site (Fig. 10.6).

To date, no study in eastern Canada has attempted to combine the two SSNM approaches. Use of MZ to optimize at-planting N management in combination with VRA during the growing season based on a measure of crop N status may be the best way to apply SSNM in potato production.

10.4 Conclusions

Fertilizer N management is an important but challenging aspect of rain-fed potato production in eastern Canada. Selection of the optimal fertilizer N rate remains one of the most important decisions for growers. Optimal fertilizer N management is necessary to achieve economic goals associated with tuber yield and size, whereas over-fertilization greatly increases the risk of environmental losses of N and of reduced tuber quality. However, large variations in crop N demand and soil N supply among fields and among years, and also within fields, make selection of an optimal fertilizer N rate problematic. Improving this will require improved predictions of crop N demand and soil N supply, both in time and in space.

Fertilizer N management can also be improved through appropriate timing of fertilizer application, fertilizer placement, and fertilizer formulation. Efficiency of N management can be improved through development of N management systems on a whole-field basis, or on a within-field basis using SSNM, where soil-based tests are used to determine at-planting N management and plant-based or soil-based tests are used for in-season N management. In addition, use of controlled release fertilizer products can be beneficial in soils where the risks of leaching losses are high.

In order to manage N efficiently and sustainably, it is important to consider N management as one component of an integrated cropping system. Sustainability of potato cropping systems can be enhanced by use of longer potato rotations, inclusion of legumes in potato rotations, use of organic amendments (Stark and Porter 2005) and reduced tillage (Carter et al. 2009). Such cropping systems can also influence soil health and populations of nematodes and soil-borne pathogens (Carter et al. 2003). Fertilizer N management can also have interactive effects with crop insect and disease management (Miller and Rosen 2005). There may also be the potential to improve efficiency of N utilization in potato production through genetic improvement of the potato crop (Errebhi et al. 1998b; Zebarth et al. 2004c).

References

Akkal-Corfini N, Morvan T, Menasseri-Aubry S, Bissuel-Bélaygue C, Poulain D, Orsini F, Leterme P (2010) Nitrogen mineralization, plant uptake and nitrate leaching following the incorporation of (15N)-labeled cauliflower crop residues (*Brassica oleracea*) into the soil: a 3-year lysimeter study. Plant Soil 328:17–26

Allen EJ, Scott RK (1992) Principles of agronomy and their application in the potato industry. In: Harris PM (ed) The potato crop: the scientific basis for improvement. Chapman and Hall, London, pp 816–881

BDO Canada (2009) Bucks in the ground: The cost to grow an acre of potatoes. http://www.bc.bdo.ca/library/publications/agriculture/articles/potatoes_cost070509.cfm. Accessed March 2011

Bélanger G, Walsh JR, Richards JE, Milburn PH, Ziadi N (2000a) Yield response of two potato cultivars to supplemental irrigation and N fertilization in New Brunswick. Am J Potato Res 77:11–21

Bélanger G, Walsh JR, Richards JE, Milburn PH, Ziadi N (2000b) Comparison of three statistical models describing potato yield response to nitrogen fertilizer. Agron J 92:902–908

Bélanger G, Walsh JR, Richards JE, Milburn PH, Ziadi N (2001a) Predicting nitrogen fertilizer requirements of potatoes in Atlantic Canada with soil nitrate determinations. Can J Soil Sci 81:535–544

Bélanger G, Walsh JR, Richards JE, Milburn PH, Ziadi N (2001b) Tuber growth and biomass partitioning of two potato cultivars grown under different N fertilization rates with and without irrigation. Am J Potato Res 78:109–117

Bélanger G, Walsh JR, Richards JE, Milburn PH, Ziadi N (2002) Nitrogen fertilization and irrigation affects tuber characteristics of two potato cultivars. Am J Potato Res 79:269–279

Bélanger G, Ziadi N, Walsh JR, Richards JE, Milburn PH (2003) Residual soil nitrate after potato harvest. J Environ Qual 32:607–612

Burton DL, Zebarth BJ, Gillam KM, MacLeod JA (2008) Effect of split application of fertilizer nitrogen on N_2O emissions from potatoes. Can J Soil Sci 88:229–239

Cambouris AN, Nolin MC, Zebarth BJ, Laverdière MR (2006) Soil management zones delineated by electrical conductivity to characterize spatial and temporal variations in potato yield and in soil properties. Am J Potato Res 83:381–395

Cambouris AN, Zebarth BJ, Nolin MC, Laverdière MR (2007) Response to added nitrogen of a continuous potato sequence as related to sand thickness over clay. Can J Plant Sci 87:829–839

Cambouris AN, Zebarth BJ, Nolin MC, Laverdière MR (2008) Apparent fertilizer nitrogen recovery and residual soil nitrate under continuous potato cropping: effect of N fertilization rate and timing. Can J Soil Sci 88:813–825

Carter MR, Kunelius HT, Sanderson JB, Kimpinski J, Platt HW, Bolinder MA (2003) Productivity parameters and soil health dynamics under long-term 2-year potato rotations in Atlantic Canada. Soil Till Res 72:153–168

Carter MR, Sanderson JB, Peters RD (2009) Long-term conservation tillage in potato rotations in Atlantic Canada: potato productivity, tuber quality and nutrient content. Can J Plant Sci 89:273–280

CRAAQ (2010) Centre de Références en Agriculture et Agroalimentaire du Québec. Fertilisation reference guide. (In French.) 2nd ed. ISBN 978-2-7649-0231-8. 473pp

De la Morena I, Guillén A, García del Moral LF (1994) Yield development in potatoes as influenced by cultivar and the timing and level of nitrogen fertilization. Am Potato J 71:165–173

Dessureault-Rompré J, Zebarth BJ, Burton DL, Sharifi M, Cooper J, Grant CA, Drury CF (2010) Relationships among mineralizable soil nitrogen, soil properties and climatic indices. Soil Sci Soc Am J 74:1218–1227

Environment Canada (2011) http://climate.weatheroffice.gc.ca/climate_normals/index_e.html. Accessed March 2011

Errebhi M, Rosen CJ, Gupta SC, Birong DE (1998a) Potato yield response and nitrate leaching as influenced by nitrogen management. Agron J 90:10–15

Errebhi M, Rosen CJ, Lauer FI, Martin MW, Bamberg JB, Birong DE (1998b) Screening of exotic potato germplasm for nitrogen uptake and biomass production. Am J Potato Res 75:93–100

Feibert EBG, Shock CC, Saunders LD (1998) Nitrogen fertilizer requirements of potato using carefully scheduled sprinkler irrigation. HortSci 33:262–265

Fiez TE, Miller BC, Pan WL (1994) Winter wheat yield and grain protein across varied landscape positions. Agron J 86:1026–1032

Gasser MO, Laverdière MR, Lagacé R, Caron J (2002) Impact of potato-cereal rotations and slurry applications on nitrate leaching and nitrogen balance in sandy soils. Can J Soil Sci 82:469–479

Gastal F, Lemaire G (2002) N uptake and distribution in crops: an agronomical and ecophysiological perspective. J Exp Bot 53:789–799

Gerendás J, Heuser F, Sattelmacher B (2007) Influences of nitrogen and potassium supply on contents of acrylamide precursors in potato tubers and on acrylamide accumulation in French fries. J Plant Nutr 30:1499–1516

Giroux M (1982) Effet des doses, des sources et du mode d'apport de l'azote sur le rendement et la maturité de la pomme de terre cultivée sur différents types de sols du Québec. (In French.) Can J Soil Sci 62:503–517

Greenwood DJ (1986) Prediction of nitrogen fertilizer needs of arable crops. Adv Plant Nutr 2:1–61

Harris PM (1992) Mineral nutrition. In: Harris PM (ed) The potato crop: the scientific basis for improvement. Chapman and Hall, London, pp 162–213

Hergert GW (1987) Status of residual nitrate-nitrogen soil tests in the United States of America. In: Brown JR (ed) Soil testing: sampling, correlation, calibration, and interpretation. SSSA, Madison, Wisconsin, USA, pp 73–78

Hutchinson C, Simonne E, Solano P, Meldrum J, Livingston-Way P (2003) Testing of controlled release fertilizer programs for seep irrigated Irish potato production. J Plant Nutr 26:1709–1723

Joern BC, Vitosh ML (1995) Influence of applied nitrogen on potato. Part I: Yield, quality, and nitrogen uptake. Am Potato J 72:51–63

Kitchen NR, Hughes DF, Sudduth KA, Birrell SJ (1995) Comparison of variable rate to single rate nitrogen fertilizer application: corn production and residual soil NO_3-N. In Robert PC, Rust RH, Larson WE (eds) Site-specific management for agricultural systems. ASA Misc. Publ., ASA, CSSA, and SSSA, Madison, WI, pp 427–439

Kleinkopf GE, Westermann DT, Dwelle RB (1981) Dry matter production and nitrogen utilization by six potato cultivars. Agron J 73:799–802

Laboski CAM, Kelling KA (2007) Influence of fertilizer management and soil fertility on tuber specific gravity: a review. Am J Potato Res 84:283–290

Lea PJ, Sodek L, Parry MAJ, Shewry PR, Halford NG (2007) Asparagine in plants. Ann Appl Biol 150:1–26

Lemaire G, Millard P (1999) An ecophysiological approach to modelling resource fluxes in competing plants. J Exp Bot 50:15–28

Li H, Parent L-É, Tremblay C, Karam A (1999) Potato response to crop sequence and nitrogen fertilization following sod breakup in a Gleyed Humo-Ferric Podzol. Can J Plant Sci 79:439–446

Li H, Parent L-É, Karam A, Tremblay C (2003) Efficiency of soil and fertilizer nitrogen of a sod-potato system in the humid, acid and cool environment. Plant Soil 251:23–26

Lobell DB (2007) The cost of uncertainty for nitrogen fertilizer management: a sensitivity analysis. Field Crops Res 100:210–217

Long CM, Snapp SS, Douches DS, Chase RW (2004) Tuber yield, storability, and quality of Michigan cultivars in response to nitrogen management and seedpiece spacing. Am J Potato Res 81:347–357

MacLean AA (1983) Sources of fertilizer nitrogen and phosphorus for potatoes in Atlantic Canada. Am Potato J 60:913–918

Maidl F-X, Brunner H, Sticksel E (2002) Potato uptake and recovery of nitrogen ^{15}N-enriched ammonium nitrate. Geoderma 105:167–177

McPharlin IR, Lancaster RA (2010) Yield and quality response of crisping potatoes (*Solanum tuberosum* L.) to applied nitrogen. J Plant Nutr 33:1195–1215

Meisinger JJ, Bouldin DR, Jones ED (1978) Potato yield reductions associated with certain fertilizer mixtures. Am Potato J 55:227–234

Milburn P, Richards JE, Gartley C, Pollock T, O'Neill H, Bailey H (1990) Nitrate leaching from systematically tiled potato fields in New Brunswick, Canada. J Environ Qual 19:448–454

Millard P, Robinson D (1990) Effect of the timing and rate of nitrogen fertilization on the growth and recovery of fertilizer nitrogen within the potato (*Solanum tuberosum* L.) crop. Fertil Res 21:133–140

Miller JS, Rosen CJ (2005) Interactive effects of fungicide programs and nitrogen management on potato yield and quality. Am J Potato Res 82:399–409

NBDAFA (2001) Crop fertilization guide. New Brunswick Dept, Agriculture, Fisheries and Aquaculture

Neeteson JJ (1989) Effect of reduced fertilizer nitrogen application rates on yield and nitrogen recovery of sugar beet and potatoes. Neth J Agric Sci 37:227–236

Neeteson JJ, Wadman WP (1987) Assessment of economically optimum application rates of fertilizer N on the basis of response curves. Fertil Res 12:37–52

Pan WL, Huggins DR, Malzer GL, Douglas CL Jr, Smith JL (1997) Field heterogeneity in soil-plant nitrogen relationships: implications for site-specific management. In: Pierce FJ, Sadler EJ (eds) The state of site-specific management for agriculture. ASA-CSSA-SSSA, Madison, WI, pp 89–99

Porter GA, Sisson JA (1993) Yield, market quality, and petiole nitrate concentration of non-irrigated Russet Burbank and Shepody potatoes in response to sidedressed nitrogen. Am Potato J 70:101–116

Robert PC (2002) Precision agriculture: a challenge for crop nutrition management. Plant Soil 247:143–149

Roberts S, Weaver WH, Phelps JP (1982) Effect of rate and time of fertilization on nitrogen and yield of Russet Burbank potatoes under center pivot irrigation. Am Potato J 59:77–86

Saffigna PG, Tanner CB, Keeney DR (1976) Non-uniform infiltration under potato canopies caused by interception, stemflow and hilling. Agron J 68:337–342

Sanderson JB, MacLeod JA, Kimpinski J (1999) Glyphosate application and timing of tillage of red clover affects potato response to N, soil N profile, and root and soil nematodes. Can J Soil Sci 79:65–72

Scharf PC, Kitchen NR, Sudduth KA, Davis JG, Hubbard VC, Lory JA (2005) Field-scale variability in optimal nitrogen fertilizer rate for corn. Agron J 97:452–461

Sharifi M, Zebarth BJ, Burton DL, Grant CA, Porter GA, Cooper JM, Leclerc Y, Moreau G, Arsenault WJ (2007) Evaluation of laboratory-based measures of soil mineral nitrogen and potentially mineralizable nitrogen as predictors of field-based estimates of soil nitrogen supply in potato production. Plant Soil 301:203–214

Sharifi M, Zebarth BJ, Burton DL, Grant CA, Porter GA (2008a) Organic amendment history and crop rotation effects on soil nitrogen mineralization potential and soil nitrogen supply in a potato cropping system. Agron J 100:1562–1572

Sharifi M, Zebarth BJ, Burton DL, Grant CA, Bittman S, Drury CF, McConkey B, Ziadi N (2008b) Response of potentially mineralizable soil nitrogen and indices of nitrogen availability to tillage system. Soil Sci Soc Am J 72:1124–1131

Sharifi M, Zebarth BJ, Porter GA, Burton DL, Grant CA (2009a) Soil mineralizable nitrogen and soil nitrogen supply under two-year potato rotations. Plant Soil 320:267–279

Sharifi M, Lynch DH, Zebarth BJ, Zheng Z, Martin RC (2009b) Evaluation of nitrogen supply rate measured by *in situ* placement of plant root simulator probes as a predictor of nitrogen supply from soil and organic amendments in potato crop. Am J Potato Res 86:356–366

Shillito RM, Timlin DJ, Fleisher S, Reddy VR, Quebedeaux B (2009) Yield response of potato to spatially patterned nitrogen application. Agric Ecosyst Environ 129:107–116

Silva GH, Chase RW, Hammerschmidt R, Vitosh ML, Kitchen RB (1991) Irrigation, nitrogen and gypsum effects on specific gravity and internal defects of Atlantic potatoes. Am Potato J 68:751–765

Simonne E, Hutchinson CM (2005) Controlled-release fertilizers for vegetable production in the era of best management practices: teaching new tricks to an old dog. HortTechnol 15:36–46

Stark JC, Porter GA (2005) Potato nutrient management in sustainable cropping systems. Am J Potato Res 82:329–338

Statistics Canada (2010) Canadian Potato Production. Catalogue no. 22-008-x

Steenvoorden J, Fonck H, Oosterom HP (1986) Losses of nitrogen from intensive grassland systems by leaching and surface runoff. In: van der Meer H, Ryden JC, Ennik GC (eds) Nitrogen fluxes in intensive grassland systems. Martinus Nijhoff Publishers, Dordrecht, The Netherlands, pp 85–97

Thompson AL, Love SL, Sowokinos JR, Thornton MK, Shock CC (2008) Review of the sugar end disorder in potato (*Solanum tuberosum* L.). Am J Potato Res 85:375–386

Vetsch JA, Malzer GL, Robert PC, Huggins DR (1995) Nitrogen specific management by soil condition: managing fertilizer nitrogen in corn. In Robert PC, Rust RH, Larson WE (eds) Site-specific management for agricultural systems. ASA Misc. Publ., ASA, CSSA, and SSSA, Madison, WI, pp 427–439

Vos J (1995) The effects of nitrogen supply and stem density on leaf attributes and stem branching in potato (*Solanum tuberosum* L.). Potato Res 38:271–279

Vos J (1997) The nitrogen response of potato (*Solanum tuberosum* L.) in the field: nitrogen uptake and yield, harvest index and nitrogen concentration. Potato Res 40:237–248

Vos J (1999) Split nitrogen application in potato: effects on accumulation of nitrogen and dry matter in the crop and on the soil nitrogen budget. J Agric Sci Camb 133:263–274

Vos J (2009) Nitrogen responses and nitrogen management in potato. Potato Res 52:305–317

Vos J, Biemond H (1992) Effects of nitrogen on the development and growth of the potato plant. 1. Leaf appearance, expansion growth, life spans of leaves and stem branching. Ann Bot 70:27–35

Vos J, van der Putten PEL (1998) Effect of nitrogen supply on leaf growth, leaf nitrogen and photosynthetic capacity in potato. Field Crops Res 59:63–72

Vos J, van der Putten PEL (2004) Nutrient cycling in a cropping system with potato, spring wheat, sugar beet, oat and nitrogen catch crops. II. Effect of catch crops on nitrate leaching in autumn and winter. Nutr Cycl Agroecosyst 70:23–31

Waddell JT, Gupta SC, Moncrief JF, Rosen CJ, Steele DD (1999) Irrigation and nitrogen management effects on potato yield, tuber quality and nitrogen uptake. Agron J 91:991–997

Wang-Pruski G, Zebarth BJ, Leclerc Y, Arsenault WJ, Botha EJ, Moorehead S, Ronis D (2007) Effect of soil type and nutrient management on potato after-cooking darkening. Am J Potato Res 84:291–299

Westermann DT, Sojka RE (1996) Tillage and nitrogen placement effects on nutrient uptake by potato. Soil Sci Soc Am J 60:1448–1453

Westermann DT, Kleinkopf GE, Porter LK (1988) Nitrogen fertilizer efficiencies on potatoes. Am Potato J 65:377–386

Wilson ML, Rosen CJ, Moncrief JF (2009) Potato response to polymer-coated urea on an irrigated coarse-textured soil. Agron J 101:897–905

Yuan F-M, Bland WL (2005) Comparison of light- and temperature-based index models for potato (*Solanum tuberosum* L.) growth and development. Am J Potato Res 82:345–352

Zebarth BJ, Milburn PH (2003) Spatial and temporal distribution of soil inorganic nitrogen concentration in potato hills. Can J Soil Sci 83:183–195

Zebarth BJ, Rosen CJ (2007) Research perspective on nitrogen BMP development for potato. Am J Potato Res 84:3–18

Zebarth BJ, Leclerc Y, Moreau G, Gareau R, Milburn PH (2003a) Soil inorganic nitrogen content in commercial potato fields in New Brunswick. Can J Soil Sci 83:425–429

Zebarth BJ, Rees H, Tremblay N, Fournier P, Leblon B (2003b) Mapping spatial variation in potato nitrogen status using the N Sensor. Acta Hort (ISHS) 627:267–273

Zebarth BJ, Leclerc Y, Moreau G, Botha E (2004a) Rate and timing of nitrogen fertilization of Russet Burbank potato: yield and processing quality. Can J Plant Sci 84:855–863

Zebarth BJ, Leclerc Y, Moreau G (2004b) Rate and timing of nitrogen fertilization of Russet Burbank potato: nitrogen use efficiency. Can J Plant Sci 84:845–854

Zebarth BJ, Tai G, Tarn R, de Jong H, Milburn PH (2004c) Nitrogen use efficiency characteristics of commercial potato cultivars. Can J Plant Sci 84:589–598

Zebarth BJ, Leclerc Y, Moreau G, Sanderson JB, Arsenault WJ, Botha EJ, Wang-Pruski G (2005a) Estimation of soil nitrogen supply in potato fields using a plant bioassay approach. Can J Soil Sci 85:377–386

Zebarth BJ, Chabot R, Coulombe J, Simard RR, Douheret J, Tremblay N (2005b) Pelletized organo-mineral fertilizer product as a nitrogen source for potato production. Can J Soil Sci 85:387–395

Zebarth BJ, Drury CF, Tremblay N, Cambouris AN (2009a) Opportunities for improved fertilizer nitrogen management in production of arable crops in eastern Canada: a review. Can J Soil Sci 89:113–132

Zebarth BJ, Arsenault WJ, Moorehead S, Kunelius HT, Sharifi M (2009b) Italian ryegrass management effects on nitrogen supply to a subsequent potato crop. Agron J 101:1573–1580
Ziadi N, Grant C, Samson N, Nyiraneza J, Bélanger G, Parent L-É (2011) Efficiency of controlled-release urea for a potato production system in Quebec, Canada. Agron J 103:60–66
Zvomuya F, Rosen CJ (2001) Evaluation of polyolefin-coated urea for potato production on a sandy soil. Hort Sci 36:1057–1060
Zvomuya F, Rosen CJ, Russelle MP, Gupta SC (2003) Nitrate leaching and nitrogen recovery following application of polyolefin-coated urea to potato. J Environ Qual 32:480–489

Chapter 11
Soil and Plant Tests to Optimize Fertilizer Nitrogen Management of Potatoes

Noura Ziadi, Bernie J. Zebarth, Gilles Bélanger, and Athyna N. Cambouris

Abstract Appropriate fertilizer nitrogen (N) management can optimize tuber yield and quality, and reduce the risk of environmental N losses. However, the optimal fertilizer N management can vary among fields and years. Plant- and soil-based tests are examined in this chapter as diagnostic tools to improve fertilizer N management in rain-fed potato production in eastern Canada. Plant-based diagnostic tests assess potato N sufficiency and can be used to guide in-season fertilizer N management. The nitrogen nutrition index (NNI) based on whole plants, the petiole nitrate concentration, and the leaf chlorophyll meter reading (SPAD) have been shown to successfully diagnose the level of potato N nutrition during the growing season in eastern Canada. The use of gene expression, a promising tool for a direct measurement of potato N sufficiency compared with chemical or optical methods, is also examined. Soil-based tests can be used to provide an estimate of soil N supply to adjust the at-planting fertilizer N rate. The use of pre-plant and in-season soil nitrate tests, ion exchange membranes, indices of soil mineralizable N, and near-infrared reflectance spectroscopy (NIRS) are examined. A combination of a soil-based test to guide at-planting fertilizer N application and a plant-based test to guide in-season N management may be most effective.

N. Ziadi (✉) • G. Bélanger
Soils and Crops Research and Development Centre, Agriculture and Agri-Food Canada, 2560 Hochelaga Blvd., Quebec City, QC, Canada G1V 2J3
e-mail: Noura.Ziadi@agr.gc.ca

B.J. Zebarth
Potato Research Centre, Agriculture and Agri-Food Canada , 850 Lincoln Rd., PO Box 20280, Fredericton, NB, Canada E3B 4Z7

A.N. Cambouris
Soils and Crops Research and Development Centre, Agriculture and Agri-Food Canada, 979 de Bourgogne Ave., Quebec City, QC, Canada G1W 2L4

Z. He et al. (eds.), *Sustainable Potato Production: Global Case Studies*,
DOI 10.1007/978-94-007-4104-1_11,

11.1 Introduction

Potato (*Solanum tuberosum* L.) crops frequently require high applications of fertilizer nitrogen (N) to achieve high tuber yield and quality. In eastern Canada, general fertilizer N recommendations vary between 125 and 200 kg N ha^{-1} (NBDAFA 2001; CRAAQ 2010). The apparent recovery of applied fertilizer N in the growing crop, however, may average less than 50% (Cambouris et al. 2008; Ziadi et al. 2011). Management of this fertilizer N is important from both economic and environmental standpoints (Zebarth et al. 2009). Nitrogen deficiency results in poor crop growth, small tuber size, and low tuber yield (Bélanger et al. 2000) while excessive N can lead to poor tuber quality, delayed crop maturity, increased N_2O emissions, and excessive nitrate leaching (Ojala et al. 1990; Bélanger et al. 2000; Burton et al. 2008). However, the optimal fertilizer N rate can vary widely among fields and among years (Zebarth et al. 2009). This variation results from variation in both the crop N demand and the soil N supply. As a result, the development of tools which predict more precisely the fertilizer N requirement on an individual field basis in potato production can be used as a strategy to optimize tuber yield and quality and to minimize the risk of N losses to the environment.

In this chapter, we examine plant- and soil-based tests which can be used as diagnostic tools to improve fertilizer N recommendations for potato production on an individual field basis. Plant-based diagnostic tests have an advantage in that they commonly assess plant N sufficiency (i.e. the balance between crop N demand and N supply), whereas soil-based tests commonly assess only soil N supply. However, plant-based tests can often only be used later in the growing season whereas soil-based tests are commonly used early in the growing season. As a result, use of a combination of soil- and plant-based tests may be most effective in optimizing fertilizer N management.

11.2 Plant-Based Diagnostic Methods

Several plant-based diagnostic methods have been developed over the last 20 years. These methods use either whole plants or specific plant parts (e.g. leaf or petiole) and they can include either chemical or optical measurements.

11.2.1 Nitrogen Nutrition Index (NNI)

The N concentration on a whole plant basis can be used as a diagnostic tool to assess crop N nutrition during the growing season. To do so, a critical N concentration (N_c), that is the minimum N concentration required for maximum crop growth, must be defined. Crop N concentration decreases over time as crop biomass increases

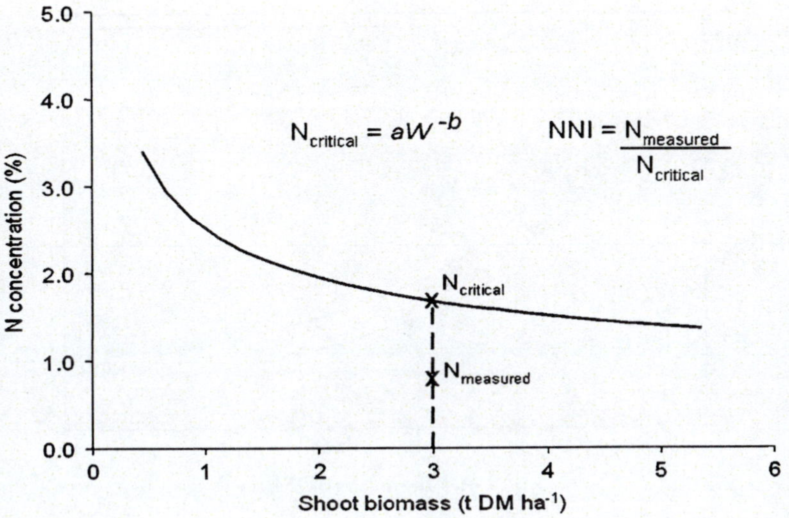

Fig. 11.1 General concept of critical N concentration. W is the total shoot biomass expressed in t dry matter (DM) ha^{-1}, $N_{critical}$ is the total N concentration in shoots expressed in % of DM, and *a* and *b* are estimated parameters. NNI is the N nutrition index

because of an increased proportion of the structural and storage components that contain little N. Consequently, N_c also decreases over time during the growing season. For that reason, N_c is commonly expressed as a function of crop biomass with critical N curves.

The concept of a critical N curve, based on the N concentration of whole plants, was first developed in France for tall fescue by Lemaire and Salette (1984) and has been successfully applied in eastern Canada to other perennial crops [timothy (Bélanger and Ziadi 2008)] and annual crops [wheat (Ziadi et al. 2010a); corn (Ziadi et al. 2008a)], including potatoes (Bélanger et al. 2001b). For the majority of crops, the N_c can be represented by the following allometric function:

$$N_c = aW^{-b} \tag{11.1}$$

where W is the total shoot biomass expressed in t dry matter (DM) ha^{-1}, N_c is the total N concentration in shoots expressed in % of DM, and *a* and *b* are estimated parameters (Fig. 11.1). The parameter *a* represents the N concentration with 1 t DM ha^{-1} and the parameter *b* represents the coefficient of dilution which describes the relationship of decreasing N concentration with increasing shoot biomass. For potatoes, the function is applied to the vines plus tubers rather than to the above-ground plant for other crop species. Therefore, the values of the parameters *a* and *b* are estimated using the combined biomass of shoots and tubers, and the N concentration of this combined biomass.

The critical N curve can then be used to calculate the N nutrition index (NNI) as the ratio between the measured N concentration of the shoot biomass and the

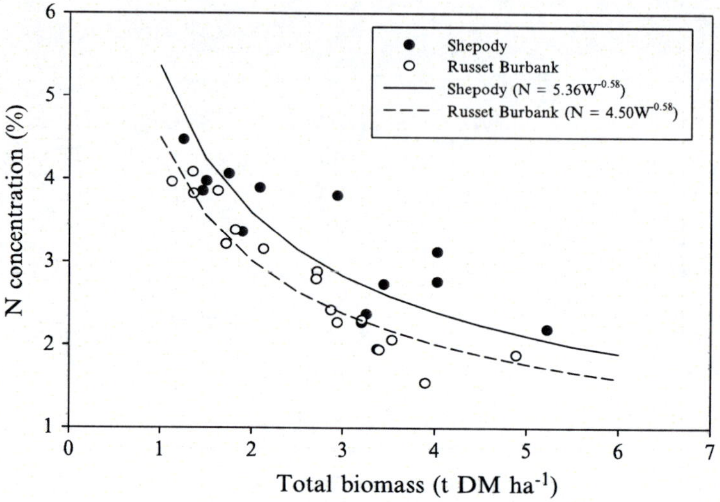

Fig. 11.2 Critical N curves for two potato cultivars under rain-fed conditions; data points correspond to maximum total biomass for each combination of site and cultivar (Bélanger et al. 2001b)

predicted N_c. This NNI describes the N nutrition status of a crop at different times during the growing season, independently of the stage of development. The critical N curve (Eq. 11.1; Fig. 11.1) discriminates three different types of N status. Data points below the curve (i.e. NNI < 1) indicate situations where N is limiting growth and additional N fertilizer would therefore increase growth. Data points above the curve (i.e. NNI > 1) indicate situations of excessive N nutrition where additional N fertilization would not increase growth. Data points located on or near the curve (i.e. NNI ≈ 1) correspond to situations where N does not limit growth and N nutrition is not excessive.

In potatoes, critical N curves were first proposed in France, Scotland, and the Netherlands (Greenwood et al. 1990; Duchenne et al. 1997). In eastern Canada, the critical N curve of potato was determined for the cultivars Russet Burbank and Shepody under rain-fed and irrigated conditions (Bélanger et al. 2001b). Critical N curves were found to be specific to cultivars and water conditions. Parameters of the critical N curves are:

$$\text{Shepody } N_c = 5.36W^{-0.58} \tag{11.2}$$

$$\text{Russet Burbank } N_c = 4.50W^{-0.58} \tag{11.3}$$

under rain-fed conditions (Fig. 11.2) and:

$$\text{Shepody } N_c = 5.04W^{-0.42} \tag{11.4}$$

$$\text{Russet Burbank } N_c = 4.57W^{-0.42} \tag{11.5}$$

under irrigated conditions.

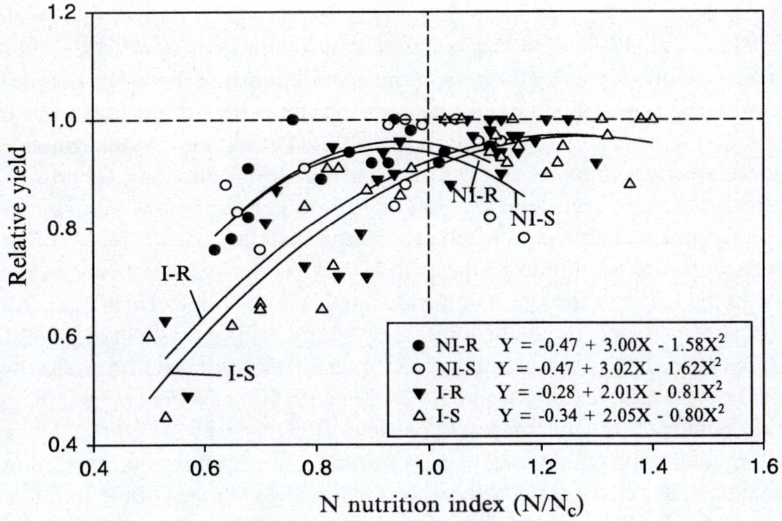

Fig. 11.3 Relationship between relative yield and the N nutrition index (NNI) of two potato cultivars (R: Russet Burbank; S: Shepody) with (I) and without (NI) irrigation (Bélanger et al. 2001b)

Using the NNI concept, relationships between potato relative yield and NNI were established for potatoes produced at six site-years in eastern Canada by Bélanger et al. (2001b) (Fig. 11.3). For a NNI equal to or greater than 1.0, the relative yield was near 1.0. In eastern Canada, there is limited evidence of yield depression at higher fertilizer N rates (Bélanger et al. 2000; Cambouris et al. 2007).With decreasing NNI below 1.0, the relative yield decreased. These results indicate that the NNI is a reliable indicator of the level of N sufficiency during the potato growing season.

The concept of N_c and the resulting NNI effectively identified situations of deficient and non-deficient N nutrition making it possible to quantify the level of potato N sufficiency. A major difficulty in using the NNI at the farm level, however, is the need to determine the actual crop biomass and its N concentration. For this reason, it may be more practical to use the NNI as a reference for calibration of simpler procedures (e.g. leaf chlorophyll measurements, petiole nitrate concentration) to determine the potato N status as described in the following sections.

11.2.2 Petiole Nitrate Concentration

Petiole nitrate concentration is one of the most widely used diagnostic tools to assess potato N sufficiency. Petiole nitrate concentration may be measured on a dry plant tissue basis or on freshly expressed petiole sap (Errebhi et al. 1998). The former is commonly done using a water extraction followed by colorimetric determination of nitrate concentration in the extract in a laboratory (Porter and Sisson 1991) whereas the latter can be measured either by a Nitrate Specific Electrode (Waterer 1997; Errebhi et al. 1998) or a combination of nitrate test strips with a hand-held reflectometer (Goffart et al. 2008). Significant relationships between petiole sap nitrate concentrations and petiole nitrate concentrations on a dry matter basis have been attained (Waterer 1997; Errebhi et al. 1998). Petiole dry matter content can vary widely among sampling dates in rain-fed potato production (Zebarth et al., unpublished data) and consequently petiole nitrate concentration on a dry tissue basis is likely more reliable in rain-fed production systems.

The concentration of nitrate in the petiole reflects the balance between nitrate reduction in the leaf and recent plant nitrate uptake from soil (Zebarth et al. 2009). Petiole nitrate concentration can be influenced by several factors including the stage of development or days after planting (DAP), fertilizer N application, water availability, and potato cultivar. Similar to the N concentration of whole plants, the petiole nitrate concentration decreases over time (Bélanger et al. 2003). Nitrogen fertilization consistently increases petiole nitrate concentration. For example in a study conducted at six sites and with two cultivars, the average petiole nitrate concentration at 63 DAP increased from 0.69% with no N applied to 2.60% when 250 kg N ha^{-1} was applied (Bélanger et al. 2003). A quadratic response to N application was reported (Porter and Sisson 1991; Bélanger et al. 2003) which is attributed to the saturation of the plant uptake capacity at high N rates.

Petiole nitrate concentration was reported to be influenced by water availability during the growing season. Insufficient water may result in the accumulation of nitrate in potato petioles (Meyer and Marcum 1998) whereas excessive water may reduce petiole nitrate concentration (Stark et al. 1993). Irrigation, however, had no consistent effect on petiole nitrate concentration in study conducted at several site-years in New Brunswick (Bélanger et al. 2003) where the level of water stress might have been insufficient to influence petiole nitrate concentration. The petiole nitrate concentration also varies with cultivars (Lewis and Love 1994; Bélanger et al. 2003). Greater petiole nitrate concentrations were reported for Shepody than for Russet Burbank on all sampling dates and all sites in a study conducted in New Brunswick (Bélanger et al. 2003).

Critical values or ranges of petiole nitrate concentrations have been suggested for potatoes in several producing areas of the world (Porter and Sisson 1991; Waterer 1997; Bélanger et al. 2003). The critical petiole nitrate concentration, that is the petiole nitrate concentration required to reach maximum yield, has most often been established using the relationship between petiole nitrate concentration and tuber yield or relative tuber yield. This relationship varies with sampling dates (DAP) and cultivars. In eastern Canada, petiole nitrate concentration increased linearly with relative yield for Russet Burbank ($R^2=0.60$) and Shepody ($R^2=0.53$) at approximately 59 DAP (Bélanger et al. 2003). However, the use of relative yield to determine the critical petiole concentration has one major limitation. Petiole nitrate concentration keeps increasing even when relative yield has reached its maximum value, that is, with no corresponding increase in tuber yield.

A novel approach to defining critical petiole concentrations was proposed by Bélanger et al. (2003) in which the NNI is used to determine critical petiole nitrate

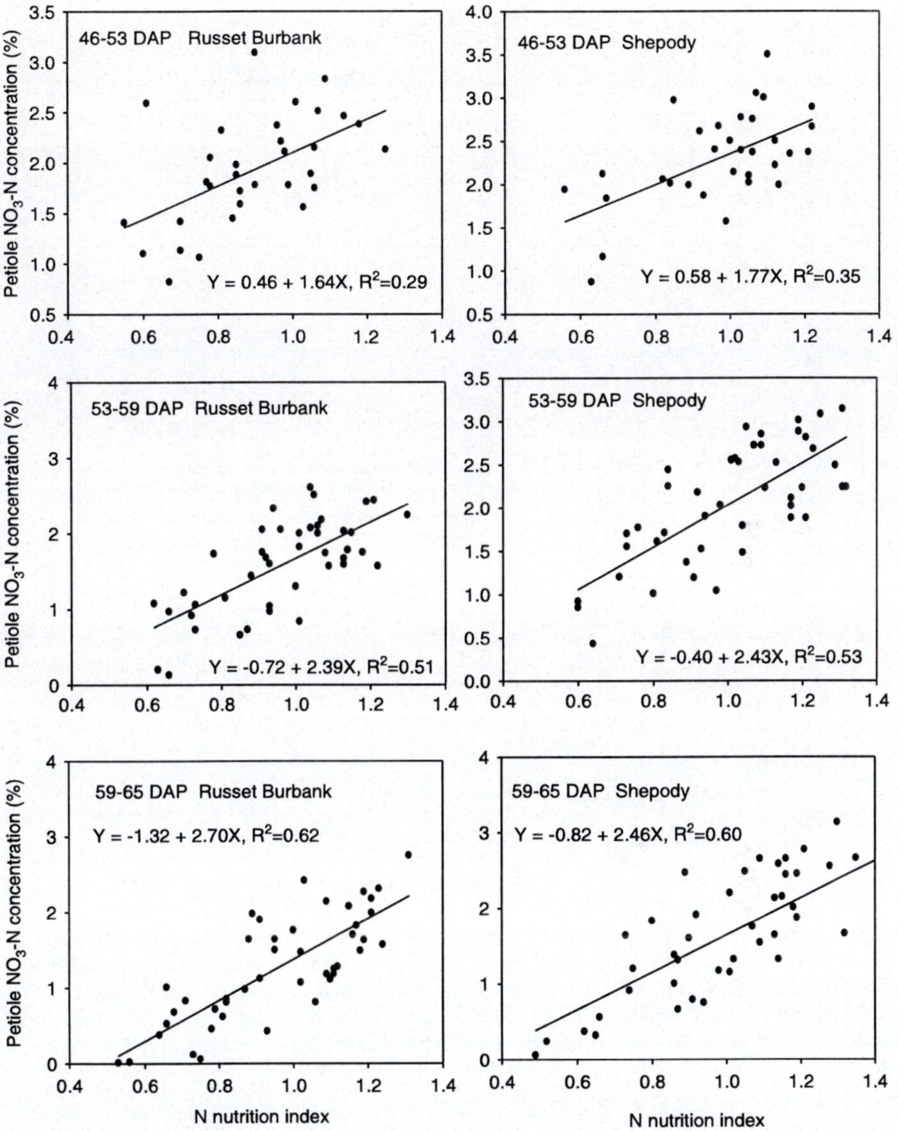

Fig. 11.4 Petiole NO_3-N concentration as a function of the N nutrition index on three sampling intervals based on the number of days after planting (DAP) and for two potato cultivars (Bélanger et al. 2003)

concentrations. They confirmed that the relationship between petiole nitrate concentration and NNI was specific to each cultivar and that it changed during the growing season (Fig. 11.4). Consequently, separate critical petiole nitrate concentrations for Russet Burbank and Shepody were proposed, taking the number of days

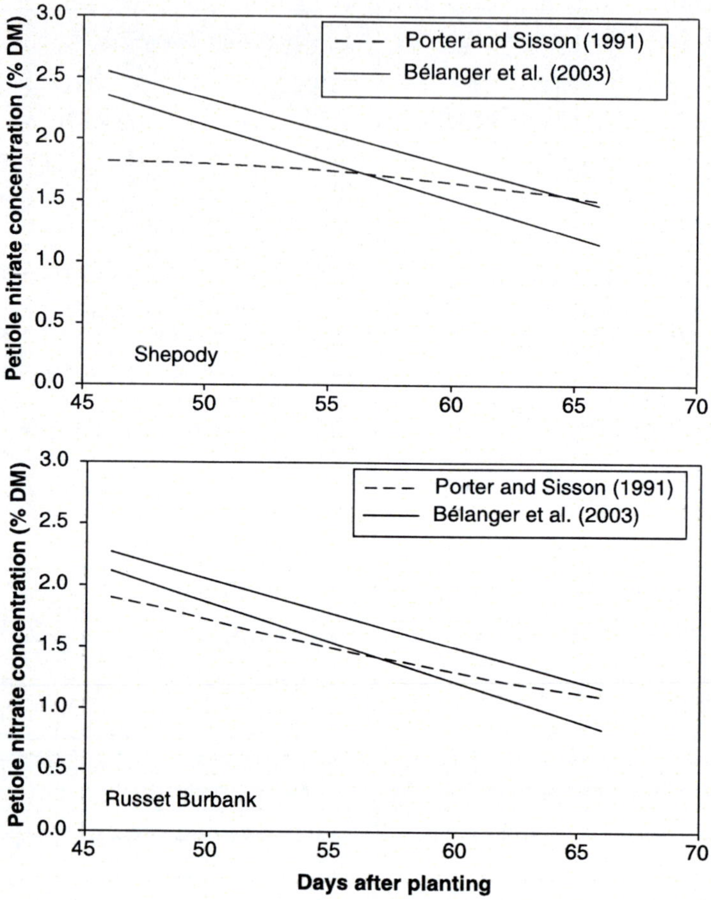

Fig. 11.5 Critical petiole nitrate concentration curves for potato cultivars Shepody and Russet Burbank as a function of the number of days after planting from two independent studies conducted in Maine (Porter and Sisson 1991) and New Brunswick (Bélanger et al. 2003). For Bélanger et al. (2003), the estimation of critical concentrations is based on the relationship with the N nutrition index (NNI) with the upper limit corresponding to NNI = 1.0 and the lower limit to NNI = 0.90

after planting into account (Fig. 11.5). These critical values are relatively close to those reported by Porter and Sisson (1991) in Maine, which were based on tuber yield (Fig. 11.5).

11.2.3 Chlorophyll Content and Chlorophyll Fluorescence

Optical methods of quantifying plant N sufficiency have been developed. Most of these methods are based on quantification of leaf chlorophyll content, which in turn is well correlated with leaf N concentration (Vos and Bom 1993). The SPAD-502 meter (Minolta Camera Co., Osaka, Japan) is the most commonly used optical instrument to measure the leaf chlorophyll content of potatoes (Vos and Bom 1993; Minotti et al. 1994). The SPAD values are the ratio of the intensities of the transmitted light at two wavelengths: red at 640 nm and near infrared at 940 nm (Spectrum Technologies Inc 2009).

Several factors, including cultivar, soil type, and climatic conditions, influence the SPAD values for potato (Gianquinto et al. 2004). This problem can be resolved by using reference plots that can be either over-fertilized or under-fertilized (i.e. no N applied). Denuit et al. (2002) concluded that over-fertilized plots were not effective for potatoes because the SPAD values were relatively insensitive to N rate at high N rates, and it was difficult to discriminate between the fertilized and over-fertilized plots. Olivier et al. (2006) compared both over-fertilized and under-fertilized reference plots and concluded that the zero-N plots discriminated well which potato fields responded to a second N application. Goffart et al. (2008) concluded that SPAD readings do not respond to potato N uptake when fertilizer N rates are above optimal and consequently can only be used to detect N deficiency.

In New Brunswick, for each date of measurements of SPAD values, relatively good positive correlations ($0.45<r<0.79$) between SPAD values and total tuber yield were obtained whereas SPAD values were poorly correlated with relative yield when all sampling dates were included (Zebarth et al. 2003b). Those results indicate that the relationship between SPAD measurements and tuber yield are specific to development stages (Zebarth and Rosen 2007). More recently, the lack of sensitivity of the SPAD values to fertilizer N rate near the optimal rate during crucial development stage for in-season N fertilization were demonstrated (Zebarth et al. 2011). Similar to petiole nitrate concentration, expressing the chlorophyll readings or the relative chlorophyll readings as a function of NNI might provide a more reasonable approach to determine critical values. This has not yet been tested in potatoes, but it has proved useful in corn (Ziadi et al. 2008b) and wheat (Ziadi et al. 2010b).

The popularity of the SPAD meter is linked to the fact that it is easier to use, faster and less costly than the current plant N tissue analyses which require destructive plant sampling. SPAD measurements are, however, still limited to small sampling areas because they require physical contact (near sensing approach) with the leaves (Botha et al. 2007). In addition, SPAD has been shown to detect N deficiency later than petiole nitrate diagnostic tool. Indeed, Wu et al. (2007) reported that N deficiency could be detected about 1 month and 2 weeks after emergence with SPAD and petiole nitrate concentrations, respectively.

Chlorophyll fluorescence analysis is another technique that can be used to determine the plant N status. It is based on the measurement of polyphenolics (Phen), which are secondary metabolites affected by stress factors (Goffart et al. 2008). A N-stressed plant has a higher content of Phen than non-stressed plants. The Phen compounds have typical ultraviolet (UV) absorption peaks in the UV-A and UV-B region (Cerovic et al. 2002) and the value of leaf UV absorbance is directly correlated with the concentration of polyphenolics in leaf tissues.

The Dualex, a portable leaf-clip tool, has been developed by Goulas et al. (2004) in France (Force-A, Orsay, France) to measure Phen contents. The Dualex provides an estimation of the absorbance by the leaf epidermis using two excitation wavelengths, one in the ultraviolet (375 nm) and one red 650 nm where the former is directly related to the concentration of Phen (Goulas et al. 2004). Cartelat et al. (2005) showed that with increasing N fertilization in wheat, leaf chlorophyll content increased and leaf polyphenolics content decreased. They further suggested that the ratio of leaf chlorophyll to polyphenolics is potentially a better indicator of leaf N concentration at the canopy level than either individual measurement. Tremblay et al. (2007) reported similar results for corn produced in eastern Canada. The Dualex has been successfully used in eastern Canada for corn (Tremblay et al. 2007), wheat (Tremblay et al. 2010) and strawberry (Fan et al. 2011). However, this technique is still under investigation for potatoes.

11.2.4 Multispectral Leaf Reflectance Measurements

Light reflectance-based measurements are an alternative approach to measuring leaf chlorophyll content, and have the advantage of being suitable for use at both the leaf and canopy scales (Botha et al. 2006, 2007). Reflectance measurements do not need a contact with the leaves and these measurements can be done with proximal or remote sensors. Reflectance measurements are therefore more suitable for measurement over larger areas. Tractor-mounted sensors such as "Greenseeker" or "Hydro N Sensor" are commercially available to map spatial variability of crop N status in a field (Zebarth et al. 2003b).

Recent studies had shown that hyperspectral leaf reflectance and transmittance measurements using a portable spectroradiometer and inverted analytical models such as PROSPECT or PROSAIL can be used to assess potato N status by estimating leaf or canopy chlorophyll contents (Botha et al. 2006, 2007). When used at the canopy level, hyperspectral reflectance measurements with the inverted PROSAIL model were most effective when the canopy structure was homogenous, and was less effective before canopy closure or after vine collapse (Botha et al. 2007). Spatial variability of potato N status in a field in New Brunswick was effectively mapped using the Hydro N Sensor (Zebarth et al. 2003b). While light reflectance-based approaches are generally effective in assessing relative potato N sufficiency, practical means of using this information to guide in-season fertilizer N management are currently lacking.

11.2.5 Use of Gene Expression

A novel approach to quantification of potato N sufficiency using gene expression is currently being evaluated. Plant responses to their environment, including abiotic stresses, are mediated through changes in gene expression (Hazen et al. 2003). Consequently, quantification of gene expression may provide a more direct measure of plant N sufficiency than current chemical or optical methods. Several studies have identified stress-specific plant gene expression profiles in response to single and combined abiotic stresses including nutrient deficiency (Hazen et al. 2003; Bohnert et al. 2006; Swindell 2006), suggesting it may be possible to use this approach to identify and distinguish among multiple abiotic stresses.

Quantitative assessment of plant N status by gene expression was first done by Li et al. (2010) using potato plants from three potato cultivars grown in a hydroponic system in the greenhouse. Although the conditions of the study were somewhat artificial, it demonstrated that a nitrate reductase gene could be used to quantitatively assess a change in potato N sufficiency within a few days of imposition of N deficiency stress. Subsequently, Zebarth et al. (2011) examined response of expression for 22 genes in leaf tissue of Shepody potatoes grown in the field at six fertilizer N rates. An ammonium transporter gene was identified which was as good as or better than petiole nitrate concentration and SPAD-502 meter readings for quantifying potato N status. While preliminary information on use of gene expression to quantify potato N status is promising, further information is required to determine the potential of this approach. In addition, practical application of this approach is currently limited by economics and by requirements for sample collection and handling protocols (Luo et al. 2011).

11.3 Soil-Based Diagnostic Methods

In most cases, soil-based tests provide an estimate of soil N supply that can be used to adjust the at-planting fertilizer N rate of a given field. Alternatively, soil-based tests can be taken in-season to estimate crop N supply (i.e. soil N supply plus applied fertilizer N). Such tests do not, however, consider crop N demand, and consequently it may be useful to utilize plant-based tests to refine in-season N management.

11.3.1 Soil Mineral Nitrogen Tests

Spring soil mineral N tests are the most commonly used soil-based diagnostic tests. In most cases, these tests are used to quantify the residual soil nitrate from the previous cropping season. Different terminology may be used to describe these tests such as the pre-plant nitrate test or the Nmin test. Such tests have been widely adopted for use in predicting fertilizer N requirements in North America (Hergert 1987) and Europe (Greenwood 1986) of several annual crops, including potatoes.

In humid regions such as eastern Canada, most residual soil nitrate from the previous growing season is lost over the autumn and winter period (Zebarth et al. 2009). Despite this, spring soil nitrate concentration is often well correlated with soil N supply because it reflects early season soil N mineralization (McTaggart and Smith 1993; Sharifi et al. 2008). Spring soil nitrate used alone, however, is not suitable as the basis for making fertilizer N recommendations for potatoes in eastern Canada (Bélanger et al. 2001a). Soil nitrate concentrations change rapidly over time when sampling would occur, and the quantity of soil nitrate in spring is relatively small compared with soil N supply (Zebarth et al. 2005). Therefore, the spring soil nitrate is not a reliable predictor of soil N supply. This is particularly true in some years when significant residual nitrate from the previous growing season is present (Zebarth et al. 2003a). As a result, it may be more appropriate in these humid environments to use spring soil mineral N as a N credit to adjust the fertilizer N recommendations (Zebarth et al. 2009).

An alternative approach is to use a mid-season nitrate test done at 32–47 DAP as a measure of crop N supply from the soil and spring-applied fertilizer to determine if supplemental N fertilizer is required. Bélanger et al. (2001a) suggested that a critical mid-season value of 80 mg NO_3-N kg^{-1} soil, measured at the 0–30 cm depth in the potato ridge following banded at-planting fertilizer application, above which additional N fertilizer may not be needed. The high spatial variability in nitrate concentration within the potato ridge/furrow system, the presence of a significant proportion of soil mineral N as ammonium at this time (Zebarth and Milburn 2003), and the variable geometry of the ridge/furrow system among grower fields may, however, complicate practical application of this approach.

11.3.2 Ion Exchange Membranes

Ion exchange membranes placed in soil have been used as an alternative to measurement of soil mineral N concentration. Both anionic and cationic exchange membranes have been used to measure nitrate and ammonium, respectively, and are commercially available as “Plant Root Simulators” (PRS). These membranes accumulate N from soils through exchange reactions by a similar mechanism to the soil-root system (Yang et al. 1991; Sharifi et al. 2009a). Thus, these membranes detect soil mineral N present at the time of insertion, plus net soil N mineralization during the period during which they are deployed, and N adsorbed on the membranes are not subject to loss through leaching or denitrification. Results from a number of field studies across Canada suggest that ion exchange membranes provide a better index of plant N availability than measurements of soil mineral N alone (Paré et al. 1995; Qian and Schoenau 1995; Ziadi et al. 1999; Nyiraneza et al. 2009). These membranes can be used to measure soil N supply when used on unfertilized plots, or crop N supply (i.e. soil N supply plus applied fertilizer N) when used on fertilized plots.

In potatoes grown in Prince Edward Island and Nova Scotia, Sharifi et al. (2009a) used PRS probes to measure soil N supply following different spring-applied organic amendments. Cumulative N supply measured over a 31 day period after planting was closely related to plant (vines plus tubers) N uptake measured at vine mechanical removal ($R^2=0.60$), and plant N uptake plus soil mineral N (0–30 cm depth) at harvest ($R^2=0.60$).

In Quebec, Ziadi et al. (2011) used anion exchange membranes (PRS-N) during three consecutive growing seasons in potatoes grown under different mineral fertilizer treatments. They concluded that PRS-N measured 40 to 50 DAP can be used as a tool to determine the need for additional N. A significant linear-plus-plateau relationship between relative yield and PRS-N was obtained indicating a critical value of 15 μg PRS-N cm^{-2} d^{-1} above which no additional N application may be required.

Ion exchange membranes can be an effective means of quantifying crop N supply, particularly in the presence of an active crop root system (Zebarth et al. 2009). Duration of deployment of the membranes should be limited to avoid the risk of saturation of the membranes (Qian and Schoenau 2002). Given the high spatial variation in soil mineral N in the potato ridge/furrow system (Zebarth and Milburn 2003), the location of placement of the ion exchange membranes should be carefully selected. In addition, the units of measurement for ion exchange membranes (i.e. flux values) cannot be converted directly to units of concentration or mass, which makes it more difficult to use them for making fertilizer N recommendations. While there has been increased interest in use of ion exchange membranes in research studies, their use in commercial potato production is limited.

11.3.3 Mineralizable Soil Nitrogen

The N mineralized from soil organic matter, organic amendments, and crop residues represents a significant proportion (between 20% and 80%) of crop N requirement (Broadbent 1984). However, estimating this source remains a challenge because of the complex soil, management and environmental controls on the N mineralization process (Dessureault-Rompré et al. 2010a, 2011a; Nyiraneza et al. 2010).

The standard laboratory-based method to quantify soil mineralizable N was developed by Stanford and Smith (1972) to estimate soil potentially mineralizable N (N_0). The N_0 is determined using a long-term aerobic incubation, and therefore this approach is not feasible for practical use. Consequently, a number of indices of soil N availability have been evaluated as predictors of N_0 (St. Luce et al. 2011). In many cases, these indices are chemical tests that target various mineralizable N pools or are biological assays of soil mineralizable N.

A number of studies have evaluated the indices of soil N availability by comparison with N_0 (Sharifi et al. 2007a; Schomberg et al. 2009). Some of the better predictors of N_0 included UV absorbance of a 0.01 *M* $NaHCO_3$ extract at 205 nm or 260 nm (Fox and Piekielek 1978; Hong et al. 1990), direct distillation with NaOH (50%) (Sharifi et al. 2009b), Illinois soil N test (ISNT) for amino sugar N (Khan et al. 2001), particulate organic matter C or N (Gregorich and Beare 2007), hot KCl extractable NH_4-N (Gianello and Bremner 1986), and hot KCl hydrolysable NH_4-N (Wang et al. 2001) (Table 11.1). However in some cases, simple soil properties, for example soil organic C or clay content, may be almost as effective in predicting N_0 as these indices of soil N availability (Simard et al. 2001).

Table 11.1 Proportion of variation in N_0 (i.e. r^2 values from linear regressions) explained by different indices of soil N availability

N availability index[Z]	r^{2Y}
$NaHCO_3$-260	0.74
NaOH-DD	0.61
ISNT	0.51
POMC	0.47
$NaHCO_3$-205	0.47
$HKCl_{HYDR}$	0.46
POMN	0.39
HKCl-NH_4	0.26
PBN_{HYDR}	0.13
PBN	0.11
MBC	0.11
Total organic C	0.60
Total organic N	0.67
Clay	0.46

Adapted from Sharifi et al. (2007a)

[Y]$r^2 \geq 0.26$ were significant $P \leq 0.001$; $n = 39$

[Z]KCl-NH_4 = extractable NH_4 with 1.7 *M* KCl; HKCl-NH_4 = extractable NH_4N with 2 *M* 100 °C KCl; $HKCl_{HYDR}$ = HKCl-NH_4 - KCl-NH_4; $NaHCO_3$-205 = UV absorbance of 0.01 *M* $NaHCO_3$ extract at 205 nm; $NaHCO_3$-260 = UV absorbance of 0.01 *M* $NaHCO_3$ extract at 260 nm; ISNT = Illinois Soil N Test for amino sugar-N; NaOH-DD = direct distillation with NaOH (50%); MBC = microbial biomass C by fumigation extraction method; PBN = direct distillation with phosphate-borate buffer (pH = 11.2); PBN_{HYDR} = PBN – (KCl-NH_4); POMC = particulate organic matter C; POMN = particulate organic matter N

Some studies compared indices of soil N availability with field-based measures of soil N supply, most commonly for corn. For example, Hong et al. (1990) found strong positive correlations between soil N supplying capacity (N uptake in the above-ground plant less 75% of starter N) with spring soil nitrate, spring soil nitrate plus hot KCl extractable NH_4-N, spring soil nitrate plus distillation with a phosphate-borate buffer solution (pH 11.2), and ultraviolet absorbance of a 0.01 *M* $NaHCO_3$ extract at 200 nm. The ISNT was highly correlated with check-plot corn yield (r = 0.79) and fertilizer N response (r = 0.82) of corn in Illinois (Mulvaney et al. 2001), however, Barker et al. (2006) concluded that the ISNT is not a good predictor of corn relative grain yield. No single index of soil N availability has gained widespread adoption.

Few studies compared indices of soil N availability with field-based measures of soil N supply in potatoes. Sharifi et al. (2007b) compared potato plant (vines plus tubers) uptake and tuber relative yield against a series of indices of N availability for sites in New Brunswick, Canada and Maine, USA under rain-fed production from 2000 to 2005. Spring soil mineral N was one of the best predictors of soil N supply, however, Sharifi et al. (2007b) recommended use of spring soil N plus Pool I (a labile pool of mineralizable N measured using a 14 day aerobic incubation) as a

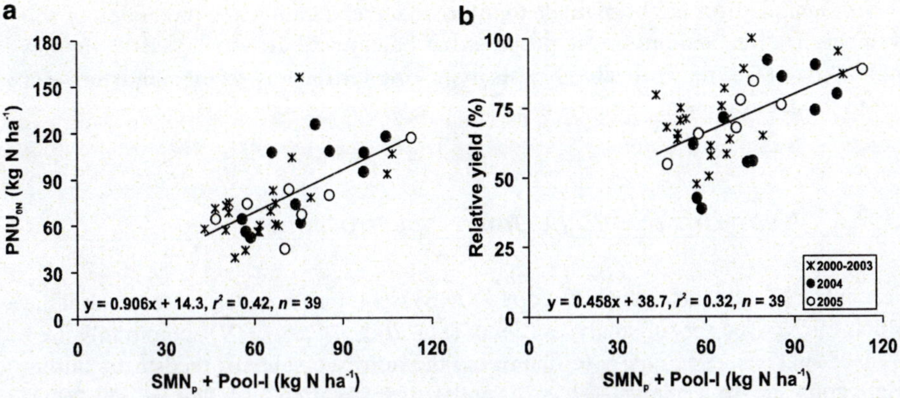

Fig. 11.6 Relationships between spring soil mineral N (0–30 cm depth) (SMN_p) plus Pool-I (a labile mineralizable N pool) and (**a**) soil N supply as estimated by plant (vines plus tubers) N uptake measured at vine desiccation with no fertilizer N application (PNU_{0N}) and (**b**) relative yield in field experiments in New Brunswick, Canada and Maine, USA in 2000–2005 (Sharifi et al. 2007b)

more robust predictor of soil N supply (Fig. 11.6). Interestingly, N_0 was a poor predictor of soil N supply. This was attributed at least in part to the exclusion of the labile mineralizable N pool in estimating the value of N_0.

Soil N availability indices provide a measure of the potential for soil N mineralization to occur, but they do not account for the effects of environmental conditions in influencing actual soil net N mineralization. One option is to predict soil N supply using simple first order kinetic models of soil N mineralization:

$$\mathrm{N}_{\min} = N_0\left[1 - \mathrm{e}^{-kt}\right] \tag{11.6}$$

where $\mathrm{N}_{\min}$ is the cumulative amount of N mineralized at time t, N_0 is potentially mineralizable N, and k is the mineralization rate coefficient (Stanford and Smith 1972; Curtin and Campbell 2007). The value of the mineralization rate constant, k, can be modified based on soil temperature (Dessureault-Rompré et al. 2010b) or soil water content (Paul et al. 2003; Dessureault-Rompré et al. 2011b) to reflect changes in environmental conditions. In some cases, satisfactory predictions of net N mineralization in the field have been achieved using a kinetic model (Stanford et al. 1977; Marion et al. 1981; Campbell et al. 1984) whereas in other cases soil N supply has been overestimated (Verstraete and Voets 1976; Griffin and Laine 1983; Cabrera and Kissel 1988; Mikha et al. 2006). In eastern Canada, Dessureault-Rompré et al. (2011a) compared estimates of soil N supply from a kinetic model with plant (vines plus tubers) N uptake in unfertilized potato plots in New Brunswick, Canada and Maine, USA. Direct application of the kinetic model significantly underestimated field measured soil N supply, however when the model considered soil mineral N and the labile mineralizable pool (i.e. Pool-I), satisfactory results were obtained. However, practical application of kinetic models is currently limited by the requirement for long-term laboratory incubations to obtain estimates of the values of N_0 and k.

Substantial effort has been made to improve understanding and prediction of soil N mineralization, and promising progress has been made. However to date, there is limited use of soil mineralizable N tests in making fertilizer N recommendations for potato production.

11.3.4 Near-Infrared Reflectance Spectroscopy

Near-infrared reflectance spectroscopy (NIRS) is a rapid, non-destructive technique which can be used for soil analyses (Dunn et al. 2002). The NIRS is commonly used in plant analysis, specifically to determine the nutritive value of feedstuffs, but its application in soil analysis is still under investigation (Malley et al. 2002; Nduwamungu et al. 2009a, b). The soil N availability as measured by NIRS was previously demonstrated to be closely related to soil N supply as measured by crop N uptake in unfertilized plots for corn ($R^2=0.49$; Fox et al. 1993) and winter wheat ($R^2=0.81$; Börjesson et al., 1999). In eastern Canada, Nduwamungu et al. (2009a) accurately predicted potentially mineralizable N calculated from soil organic matter and clay content (Simard et al. 2001) under corn production. The NIRS is a technique which merits further examination as a measure of soil N availability.

11.4 Agronomic Applications

Soil- and plant-based diagnostic tests have the potential to improve the efficiency of N utilization, and hence provide economic benefits to growers and environmental benefits to society. It is necessary for test results to be interpreted and converted into N recommendations in order for them to be effective for action (Vos 2009).

Some plant-based tests have been successfully used as a diagnostic of crop N status (e.g. petiole nitrate concentration in potatoes) or in crop models of several crops (e.g. NNI) to account for the effect of N on growth and yield. At the farm level, however, there are some limitations to their adoption by growers. Although, the NNI has been shown to have the potential to successfully diagnose the N status for different crops including potatoes, this tool requires the determination of the shoot biomass during the growing season and its N concentration, which is time-consuming for growers. Furthermore, the critical N curve is only valid for shoot biomass greater than 1.0 Mg DM ha^{-1}. The window of opportunity for a remedial action is then limited in a relatively short season. The NNI could, however, be used as a reference for simpler procedures such as the chlorophyll meter readings and petiole nitrate concentration to determine the crop N status. These simpler procedures are currently available but they are still not widely used in eastern Canada. The benefits of multi or hyperspectral measurements have not yet been demonstrated at the farm scale whereas further research is required to determine if gene expression can be used to reliably assess potato N sufficiency.

Soil tests based on residual nitrate are most commonly used world-wide, but are not as effective in eastern Canada because most residual nitrate is lost over the fall and winter period. Significant progress has been made in use of ion exchange membranes and in soil mineralizable N tests. However, further work is required before such approaches can be used as the basis of fertilizer N recommendations.

It is proposed that the most effective strategy will be the use of a combination of soil- and plant-based diagnostic tools. Soil-based tests can be used to predict soil N supply, and to adjust at-planting fertilizer N rates whereas plant-based tests can be used to assess crop N sufficiency as a guide to in-season fertilizer N management. Such an approach will facilitate the matching of the N supply to the crop N demand on an individual field basis and yield economic and environmental benefits.

References

Barker DW, Sawyer JE, Al-Kaisi MM, Lundvall JP (2006) Assessment of the amino sugar-nitrogen test on Iowa Soils: II. Field correlation and calibration. Agron J 98:1352–1358

Bélanger G, Ziadi N (2008) Phosphorus and nitrogen relationships during spring growth of an aging timothy sward. Agron J 100:1757–1762

Bélanger G, Walsh JR, Richards JE, Milburn PH, Ziadi N (2000) Yield response of two potato cultivars to supplemental irrigation and N fertilization in New Brunswick. Am J Potato Res 77:11–21

Bélanger G, Walsh JR, Richards JE, Milburn PH, Ziadi N (2001a) Predicting nitrogen fertilizer requirements of potatoes in Atlantic Canada with soil nitrate determinations. Can J Soil Sci 81:535–544

Bélanger G, Walsh JR, Richards JE, Milburn PH, Ziadi N (2001b) Critical nitrogen curve and nitrogen nutrition index for potato in eastern Canada. Am J Potato Res 78:355–364

Bélanger G, Walsh JR, Richards JE, Milburn PH, Ziadi N (2003) Critical petiole nitrate concentration of two processing potato cultivars in eastern Canada. Am J Potato Res 80:251–262

Bohnert HJ, Gong Q, Li P, Ma S (2006) Unraveling abiotic stress tolerance mechanisms – getting genomics going. Curr Opin Plant Biol 9:180–188

Börjesson T, Stenberg B, Lindén B, Jonsson A (1999) NIR spectroscopy, mineral nitrogen analysis and soil incubations for the prediction of crop uptake of nitrogen during the growing season. Plant Soil 214:75–83

Botha EJ, Zebarth BJ, Leblon B (2006) Non-destructive estimation of potato leaf chlorophyll and protein contents from hyperspectral measurements using the PROSPECT radiative transfer model. Can J Plant Sci 86:279–291

Botha EJ, Leblon B, Zebarth BJ, Watmough J (2007) Non-destructive estimation of potato leaf chlorophyll from canopy hyperspectral reflectance using the inverted PROSAIL model. Intern J Appl Earth Observ Geoinform 9:360–374

Broadbent FE (1984) Plant use of soil nitrogen. In: Hauck RD (ed) Nitrogen in crop production. ASA, CSSA, and SSSA, Madison, pp 171–182

Burton DL, Zebarth BJ, Gillam KM, MacLeod JA (2008) Effect of split application of fertilizer nitrogen on N_2O emissions from potatoes. Can J Soil Sci 88:229–239

Cabrera ML, Kissel DE (1988) Evaluation of a method to predict nitrogen mineralized from soil organic matter under field conditions. Soil Sci Soc Am J 52:1027–1031

Cambouris AN, Zebarth BJ, Nolin MC, Laverdière MR (2007) Response to added nitrogen of a continuous potato sequence as related to sand thickness over clay. Can J Plant Sci 87:829–839

Cambouris AN, Zebarth BJ, Nolin MC, Laverdière MR (2008) Apparent fertilizer nitrogen recovery and residual soil nitrate under continuous potato cropping: effect of N fertilization rate and timing. Can J Soil Sci 88:813–825

Campbell CA, Jame YW, Winkleman GE (1984) Mineralization rate constants and their use for estimating nitrogen mineralization in some Canadian prairie soils. Can J Soil Sci 64:333–343
Cartelat A, Cerovic ZG, Goulas Y, Meyer S, Lelarge C, Prioul JL, Barbottin A, Jeuffroy MH, Gate P, Agati G, Moya I (2005) Optically assessed contents of leaf polyphenolics and chlorophyll as indicators of nitrogen deficiency in wheat (*Triticum aestivum* L.). Field Crops Res 91:35–49
Cerovic ZG, Ounis A, Cartelat A, Latouche G, Goulas Y, Meyer S, Moya I (2002) The use of chlorophyll fluorescence excitation spectra for the non-destructive *in situ* assessment of UV-absorbing compounds in leaves. Plant Cell Environ 25:1663–1676
CRAAQ (2010) Centre de Références en Agriculture et Agroalimentaire du Québec. Fertilisation reference guide. (In French.) 2nd ed. ISBN 978-2-7649-0231-8, 473pp
Curtin D, Campbell CA (2007) Mineralizable nitrogen. In: Carter MC, Gregorich EG (eds) Soil sampling and methods of analysis, 2nd edn. CRC press, Boca Raton, FL, pp 599–606
Denuit JP, Olivier M, Goffaux MJ, Herman J, Goffart JP, Destain JP, Frankinet M (2002) Management of nitrogen fertilization of winter wheat and potato crops using the chlorophyll meter for crop nitrogen status assessment. Agronomie 22:847–853
Dessureault-Rompré J, Zebarth BJ, Burton DL, Sharifi M, Cooper J, Grant CA, Drury CF (2010a) Relationships among mineralizable soil nitrogen, soil properties and climatic indices. Soil Sci Soc Am J 74:1218–1227
Dessureault-Rompré J, Zebarth BJ, Georgallas A, Burton DL, Grant CA, Drury CF (2010b) Temperature dependence of soil nitrogen mineralization rate: comparison of mathematical models, reference temperatures and origin of the soils. Geoderma 157:97–108
Dessureault-Rompré J, Zebarth BJ, Burton DL, Georgallas A, Sharifi M, Porter GA, Moreau G, Leclerc Y, Arsenault WJ, Chow TL, Grant CA (2011a) Prediction of soil nitrogen supply in potato fields in a cool humid climate. Soil Sci Soc Am J 75:626–637
Dessureault-Rompré J, Zebarth BJ, Georgallas A, Burton DL, Grant CA (2011b) A biophysical water function to predict the response of soil nitrogen mineralization to soil water content. Geoderma 167–168:214–227
Duchenne T, Machet JM, Martin M (1997) Potatoes. In: Lemaire G (ed) Diagnosis of the nitrogen status in crops. Springer-Verlag, Berlin, pp 119–130
Dunn BW, Batten GD, Beecher HG, Ciavarella S (2002) The potential of near-infrared reflectance spectroscopy for soil analysis – a case study from the Riverine Plain of south-eastern Australia. Aust J Exp Agric 42:607–614
Errebhi M, Rosen CJ, Birong DE (1998) Calibration of a petiole sap nitrate test for irrigated Russet Burbank potato. Commun Soil Sci Plant Anal 29:23–25
Fan L, Fang C, Dubé C, Tremblay N, Khanizadeh S (2011) A non-destructive method to predict polyphenol content in strawberry. J Food Agric Environ 9:59–62
Fox RH, Piekielek WP (1978) A rapid method for estimating the nitrogen supplying capability of a soil. Soil Sci Soc Am J 42:751–753
Fox RH, Shenk JS, Piekielek WP, Westerhaus MO, Toth JD, Macneal KE (1993) Comparison of near-infrared spectroscopy and other soil nitrogen availability quick tests for corn. Agron J 85:1049–1053
Gianello C, Bremner JM (1986) A simple chemical method of assessing potentially available organic nitrogen in soil. Commun Soil Sci Plant Anal 17:195–214
Gianquinto G, Goffart JP, Olivier M, Guarda G, Colauzzi M, Dalla Costa L, Delle Vedove G, Vos J, Mackerron DKL (2004) The use of hand-held chlorophyll meters as a tool to assess the nitrogen status and to guide nitrogen fertilization of potato crop. Potato Res 47:35–80
Goffart JP, Olivier M, Frankinet M (2008) Potato crop nitrogen status assessment to improve N fertilization management and efficiency: past-present-future. Potato Res 51:355–383
Goulas Y, Cerovic ZG, Cartelat A, Moya I (2004) Dualex: a new instrument for field measurements of epidermal UV-absorbance by chlorophyll fluorescence. Appl Opt 43:4488–4496
Greenwood DJ (1986) Prediction of nitrogen fertilizer needs of arable crops. Adv Plant Nutr 2:1–61
Greenwood DJ, Lemaire G, Gosse G, Cruz P, Draycott A, Neeteson JJ (1990) Decline in percentage N of C3 and C4 crops with increasing plant mass. Ann Bot 66:425–436

Gregorich EG, Beare MH (2007) Physically uncomplexed organic matter. In: Carter MR, Gregorich EG (eds) Soil sampling and methods of analysis, 2nd edn. CRC Press, Boca Raton, FL, pp 607–616

Griffin GF, Laine AF (1983) Nitrogen mineralization in soils previously amended with organic wastes. Agron J 75:124–129

Hazen SP, Wu Y, Kreps JA (2003) Gene expression profiling of plant responses to abiotic stress. Funct Integr Genomics 3:105–111

Hergert GW (1987) Status of residual nitrate-nitrogen soil tests in the United States of America. In: Brown JR (ed) Soil testing: sampling, correlation, calibration, and interpretation. SSSA, Madison, pp 73–78

Hong SD, Fox RH, Piekielek WP (1990) Field evaluation of several chemical indexes of soil nitrogen availability. Plant Soil 123:83–88

Khan SA, Mulvaney RL, Hoeft RG (2001) A simple soil test for detecting sites that are nonresponsive to nitrogen fertilization. Soil Sci Soc Am J 65:1751–1760

Lemaire G, Salette J (1984) Relationship between growth and nitrogen uptake in a pure grass stand. I. Environmental effects. Agronomie 4:423–430, In French

Lewis RJ, Love SL (1994) Potato genotypes differ in petiole nitrate-nitrogen concentrations over time. HortSci 29:175–179

Li X-Q, Sveshnikov D, Zebarth BJ, Tai H, De Koeyer D, Millard P, Haroon M, Singh M (2010) Detection of nitrogen sufficiency in potato plants using gene expression markers. Am J Potato Res 87:50–59

Luo S, Tai H, Zebarth BJ, Li X-Q, Millard P, De Koeyer D, Xiong X (2011) Sample collection protocol effects on quantification of gene expression in potato leaf tissue. Plant Molec Biol Reporter 29:369–378

Malley DF, Yesmin L, Eilers RG (2002) Rapid analysis of hog manure and manure-amended soils using near-infrared spectroscopy. Soil Sci Soc Am J 66:1677–1686

Marion GM, Kummerow J, Miller PC (1981) Predicting nitrogen mineralization in Chaparral soils. Soil Sci Soc Am J 45:956–961

McTaggart IP, Smith KA (1993) Estimation of potentially mineralisable nitrogen in soil by KCl extraction. II. Comparison with soil N uptake in the field. Plant Soil 157:175–184

Meyer RD, Marcum DB (1998) Potato yield, petiole nitrogen, and soil nitrogen response to water and nitrogen. Agron J 90:420–429

Mikha MM, Rice CW, Benjamin JG (2006) Estimating soil mineralizable nitrogen under different management practices. Soil Sci Soc Am J 70:1522–1531

Minotti PL, Halseth DE, Sieczka JB (1994) Field chlorophyll measurements to assess the nitrogen status of potato varieties. HortSci 29:1497–1500

Mulvaney RL, Khan SA, Hoeft RG, Brown HM (2001) A soil organic nitrogen fraction that reduces the need for nitrogen fertilization. Soil Sci Soc Am J 65:1164–1172

NBDAFA (2001) Crop fertilization guide. Dept. Agriculture, Fisheries and Aquaculture, New Brunswick

Nduwamungu C, Ziadi N, Tremblay GF, Parent LE (2009a) Near-infrared reflectance spectroscopy prediction of soil properties: effects of sample cups and preparation. Soil Sci Soc Am J 73:1896–1903

Nduwamungu C, Ziadi N, Parent LE, Tremblay GF, Thuriès L (2009b) Opportunities for, and limitations of, near infrared reflectance spectroscopy applications in soil analysis: a review. Can J Soil Sci 89:531–541

Nyiraneza J, N'Dayegamiye A, Chantigny MH, Laverdiere MR (2009) Variations in corn yield and nitrogen uptake in relation to soil attributes and nitrogen availability indices. Soil Sci Soc Am J 73:317–327

Nyiraneza J, N'Dayegamiye A, Gasser MO, Giroux M, Grenier M, Landry C, Guertin S (2010) Soil and crop parameters related to corn nitrogen response in Eastern Canada. Agron J 102:1478–1490

Ojala JC, Stark JC, Kleinkopf GE (1990) Influence of irrigation and nitrogen management on potato yield and quality. Am Potato J 67:29–43

Olivier M, Goffart JP, Ledent JF (2006) Threshold values for chlorophyll meter as decision tool for nitrogen management of potato. Agron J 98:496–506

Paré T, Gregorich EG, Ellert BH (1995) Comparison of soil nitrate extracted by potassium chloride and adsorbed on an anion exchange membrane in situ. Commun Soil Sci Plant Anal 26:883–898

Paul KI, Polglase PJ, O'Connell AM, Carlyle JC, Smethurst PJ, Khanna PK (2003) Defining the relation between soil water content and net nitrogen mineralization. Eur J Soil Sci 54:39–47

Porter GA, Sisson JA (1991) Petiole nitrate content of Maine grown Russet Burbank and Shepody potatoes in response to varying nitrogen rate. Am Potato J 68:493–505

Qian P, Schoenau JJ (1995) Assessing nitrogen mineralization from soil organic matter using anion exchange membranes. Fert Res 40:143–148

Qian P, Schoenau JJ (2002) Practical applications of ion exchange resins in agricultural and environmental soil research. Can J Soil Sci 82:9–21

Schomberg HH, Wietholter S, Griffin TS, Reeves DW, Cabrera ML, Fisher DS, Endale DM, Novak JM, Balkcom KS, Raper RL, Kitchen NR, Locke MA, Potter KN, Schwartz RC, Truman CC, Tyler DD (2009) Assessing indices for predicting potentially nitrogen mineralization in soils under different management systems. Soil Sci Soc Am J 73:1575–1586

Sharifi M, Zebarth BJ, Burton DL, Grant CA, Cooper JM (2007a) Evaluation of some indices of potentially mineralizable nitrogen in soil. Soil Sci Soc Am J 71:1233–1239

Sharifi M, Zebarth BJ, Burton DL, Grant CA, Porter GA, Cooper JM, Leclerc Y, Moreau G, Arsenault WJ (2007b) Evaluation of laboratory-based measures of soil mineral nitrogen and potentially mineralizable nitrogen as predictors of field-based estimates of soil nitrogen supply in potato production. Plant Soil 301:203–214

Sharifi M, Zebarth BJ, Burton DL, Grant CA, Porter GA (2008) Organic amendment history and crop rotation effects on soil nitrogen mineralization potential and soil nitrogen supply in a potato cropping system. Agron J 100:1562–1572

Sharifi M, Lynch DH, Zebarth BJ, Zheng Z, Martin RC (2009a) Evaluation of nitrogen supply rate measured by in situ placement of Plant Root Simulator probes as a predictor of nitrogen supply from soil and organic amendments in potato crop. Am J Potato Res 86:356–366

Sharifi M, Zebarth BJ, Burton DL, Grant CA, Hajabbasi MA, Abbassi-Kalo G (2009b) Sodium hydroxide direct-distillation: a method for estimating total nitrogen in soil. Commun Soil Sci Plant Anal 40:2505–2520

Simard RR, Ziadi N, Nolin MC, Cambouris AN (2001) Prediction of nitrogen responses of corn by soil nitrogen mineralization indicators. In: Optimizing nitrogen management in food and energy production and environmental protection: Proceedings of 2nd international nitrogen conference on science and policy. Scientific World 1(S2):135–141

Spectrum Technologies Inc. (2009) SPAD 502 Plus Chlorophyll Meter Product Manual Item # 2900P, 2900PDL. 24pp

St. Luce M, Whalen JK, Ziadi N, Zebarth BJ (2011) Nitrogen dynamics and indices to predict soil nitrogen supply in humid temperate soils. Adv Agron 112:55–102

Stanford G, Smith SJ (1972) Nitrogen mineralization potentials of soils. Soil Sci Soc Am J 36:465–472

Stanford G, Carter JN, Westermann DT, Meisinger JJ (1977) Residual nitrate and mineralizable soil nitrogen in relation to nitrogen uptake by irrigated sugarbeets. Agron J 69:303–308

Stark JC, McCann IR, Westermann DT, Izadi B, Tindall TA (1993) Potato response to split nitrogen timing with varying amounts of excessive irrigation. Am Potato J 70:765–777

Swindell WR (2006) The association among gene expression responses to nine abiotic stress treatments in *Arabidopsis thaliana*. Genetics 174:1811–1824

Tremblay N, Wang Z, Bélec C (2007) Evaluation of the Dualex for the assessment of corn nitrogen status. J Plant Nutr 30:1355–1369

Tremblay N, Wang Z, Belec C (2010) Performance of Dualex in spring wheat for crop nitrogen status assessment, yield prediction and estimation of soil nitrate content. J Plant Nutr 33:57–70

Verstraete W, Voets JP (1976) Nitrogen mineralization tests and potentials in relation to soil management. Pedologie 26:15–26

Vos J (2009) Nitrogen responses and nitrogen management in potato. Potato Res 52:305–317

Vos J, Bom M (1993) Hand-held chlorophyll meter: a promising tool to assess the nitrogen status of potato foliage. Potato Res 36:301–308

Wang W, Smith CJ, Chalk PM, Chen D (2001) Evaluating chemical and physical indices of nitrogen mineralization capacity with an unequivocal reference. Soil Sci Soc Am J 65:368–376

Waterer D (1997) Petiole sap NO_3-N testing as a method of monitoring nitrogen nutrition of potato crops. Can J Plant Sci 77:273–278

Wu J, Wang D, Rosen CJ, Bauer ME (2007) Comparison of petiole nitrate concentrations, SPAD chlorophyll readings, and QuickBird satellite imagery in detecting nitrogen status of potato canopies. Field Crops Res 101:96–103

Yang JE, Skogley EO, Georgitis SJ, Schaff BE, Ferguson AH (1991) Phyto-availability soil test: development and verification of theory. Soil Sci Soc Am J 55:85–1365

Zebarth BJ, Milburn PH (2003) Spatial and temporal distribution of soil inorganic nitrogen concentration in potato hills. Can J Soil Sci 83:183–195

Zebarth BJ, Rosen CJ (2007) Research perspective on nitrogen BMP development for potato. Am J Potato Res 84:3–18

Zebarth BJ, Leclerc Y, Moreau G, Gareau R, Milburn PH (2003a) Soil inorganic nitrogen content in commercial potato fields in New Brunswick. Can J Soil Sci 83:425–429

Zebarth BJ, Rees H, Tremblay N, Fournier P, Leblon B (2003b) Mapping spatial variation in potato nitrogen status using the "N Sensor". Acta Hort (ISHS) 627:267–273

Zebarth BJ, Leclerc Y, Moreau G, Sanderson JB, Arsenault WJ, Botha E, Wang-Pruski G (2005) Estimation of soil nitrogen supply in potato fields using a plant bioassay approach. Can J Soil Sci 85:377–386

Zebarth BJ, Drury CF, Tremblay N, Cambouris AN (2009) Opportunities for improved fertilizer nitrogen management in production of arable crops in eastern Canada: a review. Can J Soil Sci 89:113–132

Zebarth BJ, Tai H, Luo S, Millard P, De Koeyer D, Li X-Q, Xiong X (2011) Differential gene expression as an indicator of nitrogen sufficiency in field-grown potato plants. Plant Soil 345:387–400

Ziadi N, Simard RR, Allard G, Lafond J (1999) Field evaluation of anion exchange membranes as a N soil testing method for grasslands. Can J Soil Sci 79:281–294

Ziadi N, Brassard M, Bélanger G, Cambouris AN, Tremblay N, Nolin MC, Claessens A, Parent LE (2008a) Critical nitrogen curve and nitrogen nutrition index for corn in eastern Canada. Agron J 100:271–276

Ziadi N, Brassard M, Bélanger G, Claessens A, Tremblay N, Cambouris AN, Nolin MC, Parent LE (2008b) Chlorophyll measurements and nitrogen nutrition index for the evaluation of corn nitrogen status. Agron J 100:1264–1273

Ziadi N, Bélanger G, Claessens A, Lefebvre L, Cambouris AN, Tremblay N, Nolin MC, Parent LE (2010a) Determination of a critical nitrogen dilution curve for spring wheat. Agron J 102:241–250

Ziadi N, Bélanger G, Claessens A, Lefebvre L, Tremblay N, Cambouris AN, Nolin MC, Parent LE (2010b) Plant-based diagnostic tools for evaluating wheat nitrogen status. Crop Sci 50:2580–2590

Ziadi N, Grant C, Samson N, Nyiraneza J, Bélanger G, Parent LE (2011) Efficiency of controlled-release urea for a potato production system in Quebec, Canada. Agron J 103:60–66

Chapter 12
Nitrogen Management in Organic Potato Production

Derek H. Lynch, Mehdi Sharifi, Andy Hammermeister, and David Burton

Abstract There is increasing interest in organic potato production in Canada, within a context of continuing strong growth rates for organic products globally. Using data from on-farm and station based trials, key characteristics of organic potato production in the Atlantic Canada region, notably the use of extended rotations involving leguminous crop green manures combined with organic amendments, low intensity of nitrogen and residual soil mineral N (RSMN) post harvest, and enhanced soil quality and health, are shown as sustainable outcomes of these systems. Data presented confirm nitrogen as the primary factor limiting total and marketable yields. Without additional N supplementation but following legume green manures (GMr) of red clover, or hairy vetch, potato yields and N uptake are shown to range from 30 to 35 Mg ha^{-1} and 100–125 kg N ha^{-1}, respectively, while RSMN remains low. Combining N supplementation (with composts or dehydrated manures) with GMr consistently increased total and marketable yield. The effect of N supply and GMr type on pest (wireworm, Colorado potato beetle) population dynamics is also examined. Finally, synchronizing N supply in these systems with crop demand remains challenging and the potential to use novel soil tests and plant bioassays to improve N management in organic production systems is also discussed.

D.H. Lynch (✉)
Department of plant and Animal Sciences, Nova Scotia Agricultural College, B2N 5E3 Truro, Nova Scotia, Canada
e-mail: dlynch@nsac.ca

M. Sharifi • D. Burton
Department of Environmental Science, Nova Scotia Agricultural College, B2N 5E3 Truro, Nova Scotia, Canada

A. Hammermeister
Organic Agriculture Centre of Canada, Department of Environmental Science, Nova Scotia Agricultural College, B2N 5E3 Truro, Nova Scotia, Canada

Z. He et al. (eds.), *Sustainable Potato Production: Global Case Studies*,
DOI 10.1007/978-94-007-4104-1_12,

12.1 Introduction

Organic potato (*Solanum tuberosum* L.) production is characterized by extended rotations involving leguminous crop green manure (GMr) crops often combined with organic amendments (Lynch et al. 2008a, b). These are key elements of a sustainable approach to soil and nutrient management (Stark and Porter 2005). Losses due to insects and pest damage and overall low quality, typically reduce organic potato yields to an average of 50–75% that of conventional production (Pimentel 1993; Varis et al. 1996; Mattsson and Wallen 2003; VanDelden et al. 2003; Maggio et al. 2008). In addition, inconsistent nutrient availability and the complexity of nutrient management in these systems combined with yield uncertainty increases economic risk and may even promote post-harvest N losses through leaching (Van Delden 2001; De Neve et al. 2003; Stark and Porter 2005). At the same time, strong annual growth rates have continued for retail sales of organic food products over the past 20 years and globally organic cropping acreage is estimated at 30.4 million hectares (Willer 2008). In Atlantic Canada, premiums for organic table stock potatoes have often exceeded 100–200% of conventional market prices (PEIDAF, 2000). Popularized literature and guide books for organic potato production have become more recently available (Bostock 2008).

On commercial organic farms producing field crops in Canada potatoes are typically included in rotation with small grains (wheat, barley) and leguminous forage crops or GMr. For organic vegetable producers potatoes are often included in even more complex and longer term rotations (Bostock 2008). Weeds and topkilling are managed through mechanical means. Organic potato production in Atlantic Canada currently is primarily for the tablestock rather than the processing market and varieties are typically early to mid-season in maturity (70–110 days) (PEIDAF 2000; Bostock 2008). Potato late blight (*Phytophthora infestans*) remains particularly difficult to control under organic management as synthetic fungicides, with the exception of copper products, are not allowed (Canadian General Standards Board 2009b). The reduced efficacy of copper products and concern over their continued use has led to research efforts, as yet unsuccessful, to identify either acceptable resistant varieties (Speiser et al. 2006) or fungicides (Sturz et al. 2006). However, losses due to late blight in organic potato production systems may be overestimated and the use of earlier season and earlier tuber set varieties, combined with other agronomic strategies such as chitting (or seed tuber pre-sprouting) can be effective late blight 'escape strategies' which minimize losses (Finckh et al. 2006; Möller and Reents 2007). Möller et al (2007) concluded, following a survey and data collection from 220 organic potato fields in Germany, that N availability explained 73% of the observed variability in yield, and remains the most important yield limiting factor in organic potato production. Through our on-farm and research station studies conducted in Atlantic Canada over the past 6 years, and discussed in further detail below, we have also found this to be the case (Lynch et al. 2008a, b; Liu et al. 2008, 2010)

Nitrogen is critically important for canopy development, tuber initiation and yield of potato (Porter and Sisson 1991; Belanger et al. 2001). Insufficient N supply results in lower tuber yields and tuber size, while excess N can promote abundant haulm growth, delays tuber initiation and crop maturation, and reduces tuber yield and quality (Van Delden 2001; Finckh et al. 2006). In short season humid areas such as Atlantic Canada lengthening of the vegetative period and delay of tuberization can negatively enhance risk of losses to late blight, and promote excessive post-harvest mineral N losses to the environment (Lynch et al. 2008a, b; Sharifi et al. 2009). The N requirement of high yielding potatoes ranges from 2.5 to 5.9 kg N per Mg of yield (Munoz et al. 2005), although interestingly there is some suggestion that N use efficiency can be improved under low input or organic systems (Finckh et al. 2006; Möller et al. 2007) perhaps linked to improved light use efficiency when N is limiting (Van Delden 2001). In Germany, Möller et al. (2007) reported organic potato yields of 30 Mg ha^{-1} requiring 3.5-3.7 kg NMg^{-1} (or 100–120 kg Nha^{-1}) compared to ranges closer to 4.0-6.0 kg N required per Mg^{-1} in conventional systems. However, soil and fertility management in organic systems is built around managing the complex processes of organic matter deposition and decomposition, thus managing N and synchronizing N availability with crop demand in organic potato production systems is correspondingly a challenge. The supply of N from organic amendments, crop residues plus residual soil mineral N to the crop varies with climatic conditions and among years (Sharifi et al. 2009; Zebarth et al. 2009).

Nitrogen availability and intensity may also have significant impacts with respect to potato insect pest dynamics (Alyokhin et al. 2005; Boiteau et al. 2008). With respect to defoliating insects, Colorado potato beetle (CPB) (*Leptinotarsa decemlineata* (Say)) is the most destructive insect pest in Canada and North America (Boiteau et al. 2008). Boiteau et al. (2008) found excessive rates (300 kg Nha^{-1}) of fertilization with Nutriwave (Envirem Organics Inc., Fredericton, New Brunswick), a 4-1-2 commercial pelletized organic fertilizer derived from poultry manure, promoted more rapid CPB larval development and earlier peaks of abundance of beetle larvae. The authors concluded, however, that as the influence of fertilizer on overall potato beetle populations was limited, fertility management would only have a secondary role in control of CPB. In Maine, Alyokhin et al. (2005) found CPB densities were higher on plots receiving full rates of synthetic fertilizers compared to those receiving reduced fertilizer rates combined with manure inputs. In addition to various agronomic strategies varying in effectiveness, this pest is commonly controlled in recent years in organic potato production in Canada by targeted use of foliar applications of Entrust (spinosad 80% (Dow Agrosciences, Indianapolis, Indiana)). Given this product's relatively high cost and broad spectrum activity it is typically used selectively in combination with crop scouting for CPB, and as border or perimeter sprays only (Bostock 2008). Wireworm (*Agriotes spp.*) damage from long term forage leys or pastures prior to potato planting can also reduce plant stands and /or reduce marketable yields as found by Lynch et al. (2008a,b) and Liu et al. (2010).

While the focus of the current study is on nitrogen supply in organic potato production, the importance of adequate P and K supply is often underestimated (Hagman et al. 2009). Srek et al. (2010) in the Czech Republic found that over a 53 year study soil available P (Mehlich III) concentrations below 30 mg kg^{-1} reduced tuber production. Indeed, across all organic production systems in Canada there is increasing evidence (Martin et al. 2007; Roberts et al. 2008; Knight et al. 2010) that soil phosphorus levels in particular may be becoming a critically important limitation to crop growth. Constraints with respect to P release efficacy of mined rock phosphates (Arcand et al. 2010) permitted under organic standards make this issue a further challenge. Livestock-based manure and compost sources, can in some cases, be in limited supply, and more novel organic amendments such as source separated municipal solid waste (MSW) (or 'green waste') composts may be an effective source of soil P (and N) supply to a range of crops including potatoes (Hargreaves et al. 2008; Passoni and Borin, 2009) and are discussed below. In contrast to phosphorus, potassium is much more easily supplied to meet organic potato crops needs through, in addition to manure, commercial inorganic amendments such as sulpomag, potassium sulphate or other soluble commercial sources (Lynch et al. 2008a, b, 2011; Bostock 2008; Canadian General Standards Board 2009b).

Finally, any study of productivity and nutrient dynamics in specific organic farming systems cannot be delinked from consideration of that system's sustainability, or ability to minimize environmental impacts while maintaining an economically viable production level (Hansen et al. 2001). In some jurisdictions, most notably Europe, product premiums, and consumer and government support for organic farming and its products partially reflect support for these perceived environmental benefits of organic farming systems (Lynch 2009). In Canada, as in other countries (Hansen et al. 2001), many of the key principals outlined in Canadian national standards for organic production (Canadian General Standards Board 2009a) and now enshrined in federal regulations, relate to goals associated with environmental benefits from these production systems. Environmental benefits with respect to energy use, soil quality, biodiversity and reduced off-farm nutrient impacts are closely linked to reliance on legume biological nitrogen fixation and organic matter inputs, reduced overall nutrient intensity, and increased spatial and temporal diversity associated with more complex rotations found on organic arable cropping farms (Lynch 2009, Griffiths et al. 2010; Lynch et al. 2011). In turn, these extended rotations within organic systems can enhance soil health and soil biological pools, which may contribute to soil N dynamics in these systems in ways not fully yet appreciated (Nelson et al. 2009).

While research on the topic is advancing, management of N within organic production systems remains challenging. This chapter will evaluate, primarily within the context of Atlantic Canada, how tuber yield, crop N recovery and residual soil nitrate within organic production systems are influenced by crop rotation, timing and type of green manures, and amendment with supplemental N sources (including manures, composts, and processed manure products). The potential to use soil tests to improve N management in organic production systems will also be evaluated. The effects of extended rotations and organic amendments on soil quality and soil health will also be discussed briefly.

12.2 Tuber Yield, Crop Apparent N Recovery and Residual Soil Mineral N as Affected by Rotation Design and Green Manure Type

In Atlantic Canada recommended base N requirements for conventional fertility management (i.e. from all sources) of tablestock varieties popular with organic producers such as 'Goldrush' are in the range of 190 kg N ha^{-1} (NBSCIA 2007). Earlier harvests reduce this requirement and yields by 10% or more. In general, organic potato yields in the range of 25–30 Mg ha^{-1} (or approximately 25% less than conventional) can be expected where N supplied to the crop from soil and GMr alone averages 100–130 kg N ha^{-1} from short or longer organic rotations (Sullivan et al. 2007; Liu et al. 2008; Lynch et al. 2008a, b; Maggio et al. 2008) and when pests such as wireworm (Lynch et al. 2008a, b) and late blight (Möller et al 2007) and other factors are not limiting or yield reducing. In Germany, Finckh et al. (2006) reported finding a supply of 110–130 kg N ha^{-1} supported a potential yield, under organic management, in the order of 35 Mg ha^{-1} while in Sweden, Hagman et al. (2009) reported yields of 35 Mg ha^{-1} obtained for some varieties as being high for Swedish organic production. On individual farms potential yields are affected by many factors, including environmental conditions or pest problems limiting GMr productivity (Schmidt et al. 1999) and decomposition, which may prevent farms from achieving these yield goals. Case studies of commercial organic potato farms in Ontario cited in Bostock (2008) indicated a wide range in farmer reported average yields where potatoes follow forage (typically alfalfa) of between 15 and 35 Mg ha^{-1}.

Choice of rotation design and GMr frequency and type are critical considerations for successful organic potato production. This is especially true for 'stockless' organic farming systems that are common (Bostock 2008; Schmidt et al. 1999). Total N content of a legume GMr precrop can be as high as 240 kg N ha^{-1} and GMr contribute to soil inorganic N and soil labile N pools. It must be noted also that benefits due to legume breaks in a rotation are partially due also to non-N effects such as reduction in pest incidence (Schmidt et al. 1999; Stark and Porter 2005). Schmidt et al. (1999) describes three 4-year organic potato rotations in Europe where red clover comprised 25% of the rotation phases. On commercial organic potato farms in Atlantic Canada rotations of 4–5 years duration including 2 years leguminous forage or GMr are not uncommon (Lynch et al. 2008a, b; Nelson et al. 2009). This is in contrast to much more frequent cropping of potatoes under conventional cropping practices (Angers et al. 1999). Few studies have examined mineralization of N from remaining GMr residues 2 years after incorporation, i.e. following the potato crop. While this will vary with environmental conditions and quality of GMr residue, Schmidt et al. (1999) reported utilization of GMr N by the crop following potato was small and in keeping with reported ranges of 3–15%. The effect of GMr length on potato productivity has received little attention also. In Sweden, Bath et al. 2006 found a 2-year grass-clover GMr (biomass of 107 kg N ha^{-1}), but not a 1-year GMr (biomass of only 36–44 kg N ha^{-1}) as the pre-crop increased organic potato yields. There is a need for more research to examine the net benefits to soil labile N pools of GMr of varying type and duration (1–2 years), within lengthened rotations where potatoes (and associated frequent soil disturbance) occurs less frequently, as found on organic farms (Stark and Porter 2005) . We report below in section 12.4 selected interim results from one such ongoing study at the Nova Scotia Agricultural College (NSAC) in Truro, Nova Scotia.

Practical handbooks and guides available to organic producers in Canada differ in expected N supplied from perennial forage or GMr incorporation prior to potatoes. An Ontario provincial guideline (Baute et al. 2002) suggests that plowdown of forage grass/legume stands containing 50% or more of legume content supplies 100 kg N/ha to the succeeding potato crop. In 6 years (2005–2011) of research trials on organic potato fertility in Atlantic Canada, we have typically obtained potato whole crop N recovery at topkilling (i.e. PNU_{0N}) of over 100 kg N ha^{-1} following fall plowing of a red clover (*Trifolium pratense*) GMr crop. Exceptions were during unseasonably dry conditions. In a study to examine productivity and N dynamics under extended rotation characteristic of production practices on commercial organic potato farms in Atlantic Canada, conducted in Prince Edward Island (PEI) and at an experimental site at NSAC, relatively high tuber yields (~30 Mg ha^{-1}) were achieved without supplemental N application, in 2 of 3 years when seasonal moisture levels were non-limiting. PNU_{0N} in un-amended soils, but following a red clover GMr, averaged 112 kg N ha^{-1} (Lynch et al. 2008a). In contrast, Sharifi et al. (2008), reporting on data from a 13 year University of Maine research trial in Presque Isle, Maine, found for PNU_{0N} a 1-year mixed alfalfa/timothy crop (within a 4 year potato-soybean-barley-alfalfa/timothy sequence) supplied only 99 kg N ha^{-1}. Background soil fertility undoubtedly also influences the degree of response to a leguminous GMr. Liu et al. (2008, 2010) conducted a four organic rotation study in Truro, Nova Scotia, where various farming systems (stockless, ruminant and monogastric) were reflected in choice of crop sequence (including 0, 1 or 2 years red clover forage) and amendment. While soil N supply in the potato year (year 4) increased by 15–22% with increasing forage frequency, limited PNU_{0N} differences (average 103 kg N ha^{-1}) among rotation treatments was attributed to the high soil fertility from a previous long term pasture at the experimental site. In PEI, Sanderson and MacLeod (1994) report a N replacement value (determined from tubers alone) of 53 kg N ha^{-1} following a late fall incorporated lupin (*Lupinus albus*) whole plant GMr and no other N source applied. That study was not managed organically, however.

Green manure options for organic producers are not limited to perennial leguminous forage crops (Munoz et al. 2005) and variability in PNU_{0N} response to a legume precrop may also be attributed to type and productivity of a GMr and timing (discussed below) of it's incorporation. In Italy, Campiglia et al. (2009) grew a range of overwintering cover crops before organic potatoes and produced remarkably high marketable yields of 48.5 Mg ha^{-1}matching that from N-P-K fertilizer following GMrs of subclover (*Trifolium subterraneum L.*) or hairy vetch (*Vicia villosa Roth.*), which produced 169 and 147 kg N ha^{-1} in aboveground biomass alone. Vetch as GMr is popular as an annual GMr for organic vegetable production in the eastern seaboard of the US and has been part of an organic systems trial comparing GMrs at NSAC over the past 5 years, and discussed below in section 12.4. Provincial guides to cover crops for organic production in Eastern Canada (OMAFA 2011a, b) and general organic potato production guides (Bostock 2008) suggest, in addition to perennials red clover or alfalfa (*Medicago sativa* L.), GMr such as hairy vetch, sweetclover (*Melilotus* spp.) and field peas (*Pisum sativum* L.).

Timing of N release from GMrs may not always synchronize with crop demand. In Maine USA, Porter et al. (1999) found that while a leguminous green manure (oats-pea-vetch mixture), compared to oats, increased soil nitrate levels throughout the season and late season haulm growth, this was not reflected in a yield benefit, as treatments in that study also received N fertilizer (at 134–202 kg N ha^{-1}). In Michigan, USA, Griffin and Hesterman (1991) reported on use of GMrs of alfalfa, clover, sweetclover, hairy vetch or birdsfoot trefoil in short (2-year) rotations with potatoes conventionally grown and fertilized with N at 0–225 kg N ha^{-1}. PNU_{0N} was increased by 75–145% by these GMrs which contained as much as 165–230 kg biomass N ha^{-1}. The lack of total and marketable yields response, however, was attributed to a poor synchrony of GMr-N release and potato N demand. In PEI, Sanderson and MacLeod (1994) attributed their finding that N from lupin GMr contributed more to late season N uptake , but not yield, and was partly due to their use of an indeterminate variety, Russet Burbank, compared to the determinate varieties ('Shepody' and 'Atlantic') used by Griffin and Hesterman (1991). As noted by Möller and Reents (2007) selection of an appropriate potato variety is a key agronomic strategy to take advantage of N status and avoid late blight under organic production.

In eastern Canada an increasing acreage of agricultural soils over the past 20 yr has been classified as at high risk of being a source of nitrate losses to water (de Jong et al. 2007). In humid regions of Atlantic Canada, most leaching losses of nitrates from agricultural soil occur between growing seasons and in major conventional potato growing areas such as PEI, losses of nitrates to groundwater is currently a major concern (Lynch 2009). Residual soil mineral N (RSMN) after harvest of conventionally managed potatoes of up to 110 kg NO_3-N ha^{-1} are not unknown globally (VanDelden et al. 2003) but regionally have averaged closer to 60 kg NO_3-N ha^{-1} under conventional management. In organic systems, optimizing management of GMr to synchronize with crop demand and prevent excessive N release and loss from the systems is a continuing challenge for organic systems, particularly in humid regions (Pimentel et al. 2005; Lynch 2009). In studies conducted on commercial organic potato farms in PEI and New Brunswick (Lynch et al. 2008a,b), much lower RSMN (<25 kg NO_3-N ha^{-1}) in the soil following organic potato harvest was found than for more intensive conventional potato systems in the region (Zebarth et al. 2003; de Jong et al. 2008). While some of these organic farms are mixed livestock and cropping operations, it should be noted that manure is often in limited supply.

Hansen et al. (2001) in Europe attributed the reduced nitrate leaching from organic farming systems in general, (in the range of 8 to 34 kg N ha^{-1} yr^{-1}), including arable cropping systems, to be partially due to the greater use of catch crops in fall and winter under organic management. In the short, cool, humid growing season of Atlantic Canada, there is a limited post-harvest window after a potato crop in which a catch crop can rapidly produce the extensive root system required to reduce RSMN and prevent leaching.

Finally, GMrs and extended rotations used in organic production can contribute not only to maintenance of soil organic matter and turnover of labile pools of crop residue, but also promote soil health and associated biological activity in soil. In a study conducted on four commercial potato farms in Atlantic Canada over 2 years, Nelson et al. (2009) found microbial quotient (microbial biomass as a fraction of total organic carbon (TOC)) and earthworm abundance and biomass increased continuously during the 5 year forage legume green manure-grain-potato rotation at

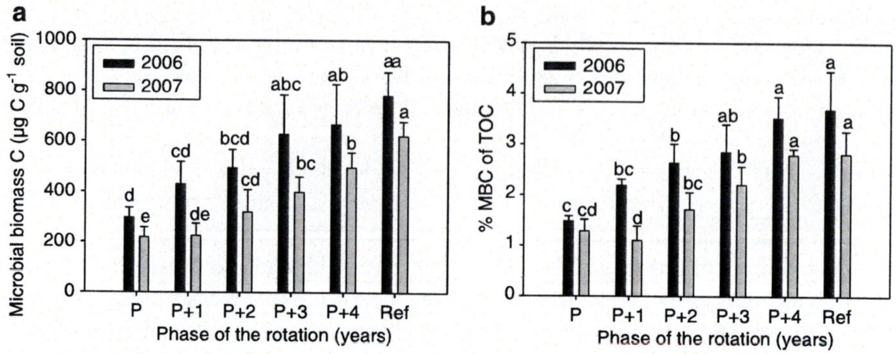

Fig. 12.1 Microbial biomass C amount (**a**) and microbial quotient (**b**) for each phase of 5-year organic potato rotations and adjacent undisturbed reference fields for years 2006 and 2007. Values are means of n=4 sites. Error bars represent standard error of the means and treatments sharing the same letter are not significantly different according to Fisher's LSD at $p<0.05$. P, P+1, P+2, P+3, P+4 and ref refer to potato phase, 1 year after potato, 2 years after potato, 3 years after potato, 4 years after potato and reference field. Reprinted from Agriculture, Ecosystems and Environment, 131, Nelson, KL, Lynch, DH and Boiteau, G, Assessment of changes in soil health throughout organic potato rotation sequences, 220–228, 2009, with permission from Elsevier

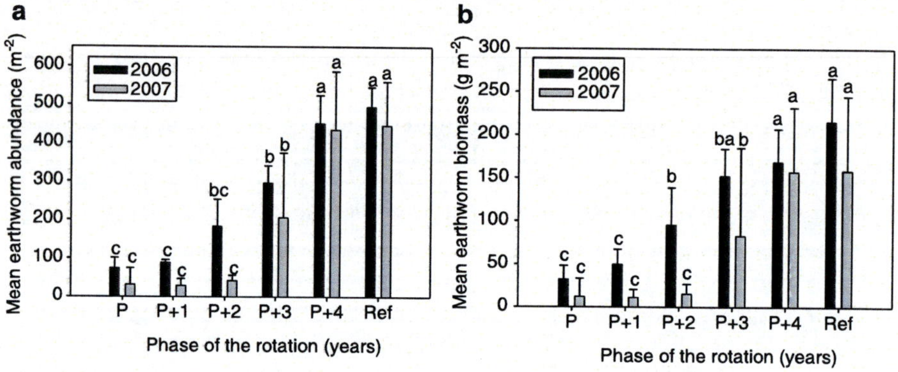

Fig. 12.2 Earthworm abundance (**a**) and biomass (**b**) for each phase of 5-year organic potato rotations and adjacent undisturbed reference fields for 2006 and 2007. Values are means of n=4 sites. Error bars represent standard error of the means and treatments sharing the same letter are not significantly different according to Fisher's LSD at $p<0.05$. P, P+1, P+2, P+3, P+4 and ref refer to potato phase, 1 year after potato, 2 years after potato, 3 years after potato, 4 years after potato and reference field. Reprinted from Agriculture, Ecosystems and Environment, 131, Nelson, KL, Lynch, DH and Boiteau, G, Assessment of changes in soil health throughout organic potato rotation sequences, 220–228, 2009, with permission from Elsevier

each site (Figs. 12.1a,b and 12.2a,b). Microbial quotient and earthworm abundance recovered to levels (3–4% of TOC and 400–500 m^2, respectively) found in adjacent undisturbed (pasture) fields and was greatest immediately prior to potato planting at all sites. How this enhanced biological activity and turnover specifically contributes to N (and P) dynamics in these rotations remains largely unexplored.

12.3 Tuber Yield, Crop Apparent N Recovery and Soil Mineral N as Affected by Timing of Green Manure Incorporation

With some exceptions, fall incorporation of legumes is standard practice in conventional and organic potato rotations in Atlantic Canada, in order to prepare an adequate seed bed for subsequent potato planting (Sanderson et al. 1999). Soil conditions on some organic farms, however, or type of GMr, can allow for spring rather than fall tillage. In Maine, Porter et al. (1999) found a GMr of oats-peas-hairy vetch could be left to overwinter before tillage prior to potatoes. As part of a larger study on N dynamics and GHG emissions in organic potato production at NSAC (Lynch et al. 2008b), we compared the effect of timing of tillage (spring vs fall) of a timothy or red clover (2 year stand) GMr, and potato N fertility regime (with or without added ammonium nitrate fertilizer N at 90 and 140 kg N ha^{-1} following red clover and timothy respectively, banded in the potato hill at planting) on yield and PNU_{0N}. Weed and pest management followed required organic certification protocols. Clover GMr dry matter production averaged 4.4 Mg ha^{-1}, while timothy forage DM yields increased with timothy N fertilization (120 kg N ha^{-1}) from 3.8 to 6.1 Mg ha^{-1}. Tuber yield and average size, and PNU_{0N} was greatest when forages were spring rather than fall plowed, and when supplemented with N fertilizer (Table 12.1). Total yields following clover consistently averaged in the 30–35 T ha^{-1} range and benefited less from added N supplementation. In unfertilized potato plots, PNU_{0N} increased from an average of 77.7 kg N ha^{-1} following timothy plowed in fall or spring to an average of 113.1 kg N ha^{-1} following clover and was highest (145.4 kg N ha^{-1}) for spring incorporated clover + N fertilizer. For both GMr, spring plowing increased PNU_{0N} by 20–30 kg N ha^{-1} suggesting overwinter losses of nitrate N occurred following fall plowing. However, Sanderson et al. (1999) in PEI, found spring incorporation of a 2 year red clover stand (with no additional N fertilization but otherwise under conventional management) only produced greater potato yields (47 Mg ha^{-1} compared

Table 12.1 Potato tuber dry matter yield plus tuber and whole plant N uptake as affected by type of green manure precrop, timing (fall vs spring) of green manure tillage, and supplemental fertilizer N applied to potato. Values are means (n = 4). Adapted from Lynch et al. 2008b

Potato fertilizer rate (kg N ha^{-1})	Previous crop	Forage tillage date	Tuber dry matter yield (Mg ha^{-1})	Tuber N uptake (kg N ha^{-1})	Whole plant N uptake (kg N ha^{-1})
0	Timothy	Spring	5.67[1]	69.2[1]	94.3[1]
140	Timothy	Spring	6.10	80.0	112.9
0	Timothy	Fall	4.70	47.0	61.1
140	Timothy	Fall	6.04	70.8	93.6
0	Clover	Spring	7.19	89.9	127.5
90	Clover	Spring	6.91	101.9	145.4
0	Clover	Fall	6.69	72.7	98.7
90	Clover	Fall	7.13	92.0	127.3

to 42–43 Mg ha^{-1}) than fall incorporation in one out of 3 years. Where fall incorporation of GMr is a necessity, delaying incorporation can reduce overwinter N losses. Sanderson and MacLeod (1994) in PEI, found late fall (October 1st) compared to early fall (September 1st) incorporation of a lupin GMr benefited subsequent potato tuber yield (35.1 vs 32.4 Mg ha^{-1}). This was due to greater N (190–265 kg N ha^{-1}) in lupin GMr biomass in late compared to early fall (160–250 kg N ha^{-1}) and that, as evidenced by enhanced fall and lower spring soil nitrate levels, earlier incorporation in the fall provided more opportunity for mineralization of GMr residue and subsequent overwinter losses. Sanderson et al. (1999) similarly found late fall (mid-October) compared to early fall (mid-September) tillage of red clover increased subsequent potato yields and reduced nitrate N in soil in late November following GMr incorporation. Petiole sampling was used as the index of potato crop N status and petiole nitrate-N generally increased as tillage of red clover was delayed.

12.4 Tuber Yield and Quality, Crop Apparent N Recovery and Soil Mineral N as Affected by Rotation Design and Use of Supplemental Organic Amendments

Organic potato production relies on well designed, extended rotations, ideally including 25% or more of leguminous GMr typically combined with some additions of manures and other organic amendments (Finckh et al. 2006; Lynch et al. 2008a). Möller et al. (2007) in Germany characterized, from among 220 organic potato fields, broad N fertility groupings of 'low' (precrop of cereals with manure applied at a maximum of 40 kg N ha^{-1}), 'intermediate' (precrop of pea or cereals with manure 40–100 kg N ha^{-1}) and 'high' (precrop of grass-clover or cereals and manure at or above 100 kg N ha^{-1}). Yield potential ranges from 20–25 Mg ha^{-1} ('low') to 30–40 Mg ha^{-1} ('high') compared with a range of 50–60 Mg ha^{-1} for conventional production. While laboratory analyses can provide some indication of the average potential nutrient contributions from manures and composts, the availability and timing of nutrient release from these materials in the field is highly variable and remains hard to predict (Stark and Porter 2005; Lynch et al. 2004; Zebarth et al. 2009). Routine use of manures for N supply in organic potato production also runs the risk of oversupply of P if manure is relied on as the primary N source (Lynch et al. 2004; Stark and Porter 2005). In an effort to improve potato yields and reduce P accumulation in organic farming systems, VanDelden et al. (2003) modeled N uptake, tuber yield and RSMN over 30 years as affected by (i) N:P ratio of the manure (pig (4.0) vs cattle (6.2)), (ii) time of manure (slurry) application, (iii) historical N use, and (iv) cultivar maturity. Manure application rates varied from 0 to 490 kg N ha^{-1}. The model indicated that when slurry is spring applied at the maximum rate (128 kg slurry N ha^{-1}) to avoid soil P accumulation (based on estimated crop export of 21 kg P ha^{-1}) potato yields would average 77% of that obtainable at the maximum slurry application rate (PNU_{0N} for unamended controls were estimated at 95 kg N ha^{-1}). As choosing a manure with a lower N:P reduced allowable application rates by a further one third, it was concluded leguminous crops would be needed in the rotation to avoid N limitations. While regional validation would be required, such studies provide useful tools for balancing N and P inputs in organic potato production systems using manure resources.

In Maine, USA, Porter et al. (1999) compared the effect of two GMrs (oats or oats-pea-vetch (OPV)) precrops, with or without amendment (beef or dairy manure vs. potato waste compost), or irrigation, on soil properties and yield and quality of 'Superior' potatoes. No interaction of amendments with GMr occurred but amendments increased yields by 4.0 to 8.6 Mg ha^{-1} across all rotations, although supplemental N fertilizer (at 134 to 202 kg N ha^{-1}) was also applied to all treatments. Similarly to GMr in rotation, organic amendments, through their effect on soil structure and organic matter, water holding capacity and root development have also been found to benefit potato yields and yield stability though 'non-nutrient' benefits as well as through direct nutrient effects. These benefits can persist for many years after a single amendment application (Opena and Porter 1999; Carter 2007; Mallory and Porter 2007). In contrast to GMr, organic amendments can also produce a substantial residual nutrient effect with substantial continued N mineralization after harvest (Mallory et al. 2010) and beyond the succeeding potato year, and these effects may mask rotation effects.

There is an increasing availability of commercially produced organic amendments approved for use in organic production, such as composts and dehydrated pelletized manures, which are changing options for producers and possibly the intensity of N use in organic potato production systems (Lynch et al. 2008a). Under certain strict quality control criteria with respect to potential contaminants, composts and byproducts from municipalities and industry may also be permitted within Canadian organic systems (Canadian General Standards Board 2009b). In Nova Scotia all municipalities are required to produce compost from food and yard waste source separated and collected from residences. The efficient use of all such sources of N in organic potato production is required to minimize negative impacts on water and air quality and increase the economic return on the crop.

Supplemental organic N sources with a substantial proportion of mineral or readily mineralizable N applied at planting or shortly after emergence can improve organic potato yields (VanDelden et al. 2003). Given current premiums for certified organic potatoes in Canada, improving yields through application of amendments supplying moderate rates of N or organic matter appears warranted. However, excessive N fertilization with organic amendments can result in excessive N supply, haulm growth and RSMN levels, delay tuber initiation and crop maturation, and reduces tuber yield and quality (Van Delden 2001; Neuhoff and Kopke 2002; Finckh et al. 2006; Möller et al. 2007; Lynch et al. 2008a). Delayed crop maturation also favours increased risk of infection with late blight (Finckh et al. 2006). Lynch et al. (2008a) examined, at two sites in Atlantic Canada (a commercial farm in Winslow, PEI, and a research site at NSAC in Nova Scotia), the impacts of contrasting organic amendments (compost and dehydrated poultry manure) on Shepody potato yield, quality and soil mineral nitrogen dynamics under organic management. Application of a commercial pelletized poultry manure product (NW; analyses 4-1-2; C:N ~9:1) at 300 kg total N ha^{-1} (broadcast applied at planting to supply an estimated 112 kg plant available N (PAN) ha^{-1}) promoted gains in yields (+ 5.8 Mg ha^{-1} average above ~30 Mg ha^{-1} for plots relying on GMr-N alone) and marketable yields (+ 7.0 Mg ha^{-1} average) of Shepody, but RSMN levels rose to 61 kg N ha^{-1} from 25 kg NO_3-N ha^{-1} for unamended and compost amended plots. At very higher rates of NW application (600 kg total N ha^{-1}), no yield response was obtained and excessive haulm growth and delayed senescence was clearly visible. While tuber number differed little between NW treatments average tuber size for the highest NW rate was much smaller than the unamended control as a result of delayed plant maturity and tuber fill (Möller et al. 2007). Tuber N uptake as a proportion of total plant at harvest also dropped from an average of 62% for the unamended treatments to 49%, and RSMN values rose to 141 kg NO_3-N ha^{-1} for the highest rate of NW. In Atlantic Canada, most leaching losses of NO_3-N occur between planting seasons, and such excessive levels of RSMN must be avoided. The moderate rates of manure N applied here falls within the range (0 to 350 kg manure N ha^{-1}) reported from a survey of 115 farms across seven European countries (Finckh et al. 2006). Similar data on on-farm manure availability on organic farms is not currently available in Eastern Canada.

Compost (hog manure and sawdust; C:N ~18:1) resulted in higher total yields than unamended treatments in one of three site-years. Apparent mineralized N and crop recovery of N from compost was negligible and yield benefits were attributed to factors other than N availability. The in-season dynamics of soil (0-30 cm) mineral N (NH_4 and NO_3), averaged across three site years and as affected by amendments, are shown in Fig. 12.3 (Sharifi et al. 2009). In the unamended control, apparent soil N mineralization over the growing season averaged 141 kg N ha^{-1} while NW treatments added an extra 115 and 195 kg N ha^{-1}. Much of the readily mineralizable N from NW was released within the first 10 days of application, with nitrification of the majority of this mineral N present as NH_4 by 31 days after planting. As discussed below, improved use of plant bioassays and soil tests to generate site specific information for organic producers on the N supplying ability of their rotations sequences and GMr management strategies will help refine the targeted use of supplementary organic N sources.

As noted above for the compost treatment, organic amendments, particularly when applied at moderate supplementary rates, do not always contribute to RSMN. Liu et al. (2010), in Nova Scotia, found N supply from two manure based composts (from beef or poultry) in the field was much lower than could be predicted from controlled environment mineralization studies and suggests this limits the effectiveness of composts as a means of addressing short term N deficiencies in potato based rotations. Lynch et al. (2004) suggests using the legume component of the rotation as a 'N buffer' to accommodate, through biological nitrogen fixation, this variability and low N supplying ability of composts when building a labile N pool for a succeeding crop such as potatoes of high N demand. Hargreaves et al. (2008) provides a review on use of MSW composts in agriculture. To provide a high input of inorganic N, studies have shown application rates have to be very high, i.e. 40–50 Mg compost ha^{-1}. In contrast significant increases in plant available P have been more consistently achieved with these composts. Fahmy et al. (2010) in New Brunswick

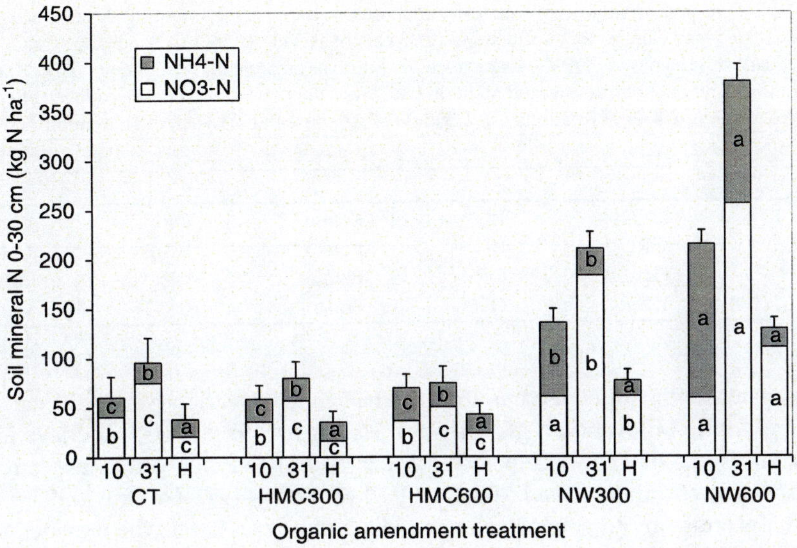

Fig. 12.3 Soil NH_4-N and NO_3-N at 10 days after planting (DAP), 31 DAP and harvest (H) as affected by organic amendment source (non-amended control (CT), hog manure-sawdust compost (HMC) and pelletized dehydrated poultry manure (NW)) and application rate (300 and 600 kg N ha^{-1}). The means comparison was done separately for each sampling date and for NO_3-N and NH_4-N. Values with the same letter are not significantly different at 0.05 probability level using Fisher's protected least significant difference test. Bars represent standard error for soil mineral N. Reprinted with permission from American Journal of Potato Research, 86, Sharifi et al., Evaluation of nitrogen supply rate measured by in situ placement of Plant Root Simulator™ probes as a predictor of nitrogen supply from soil and organic amendments in potato crop, 356–366, 2009

reported that composted pulp fibre residues was a good source of plant available P and K to potatoes.

In an organic cropping systems study commenced in 2006 at NSAC in Truro, the impact of GMr type and frequency in rotation, and with or without organic amendment or N fertilizer supplement, on potato yield and marketable yield, soil quality, soil N dynamics, greenhouse gas emissions and overwinter N losses are being evaluated. Interim (first 3 years until the first potato phase) data on crop and soil mineral N dynamics are presented here. The soil at the site is a Pugwash sandy loam classified as Orthic Humo-Ferric Podzols in Canadian soil classification (Webb et al. 1991). The experiment consists of a split-plot arrangement of treatments in a randomized complete block design with three replications. Main plots (i.e. rotation sequences) differed in green manure type (red clover vs. oats/pea/vetch mixture) and frequency (Table 12.2). Data is presented here for three sequences; (C1) ORC-RC (oats (*Avena sativa* L.) underseeded with red clover-red clover) (C3) carrots-OPV (carrots (*Daucus carota* subsp. sativus)- oats/pea /hairy vetch mixture); and (C4) BBu-OPV (beans (*Phaseolus lunatus* L.) followed by buckwheat (*Polygonum convolvulus* L.)- Oats/pea/vetch mixture). The experiment is partially phased such that three phases of each 5 year rotation sequence are present in each year.

Table 12.2 Five-year organic crop rotation sequences in the organic cropping systems study at NSAC, Truro, Nova Scotia. RC, solid stand (2nd year) of red clover as green manure; ORC, oats underseeded with red clover; OPV, oats/pea/vetch mixture as an annual green manure; BBu, yellow beans followed by buckwheat as green manure. Data from the potato crop phase produced after the first 3 years (2006–2008) of rotations C1, C3 and C4 are presented. (Lynch et al. unpublished)

	Year				
Rotation	1	2	3	4	5
C1	ORC	RC	Potato	ORC	Carrots
C2	ORC	RC	Potato	BBu	Carrots
C3	Carrots	OPV	Potato	ORC	BBu
C4	BBu	OPV	Potato	ORC	Carrots

Seeding rates for the rotation crops are 70 kg ha^{-1} for oats (variety(var.) AC Francis), 12 kg ha^{-1} for Red Clover (var. AC Christie), 3.20 kg ha^{-1} for carrots (var. Maverick), 60, 60, and 34 kg ha^{-1} for oats, pea (var. Mozart), and common vetch, respectively in oats/pea/vetch mixture, and 108.4 kg ha^{-1} for beans (var. Goldrush). The RC plots that were going to potatoes were moldboard plowed late in the previous fall. The OPV GMr was incorporated by mowing and discing late in the fall. Beans residues were disced and plowed in late July to early August, and buckwheat was seeded and allowed to grow until mid-October then incorporated.

Fertility treatments (applied in potato years only) comprise the subplots and include (i) a control (unamended), (ii) fertilized with mineral N and P at recommended rates based on soil test for P and regional N credits for leguminous green manures (FERT), (iii) source separated MSW compost (12 Mg ha^{-1} wet weight; dry weight = 60%) and (iv) composted paper mill biosolids (PMB) (30 Mg ha^{-1} wet weight; dry weight = 39%). Compost application rates were designed to provide the potato crop with a current-season average of 60 kg ha^{-1} of plant available P_2O_5 based on compost analyses and assuming 50% of total P is plant available in the year of application. The MSW compost is of higher quality, i.e. higher N content (2.1% vs 1.6%) and lower C:N (7.0 vs 14.3), than that of the PMB compost (Lynch et al. unpublished data). Composts thus also provided 82 kg total N ha^{-1} (MSW) and 121 kg total N ha^{-1} (PMB), respectively. Potato N and P needs in FERT subplots were met with 80 kg N ha^{-1} as ammonium nitrate and 60 kg P_2O_5 ha^{-1} as triple super phosphate fertilizer. The N rate was adjusted for a regional estimated N credit of 50 kg N ha^{-1} from legumes (Zebarth and Rosen 2007). Composts were broadcasted on subplots mid-May and hills and furrows were formed approximately 4 weeks after planting using a tool carrier. Nitrogen and P fertilizers were banded 5 cm below and 5 cm to the side of the seed pieces at planting in fertilized subplots. Potassium (K) was broadcasted each year on all experiment plots as Sulpomag (0-0-22) as required by soil test. Potatoes (cv. Goldrush) were planted by hand on June 6–9 using hand-cut seed pieces (≈50 ± 3 g each) in rows 83 cm apart, at 31 cm within-row spacing and at a depth of 5–10 cm. Weeds were controlled by cultivation and hand. Late blight and Colorado potato beetle were controlled with the application of copper hydroxide (Parasol) and Entrust®, respectively, as required. Potatoes were harvested on 25th September 2008. No supplemental irrigation was applied.

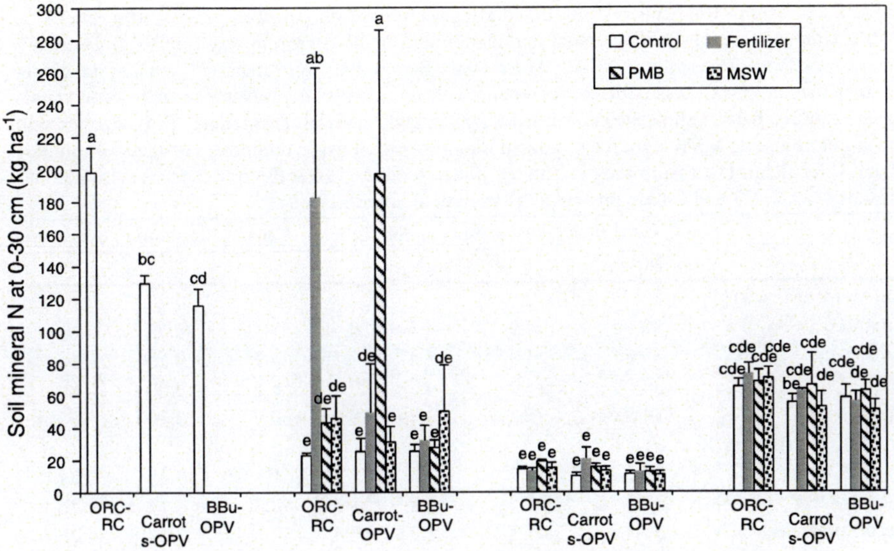

Fig. 12.4 Seasonal changes in soil mineral N as influenced by rotation sequences and fertility treatments (composts or inorganic N fertilizer) in 2008 (potato phase) of the organic cropping systems study at NSAC. Soil mineral N was measured at pre-plant, tuber initiation stage, tuber bulking stage and post-harvest in 2008. Error bars represent standard error of the means and treatments sharing the same letter are not significantly different according to Fisher's LSD at 0.05 probability level. MSW, source separated municipal food waste compost; PMB, paper mill biosolids compost; BBu, yellow beans followed by buckwheat as green manure; ORC, oats underseeded with red clover; OPV, oats/pea/vetch mixture as an annual green manure; RC, solid stand of red clover as green manure. (Lynch et al. unpublished)

Soil sampling was carried out prior to planting up to 45 cm (0–15 cm, 15–30 cm and 30–45 cm), at tuber initiation stage (~35 days after planting (DAP)) up to 15 cm, at tuber bulking stage (~80 DAP) up to 30 cm (0–15 cm and 15–30 cm) and at post-harvest up to 60 cm (0–15 cm, 15–30 cm and 30–60 cm).

Soil mineral N (SMN) dynamics are presented in Fig. 12.4. At planting, SMN for rotation C1 (198 kg NO_3-N ha^{-1}) was 62% ($P<0.05$) greater than C3 and C4 rotations, but SMN did not significantly differ between C3 and C4 (130 and 116 kg NO_3-N ha^{-1}, respectively). Ammonium N comprised less than 1 kg N ha^{-1}. Spring soil N supply rate measured *in-situ* by plant root simulator probes (PRS™) (Western Ag. Innovations Inc., Sakatoon, SK) ion exchange membranes was also 58% greater in C1 compared with C3 (data not shown). Rotation did not influence soil total organic C and N but a substantial proportion of total organic N (41%) was in the form of particulate (>53 μm) organic N of high quality (C:N of 6.0 to 8.3) which is highly sensitive to mineralization. The high SMN values at planting support the importance of GMrs as a source of N for organic potato crops in this region (Lynch 2009). At the tuber initiation stage, higher SMN was measured under FERT/C1 and under PMB/C4 in comparison with other treatments (Fig. 12.4). Overall, greater than a nine-fold decrease in SMN was measured for all amendments and fertility

Table 12.3 Potato total tuber yield, dry matter accumulation, specific gravity, and tuber size distribution as affected by amendment application and rotation crop during the first 3 years of a organic potato rotation study at NSAC, Truro, Nova Scotia. RC, solid stand (2nd year) of red clover as green manure; ORC, oats underseeded with red clover; OPV, oats/pea/vetch mixture as an annual green manure; BBu, yellow beans followed by buckwheat as green manure. PMB = paper mill biolsolids compost; MSW = source separated municipal food waste compost; Fertilizer = inorganic N and P fertilizer. Data from the potato crop phase produced after the first 3 years (2006–2008) of rotations C1, C3 and C4 are presented. (Lynch et al. unpublished)

		Tuber yield	Dry matter	Culls	Small	Canada #1	Jumbo	Wireworm
		Mg ha^{-1}				%		
Amendment ($n=9$)								
Control		32.3b	24.9	4.8	24.1a	57.4	0.23	26.6
Fertilizer		39.8a	24.4	3.4	17.5b	64.3	0.66	20.6
PMB		33.2b	24.7	4.7	23.1a	56.0	0.00	28.8
MSW		34.9ab	25.2	3.7	19.8ab	60.9	0.16	26.3
Proceeding crop ($n=12$)								
Oats/RCl-RCl		35.6	24.7	3.8	21.7	61.8	0.45	50.3a
Carrots-OPV		34.3	24.4	4.8	21.3	58.8	0.27	9.7b
Bean/Buckwheat-OPV		35.3	25.2	3.8	20.5	58.3	0.07	16.7b
Source of variation	df							
Block	2	NS	NS	NS	NS	***	NS	***
Proceeding crop (PC)	2	NS	NS	NS	NS	NS	NS	***
Amendment (A)	3	*	NS	NS	*	NS	NS	NS
A×PC	6	NS	NS	NS	NS	NS	NS	NS
EMS	22	NS	NS	NS	NS	NS	NS	NS
CV %	–	15	3	38	20	17	321	38

treatments from planting to tuber bulking. More than 80% of the decrease in SMN from planting to harvest (average 133 kg N ha^{-1}) can be attributed to plant N uptake (average 109 kg N ha^{-1}). Rotation affected PNU_{0N} and was greatest for C1 (124.4 kg N ha^{-1}) compared to C3 and C4 (106.2 and 99.7 kg N ha^{-1}, respectively). These values are within the range of PNU, of 100–120 kg N ha^{-1} obtained for fields of 'high' N input status among 220 commercial organic fields in Germany by Möller et al (2007). PNU increased to 149 kg N ha^{-1} under the FERT treatment (data not shown).

Yields did not differ by rotation (range 34.3 – 35.6 Mg ha^{-1}) but were influenced by amendment treatments (Table 12.3). Yields for the unamended control and PMB treatment were similar (average 32.5 Mg ha^{-1}) but were significantly lower than those obtained for MSW (34.9 Mg ha^{-1}) and FERT (39.8 Mg ha^{-1}) treatments. Tuber size distribution also benefited from these latter treatments, with % small sized tubers decreasing from an average of 23.5% (Cont and PMB) to 19.8% (MSW) and 17.5% (FERT). Nitrogen use efficiency decreased, however, from approximately 30 kg N uptake required per 10 Mg of yield for C3 and C4, to 35 kg N per 10 Mg for C1 and 37 kg N per 10 Mg for the FERT treatment (data not shown), in general agreement with Möller et al (2007). Much greater incidence of wireworm damage for rotation C1(~50% of tubers) compared to C3 and C4 (average 11.7%) was the major factor influencing loss of marketable yield. These results suggest a GMr of hairy vetch can be an effective substitute for red clover for organic potato production in the Eastern Canada region, especially when combined with supplemental organic N sources. Once elaborated, effective and productive rotations such as these can be combined with further agronomic strategies to take full advantage of fertility status. Möller and Reents (2007) found, particularly under conditions of high N availability for organic systems, selection of varieties with mid-early tuber initiation and tuber pre-sprouting further increased yields by 18-23%.

The measured RSMN values to 30 cm depth post harvest of ~60 kg NO_3-N ha^{-1} (Fig. 12.4), while above those found on commercial organic potato farms in the region noted above, were similar to the level (61 kg N ha^{-1}) reported above following 300 kg N ha^{-1} pelletized poultry manure application to organic potatoes (Lynch et al. 2008a, b). The RSMN values are in the lower end of the range reported by Zebarth et al. (2003) for 228 commercial conventionally managed potato fields in New Brunswick Canada (3 to 250 kg N ha^{-1}) and by Cambouris et al. (2008) in 2 sites for 3 years in Quebec, Canada (52–114 kg N ha^{-1}).

In Italy, Canali et al. (2010) assessed the contribution of contrasting organic amendments (farmyard manure or municipal 'green waste' compost) in combination with a subterranean clover cover crop on organic potato (var. Monna Lisa) N nutrition. Amendments were applied at 0, 50 and 100 kg total N ha^{-1} to cultivated or uncultivated GMr main plots. Incorporation of GMr or farmyard manure alone increased tuber yields by 22–25%. When GMr and amendments were combined yields increased by 43%, but RSMN did not increase significantly as found in our cropping systems study discussed above. In Sweden, Bath et al. (2006) found the combination of a clover-grass GMr precrop and fermented manure slurry (containing between 37 and 94 kg NH_4-N ha^{-1}) increased organic potato N uptake by approximately 50% and yields by approximately 40%, on a site with poorer soil but not on a site with more fertile soil. In late fall (November), RSMN (0–0.9 m depth) was lower at the poor soil site (25–60 kg N ha^{-1}) than under more fertile soil conditions (50–90 kg N ha^{-1}). The results of these studies suggest effective combinations of GMrs and supplementary organic amendments can be developed, adapted to regional conditions and production systems, which enhance organic potato productivity without compromising on maintaining a low environmental footprint of these systems.

12.5 Use of Plant Bioassays and Soil Tests to Improve N Management in Organic Potato Production

Unlike phosphorus and potassium, there is currently no standard soil test for N to use in making N recommendations in humid environments such as Atlantic Canada. In this region, the loss of soil residual mineral N over late fall and winter results in N supply being dominated by in-season organic N mineralization (Sharifi et al. 2007; Zebarth et al. 2009). Advances in, and potential of, soil and plant tests to improve N use in potato production systems are reviewed in this text and elsewhere by Zebarth et al. (2009), Sharifi et al. (2007) and Sharifi et al. (2008). Many of these approaches, particularly with respect to a reliable lab test for potentially mineralizable soil N, would be important tools to help refine organic production systems. A few additional approaches of note are summarized below.

Sullivan et al. (2007), using a whole crop bioassay approach sampled just before vine kill from a zero supplemental N plot (PNU_{0N}) reports on plant-available N on six organic farms in Oregon, in the northeastern USA. This approach has the advantage of being conducted under field conditions where temperature, moisture, and aeration are representative of that experienced by the crop. It also incorporates the root interactions of the specific crop species, and is therefore expected to provide a better estimate of the crop specific soil N supply (Zebarth et al. 2005). Soil N supply also can be estimated as the sum of PNU_{0N} plus RSMN after crop harvest. Most organic producers in Atlantic Canada have not attempted to gauge the N supply from soil and GMr residues as affected by their management system, and this simple approach could be more widely promoted as suggested by Sullivan et al. (2007). The relative recovery of N in tubers compared to haulms can also be used to gauge N oversupply.

The *in situ* use of ion-exchange resins and membranes has been explored as a means to assess plant available N and as an alternative to a static measurement of SMN for timothy (Ziadi et al. 1999) and canola (Qian and Schoenau 2005). When buried *in situ*, nutrient adsorption to the ion-exchange membrane is influenced by the same edaphic factors that affect nutrient uptake by plant roots and the membrane thus measures the actual N flux over time in soils. The measured N flux can be influenced by duration of burial in the soil, soil temperature, soil moisture, competing sinks (microbial and roots), and crop N uptake pattern (Ziadi et al. 1999). Johnson et al. (2005) reported that PRS-N results are more sensitive to soil moisture, but less sensitive to temperature, compared with SMN. As part of the study described above and in Lynch et al. (2008a), Sharifi et al. (2009) examined the relationship between spring PRS-N flux and PNU_{0N} or PNU_{0N} and residual soil mineral N left at harvest (SMN_h) (Fig. 12.5a,b). The cumulative PRS-N flux for a burial period of 31 days after planting (DAP), soil mineral N at 10 days DAP and soil NO_3-N at 31 DAP successfully ($r=0.71$-0.81) predicted N supply from soil and organic amendments to the potato crop. The PRS-N data was less variable (i.e. lower CV) compared with in-season soil mineral N data, but use of the sequential multi-burial approach can make it more costly and labour intensive. In our rotation trial described in section 4 above, we are continuing to test the effectiveness of *in situ* PRS as a measure of N availability for organic potato production systems in Atlantic Canada.

12.6 Conclusion

On farm and station based trials in Atlantic Canada, suggest effective combinations of GMrs and supplementary organic amendments can be developed, adapted to regional conditions and production systems, which enhance organic potato pro-

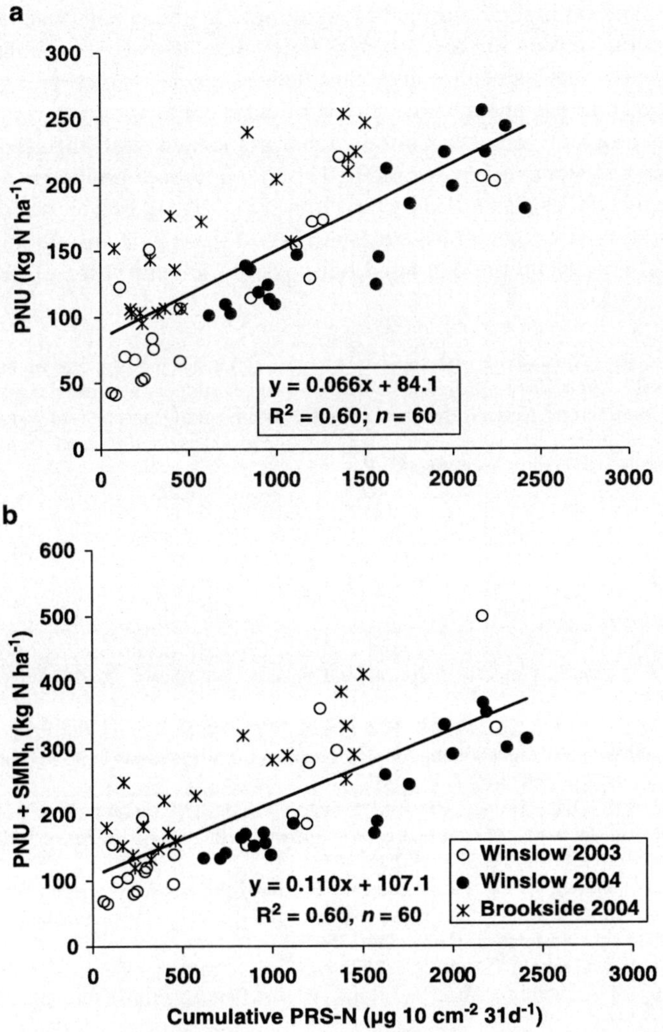

Fig. 12.5 Relationship between N supply rate measured by in situ burial of Plant Root SimulatorTM probes (PRS-N) for a cumulative burial period of 31 days after planting and N supply from soil and organic amendments as estimated by total plant uptake taken at potato vine topkilling (mechanical removal) (PNU) (**a**) or PNU plus soil (0-30 cm depth) mineral N at harvest (PNU + SMN_h) (**b**). All regression coefficients are significant at 0.001 probability level. Reprinted with permission from American Journal of Potato Research, 86, Sharifi et al., Evaluation of nitrogen supply rate measured by in situ placement of Plant Root Simulator™ probes as a predictor of nitrogen supply from soil and organic amendments in potato crop, 356–366, 2009

ductivity without compromising on maintaining a low environmental footprint of these systems. There is a need for more research to examine the net benefits to soil labile N pools of GMr of varying type and duration (1–2 years), within lengthened rotations where potatoes (and associated frequent soil disturbance) occur less frequently, as found on organic farms. Practical tools to gauge soil N supply within these production systems are also needed. Innovative approaches to management of green manures that reduce reliance on tillage, such as strip cropping systems, would also be an important advance. In the broader context a greater understanding of the ecological versus economic risks associated with different organic potato cropping system designs would be beneficial to both producers and policy makers. Finally, and as also noted by Griffiths et al. (2010), greater understanding of the potential environmental tradeoffs at play in these systems associated with improved soil quality on thc onc hand but possibly greater N losses to leaching and GHG is needed.

Acknowledgements The authors wish to acknowledge support for the principal author provided by the Canada Research Chairs program. Individual project support was provided through the Nova Scotia Department of Agriculture, and Prince Edward Island Department of Agriculture and Forestry. The keen interest and assistance of many organic producers and research technicians, too numerous to list, is also greatly acknowledged.

References

Alyokhin A, Porter G, Groden E, Drummond F (2005) Colorado potato beetle response to soil amendments: a case in support of the mineral balance hypothesis? Agric Ecosyst Environ 109:234–244

Angers DA, Edwards LM, Sanderson JB, Bissonnette N (1999) Soil organic matter quality and aggregate stability under eight potato cropping sequences in a fine sandy loam of Prince Edward Island. Can J Soil Sci 79:411–417

Arcand MM, Lynch DH, Voroney RP, van Straaten P (2010) Residues from a buckwheat (*Fagopyrum esculentum*) green manure crop grown with phosphate rocks influence soil phosphorus dynamics. Can J Soil Sci 90:257–266

Bath B, Ekbladh G, Ascard J, Olsson K, Andersson B (2006) Yield and nitrogen uptake in organic potato production with green-manures as pre-crop and the effect of supplementary fertilization with fermented slurry. Biol Agric Hortic 24:135–148

Baute T, Hayes A, McDonald I, Reid K (2002) Agronomy guide for field crops. Publication 811. Ministry of Agriculture, Food and Rural Affairs, Queens Printer for Ontario, Toronto, ON, Canada

Belanger G, Walsh JR, Richards JE, Milburn PH, Ziadi N (2001) Critical nitrogen curve and nitrogen nutrition index for potato in eastern Canada. Am J Potato Res 78:355–364

Boiteau G, Lynch DH, Martin RC (2008) Influence of fertilization on the Colorado potato beetle, Leptinotarsa decemlineata (Say), in organic potato production. Econ Entomol 37:575–585

Bostock M (2008) Growing potatoes organically from market garden to field crop: a COG practical skills handbook. Canadian Organic Growers, Ottawa

Cambouris AN, Zebarth BJ, Nolin MC, Laverdiere A (2008) Apparent fertilizer nitrogen recovery and residual soil nitrate under continuous potato cropping: effect of N fertilization rate and timing. Can J Soil Sci 88:813–825

Campiglia E, Paolini R, Colla G, Mancinelli R (2009) The effects of cover cropping on yield and weed control of potato in a transitional system. Field Crops Res 112:16–23

Canadian General Standards Board (2009a) Organic production systems: general principles and management standards. Canadian General Standards Board, Gatineau, Canada

Canadian General Standards Board (2009b) Organic production systems: permitted substances lists. Canadian General Standards Board, Gatineau, Canada

Canali S, Ciaccia C, Antichi D, Barberi P, Montemurro F, Tittarelli F (2010) Interactions between green manure and amendment type and rate: effects on organic potato and soil mineral N dynamic. J Food Agric Environ 8:537–543

Carter MR (2007) Long-term influence of compost on available water capacity of a fine sandy loam in a potato rotation. Can J Soil Sci 87:535–539

de Jong R, Yang JY, Drury CF, Huffman EC, Kirkwood V, Yang XM (2007) The indicator of risk of water contamination by nitrate-nitrogen. Can J Soil Sci 87:179–188

de Jong R, Qian B, Yang JY (2008) Modelling nitrogen leaching in Prince Edward Island under climate change scenarios. Can J Soil Sci 88:61–78

De Neve S, Dietjens I, Moreels E, Hofman G (2003) Measured and simulated nitrate leaching on an organic and conventional mixed farm. Biol Agric Hortic 21:217–229

Fahmy SH, Sharifi M, Hann SWR, Chow TL (2010) Crop productivity and nutrient availability in a potato-based three-year rotation as affected by composted pulp fibre residue application and supplemental irrigation. Comm Soil Sci Plan 41:744–756

Finckh MR, Schulte-Geldermann E, Bruns C (2006) Challenges to organic potato farming: disease and nutrient management. Potato Res 49:27–42

Griffin TS, Hesterman OB (1991) Potato response to legume and fertilizer N sources. Agron J 83:1004–1012

Griffiths BS, Ball BC, Daniell TJ, Hallett PD, Neilson R, Wheatley RE, Osler G, Bohanec M (2010) Integrating soil quality changes to arable agricultural systems following organic matter addition, or adoption of a ley-arable rotation. Appl Soil Ecol 46:43–53

Hagman JE, Mårtensson A, Grandin U (2009) Cultivation practices and potato cultivars suitable for organic potato production. Potato Res 52:319–330

Hansen B, Alrøe HF, Kristensen ES (2001) Approaches to assess the environmental impact of organic farming with particular regard to Denmark. Agric Ecosyst Environ 83:11–26

Hargreaves JC, Adl MS, Warman PR (2008) A review of the use of composted municipal solid waste in agriculture. Agric Ecosyst Environ 123:1–14

Johnson DW, Verburg PSJ, Arnone JA (2005) Soil extraction, ion exchange resin, and ion exchange membrane measures of soil mineral nitrogen during incubation of a tallgrass prairie soil. Soil Sci Soc Am J 69:260–265

Knight JD, Buhler R, Leeson JY, Shirtliffe SJ (2010) Classification and fertility status of organically managed fields across Saskatchewan, Canada. Can J Soil Sci 90:667–678

Liu K, Hammermeister AM, Patriquin DG, Martin RC (2008) Assessing organic potato cropping systems at the end of the first cycle of four-year rotations using principal component analysis. Can J Soil Sci 88:543–552

Liu K, Hammermeister AM, Entz MH, Astatkie T, Warman PR, Martin RC (2010) Nitrogen availability in an organic potato crop following 3-year transition under contrasting farming systems. J Sustain Agr 34:821–835

Lynch DH (2009) Environmental impacts of organic agriculture: a Canadian perspective. Can J Plant Sci 89:621–628

Lynch DH, Voroney RP, Warman PR (2004) Nitrogen availability from composts for humid region perennial grass and legume-grass forage production. J Environ Qual 33:1509–1520

Lynch DH, Zheng Z, Zebarth BJ, Martin RC (2008a) Organic amendment effects on tuber yield, plant N uptake and soil mineral N under organic potato production. Renew Agr Food Syst 23:250–259

Lynch DH, Clegg E, Owen J, Burton D (2008b) Greenhouse gas emissions from organic crop management in humid region eastern Canada. Proceedings of 'Organic Agriculture and Climate Change', April 17–18th, Enita Clermont, Lempdes, France

Lynch DH, MacRae RJ, Martin RC (2011) The carbon and global warming potential impacts of organic farming: does it have a significant role in an energy constrained world? Sustainability 3:322–362

Maggio A, Carillo P, Bulmetti GS, Fuggi A, Barberi G, De Pascale S (2008) Potato yield and metabolic profiling under conventional and organic farming. Eur J Agron 28:343–350

Mallory EB, Porter GA (2007) Potato yield stability under contrasting soil management strategies. Agron J 99:501–510

Mallory EB, Griffin TS, Porter GA (2010) Seasonal nitrogen availability from current and past applications of manure. Nutr Cycl Agroecosys 88:351–360

Martin RC, Lynch DH, Frick B, VanStraaten P (2007) Phosphorous status on Canadian organic farms. J Sci Food Agr 87:2737–2740

Mattsson B, Wallen E (2003) Environmental life cycle assessment (LCA) of organic potatoes. Acta Hortic 619:427–435

Möller K, Reents HJ (2007) Impact of agronomic strategies (seed tuber pre-sprouting, cultivar choice) to control late blight (*Phytophthora infestans*) on tuber growth and yield in organic potato (*Solanum tuberosum* L.) crops. Potato Res 50:15–29

Möller K, Habermeyer J, Zinkernagel V, Reents HJ (2007) Impact and interactions of nitrogen and Phytophthora infestans as yield-limiting factors and yield reducing factors in organic potato (*Solanum tuberosum* L.) crops. Potato Res 49:281–301

Munoz F, Mylavarapu RS, Hutchinson CM (2005) Environmentally responsible potato production systems: a review. J Plant Nutr 28:1287–1309

NBSCIA (2007) Nitrogen management for potato: general fertilizer recommendations. A factsheet prepared by the New Brunswick Soil and Crop Association, pp 4

Nelson KL, Lynch DH, Boiteau G (2009) Assessment of changes in soil health throughout organic potato rotation sequences. Agric Ecosyst Environ 131:220–228

Neuhoff D, Kopke U (2002) Potato production in organic farming: effects of increased manure application and different cultivars on tuber yield and quality. Pflanzenbauwissenschaften 6:49–56

OMAFA (2011a) 2011 Field crop budgets. Publication 60. Ontario Ministry of Food, Agriculture and Rural Affairs. http://www.omafra.gov.on.ca/english/busdev/facts/pub60.htm#2010. Accessed 22 Jan 2011

OMAFA (2011b) Cover crops. Ontario Ministry of Food, Agriculture and Rural Affairs. http://www.omafra.gov.on.ca/english/crops/facts/cover_crops01/covercrops.htm. Accessed 22 Jan 2011

Opena GB, Porter GA (1999) Soil management and supplemental irrigation effects on potato: II. Root growth. Agron J 91:426–431

Passoni M, Borin M (2009) Effects of different composts on soil nitrogen balanceand dynamics in a biennial crop succession. Comp Sci Util 17:108–116

(PEIDAF) Prince Edward Island Department of Agriculture and Forestry (2000) Agricultural business profile on organic potatoes. Factsheet 161/00. PEIDAF, Charlottetown, Prince Edward Island, Canada

Pimentel D (1993) Economics and energetics of organic and conventional farming. J Agr Environ Ethic 6:53–60

Pimentel D, Hepperly P, Hanson J, Douds D, Seidel R (2005) Environmental, energetic, and economic comparisons of organic and conventional farming systems. Bioscience 55:573–582

Porter GA, Sisson JA (1991) Response of Russet Burbank and Shepody potatoes to nitrogen fertilizer in two cropping systems. Am Potato J 68:425–443

Porter GA, Opena GB, Bradbury WB, McBurnie JC, Sisson JA (1999) Soil management and supplemental irrigation effects on potato: I. Soil properties, Tuber Yield and quality. Agron J 91:416–425

Qian P, Schoenau JJ (2005) Use of ion-exchange membrane to assess nitrogen-supply power of soils. J Plant Nutr 28:2193–2200

Roberts CJ, Lynch DH, Voroney RP, Martin RC, Juurlink SD (2008) Nutrient budgets of Ontario organic dairy farms. Can J Soil Sci 88:107–114

Sanderson JB, MacLeod JA (1994) Soil nitrate profile and response of potatoes to fertilizer N in relation to time of incorporation of lupin (Lupinus albus). Can J Soil Sci 74:241–246

Sanderson JB, MacLeod JA, Kimpinski J (1999) Glyphosphate application and timing of tillage of red clover affects potato response to N, soil N profile, and root and soil nematodes. Can J Soil Sci 79:65–72

Schmidt H, Philipps L, Welsh JP, Fragstein PV (1999) Legume breaks in stockless organic farming rotations: nitrogen accumulation and influence on the following crops. Biol Agric Hortic 17: 159–170

Sharifi M, Zebarth BJ, Burton DL, Grant CA, Porter GA, Cooper JM, Leclerc Y, Moreau G, Arsenault WJ (2007) Evaluation of laboratory-based indices of nitrogen availability as predictors of field-based estimates of soil nitrogen supply in potato production. Plant Soil 301:203–214

Sharifi M, Zebarth BJ, Burton DL, Grant CA, Porter GA (2008) Organic amendment history and crop rotation effects on soil nitrogen mineralization potential and soil nitrogen supply in a potato cropping system. Agron J 100:1562–1572

Sharifi M, Lynch DH, Zebarth BJ, Zheng Z, Martin RC (2009) Evaluation of nitrogen supply rate measured by in situ placement of Plant Root Simulator probes as a predictor of nitrogen supply from soil and organic amendments in potato crop. Am J Potato Res 86:356–366

Speiser B, Tamm L, Amsler T, Lambion J, Bertrand C, Hermansen A, Ruissen MA, Haaland P, Zarb J, Santos J, Shotton P, Wilcockson S, Juntharathep P, Ghorbani R, Leifert C (2006) Improvement of late blight management in organic potato production systems in Europe: field tests with more resistant potato varieties and copper based fungicides. Biol Agric Hortic 23:393–412

Srek P, Hejcman M, Kunzova E (2010) Multivariate analysis of relationship between potato (*Solanum tuberosum* L.) yield, amount of applied elements, their concentrations in tubers and uptake in a long-term fertilizer experiment. Field Crops Res 118:183–193

Stark JC, Porter GA (2005) Potato nutrient management in sustainable cropping systems. Am J Potato Res 82:329–338

Sturz AV, Lynch DH, Martin RC, Driscoll AM (2006) Influence of compost tea, powdered kelp and Manzate 7 75 on the community composition and antibiosis against Phytophthora infestans of bacteria in the potato phylloplane. Can J Plant Pathol 28:52–62

Sullivan DM, McQueen JPG, Horneck DA (2007) Estimating Nitrogen mineralization in organic potato production. Factsheet produced by Oregon State University Extension Service, 8 pp

Van Delden A (2001) Yield and growth components of potato and wheat under organic nitrogen management. Agron J 93:1370–1385

VanDelden A, Schroder JJ, Kropff MJ, Grashoff C, Booij R (2003) Simulated potato yield, and crop and soil nitrogen dynamics under different organic nitrogen management strategies in The Netherlands. Agric Ecosyst Environ 96:77–95

Varis E, Pietils L, Koikkalainen K (1996) Comparison of conventional, integrated and organic potato production in field experiments in Finland. Acta Agr Scand 46:41–48

Webb KT, Thompson RL, Beke GJ, Nowland JL (1991) Soils of Colchester County, Nova Scotia. Report No. 19. Nova Scotia Soil Survey. Research Branch, Agriculture Canada, Ottawa, Ontario, Canada

Willer H (2008) Organic agriculture worldwide: current statistics. In: Wliier H, Yussefi M, Sorensen N (eds) The world of organic agriculture: statistics and emerging trends 2008. International Federation of Organic Agriculture Movements, Bonn, Germany

Zebarth BJ, Rosen CJ (2007) Research perspective on nitrogen BMP development for potato. Am J Potato Res 84:3–18

Zebarth BJ, Leclerc Y, Moreau G, Gareau R, Milburn PH (2003) Soil inorganic nitrogen content in commercial potato fields in New Brunswick. Can J Soil Sci 83:425–429

Zebarth BJ, Leclerc Y, Moreau G, Sanderson JB, Arsenault WJ, Botha E, Wang-Pruski G (2005) Estimation of soil nitrogen supply in potato fields using a plant bioassay approach. Can J Soil Sci 85:377–386

Zebarth BJ, Drury CF, Tremblay N, Cambouris AN (2009) Opportunities for improved fertilizer nitrogen management in production of arable crops in eastern Canada: a review. Can J Soil Sci 89:113–132

Ziadi N, Simard RR, Allard G, Lafond J (1999) Field evaluation of anion exchange membranes as a N soil testing method for grasslands. Can J Soil Sci 79:281–294

Chapter 13
Nitrate Leaching from Potato Production in Eastern Canada

Yefang Jiang, Bernie J. Zebarth, George H. Somers, John A. MacLeod, and Martine M. Savard

Abstract Nitrate leaching from potato production systems is of economic importance, and also of concern for both drinking water quality and the health of aquatic ecosystems. We examine the processes, timing and magnitude of nitrate leaching, and examine practices developed to reduce nitrate leaching from potato production systems, with a particular focus on Prince Edward Island. Results from tile-drain experiments indicate that nitrate leaching occurs primarily during late autumn winter and early spring when crop uptake diminishes, and elevated nitrate concentrations coexist with water movement from the root zone, with the timing of nitrate leaching generally corresponding with major recharge events. Based on stable isotopic signatures, nitrate leached during the growing season is primarily from mineral fertilizers and mineralization of soil organic N whereas nitrate leached outside of the growing season

Y. Jiang (✉)
Agri-Environment Services Branch, Agriculture and Agri-Food Canada,
440 University Avenue, Charlottetown, PEI, Canada C1A 4N6
e-mail: Yefang.jiang@agr.gc.ca

B.J. Zebarth
Potato Research Centre, Agriculture and Agri-Food Canada, 850 Lincoln Rd.,
Fredericton, NB, Canada E3B 4Z7
e-mail: bernie.zebarth@agr.gc.ca

G.H. Somers
PEI Dept. Environment, Energy and Forestry , 11 Kent St.,
Charlottetown, PE, Canada C1A 7N8
e-mail: ghsomers@gov.pe.ca

J.A. MacLeod
Crops and Livestock Research Centre, Agriculture and Agri-Food Canada,
440 University Avenue, Charlottetown, PEI, Canada C1A 4N6
e-mail: johnmacleod.pei@gmail.com

M.M. Savard
Geological Survey of Canada, Natural Resources Canada, 490 rue de la Couronne,
Quebec, QC, Canada G1K 9A9
e-mail: msavard@nrcan.gc.ca

Z. He et al. (eds.), *Sustainable Potato Production: Global Case Studies*,
DOI 10.1007/978-94-007-4104-1_13,

originates primarily from mineralized N (including microbially-modified fertilizer N). Nitrate concentrations in tile-drain water from potato plots were commonly above the 10 mg NO_3-N L^{-1} drinking water quality guideline, and were higher from potatoes than from red clover or cereal plots. Simulations using LEACHN predicted that >80% of nitrate leaching occurred outside of the growing season, a finding consistent with water balance calculations in intensively cultivated watersheds. Nitrate leaching under conventional potato-barley-red clover rotations were predicted to be 27–91 kg N ha^{-1} (average 56 kg N ha^{-1}), corresponding to concentrations of nitrate in drainage of 5.6–18 mg N L^{-1} (average 10.7 mg N L^{-1}). Nitrate leaching was predicted to increase with fertilizer N rate for potato, and decreased with rotation length. Nitrate leaching can be reduced through implementation of practices which reduce the accumulation of nitrate in the root zone, particularly at potato harvest. These practices may include improved crediting of soil N mineralization from soil organic matter and crop residues, use of in-season measures of crop N status to guide in-season N management, and introducing potato cultivars with lower fertilizer N requirements or higher N use efficiency. The requirement for high N inputs to obtain economic tuber yields, in combination with the high risk of nitrate loss from the root zone, present significant challenges to growers with respect to N management. Reducing the frequency of potato in the crop rotation, and use of practices to reduce residual nitrate in potato production, will be most effective in reducing nitrate leaching losses.

13.1 Introduction

Nitrogen (N) losses to groundwater from agricultural production systems are a common environmental issue world-wide (Power and Schepers 1989; Spalding and Exner 1993; Böhlke 2002). In agricultural settings, nitrate leached below the crop root zone can impair the quality of underlying groundwater (Nolan et al. 2002; McMahon et al. 2007; Puckett et al. 2011). In addition, the discharge of nitrate-enriched groundwater as base flow delivers high loads of N to surface waters, which impair aquatic ecosystems (Mitsch et al. 1999; Keith and Zhang 2004).

High concentrations of nitrate are of concern for drinking water quality, and drinking water guidelines for nitrate are commonly set at 10 mg NO_3-N L^{-1} (United States Environmental Protection Agency 2009; Health Canada 2010). High nitrate concentrations are also of concern in fresh surface waters and in estuarine environments and may contribute to eutrophication and deterioration of aquatic habitat (Pionke and Urban 1985; Bachman et al. 1998). In Canada, the guideline for the protection of freshwater aquatic habitat is 2.9 mg NO_3-N L^{-1} (CCME 2003). Thus, N losses from agricultural production systems are of concern both for drinking water quality and the health of aquatic ecosystems.

Nitrogen is the nutrient required in the largest quantities for crop growth and almost all non-leguminous crops need N inputs as mineral or organic fertilizer or biological N fixation for optimal production. Agricultural systems are, however, inherently "leaky", and a certain amount of N within the system is subject to losses. Given the important role of N in crop production, these N losses can also have important economic as well as environmental implications. Nitrate leaching is a major pathway for N loss in humid regions and in irrigated agricultural systems (Jemison and Fox 1994; Baker 2001).

Nitrate leaching losses from potato (*Solanum tuberosum* L.) production are of particular concern, and there is evidence of increased groundwater nitrate contamination associated with potato production (Hill 1986; Richards et al. 1990; Benson et al. 2006). The potato crop usually receives high fertilizer N inputs in order to meet industry tuber yield and size requirements (Zebarth and Rosen 2007), whereas apparent recovery of applied fertilizer N in the potato crop commonly ranges from 40–60% in Eastern Canada (Zebarth and Rosen 2007) and Western Europe (Vos 2009). In drier regions, where not all nitrate lost from the root zone over the winter period, the risk of nitrate leaching losses may be reduced by growing potatoes in rotation with deeper-rooted crops, such as barley (Delgado et al. 1999, 2001b; Dabney et al. 2001). Estimates of nitrate leaching losses from commercial potato fields in Eastern Canada ranged from 10–171 kg N ha^{-1} (Milburn et al. 1990; Gasser et al. 2002). Nitrate concentrations in leachate from potato fields commonly exceed the 10 mg NO_3-N L^{-1} drinking water guideline for nitrate (Milburn et al. 1990; Vos and van der Putten 2004).

This chapter examines nitrate leaching from rain-fed potato production in Eastern Canada, and considers the effects of climatic conditions, soil properties and management practices in influencing nitrate leaching. The chapter has a primary focus on potato production in Prince Edward Island (PEI), and uses specific examples of work done in PEI to illustrate these controls on the leaching process.

13.2 Groundwater Nitrate Contamination in PEI

PEI is the smallest province in Canada, yet it produces about one fourth of the Canadian potato crop (Statistics 2009). Agricultural land accounts for 40% of the island's land mass, about half of which is in potato rotations. Potatoes are commonly grown in a 3-year rotation with barley (*Hordeum vulgare* L.) and forage crops (red clover (*Trifolium pratense* L.) or mixture of red clover and perennial grasses such as timothy (*Phleum pratense* L.)). Russet Burbank is the main cultivar grown, and the main market for the potato crop is for processing (French fry) purposes. Barley is grown primarily for livestock feed. Traditionally the forage crops were grown in a system where the forage was harvested mid-season for animal feed and ploughed down at the end of the growing season as a green manure. Inclusion of forage crops, particularly legumes, in potato rotations are important in maintaining the productivity of these low organic matter soils (Stark and Porter 2005).

PEI has a cool maritime climate, humid soil moisture regimes, cool wet spring conditions and short growing seasons. The short growing season limits the potential for use of cover crops to take up residual nitrate after potato harvest, and most nitrate is lost from the root zone over the winter period (Zebarth et al. 2009).

The soils used for potato production were derived from continental glacial till and are sandy and well-drained, however subsoils may be prone to compaction and have low soil pH, and consequently root penetration is frequently limited. The soils are underlain by an unconfined and semi-confined fractured-porous sandstone aquifer, which supplies all of the drinking water, and a large majority of industrial water and base flow to freshwater streams, on the island.

The combination of a humid climate, sandy soils and intensive potato production systems over an unconfined and semi-confined aquifer creates a favourable situation for nitrate leaching in PEI. As a result, groundwater nitrate contamination is prevalent. Nitrate concentrations in well water in PEI averaged 3.7 mg NO_3-N L^{-1}, which is 2 to 3 mg NO_3-N L^{-1} higher than background groundwater nitrate concentrations (Jiang and Somers 2009). In addition, 15–20% of domestic wells had nitrate concentrations above the 10 mg NO_3-N L^{-1} drinking water guideline in watersheds with a large proportion of land in potato production (Jiang and Somers 2009).

Nitrate-enriched groundwater discharged to the local streams and associated estuaries has been suggested as one of the factors implicated with the anoxia events prevailing in some estuaries in PEI (Young et al. 2002). Isotopic analyses of nitrate in groundwater and surface waters of the Wilmot River watershed show essentially identical characteristics suggesting base flow is a primary source of the nitrate in surface waters (Savard et al. 2007). Island-wide monitoring data indicates that nitrate concentrations of stream water have increased over time, and in some cases have increased several-fold since the 1960s (Somers 1998). Isotopes of nitrate in both groundwater and surface water in the Wilmot River watershed in PEI suggest that nitrate originates in approximately equal proportions from mineral fertilizers and from soil organic materials during the growing season, whereas outside of the growing season, N was derived primarily from soil organic N (including microbially-modified fertilizer N) (Savard et al. 2010).

13.3 Controls on Nitrate Leaching from Potato Production

Nitrate leaching occurs when moving water and nitrate coexist in the soil (Meisinger and Delgado 2002). The movement of water, and thus nitrate, below the root zone occurs when precipitation rate or irrigation exceeds the evapo-transpiration rate and soil water content exceeds field capacity. The quantity of nitrate in the soil which is available for leaching depends on the complex interaction of several processes in the C (carbon) and N cycles. These processes and transformations that control nitrate availability occur simultaneously with nitrate transport within the soil profile. The net effect of these processes, and thus the potential for nitrate leaching from the root zone to groundwater, depends on climatic conditions, soil properties and management practices.

13.3.1 Soil and Climatic Controls on Water Movement

Soil and climatic conditions play an important role in influencing the amount of drainage. Irrigation is limited in PEI, and consequently precipitation and evapo-transpiration are the primary factors that influence the magnitude and timing of drainage.

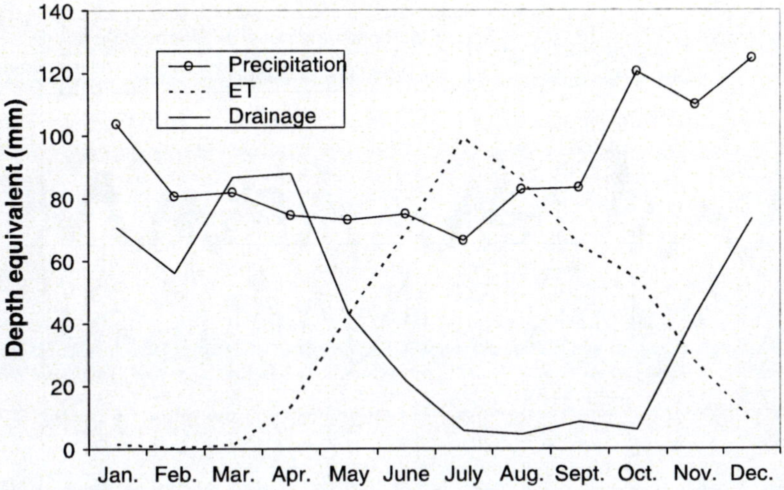

Fig. 13.1 Monthly precipitation, drainage and evapo-transpiration (ET) from a typical potato production system in PEI with a barley-red clover-potato rotation averaged over the period 2000–2008. Drainage and ET values are based on LEACHN predictions using Charlottetown soil series information and climate data from the Environment Canada weather station at the Charlottetown Airport

Timing of drainage is illustrated by a simulation (Jiang et al. 2011) performed using LEACHN (Hutson 2003) for a potato production system for the period 2000–2008 (Fig. 13.1). Average annual precipitation during this period (1075 mm) was very close to the long-term average (1100 mm). Average annual evapo-transpiration and drainage were predicted to be 416 and 506 mm, respectively. Drainage in March and April was predicted to be greater than precipitation due to snowmelt. It was predicted that 82% of the drainage occurred from November to April, emphasizing that most drainage occurs outside of the May to October growing season. Monthly evapo-transpiration was similar to or exceeded precipitation from June to September, and consequently drainage during this time would primarily occur in response to significant rainfall events.

These simulations are consistent with results from tile-drain experiments in PEI and New Brunswick. High tile-drainage effluent discharges were measured during the October to April period and near zero or very limited drainage effluent was measured during the May to September period (Milburn et al. 1990, 1997). The simulations are also consistent with the response of shallow water table and stream discharge to recharge events on PEI. For example, water table elevations at Sleepy Hollow monitoring well near Charlottetown generally increased in spring over the March to May period each year in response to snowmelt recharge events and declined over the June to October period when evapo-transpiration exceeded precipitation and net drainage (i.e. recharge) was limited (Fig. 13.2). The water table elevation subsequently increased over the November to February period because evapo-transpiration diminished and net drainage from snowmelt and/or rainfall infiltration recharged the water table. Similar responses can be observed at the other

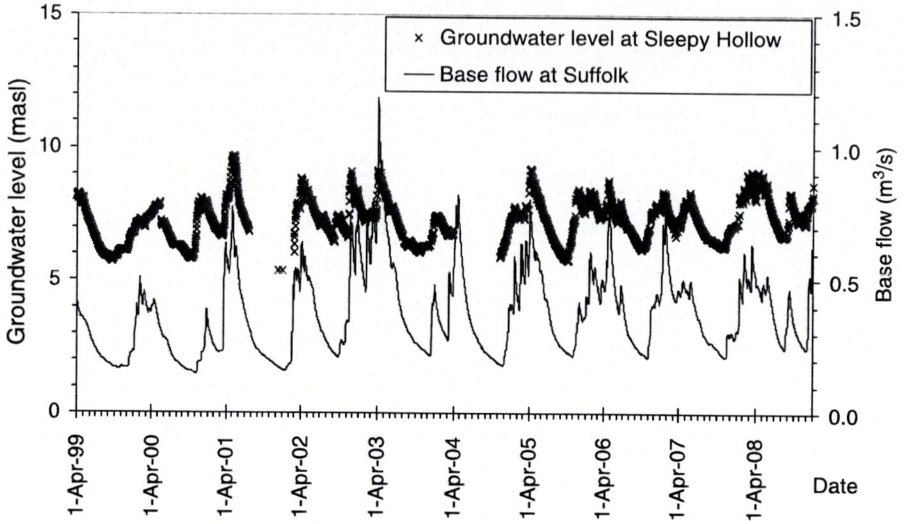

Fig. 13.2 Daily groundwater level elevation (masl-meters above sea level) and base flow (separated from measured stream discharges) from Sleepy Hollow and Suffolk PEI, respectively. Recharge in late autumn winter and spring are reflected in both groundwater level elevations and base flow

shallow monitoring wells on the island (PEIEEF 2011). The calculated base flows at Suffolk (Fig. 13.2), as well at the other hydrometric stations on PEI, demonstrate similar seasonal rising and falling responses.

Nitrate leaching is also influenced by soil properties. The risk of leaching is increased in sandy soils and soils with low organic matter content due to reduced soil water holding capacity. In coarse-textured soils, water infiltration due to rainfall events can rapidly move nitrate below the surface soil, where it is no longer accessible to the growing crop. Shallow unconfined aquifers underlying sandy soils are particularly vulnerable to nitrate contamination (Dubrovsky and Hamilton 2010). Soils in PEI are commonly sandy and have relatively low (15 to 35 $g\,kg^{-1}$) organic matter content, which favours nitrate leaching. Elevated nitrate in the underlying shallow unconfined aquifer was highly spatially correlated with intensity of potato cropping (Benson et al. 2006).

In a 2 year geochemical study in the Wilmot River watershed of PEI, a strong seasonal association was observed between $\delta^{18}O$ ratios (i.e. the ratio of $^{18}O/^{16}O$ in water samples compared with the standard) in precipitation and in nitrate dissolved in shallow groundwater samples. The relationship has as its basis in the approximation that during nitrification, two-thirds of the oxygen incorporated into the nitrate molecule is from soil water, bearing the isotopic characteristics of recent precipitation, and one-third is from the atmosphere with a constant $\delta^{18}O$ ratio (Snider et al. 2010). The seasonally distinct $\delta^{18}O$ ratios in nitrate observed in shallow groundwater therefore provide a geochemical marker for the timing of nitrification and of recharge to the aquifer. Combined with estimated daily recharge rates, these results indicate that winter and spring periods provide high rates of N transfer to the aquifer relative to the growing season (Savard et al. 2007).

It is important to have information on the magnitude and timing of leaching from the root zone, however direct measurement of leaching from the root zone to the underlying groundwater is a challenging task (Powlson 1993). Tile drain effluent has been used by some researchers to estimate leaching (Jemison and Fox 1994), however this may significantly underestimate recharge. For example, annual tile drainage was reported to be 86 to 132 mm yr^{-1} in PEI during 1989–1992 (Milburn et al. 1997) whereas annual recharge during the same time period was estimated to be as high as 310–490 mm yr^{-1} during the same timer period using a local empirical recharge coefficient of 30–40% of annual precipitation (Francis 1989) and annual precipitation of 1056–1232 mm for the same period. This suggests that a large proportion of recharge may bypass the tile-drain systems.

Crude approximations of N fluxes may be made at a watershed scale, and have been used for the Wilmot River watershed to estimate the relative magnitude of N delivered to the aquifer during the growing and non-growing seasons (Somers and Savard 2008). Combining daily recharge rates determined using the approach of Healy and Cook (2002) and seasonal groundwater nitrate concentrations, N flux to the aquifer is estimated to be 25% during the growing season and 75% during the non-growing season.

The potential to estimate leaching through water balance calculations is limited because not all terms within the water balance equation can be readily and accurately characterized (Itier and Brunet 1996). Consequently, because of the many physical, chemical and biological processes affecting leaching, simulation models are commonly used to predict leaching (Addiscott and Whitmore 1991). Many process-based models have been developed to simulate nitrate leaching below the crop root zone as a function of soil properties, climatic conditions and management practices. Examples of these include SOILN (Johnsson et al. 1987), LEACHM (Wagenet and Hutson 1989; Hutson 2003), NLEAP (Shaffer et al. 1991) and HYDRUS (Simunek et al. 2008). Most of the models are one dimensional and subsequently do not capture the heterogeneities (especially in the horizontal extent) in soil, climate, management and topography. These models commonly require site-specific soil and climate data and management-specific data as input, and many of the data were not collected in the field if model calibration was not the major intent. Lack of input data creates a barrier for their practical application to agricultural systems (Shaffer 1995).

13.3.2 Soil, Climatic and Management Controls on Nitrate Availability

Soil nitrate originates primarily from mineral fertilizers, organic amendments, crop residues, biological N fixation and soil organic matter. Soil nitrate present in the soil may be taken up by plants, assimilated by soil microorganisms, or lost to the atmosphere by denitrification. The balance between these processes controls the availability of nitrate in soil which has the potential to be lost by leaching.

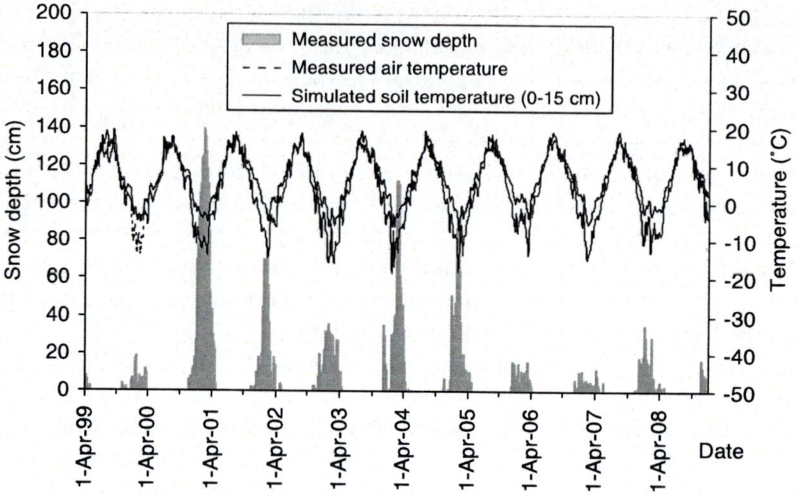

Fig. 13.3 Seasonal variation in soil temperature and drainage simulated using LEACHN and using observed air temperature and snow depth at the Charlottetown Airport, PEI

The mineralization of organic N in soil is controlled primarily by the amount and quality of soil mineralizable N and environmental conditions, primarily soil temperature and water content (Zebarth et al. 2009). Soil mineralizable N is influenced by organic amendment use (Sharifi et al. 2008a), addition of crop residues (Sharifi et al. 2009), tillage (Sharifi et al. 2008b), soil properties and climatic zone (Dessureault-Rompré et al. 2010). The C/N ratio, composition and particle size of crop residues affect the mineralization process (Kumar and Goh 2000).

Soil temperature and water content control rates of microbial growth and activity, and consequently influence the processes of mineralization, nitrification and denitrification (Johnsson et al. 1987; Hutson 2003). Therefore climate, and the annual variations in climatic conditions, influences both the availability of nitrate for leaching and the potential for leaching to occur. Most microbial activity occurs under warmer soil temperatures during the growing season. However, it is common for soils to remain unfrozen over much of the winter period. For example, simulated soil temperatures for 0–15 cm depth could be above zero for much of the late autumn winter and spring despite some sub-zero air temperatures (Fig. 13.3). As a result, soil N and C cycling can occur, albeit at a slower rate, outside of the crop growing season.

An isotopic approach was used to investigate N dynamics at a watershed scale, and to quantify seasonal N fluxes to groundwater in the Wilmot River watershed in PEI. Land use in the watershed is predominantly agricultural (74% of total land area) dominated by potato production. As such it provides a good example of the impact of intensive potato production on water resources at a watershed scale. A source apportionment model was developed using $\delta^{15}N$ and $\delta^{18}O$ ratios in groundwater and for principle N sources (manure and sewage, inorganic fertilizers and

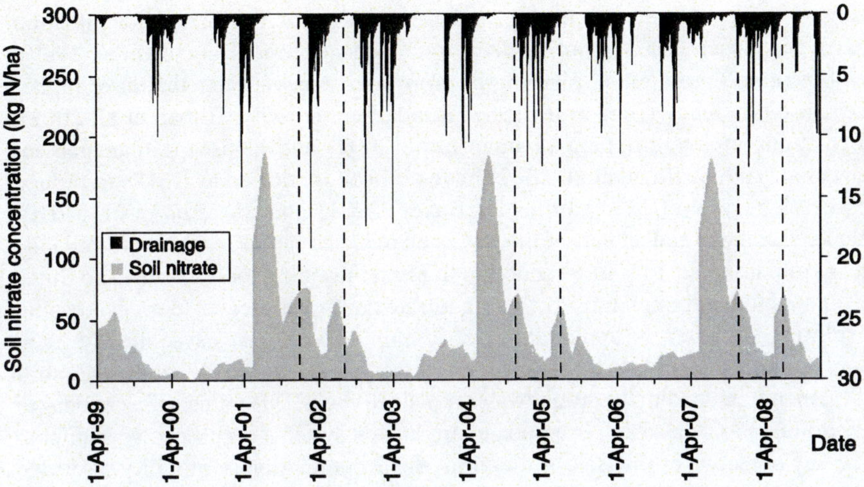

Fig. 13.4 Soil nitrate concentration and drainage under a conventional barley-red clover-potato production system in PEI simulated using LEACHN. A potato crop was present in 2001, 2004 and 2007 (The dash boxes highlight the periods of high nitrate leaching risk where potatoes are harvested, and high soil nitrate concentration and drainage coexist)

soil organic matter) (Savard et al. 2010). It was concluded that nitrate loading to groundwater during the growing season originated about equally from mineral fertilizer N and N mineralized from soil organic matter. In comparison, nitrate was derived primarily from mineralization of soil organic matter (including microbially-modified fertilizer N) outside of the growing season. Nitrate derived from manure or septic systems constituted only a minor part of the total N flux to groundwater in the watershed. The relationship between $\delta^{15}N$ and $\delta^{18}O$ in groundwater samples indicated that denitrification is occurring to a limited extent and does not significantly deplete nitrate within the aquifer.

Similar to the isotopic results, simulations using LEACHN (Jiang et al. 2011) predicted occurrence of high drainage and soil nitrate concentration following potato harvest under a conventional barley-red clover-potato rotation system in the autumn winter-spring period in PEI (Fig. 13.4). Both the isotopic and modeling results highlighted that nitrate leaching from potato production occurred primarily outside of the growing season in PEI.

13.4 Management of Nitrate Leaching in Potato Production

Management strategies that minimize nitrate accumulation during time periods when the risk of drainage is high will reduce the risk of nitrate loss to the groundwater. An array of Beneficial Management Practices (BMPs) has been developed to help minimize nitrate leaching from crop production within different cropping systems. Choice of the rate, timing and formulation of fertilizer N application (Zebarth and Rosen 2007), planting deep-rooted cover crops (Delgado et al. 1999), development and preferential planting of crops and crop varieties that have higher nutrient-use efficiency (Bergstrom 1987; Randall et al. 1997; Tilman et al. 2002), crop rotations or intercropping (Tilman et al. 2002) and landscape management (Chow et al. 1999; Tilman et al. 2002) have all been identified as BMPs to reduce N losses. A review of field studies by Baker (2001) and Shrestha et al. (2010) concluded that the overall effects of the BMPs on reducing nitrate leaching are variable, ranging from no effect to a 30% reduction in nitrate leaching losses. Kraft and Stites (2003) noted that although most orthodox nitrate control strategies (e.g., decreasing and splitting fertilizer N applications, irrigation scheduling) have already been implemented in the irrigated potato production systems in the Wisconsin Central Sand Plain, nitrate loading to groundwater remained high. This reflects the challenges for controlling nitrate leaching from potato production systems over sensitive aquifers.

The key objective of the BMPs is to limit the accumulation or mobility of nitrate in the soil for the period when crop uptake is low or absent and drainage is occurring (Meisinger and Delgado 2002). The challenge is that almost all non-leguminous crops need N inputs as mineral or organic fertilizer, or require biological N fixation within the crop rotation, to achieve economic crop yields. This is particularly true for the potato crop where high fertilizer N rates are commonly required to achieve tuber size requirements (Zebarth and Rosen 2007). Furthermore, growers may apply additional N as “insurance N” to avoid the risk of yield loss (Mitsch et al. 1999). However, it is inevitable that some loss of N occurs from agricultural systems, and the risk of loss increases rapidly as N rate increases above the optimum (van Es et al. 2002).

In Eastern Canada, a number of BMPs have been evaluated to improve the fertilizer N management in the potato crop. These include choice of rate, timing and form of N applied (Chap. 10) and use of soil- and plant-based diagnostic tests to improve at-planting and in-season N management (Chap. 11). These BMPs were evaluated with respect to potato crop response (tuber yield and quality and plant N uptake) and residual soil nitrate, however no estimates of nitrate leaching were obtained.

Tile drainage experiments were performed to evaluate the ability of wheat straw mulch or an autumn seeded cover crop to reduce nitrate leaching following an early harvested potato crop (cultivar Superior) from spring 1989 to spring 1993 (Milburn and MacLeod 1991; Milburn et al. 1997). Application of a wheat straw mulch at a rate of 3000 kg ha^{-1} and lightly incorporated into the soil with a single pass of a disc harrow after early harvested potato reduced annual flow-weighted leached nitrate concentrations by 16% and 35% in the 1989–1990 and 1991–1992 leaching seasons, respectively, compared with conventional fallow practice (Milburn et al. 1997; MacLeod and Sanderson 2002). The reduction in nitrate leaching was attributed to net immobilization of soil nitrate by the high C/N ratio wheat straw. Autumn seeded winter wheat following early harvested potato was shown to reduce annual flow-weighted leached nitrate concentrations by 31% and 13% in the 1989–1990 and 1991–1992 leaching seasons, respectively, compared with conventional fallow practice (Milburn et al. 1997; MacLeod and Sanderson 2002). However, the use of cover crops or straw mulch may not always work efficiently. The growing season in eastern Canada is relatively short and the commonly-grown potato cultivar (Russet Burbank) is harvested late in the year. As a result there may not be sufficient time for N uptake by an autumn seeded cover crop, or for significant immobilization of N by a straw mulch, before the winter season in most years.

Tile drainage experiments and soil nitrate measurements were also performed to examine the effect of the timing of red clover plough-down on nitrate leaching from barley-red clover-potato rotation systems. When red clover was incorporated in early autumn nitrate concentration in tile-drain effluent was higher, and soil nitrate in the subsequent spring was lower, than when incorporation was delayed until late autumn or spring (Sanderson et al. 1999; Sanderson and MacLeod 2002). When the rate of N fertilizer application to the potato crop was not reduced to account for the additional N retained following late incorporation of the red clover crop, the additional N resulted in increased nitrate leaching following the subsequent potato harvest.

Mean flow-weighted nitrate concentrations of tile drainage over the autumn winter and spring following potato harvest from 1989 to 1992 under a barley-potato rotation were consistently above 10 mg NO_3-N L^{-1} (average 17 mg NO_3-N L^{-1}) except the concentration (8.8 mg NO_3-N L^{-1}) from the treatment of winter wheat in 1989 (Milburn et al. 1997). This was the case regardless of wheat straw mulch or autumn seeded cover crop treatments. In comparison, flow-weighted nitrate concentrations in tile drainage following barley in the potato-barley rotation systems averaged 5.5–6.5 mg NO_3-N L^{-1}. Mean flow-weighted nitrate concentrations of tile drainage over the autumn winter and spring following potato harvest from 1993 to 2004 under a conventional barley-red clover-potato rotation were consistently above 10 mg NO_3-N L^{-1} (average 19.2 mg NO_3-N $L^{-1)}$ while the nitrate concentrations in tile drainage following barley and red clover averaged 7.1 and 8.2 mg NO_3-N L^{-1}, respectively (MacLeod, unpublished data). In contrast, nitrate concentrations from wells in forested areas, where natural background concentrations of nitrate are present, were typically less than 2 mg NO_3-N L^{-1} (Jiang and Somers 2009). These findings suggest that potato productions systems have the potential to result in nitrate loading well above background concentrations, and that the risk of nitrate leaching is primarily associated with the potato phase of conventional barley-red clover-potato rotations.

Nitrate leaching from conventional potato production systems was also simulated using LEACHN (Jiang et al. 2011). Model inputs related to soil, climate and managements were derived from measurements at the above tile drainage facility. The model was calibrated and verified against measured nitrate concentrations of tile drainage and water levels for the period 1999–2008 through coupled LEACHN and MODFLOW simulations. Simulations were conducted for multiple cycles of barley-potato (2-year) and barley-red clover-potato (3-year) given recommended fertilizer N rates for potato, barley and red clover as 200, 60 and 0 kg N ha^{-1}, respectively, over the period of 1999 and 2008. Simulations using fertilizer N rates of 0, 50, 100, 150 and 200 kg N ha^{-1} for the potato crop, holding N management for the barley and red clover crops constant, were also performed.

Annual nitrate leaching and leached nitrate concentration were predicted to increase with increasing fertilizer N rate for potato crop, and to be greater from

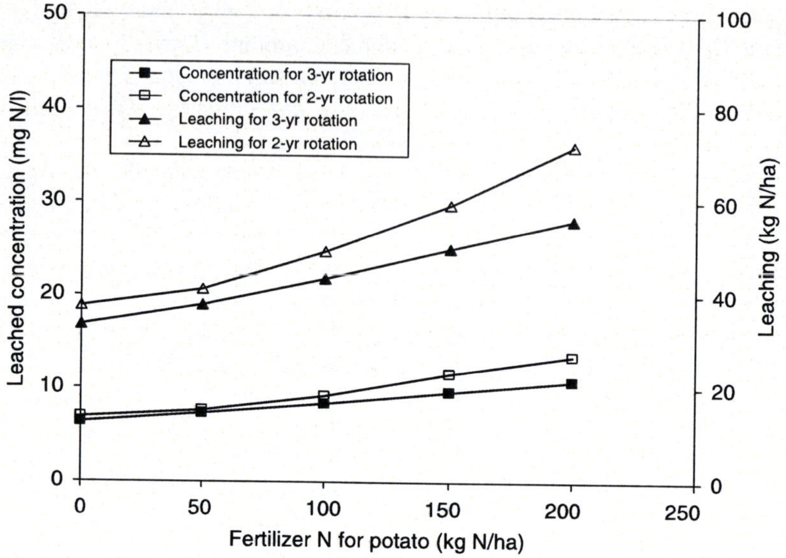

Fig. 13.5 Long-term average annual nitrate leaching from barley-red clover-potato (3-year) and barley-potato rotation (2-year) production systems as affected by fertilizer N for potato crop, simulated using LEACHN. Fertilizer N rate for barley and red clover were held constant at 60 and 0 kg N ha^{-1}, respectively

the 2-year potato rotation compared with the 3-year potato rotation (Fig. 13.5). For current typical fertilizer N application rates (150–200 kg N ha^{-1}), annual average nitrate leaching for 2-year and 3-year rotation systems were predicted to be 60–72 and 50–56 kg N ha^{-1}, respectively, and the corresponding leached concentrations were predicted to be 11.6–13.5 and 9.6–10.7 mg NO_3-N L^{-1}, respectively. Even when the fertilizer N rate for potato was reduced to 0 kg N ha^{-1}, the model predicted annual average nitrate leaching of 38 and 34 kg N ha^{-1} (leached nitrate concentrations of 6.9 and 6.4 mg NO_3-N L^{-1}) for 2- and 3-year rotation systems respectively. This suggests that significant nitrate leaching can occur from potato production systems even under low fertilizer N input for potato crops but recommended N application for barley, at least over limited time periods. When the potato crop received fertilizer N application rates of 150–200 kg N ha^{-1}, approximately 50–60% of the nitrate leaching losses were associated with the barley and red clover phases of the rotation. This suggests there is the potential to reduce nitrate leaching not just under potato production, but also under potato rotation crops.

Nitrate leaching from the representative 3-year rotation system barley-red clover-potato was simulated using LEACHN (Fig. 13.6). Annual nitrate leaching was predicted to range from 27–91 kg N ha^{-1} with an average of 56 kg N ha^{-1}, depending on crop species, management practices and climatic conditions. Plant N uptake and N harvest index for each crop were assumed to be constant over the simulation period, and therefore the annual variation of nitrate leaching primarily reflected the effects of variation in climatic conditions. Nitrate leaching was generally higher

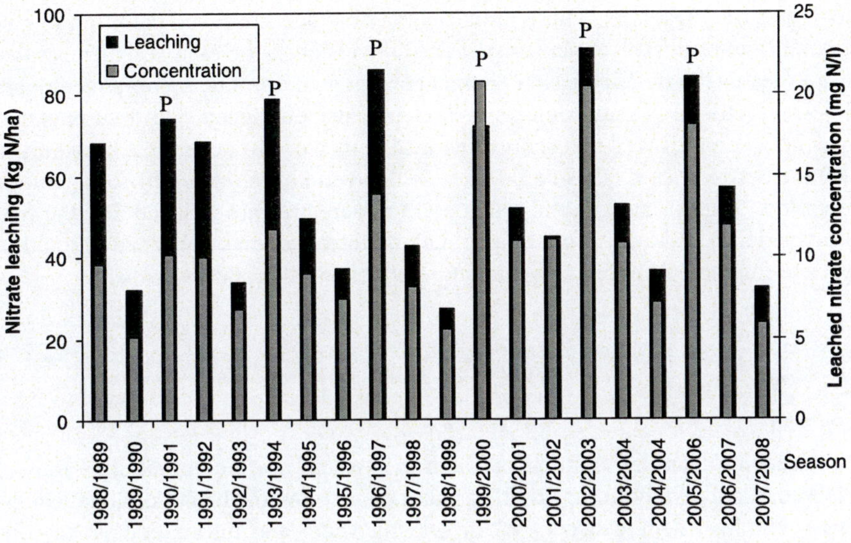

Fig. 13.6 Annual nitrate leaching for the October to April period from a conventional barley-red clover-potato rotation in PEI simulated using LEACHN. All the input parameters except weather were held constant for each cycle of the rotation(The bars marked with "P" represent overall nitrate leaching and flow-weighted leached nitrate concentration during the period after potato harvest and before planting barley crop)

following a potato crop (72–91 kg N ha^{-1}, average 81 kg N ha^{-1}) compared with the rotation crops of barley (43–69 kg N ha^{-1}, average 56 kg N ha^{-1}) or red clover (32–45 kg N ha^{-1}, average 35 kg N ha^{-1}). The model predicted that nitrate leaching outside of the growing season accounted for between 72% and 98% of the annual leaching, with an average of 86%. The predicted magnitude and timing of nitrate leaching generally agreed with the isotopic evidence discussed above and those reported by Delgado et al. (2001a), De Neve et al. (2003), Peralta and Stockle (2001) and Vos and van der Putten (2004). In comparison, predicted N leaching as NH_4^+ ranged from 0–0.8 kg N ha^{-1} with an average of 0.4 kg N ha^{-1}. Annual N losses through denitrification were predicted to vary from 0.2–8.3 kg N ha^{-1} with an average of 2.4 kg N ha^{-1}, and fell in the ranges of values reported by Hoffmann and Johnsson (2000) and Zebarth and Rosen (2007). As expected, N losses through NH_4^+ leaching and by denitrification were predicted to be much less significant than nitrate leaching.

These findings suggest that three general approaches may be most effective in reducing nitrate leaching losses from potato production systems in PEI. First, nitrate leaching occurs primarily from the potato phase of potato rotations. Consequently, increasing the length of potato rotations, for example from current 3-year to 4-year rotations, and reducing the proportion of land in potato production within a watershed, should reduce the overall nitrate loading within a given watershed. Second, N management practices and BMPs that reduce the risk of nitrate accumulation in soil during the potato rotation, especially residual nitrate at potato harvest, may reduce the risk of nitrate leaching. These practices may include improved crediting of soil N mineralization from soil organic matter and crop residues, use of in-season measures of crop N status to guide in-season N management, and introducing potato cultivars with lower fertilizer N requirements or higher N use efficiency. Third, N management practices and BMPs that reduce the availability of nitrate for leaching during the rotation crop phases. This could include delayed plough-down of forage crops until spring, or until late autumn when soil temperatures are reduced and the rate decomposition is reduced. It could also include introduction of alternative rotation crops, and implementing N management practices for thc rotation crops.

13.5 Conclusions

Nitrate leaching from agricultural production systems is of economic importance, and is also of concern both for drinking water quality and the health of aquatic ecosystems. Nitrate leaching occurs when moving water and nitrate coexist in soil. Intensive agricultural crop production systems with high N inputs in coarse-textured soil and under humid climatic conditions present a high risk of nitrate leaching. Elevated nitrate nitrogen in groundwater and associated surface water in PEI was attributed to the configuration of intensive potato cropping in sandy soil over a shallow unconfined and semi-confined sandstone aquifer under a humid climate. Nitrate leaching from conventional barley-red clover-potato rotation systems primarily occurred during autumn winter and spring when crop uptake diminishes, and nitrate from mineralization and fertilizer residual and excessive moisture from rainfall and snowmelt infiltration coexists in the soil in PEI. This finding was supported by tile drainage experiments, isotopic evidences, long-term hydrological monitoring and LEACHN simulations. The LEACHN model predicted that about 50–60% of the nitrate leaching losses were associated with the barley and red clover phases, suggesting there is the potential to reduce nitrate leaching not just in the potato phase, but also in the rotation crop phases. The requirement for high N inputs to obtain economic tuber yields, in combination with the high risk of nitrate loss from the root zone, present significant challenges to growers with respect to N management. Reducing the frequency of potato in the crop rotation, and use of practices to reduce residual nitrate in potato phase as well as the rotation crop phases, will be most effective in reducing nitrate leaching losses.

Future studies should investigate the opportunities of minimizing nitrate accumulation in the soil both during and outside of the growing season both in the potato phase and the rotation crop phases. Development of improved practices with respect to the selection of rate, timing and form of fertilizer N products (Chap. 10) and development of soil- and plant-based diagnostic tests to guide at-planting and in-season fertilizer N management (Chap. 11) will be important in reducing the risk of nitrate leaching. Development of potato cultivars with low N input requirements and high N use efficiency may also reduce the risk of nitrate leaching (Zebarth and Rosen 2007). The effects of BMPs developed to minimize nitrate leaching losses from potato production systems outside of the growing season, such as incorporation of wheat straw, planting cover crops following potato crops, and delayed red clover plough-down, have been tested with various degrees of success in PEI. The potential effects of these BMPs on reducing nitrate leaching losses should be evaluated through well-designed field tests and modeling at field and watershed scales in eastern Canada. In addition, the possibilities of growing other high value and low N input crops and removing legume crops, such as red clover, out of potato rotation systems should be explored.

References

Addiscott TM, Whitmore AP (1991) Simulation of solute leaching in soils of differing permeabilities. Soil Use Manage 7:94–102

Bachman LJ, Lindsey B, Brakebill J, Powars DS (1998) Ground-water discharge and base-flow nitrate loads of national streams, and their relation to a hydrogeomorphic classification of the Chesapeake Bay watershed, Middle Atlantic Coast. USGS Water-Resources Investigations Report, 98–4059

Baker JL (2001) Limitations of improved nitrogen management to reduce nitrate leaching and increase use efficiency. In: Optimizing nitrogen management in food an energy production and environmental protection. Proceedings of the 2nd international nitrogen conference on science and policy. The Scientific World 1(S2):10–16

Benson VS, VanLeeuwen JA, Sanchez J, Dohoo IR, Somers GH (2006) Spatial analysis of land use impact on ground water nitrate concentrations. J Environ Qual 35:421–432

Bergstrom L (1987) Nitrate leaching and drainage from annual and perennial crops in tile-drained plots and lysimeters. J Environ Qual 16:11–18

Böhlke JK (2002) Groundwater recharge and agricultural contamination. Hydrogeol J 10:153–179

Canadian Council of Ministers of the Environment (CCME) (2003) Canadian Water Quality Guidelines for Protection of Aquatic Life. Environment Canada, National Guidelines and Standards Office, Hull, Quebec

Chow TL, Rees HW, Daigle JL (1999) Effectiveness of terraces/grassed waterway systems for soil and water conservation: a field evaluation. J Soil Water Conserv 54:577–583

Dabney SM, Delgado JA, Reeves DW (2001) Using winter cover crops to improve soil and water quality. Commun Soil Sci Plant Anal 32:1221–1250

De Neve S, Dieltjens I, Moreels E, Hofman G (2003) Measured and simulated nitrate leaching on an organic and a conventional mixed farm. Biol Agric Hortic 21:217–229

Delgado JA, Sparks RT, Follett RF, Sharkoff JL, Riggenbach RR (1999) Use of winter cover crops to conserve soil and water quality in the San Luis Valley of South Central Colorado. In: Lal R (ed) Soil quality and soil erosion. CRC Press, Boca Raton, pp 125–142

Delgado JA, Riggenbach RR, Sparks RT, Dillon MA, Kawanabe LM, Ristau RJ (2001a) Evaluation of nitrate-nitrogen transport in a potato-barley rotation. Soil Sci Soc Am J 65:878–883

Delgado JA, Ristau RJ, Dillon MA, Duke HR, Stuebe A, Follett RF, Shaffer MJ, Riggenbach RR, Sparks RT, Thompson A, Kawanabe LM, Kunugi A, Thompson K (2001b) Use of innovative tools to increase nitrogen use efficiency and protect environmental quality in crop rotations. Commun Soil Sci Plant Anal 32:1321–1354

Dessureault-Rompré J, Zebarth BJ, Burton DL, Sharifi M, Cooper J, Grant CA, Drury CF (2010) Relationships among mineralizable soil nitrogen, soil properties and climatic indices. Soil Sci Soc Am J 74:1218–1227

Dubrovsky NM, Hamilton PA (2010) Nutrients in the nation's streams and groundwater: national findings and implications. USGS fact sheet 2010–3078, 6pp

Francis RM (1989) Hydrogeology of the Winter River Basin, Prince Edward Island. Department of the Environment, Prince Edward Island

Gasser MO, Laverdière MR, Lagacé R, Caron J (2002) Impact of potato-cereal rotations and slurry applications on nitrate leaching and nitrogen balance in sandy soils. Can J Soil Sci 82:469–479

Health Canada (2010) Guidelines for Canadian drinking water quality at http://www.hc-sc.gc.ca/ewh-semt/pubs/water-eau/2010-sum_guide-res_recom/index-eng.php. Accessed 4 Aug 2011

Healy RW, Cook PG (2002) Using groundwater levels to estimate recharge. Hydrogeol J 10:91–109

Hill AR (1986) Nitrate and chloride distribution and balance under continuous potato cropping. Agric Ecosyst Environ 15:267–280

Hoffmann M, Johnsson H (2000) Nitrogen leaching from agricultural land in Sweden. Ambio 29:67–73

Hutson JL (2003) Leaching estimation and chemistry model-LEACHM: model description and user's guide. School of Chemistry, Physics and Earth Sciences, The Flinders University of South Australia, GPO Box 2100, Adelaide, SA5001 at http://www.scieng.flinders.edu.au/cpes/people/hutson_j/leachweb.html. Accessed 10 Aug 2007

Itier B, Brunet Y (1996) Recent developments and present trends in evaporation research: a partial survey. In: Evapotranspiration and irrigation scheduling. Proc Int Conf, San Antonio, Texas. ASCE, pp 1–20

Jemison JM, Fox RH (1994) Nitrate leaching from nitrogen-fertilized and manured corn measured with zero-tension pan lysimeters. J Environ Qual 23:337–343

Jiang Y, Somers GH (2009) Modeling effects of nitrate from non-point sources on groundwater quality in an agricultural watershed in Prince Edward Island, Canada. Hydrogeol J 17:707–724

Jiang Y, Zebarth BJ, Love J (2011) Long-term simulations of nitrate leaching from potato production systems in Prince Edward Island, Canada. Nutr Cycl Agroecosyst 91:307–325

Johnsson H, Bergstorm L, Jansson P (1987) Simulated nitrogen dynamics and losses in a layered agricultural soil. Agric Ecosyst Environ 18:333–356

Keith S, Zhang Y (2004) Base flow contribution to nitrate-nitrogen export from a large, agricultural watershed, USA. J Hydrol 295:305–316

Kraft G, Stites W (2003) Nitrate impacts on groundwater from irrigated-vegetable systems in a humid north-central US sand plain. Agric Ecosyst Environ 100:63–74

Kumar K, Goh KM (2000) Crop residues and management practices: effects on soil quality, soil nitrogen dynamics, crop yield, and nitrogen recovery. Adv Agron 68:197–319

MacLeod JA, Sanderson JB (2002) Fall cover crops for reducing nitrate loss. In: Proceedings of the national conference on agricultural nutrients and their impact on rural water quality, 28–30 April 2002 Waterloo, Ontario, Canada

McMahon PB, Bohlke LK, Kauffman LJ, Kipp KL, Landon MK, Crandall CA, Burow KR, Brown CJ (2007) Source and transport controls on the movement of nitrate to public supply wells in selected principal aquifers of the United States. Water Resour Res 44:W04401

Meisinger JJ, Delgado JA (2002) Principles for managing nitrogen leaching. J Soil Water Conserv 57:485–498

Milburn P, MacLeod J (1991) Considerations for tile drainage water quality studies in temperate regions. Appl Eng Agric 7:209–215

Milburn P, Richards JE, Gartley C, Pollock T, O'Neill H, Bailey H (1990) Nitrate leaching from systematically tiled potato fields in New Brunswick, Canada. J Environ Qual 19:448–454

Milburn P, MacLeod JA, Sanderson JB (1997) Control of fall nitrate leaching from early harvested potatoes on Prince Edward Island. Can Agric Eng 39:263–271

Mitsch WJ, Day JW, Gilliam JW, Groffman PM, Hey DL, Randall GW, Wang N (1999) Reducing nutrient loads, especially nitrate-nitrogen, to surface water, ground water and the Gulf of Mexico. Topic 5 Report for the integrated assessment on Hypoxia in the Gulf of Mexico. NOAA coastal Ocean program, decision analysis series No. 19

Nolan BT, Hitt KJ, Ruddy BC (2002) Probability of nitrate contamination of recently recharge groundwaters in the coterminous United States. Environ Sci Tech 36:2138–2145

PEIEEF (Prince Edward Island Department of Environment, Energy and Forestry) (2011) Groundwater level monitoring data at http://www.gov.pe.ca/envengfor/groundwater/app.php

Peralta JM, Stockle CO (2001) Dynamics of nitrate leaching under irrigated potato rotation in Washington State: a long-term simulation study. Agric Ecosyst Environ 88:23–34

Pionke HB, Urban JB (1985) Effect of agricultural land use on ground-water quality in a small Pennsylvania watershed. Ground Water 23:68–80

Power JF, Schepers JS (1989) Nitrate contamination of groundwater in North America. Agric Ecosyst Environ 26:165–187

Powlson DS (1993) Understanding the soil nitrogen cycle. Soil Use Manage 9:86–94

Puckett LJ, Tesoriero AJ, Dubrovsky NM (2011) Nitrogen contamination of surficial aquifers – a growing legacy. Environ Sci Technol 45:839–844

Randall GW, Huggins DR, Russelle MP, Fuchs DJ, Nelson WW, Anderson JL (1997) Nitrate losses through subsurface tile drainage in conservation reserve program alfalfa and row crop systems. J Environ Qual 26:1240–1247

Richards JE, Milburn PH, MacLean AA, Demerchant G (1990) Intensive potato production effects on nitrate-N concentrations of rural New Brunswick well water. Can Agric Eng 32:189–196

Sanderson JB, MacLeod JA (2002) Maximizing legume N use and minimizing nitrate leaching. In: Proceedings of the national conference on agricultural nutrients and their impact on rural water quality, 28–30 April 2002, Ontario, Canada

Sanderson JB, MacLeod JA, Kimpinski J (1999) Glyphosate application and timing of tillage of red clover affects potato response to N, soil N profile, and root and soil nematodes. Can J Soil Sci 79:65–72

Savard MM, Paradise D, Somers GH, Liao S, van Bochove E (2007) Winter nitrification contributes to excess NO_3^- in groundwater of an agricultural region: a dual-isotope study. Water Resour Res 43:W06422

Savard MM, Somers G, Smirnoff A, Paradis D, van Bochove E, Liao S (2010) Nitrate isotopes unveil distinct seasonal N-sources and the critical role of crop residues in groundwater contamination. J Hydrol 381:134–141

Shaffer MJ (1995) Fate and transport of nitrogen: what models can and cannot do? At http://www.nrcs.usda.gov/technical/rca/wp11text.html. Accessed 6 Jan 2010

Shaffer MJ, Halvorson AD, Pierce FJ (1991) Nitrate leaching and economic analysis package (NLEAP): model description and application. In: Follett RF, Keeney DR, Cruse RM (eds) Managing nitrogen for groundwater quality and farm profitability. Soil Science Society of America, Madison, 285pp

Sharifi M, Zebarth BJ, Burton DL, Grant CA, Porter GA (2008a) Organic amendment history and crop rotation effects on soil nitrogen mineralization potential and soil nitrogen supply in a potato cropping system. Agron J 100:1562–1572

Sharifi M, Zebarth BJ, Burton DL, Grant CA, Bittman S, Drury CF, McConkey B, Ziadi N (2008b) Response of soil potentially mineralizable N and indices of N availability to tillage system. Soil Sci Soc Am J 72:1124–1131

Sharifi M, Zebarth BJ, Porter GA, Burton DL, Grant CA (2009) Soil mineralizable nitrogen and soil nitrogen supply under two-year potato rotations. Plant Soil 320:267–279

Shrestha RK, Cooperband LR, MacGuidwin AR (2010) Strategies to reduce nitrate leaching into groundwater in potato grown in sandy soil: case study from North Central USA. Am J Potato Res 87:229–244

Simunek J, van Genuchten MT, Miroslav S (2008) Development and applications of the HYDRUS and STANMOD software packages and related codes. Vadose Zone J 7:587–600

Snider DM, Spoelstra J, Schiff SL, Venkiteswaran JJ (2010) Stable oxygen isotope ratios of nitrate produced from nitrification: ^{18}O-labelled incubations of agricultural and temperate forest soils. Environ Sci Tech 44:5358–5364

Somers GH (1998) Distribution and trends for occurrence of nitrate in PEI groundwater. Paper presented in the workshop on "Nitrate-agricultural sources and fate in the Environment-Perspectives and Directions", Charlottetown, PEI, 26 February 1998

Somers G, Savard MM (2008) Dynamics of nitrate transfer from agricultural soils to aquifers inferred from stable isotopes. In: Proceedings of the 2008 American Water Works Association

Inorganic Contaminants Workshop, Albuquerque, NM. January 27–29. American Water Works Association, Denver, CO
Spalding RF, Exner ME (1993) Occurrence of nitrate in groundwater – a review. J Environ Qual 22:392–402
Stark JC, Porter GA (2005) Potato nutrient management in sustainable cropping systems. Am J Potato Res 82:329–338
Statistics Canada, 2009 Canadian Potato Production. Catalogue no.22-008-X, Vol. 6(3)
Tilman D, Cassman KG, Matson PA, Nalor R, Polasky S (2002) Agricultural sustainability and intensive production practices. Nature 418:671–677
United States Environmental Protection Agency (2009) National Primary Drinking Water Protection Regulations, EPA 816-F-09-004
van Es HM, Czymmek KJ, Ketterings QM (2002) Management effects on nitrogen leaching and guidelines for a nitrogen leaching index in New York. J Soil Water Conserv 57:499–504
Vos J (2009) Nitrogen responses and nitrogen management in potato. Potato Res 52:305–317
Vos J, van der Putten PEL (2004) Nutrient cycling in a cropping system with potato, spring wheat, sugar beet, oats and nitrogen catch crops. II. Effect of catch crops on nitrate leaching in autumn and winter. Nutr Cycl Agroecosyst 70:23–31
Wagenet RJ, Hutson JL (1989) LEACHM: leaching estimation and chemistry model. A process based model of water and solute movement, transformations, plant uptake, and chemical reactions in the unsaturated zone. Water Resources Institute, Cornell University, Ithaca
Young JJ, Somers GH, Raymond BG (2002) Distribution and trends for nitrate in PEI groundwater and surface waters. Paper presented in the National Conference on Agricultural Nutrients and their Impact on Rural Water Quality, Waterloo, Ontario, 28–30 April 2002
Zebarth BJ, Rosen CJ (2007) Research perspective on nitrogen BMP development for potato. Am J Potato Res 84:3–18
Zebarth BJ, Drury CF, Tremblay N, Cambouris AN (2009) Opportunities for improved fertilizer nitrogen management in production of arable crops in eastern Canada: a review. Can J Soil Sci 89:113–132

Chapter 14
Nitrous Oxide Emissions from Potato Production and Strategies to Reduce Them

David L. Burton, Bernie J. Zebarth, John A. McLeod, and Claudia Goyer

Abstract The potato production system provides an interesting environment to examine denitrification and nitrous oxide (N_2O) emissions. Tillage is used to form ridges of topsoil in which the potato plant and fertilizer are placed. The tillage operation results in a low bulk density within this ridge resulting in a well-aerated environment rich in carbon (C) and nitrate (NO_3^-) in contrast with the compact furrow that remains between the ridges. We review the soil factors that control denitrification and N_2O emissions within the context of rain-fed potato production systems in Atlantic Canada. Further, we examine the role of water content in controlling denitrification and N_2O emissions at different spatial scales: the ridge-furrow complex, soil profile and landscape scales. While denitrification is found to predominate in the compact and wet furrow environment, N_2O emissions are greatest in the potato ridge as a result of the greater C and NO_3^- concentrations found there. Water movement at larger spatial scales results in a shift of N_2O production to deeper in the soil profile and to lower slope positions in the landscape. Beneficial management practices (BMPs) for reducing N_2O emissions from rain-fed potato production include improved fertilizer N management (placement and timing), choice of crop rotation, manure management and drainage. It is important to realize, however, that because of the complex controls on denitrification and N_2O emissions, the effects of different BMPs on N_2O emissions can vary with soil properties and climatic conditions.

D.L. Burton (✉)
Nova Scotia Agricultural College, P.O. Box 550, Truro, NS, Canada B2N 5E3
e-mail: dburton@nsac.ca

B.J. Zebarth • C. Goyer
Potato Research Centre, Agriculture and Agri-Food Canada, , 850 Lincoln Rd., P.O. Box 20280, Fredericton, NB, Canada E3B 4Z7

J.A. McLeod
Crops and Livestock Research Centre, Agriculture and Agri-Food Canada, 440 University Ave., Charlottetown, PE, Canada C1A 4N6

Z. He et al. (eds.), *Sustainable Potato Production: Global Case Studies*, DOI 10.1007/978-94-007-4104-1_14, © Springer Science+Business Media B.V. 2012

14.1 Introduction

Changes in the global climate system have been linked to anthropogenic emissions of greenhouse gases (GHG) (IPCC 2007). In 2008, 8.4% of the total GHG emissions reported in Canada resulted from agricultural activities (Environment 2010). Nitrous oxide (N_2O) from agricultural soils accounted for 50% of the anthropogenic GHG emissions from Canadian agroecosystems. In addition to being a greenhouse gas, N_2O is also responsible for stratospheric ozone depletion (Crutzen and Ehhalt 1977). Nitrogen fertilizer, the major source of agricultural N_2O emissions, has been instrumental in increasing yield from arable cropping systems over the past 50 years (Matson et al. 1997; Tillman et al. 2002). Potato is the dominant arable crop in Atlantic Canada. Large fertilizer N inputs, frequently 200 kg N ha^{-1} or more, are applied to the potato crop to achieve the tuber yield and size distribution demanded by the potato processing industry (Zebarth et al. 2003; Zebarth and Rosen 2007). These large N inputs raise concerns as to the potential for increased N_2O emissions (Zebarth et al. 2009a).

The potato production system is an interesting one from the perspective of N_2O emissions. As with other ridge/furrow management production systems, soil management practices cause distinct distributions of soil physical, chemical and biological characteristics over relatively short distances. The potato ridge, formed as a result of tillage, provides a low bulk density, well-aerated environment suitable for the formation of tubers (Fig. 14.1). In Atlantic Canada, the ridge is commonly formed when the plants are 15–30 cm tall, but there is a trend towards earlier ridge formation that can occur anytime from planting to emergence. The majority of the fertilizer application in Atlantic Canada is as a band application at planting placed adjacent to the seed-pieces in the potato ridge. The potato furrows by contrast have a greater soil bulk density and therefore less aerobic environment where carbon and nitrate (NO_3^-) supplies are reduced relative to the ridge. In Atlantic Canada the majority of the A horizon, the horizon in which organic carbon accumulates, is used to form the ridge. The furrow consists of a thin layer of A horizon (maximum of 5 cm) that remains after tillage overlying a B horizon or in some cases C horizon. In Atlantic Canada the soils tend to be rather shallow soil with low pH and frequently have compact subsurface horizons making the distinction between ridge and furrow even more dramatic. The furrow is also a region of water accumulation as a result of the runoff of precipitation. This substantial variation in soil physical and chemical properties over a short distance has the potential to result in markedly different soil N processes and microbial communities. In this chapter we will use the results of studies we have conducted in rain-fed potato production in Atlantic Canada to examine the controls on N_2O emissions from potato production systems and strategies for managing these emissions.

14.2 Factors Controlling N_2O Emissions

The biological production of N_2O in agricultural soils occurs primarily as a result of the processes of nitrification and denitrification (Firestone and Davidson 1989; Mosier 1998). The conditions that favor these processes are very different. Nitrification, the

Fig. 14.1 The ridge-forming operation results in the formation of a low bulk density ridge that contains higher NO_3^- due to banded fertilizer application, and higher organic carbon concentrations because most of the C-rich surface horizon is located in the ridge, compared with the furrow. The hilling operation can occur at different times, but is commonly done in eastern Canada when the plants are 15–30 cm high

oxidation of ammonium (NH_4^+) to NO_3^-, occurs as a result of the activity of autotrophic organisms in aerobic soils. Denitrification is a respiratory process whereby heterotrophic organisms respire oxides of N in the absence of O_2, producing N_2 as the final product. As a result, denitrification generally occurs in moist soils where O_2 supply is reduced. For this, an energy source, primarily C, is required by the microorganisms for the process to proceed. In humid climates such as those found in Atlantic Canada[1], denitrification is the predominant source of N_2O (Firestone and Davidson 1989; Granli and Bøckman 1994; Mosier 1998) and will be the primary focus of our

[1] Atlantic Canada experiences a cool humid Maritime climate with annual rainfall averaging between 900 and 2,000 mm annually, growing season temperatures averaging 8.5-12.5 °C, 170–180 frost free days and 2,000–2,700 crop heat units.

discussion here. Since denitrification is a respiratory process the primary factors that control denitrification relate to the supply of preferred terminal electron acceptor (O_2 availability or aeration), the demand for terminal electron acceptors (carbon availability) and the relative supply of alternate terminal electron acceptors, primarily NO_3^-availability. Other factors (e.g., temperature, pH, texture) also influence the denitrification process (Firestone and Davidson 1989; Granli and Bøckman 1994) but are less impacted by management practices in the potato production system.

14.2.1 Aeration

Denitrification commonly occurs at water-filled pore space (WFPS) of 0.70 m^3 m^{-3} and greater (Bateman and Baggs 2005). The rate of gaseous diffusion through the soil depends on water content and bulk density (Ball et al. 1999, 2008). Soils with greater bulk density have smaller pore volume. Similarly, greater soil water content reduces air-filled porosity. Both factors result in reduced and less continuous air-filled porosity, decreasing the rate of O_2 movement through the soil (Focht 1992). Increasing the percentage of WFPS has been shown to increase denitrification and N_2O emissions (Ruser et al. 1998, 2006; Ball et al. 1999, 2008; Flessa et al. 2002).

In C-rich environments, increased microbial activity can also result in reduced availability of O_2. For example, Gillam et al. (2008) found that even in relatively well-aerated soils (WFPS 0.60 m^3 m^{-3}), increased C availability can substantially increase N_2O emissions in soils through its influence on O_2 consumption. In comparison, N_2O emissions were negligible at WFPS below 0.45 m^3 m^{-3} due to enhanced aeration, or above 0.75 m^3 m^{-3} because restricted gaseous diffusion limits the diffusion of N_2O away from the site of denitrification thereby favoring the further reduction of N_2O to N_2, and consequently C availability has a more limited effect on N_2O emissions under these conditions.

The WFPS varies in ridges and furrows. It is consistently greater in the furrow than in the ridge (Burton et al. 2008a; Dandie et al. 2008; Haile-Mariam et al. 2008). This difference is due in large part to differences in soil bulk density. For example, soil bulk density for 0–15 cm depth in the ridge and furrow were commonly 1.0 and 1.2 Mg m^{-3}, respectively, in the study of Burton et al. (2008a). Greater WFPS in the furrow can also reflect the redistribution of water after rainfall events, and the greater loss of water from the ridge due to transpiration (Zebarth and Milburn 2003). The greater total pore volume and preponderance of large pores in the ridge results in a soil in which the pores are largely filled with air. The presence of continuous air-filled porosity greatly enhances the movement of O_2 into the soil (Focht 1992). The enhanced supply of O_2 to the microbial community suppresses anaerobic processes such as denitrification (Robertson and Tiedje 1987; Focht 1992; Granli and Bøckman 1994). This is consistent with reports of increased N_2O emissions in furrows that were compacted as a result of wheel traffic (Ruser et al. 1998; Thomas et al. 2004) and with increased denitrification in furrows compared with ridges (Dandie et al. 2008).

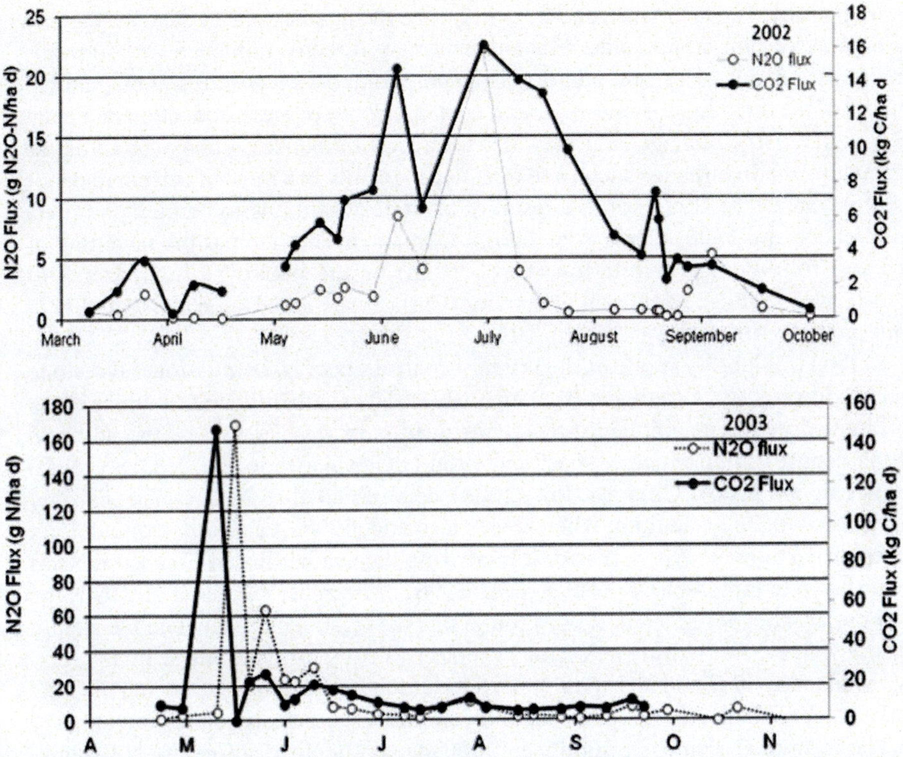

Fig. 14.2 Synchrony of nitrous oxide (open symbols; g N_2O-N ha^{-1} d^{-1}) and carbon dioxide (closed symbols; kg CO_2-C ha^{-1} d^{-1}) emissions in the 2002 and 2003 growing seasons from a potato crop in New Brunswick (Burton et al. 2008a)

14.2.2 Carbon Availability

Denitrification is a respiratory process where nitrogen oxides (NO_3^-, NO_2^-, NO and N_2O) are utilized as terminal electron acceptors by microorganisms in the absence of oxygen. Organic C substrates serve as a source of energy for heterotrophs in this process (Beauchamp et al. 1989). Carbon also has an indirect influence on denitrification by increasing the consumption of O_2 during aerobic respiration stimulating the formation of anoxic sites within the soil (Smith 1980; Hojberg et al. 1994). Sources of soil C inputs include soil organic matter, the potato rhizosphere, residues from the preceding crop, senescing leaves and roots from the current crop, and organic amendments added to the soil. In most cases soil management results in all of these sources being concentrated in the potato ridge, either as a result of ridge formation or as a result of the location of the potato plant. The greater concentration of the organic carbon in the potato ridge results in Greater microbial activity in these locations and greater N_2O and CO_2 emissions (Fig. 14.2; Haile-Mariam et al. 2008).

Management and climatic conditions influence the quantity, quality and availability of organic carbon compounds. Management may directly influence carbon availability as a result of carbon addition (manure, compost or crop residues), altered quality of the organic substrates (Zebarth et al. 2009b, c) or the introduction of a plant (Sey et al. 2010). Management may also indirectly influence carbon availability as a result of soil disturbance such as tillage, which results in a flush of carbon metabolism (Alvarez et al. 2001), or as a result of plant desiccation with herbicide (Tenuta and Beauchamp 1996; Flessa et al. 2002). Climatic events such as the rewetting of dry soil following precipitation (Birch 1958) or thc thawing of a frozen soil also result in a flush of microbial activity (Christensen and Tiedje 1990) and N_2O emissions (Burton and Beauchamp 1994).

Carbon availability is one of the primary controllers of N_2O emissions (Firestone and Davidson 1989; Granli and Bøckman 1994). The availability of C sources in the soil influences the magnitude of the denitrification process as well as the products of the denitrification process, specifically the N_2O molar ratio (i.e. $N_2O{:}N_2+N_2O$) (Baggs et al. 2000). Gillam et al. (2008) determined that N_2O emissions only increased with NO_3^- addition when C was also added; NO_3^- addition alone did not necessarily stimulate N_2O emissions. Increasing the availability of C substrates has been shown to increase (Dendooven et al. 1996) or decrease (Weier et al. 1993) the N_2O molar ratio. Miller et al. (2008) found that the relationship between the supply of, and demand for, terminal electron acceptors (TEAs), as determined by the relative availability of C and NO_3^-, influenced the amount of denitrification and the N_2O molar ratio for both simple (glucose) and complex C sources (plant residues).

The form of C also has a significant influence on denitrification and N_2O emissions. Gillam et al. (2008) found ground red clover was equivalent to glucose in its ability to stimulate N_2O emissions per unit of available C added, but the greater C:N ratio of barley straw resulted in less CO_2 and N_2O emissions (Gillam et al. 2008). Similarly, Miller et al. (2008) found that C from ground red clover was equivalent to glucose C in stimulating the microbial population. Furthermore, Miller et al. (2008, 2009) found N_2O emissions to be linearly related to C utilization by microbes as indicated by soil respiration across a wide range of carbon sources (glucose, plant residues, manures).

14.2.3 Nitrate Availability

Nitrate has been found to limit denitrification when NO_3^- concentrations are low ($<$ 5–10 mg N kg^{-1} soil; Ryden 1983), a situation not commonly found in soils cropped to potato. In agricultural soils where NO_3^- often accumulates to $>$ 10 mg N kg^{-1} soil, the availability of NO_3^- was found to have no effect on denitrification but a significant effect on the N_2O molar ratio (Gillam et al. 2008). The influence of NO_3^- on the N_2O molar ratio is due to NO_3^- being a more favorable electron acceptor than is N_2O (Betlach and Tiedje 1981; Cho et al. 1997), and therefore its presence in soil suppresses both the expression (Firestone et al. 1979) and activity (Betlach and Tiedje 1981) of nitrous oxide reductase enzyme. The N_2O molar ratio is influenced by the relative availability of N-oxides and C, as simple or complex sources (Weier et al. 1993; Miller et al. 2008). Concentrations of NO_3^--N ranging from 100 to 300 kg N ha^{-1} soil can inhibit N_2O reductase activity (Blackmer and Bremner 1978; Weier et al. 1993), reducing the conversion of N_2O to N_2 and thereby making N_2O the principle product of denitrification (Blackmer and Bremner 1978; Firestone et al. 1979; Weier et al. 1993).

In potato production systems in Atlantic Canada, Dandie et al. (2008) found that in the ridge, the magnitude of N_2O emissions was comparable with the denitrification rate suggesting that most gaseous emissions occurred as N_2O. In contrast in the furrow, the denitrification rate was greater than N_2O emissions, suggesting there was significant conversion of N_2O to N_2. This was attributed to the N fertilizer being concentrated in the potato ridge as a result of banded fertilizer application, and the increased NO_3^- concentration decreasing N_2O reduction, increasing the N_2O molar ratio of denitrification.

14.2.4 Denitrifier Community

Denitrifiers are mainly aerobic heterotrophs (utilize C as an energy source) and are not necessarily controlled by the presence of anaerobic conditions or NO_3^- availability in the soil (Murray et al. 1990). Several microcosm studies using agricultural soils from potato production systems investigated the effect of organic C availability and source under completely anoxic conditions. Miller et al. (2008, 2009) found that C from plant residues or manures applied at 0–500 mg C kg^{-1} soil did not increase the abundance of broad groups of denitrifiers but did increase denitrification rate and N_2O emissions. In contrast, the abundance of a broad group of denitrifiers, N_2O emissions and denitrification all increased in soil amended with plant residues at 1,000 mg C kg^{-1} soil (Henderson et al. 2010). Plant residues with different C/N ratios increased similarly the abundance of a broad group of denitrifiers (Henderson et al. 2010). These findings suggest that abundance of broad groups of denitrifiers increase when availability of organic carbon is greater than 1000 mg C kg^{-1} soil but that the source of organic carbon had little influence on growth.

Nitrate and reduced intermediate products induce denitrification gene expression under anoxic conditions (Saleh-Lakha et al. 2009). Denitrification gene expression in a smaller group of denitrifiers, *Pseudomonas mandelii* and related species (Dandie et al. 2007) increased with nitrate addition in soil without organic C amendment (Shannon et al. 2011). In contrast, there was no increase in denitrification gene expression in response to soil NO_3^- addition following organic amendment even despite there being an increase in denitrification rate (Shannon et al. 2011).

It is not clear whether the abundance of the denitrifier community and/or their denitrification gene expression play a role in controlling denitrification and N_2O emissions. Several studies found no relationship between abundance of denitrifiers (Dandie et al. 2007; Miller et al. 2008, 2009; Henderson et al. 2010; Shannon et al. 2011) or denitrification gene expression (Henderson et al. 2010; Shannon et al. 2011) and N_2O emissions or denitrification rate. These findings suggest that the denitrification activity was decoupled from the abundance and denitrification gene expression of denitrifiers and that control over the magnitude of denitrification might be occurring primarily at the level of the denitrifying enzymes. Absence of relationships between abundance and gene expression could also be explained from the lack of information as to which denitrifier species are active in a particular environment and from limitations in the tools used to quantify the denitrifiers.

14.3 Temporal and Spatial Variation in N_2O Emissions

14.3.1 Temporal Variation

The interaction between the controlling factors discussed above result in highly episodic N_2O emissions. Despite this episodic nature, seasonal patterns in N_2O emissions are frequently observed. Following spring thaw and during the early part of the growing season soil moisture conditions, the release of organic carbon and the availability of NO_3^- often combine to result in large N_2O emissions (Christensen and Tiedje 1990). This has often been observed in potato production systems in Atlantic Canada (Fig. 14.2) but is not always the case. N_2O emissions are, however, seldom monitored during the non-cropping period (November to May). In studies that have examined the non-cropping period in temperate climates, the emissions can be significant in the autumn and spring periods (Wagner-Riddle and Thurtell 1998; Ruser et al. 2006). This may be particularly true in the autumn to spring period following potato production where residual NO_3^- after harvest is large (Zebarth et al. 2003).

The location, timing and intensity of C availability combine to determine the magnitude of N_2O emissions (Ball et al. 2008). This can be seen in the similarity in patterns of N_2O and CO_2 emissions (Fig. 14.2). The spring flush of CO_2 and N_2O emissions is often attributed to the release of C following soil thawing (Christensen and Tiedje 1990; Burton and Beauchamp 1994) or as a result of tillage operations (Alvarez et al. 2001). The rewetting of dry soil can also result in a flush of respiration and has been shown to result in N_2O emission events (Zebarth et al. 2008a). Snowden (2010) demonstrated that the preceding crop significantly affected denitrification and N_2O emissions from the potato crop. While there were no significant differences in N_2O or CO_2 emissions during the growth of the non-potato rotational crops, the subsequent potato crop had greater rates of denitrification and N_2O emissions where the preceding crops were forages (red clover, Italian ryegrass). This effect was attributed to the quality of the forage crop as a substrate driving denitrification. Nitrate exposure (a temporally integrated measure of NO_3^- concentration), CO_2 emissions and WFPS were able to explain 60% of the variation in cumulative growing season N_2O emissions from the potato crop (Snowden 2010).

14.3.2 *Spatial Variations in N_2O Emissions and Denitrification*

The spatial distribution of water is one the primary factors that govern where denitrification and N_2O emissions occur. The movement of water falling on the soil as precipitation through the soil profile and along landscape features results both in the transport of C and N compounds, particularly NO_3^-, as well as influencing the water content and therefore aeration status of the soil environment. Consideration of the influence of hydrology on the spatial distribution of N_2O emissions must be addressed at several scales: the ridge-furrow complex, the soil profile, and the landscape.

14.3.2.1 Ridge-Furrow Complex

Water is not uniformly distributed in the potato ridge-furrow complex. Water applied to a potato field, whether as rainfall or irrigation, tends to accumulate in the furrow due to runoff from the ridge and leaf-drip from the outer foliage system (Saffigna et al. 1976). There is also greater soil wetting at the base of the plant due to stemflow. Zebarth and Milburn (2003) reported lower soil water content in the ridge than in the furrow in a potato field, a finding attributed to greater loss of water from the ridge by transpiration because no difference in water content was measured when no plants were present.

The architecture of the potato ridge-furrow complex is important in influencing both denitrification and N_2O emissions. The furrow locations systematically maintain a greater percentage of water-filled pores than do ridges (Table 14.1). The greater WFPS results in the impeded diffusion of O_2 into the soil, favoring denitrification (Table 14.1), and impeded diffusion of N_2O away for the site of denitrification, resulting in a lower N_2O molar ratio. The lower N_2O molar ratio is also favored by the low NO_3^- concentration in the furrow because fertilizer N is banded in the ridge. By comparison, the ridge has lower WFPS due to lower soil water content and soil bulk density, and consequently is less favorable for denitrification (Table 14.1). Despite the greater potential of gas exchange, denitrification does occur in the ridge due to the presence of organic matter rich soil and the potato rhizosphere resulting in a large rate of O_2 consumption. In addition, the presence of large concentrations of NO_3^- in the ridge ensure that when O_2 supply cannot meet demand, denitrification will result in N_2O being the primary end product and the subsequent reduction of N_2O to N_2 will be limited due to the greater substrate affinity of the nitrate reductase compared to the N_2O reductase (Betlach and Tiedje 1981).

The end result of this architecture and the placement of N fertilizer in the ridge is that the potato ridge has a lower rate of denitrification, but greater N_2O emissions, compared with the furrow (Table 14.1; Burton et al. 2008a; Dandie et al. 2008). Where fertilizer N is broadcast, resulting uniform distribution of NO_3^-, greater N_2O emissions occur from the furrow, particularly compacted furrows, and over-all N_2O emissions from the field are greater (Ruser et al. 1998).

Table 14.1 Influence of row location (ridge, furrow) on cumulative growing season N_2O emissions, cumulative CO_2 emissions, mean denitrification rate and mean water-filled pore space (WFPS) as influenced fertilizer N management (N0=0 N; N200=200 kg N ha^{-1} at planting; N120+80=120 kg N ha^{-1} at planting plus 80 kg N ha^{-1} at final ridge formation in a potato crop located in New Brunswick (Burton et al. 2008a)

	2002			2003	
Treatment	Ridge	Furrow		Ridge	Furrow
Cumulative N_2O emissions	(kg N ha^{-1})				
N0	0.19	0.19		0.64	0.62
N200	0.98	0.32		3.09	1.32
N120+80	0.59	0.32		2.15	0.94
SEM (n=4, df=8)			0.35		
Cumulative CO_2 emissions	(Mg C ha^{-1})				
N0	1.18	1.01		1.16	1.12
N200	1.38	0.89		1.09	0.87
N120+80	1.16	0.80		1.31	0.92
SEM (n=4, df=8)			0.27		
Mean denitrification	(kg N ha^{-1})				
N0	1.7	6.5		2.8	41
N200	15.1	22.2		45.2	24.3
SEM (n=4, df=12)			10.5		
Mean WFPS	(m^3 m^{-3})				
N0	0.39	0.64		0.35	0.52
N200	0.40	0.65		0.38	0.59
SEM (n=4, df=12)			0.02		

SEM standard error of the mean

The distinct soil environments within the ridge-furrow complex also result in distinguishable changes in the abundance and diversity of the soil denitrifier community. Distinct differences in denitrifier community abundance (Dandie et al. 2008) and diversity (Wertz et al. 2009) were measured between the ridge and furrow, and also between locations within the ridge that were adjacent to or more distal from the plant. However the spatial distribution of the denitrifier communities within the ridge-furrow complex did not match the spatial distribution of denitrification activity. Thus, as indicated above, denitrifier abundance is de-coupled from denitrification activity. This shows that soil environmental conditions influence the activity of denitrifier enzymes and that greater denitrification is not necessarily the result of greater gene abundance (Dandie et al. 2008; Wertz et al. 2009).

14.3.2.2 Soil Profile

The ridge-furrow complex is hydrologically connected to the underlying soil as a result of the water draining through the soil profile. While the most biologically active region of the soil occurs at the soil surface, microbial processes also occur in the sub-surface. Water not only influences the potential for denitrification through its impacts on aeration, but also as the primary means by which C and NO_3^- are

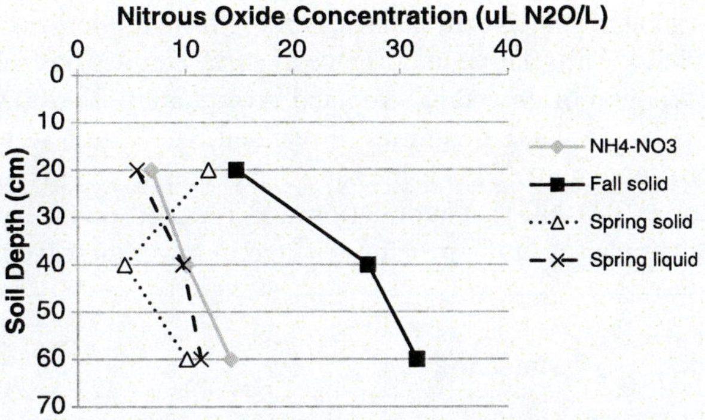

Fig. 14.3 Influence of N source on the N_2O concentration of the soil profile of a potato–barley–red clover rotation. Sampling was done in the spring prior to the potato phase of the rotation following fall manure application but prior to spring manure applications

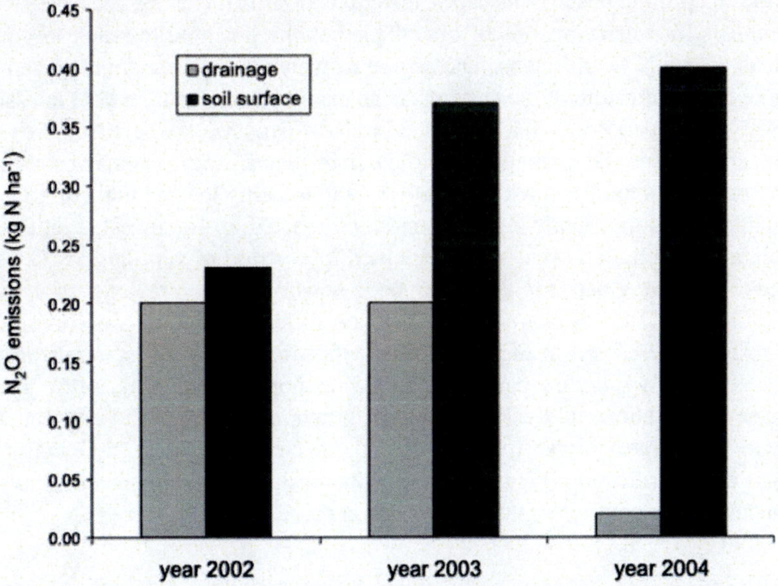

Fig. 14.4 Annual emissions of N_2O dissolved in tile drainage water (drainage) or emitted from the soil surface (soil surface) in a soil cropped to barley (2002), red clover (2003), and in the spring period following red clover plow-down (2004)

delivered to the sub-surface. Examination of the N_2O concentration in the soil profile of a potato – barley – red clover rotation revealed that the greatest concentrations of N_2O were often not found in the biologically active surface layer but at depth (Fig. 14.3). Fall application of solid swine manure resulted in an increase in N_2O concentration throughout the profile and in particular at 60 cm depth. This result points to the role of leaching of C and N, even if applied at the soil surface, in influencing the magnitude and location of N_2O production (Burton and Beauchamp 1994) and denitrification (Paul et al. 1997).

In Atlantic Canada many potato producers have installed tile drainage as a means of removing excess water from their fields during the spring period. Examination of N_2O emitted in tile drainage water indicated that dissolved N_2O emissions could be as great in some years as those coming from the soil surface (Fig. 14.4; Burton et al. 2008b). The relative importance of surface emissions relative to dissolved N_2O was a function of preceding crop and precipitation. In 2002, the year following potato production, large quantities of residual NO_3^- and a wet autumn combined to favor the production of N_2O in the sub-surface and its loss in tile-drainage water (Fig. 14.4). Lower soil NO_3^- accumulation and reduced precipitation in the barley and red clover phases of the rotation resulted in lower dissolved N_2O emissions.

14.3.2.3 Landscape

The influence of water on the distribution of denitrification and its end products extends beyond the soil profile. In soil landscapes, lower positions in the landscape have greater water contents for longer periods of time. This has the potential to result in greater denitrification and N_2O emissions. The degree to which this potential is realized is a function of the architecture of the system. As an example, consider the N_2O emissions from a soil landscape located in a single field in New Brunswick where soils vary primarily in terms of internal drainage (Fig. 14.5). In this landscape, a catena of soils exist along the slope with the Research Station Association being located in the uppermost slope positions and the Fundy Association being located at the lowest positions. In 2002, a large precipitation event in July resulted in a period of very large N_2O emissions and these were greatest in the lower slope positions. The remainder of the year had little to no N_2O emissions and there was no difference among slope positions. Similar patterns were seen in the rate of denitrification with greatest denitrification occurring on the July sampling date and at the lower slope position (Fig. 14.6). In contrast, lower precipitation and the lack of a large growing season rainfall event in 2003 resulted in lower rates of N_2O emissions and no clear pattern in N_2O emissions (Fig. 14.5) or denitrification (Fig. 14.6) with slope position. This emphasizes the situational nature of the controls of the denitrification process and its products.

14.4 Implications of Management on N_2O Emissions

14.4.1 Crop Species

One of the most profound influences on soil biology and its activity is the choice of crop species grown. Crop species differ in the nature and extent of the plant root system and its associated rhizosphere, fertility requirements, residue quality (C/N ratio, lignin content), and soil and water management practices. These conditions also influence the potential for N_2O emissions. Potato is often grown in rotation with other crops, often in 2- or 3-year rotations. The potential for N_2O production during the non-potato phases should also be considered in the overall N_2O emission profile of the potato production system and may offer an opportunity to reduce the N_2O emissions from this production system. Snowden (2010) demonstrated that the preceding crop has an impact on the N_2O emissions occurring during potato

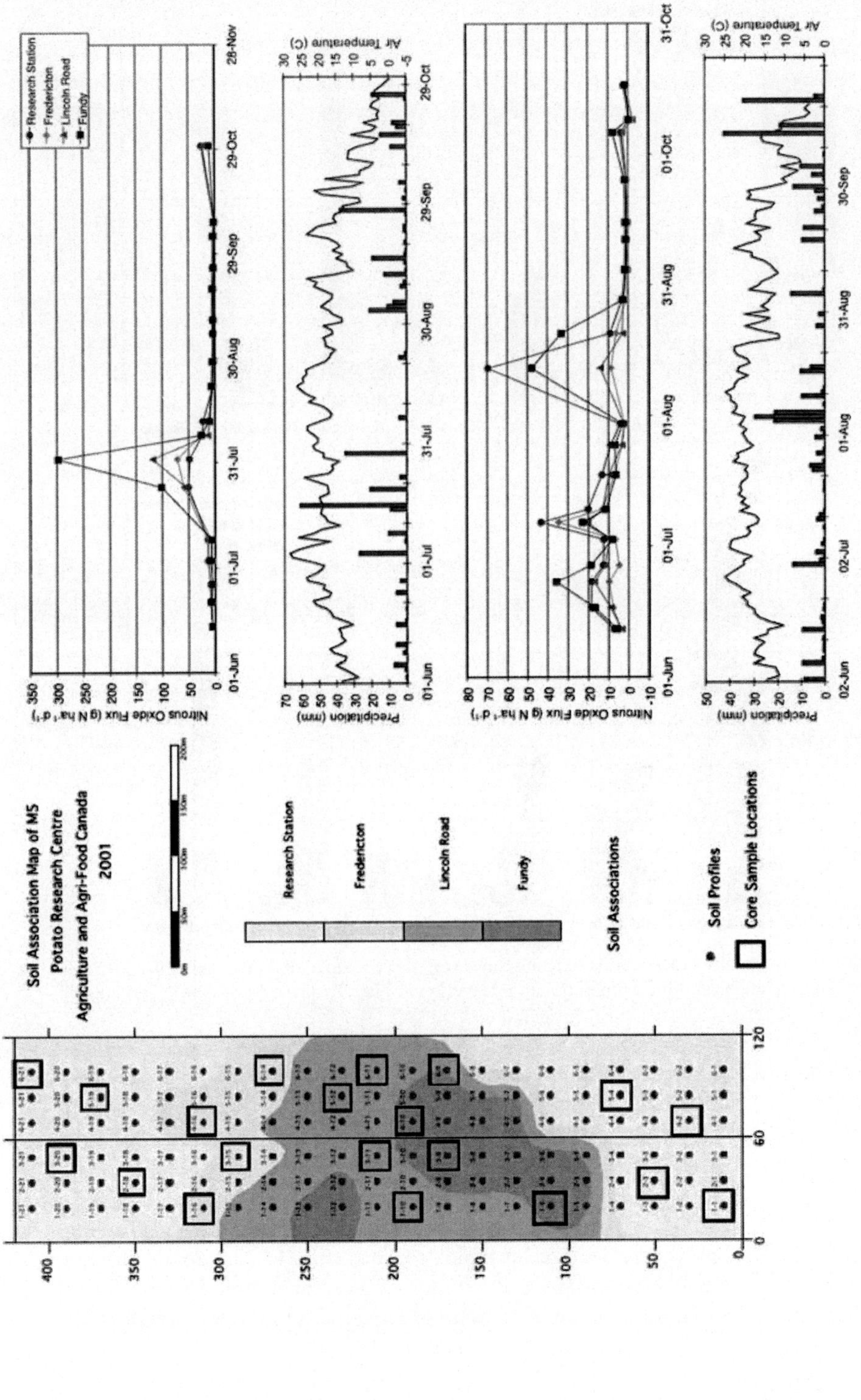

Fig. 14.5 Influence of precipitation on the timing of N_2O emission events as a function of landscape. The Research Station soil association is located at the top of the slope and Fundy at the bottom. The relative internal drainage of the associations decrease in the order Research Station > Fredericton > Lincoln Road > Fundy

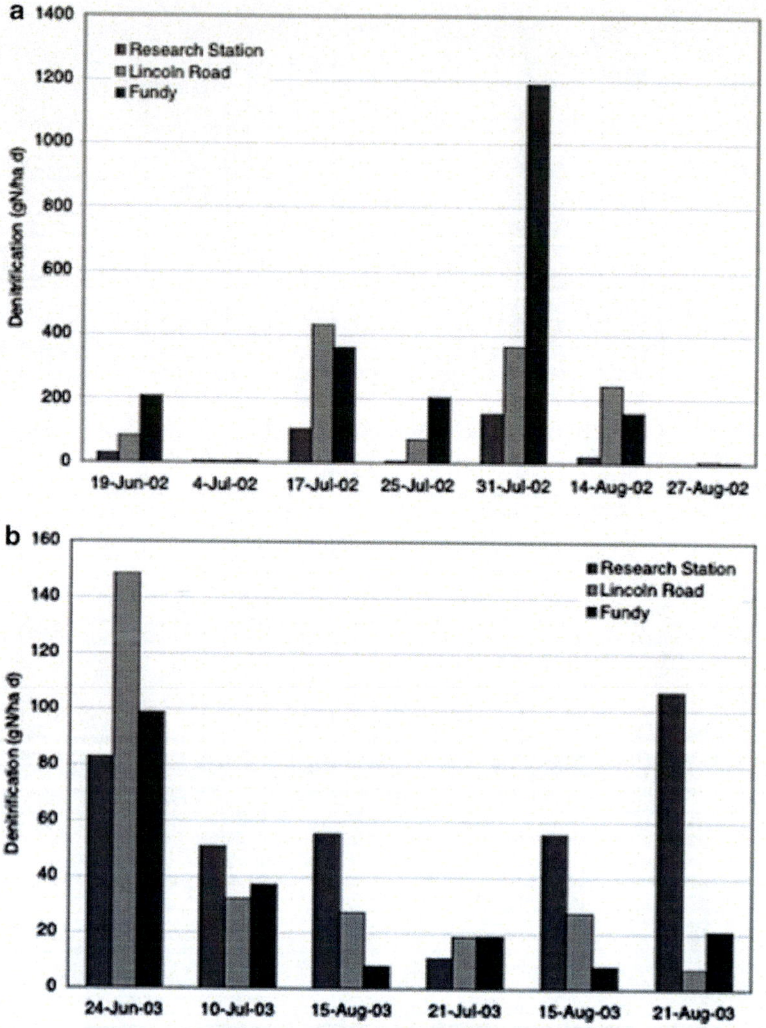

Fig. 14.6 Denitrification rate measured using the acetylene blockage method as influenced by soil association in a landscape study under potato production in New Brunswick: (**a**) 2002, (**b**) 2003

growth and attributed this to differences in the C/N ratio and quantity of crop residue returned and the subsequent impacts on C and NO_3^- availability. An examination of the N_2O emissions associated with each phase of a potato–barley–red clover rotation indicated that N_2O emissions were generally related to the accumulation of NO_3^- in the soil (Fig. 14.7; Burton 2005). Nitrate exposure, an integrated measure of NO_3^- concentration over time, explained 65% of the variation in cumulative growing season N_2O emissions across the three cropping seasons (Fig. 14.7). Periods of large emissions were generally associated with periods of NO_3^- accumulation as a result of N fertilization or the plow-down of N-rich red clover residues. One exception to

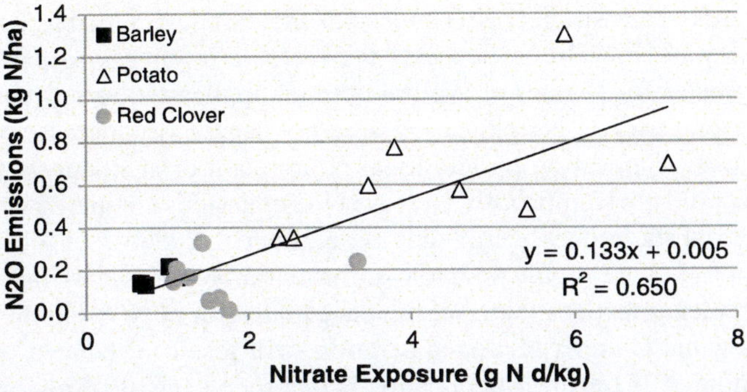

Fig. 14.7 Relationship between nitrate exposure and cumulative growing season N_2O emissions as measured in a potato-barley-red clover crop rotation located at the Crops and Livestock Research Station, Charlottetown, PE (Burton 2005)

this generalization was large N_2O emissions in the spring of the red clover phase that was attributed to precipitation events. Selection of rotational crops that maintain small soil NO_3^- concentrations through plant uptake or smaller rates of N application will generally result in smaller N_2O emissions. For potatoes grown in 2-year rotations, differences in cumulative N_2O emissions during the potato phase could be explained primarily by the effect of the preceding crop on soil nitrate availability (measured as nitrate exposure). The exception was where the preceding crop was a legume or non-legume forage crop, where N_2O emissions were greater than would be expected based on nitrate exposure alone, a finding attributed to increased N_2O emissions early in the growing season due to enhanced respiration (Snowden 2010).

14.4.2 Fertilizer N Management

Emissions of N_2O can be influenced by the rate, timing, placement and form of fertilizer N applied. In general, N_2O emissions increase as a function of N application rate, and this has also been reported for potato production (Ruser et al. 2001). This concept forms the basis of the IPCC coefficient for N_2O emissions associated with fertilizer application that assumes that 1% of the applied N fertilizer is emitted as N_2O on an annual basis (IPCC 2007).

While research is limited on the effect of fertilizer N rate on N_2O emissions in potato production in Atlantic Canada, greater N_2O emissions were measured for potatoes grown with 200 kg N ha^{-1} applied than with an unfertilized control due to increased soil NO_3^- concentration (Burton et al. 2008a). This concept is also demonstrated in the relationship between nitrate exposure and cumulative growing season N_2O emissions (Fig. 14.7). Nitrate exposure increases with increasing N rate (Burton et al. 2008a; Zebarth et al. 2008a) and also results in increased N_2O

emissions. In other crop species, N_2O emissions increased with increasing fertilizer N rate on barley in New Brunswick, and the increase in emissions per unit of N applied was greater for N rates above the optimal application rate (Zebarth et al. 2008a). However, fertilizer N application does not necessarily increase N_2O emissions in all cases. Application of additional N pre-plant or at sidedress to corn in New Brunswick did not significantly increase N_2O emissions compared with a treatment that received starter fertilizer N only (Zebarth et al. 2008b).

Emissions of N_2O are highly sporadic and event driven. It seems reasonable to expect management practices that avoid thc accumulation of NO_3^-, in the presence of available organic C under restricted aeration, will result in reduced N_2O emissions. The timing of N fertilizer application is one such example. Split application of NH_4NO_3 has been shown to result in reduced N_2O emissions (Burton et al. 2008a), but this only occurred in a year when there was significant precipitation between the initial application of N fertilizer and the split application. Thus, while split application of N fertilizer has the potential to reduce N_2O emissions and increase N use efficiency, this practice will only do so in the years where the conditions are favorable for N losses during the timing between planting and the split application of N.

The placement of N fertilizer in the potato ridge is a common practice in Atlantic Canada. Earlier in this chapter we described why this results in an increase in the N_2O molar ratio, resulting in more N_2O emissions from the ridge than if no NO_3^- was present. Other studies have shown that the broadcast of N fertilizer results in an increase in the N_2O molar ratio in both ridge and furrow (Ruser et al. 1998). Placement of the N fertilizer in the potato ridge lowers the soil NO_3^- concentration in the furrow, the area of most active denitrifying activity (Table 14.1; Burton et al. 2008a) and therefore results in lower overall N_2O emissions.

Controlled release N fertilizer products have been shown to increase nitrogen-use efficiency (Shoji et al. 2001) and reduce N_2O emissions (Hyatt et al. 2010). Plant N uptake can be greater with controlled release N fertilizer products than conventional products, particularly on sandy soils (Shoji et al. 2001; Wilson et al. 2009; Ziadi et al. 2011). Hyatt et al. (2010) reported reduced N_2O emissions with application of a polymer-coated urea product compared with conventional fertilizer N products in irrigated potato production on sandy soils in Minnesota. Controlled-release products are most likely to be most beneficial in reducing N_2O emissions when the rates of application are reduced to reflect the increased efficiency of delivery of the controlled release product.

14.4.3 Influence of Tile Drainage

Tile-drainage influences N_2O emissions in several ways. The most obvious is by reducing the extent and duration of water accumulation within the soil profile. Greater water accumulation increases the potential for anoxic processes such as denitrification, potentially resulting in greater N_2O production. Alternatively, greater water accumulation may also result in a reduced N_2O molar ratio under very wet conditions, potentially reducing N_2O emissions. As a result, the introduction of tile drainage may increase or decrease N_2O emissions. The relative importance of these two opposing influences will vary from year-to-year and from site-to-site.

The less obvious implication of tile drainage is its role in transporting NO_3^- and dissolved organic carbon into the soil subsurface and for the production of N_2O and its loss as dissolved N_2O in tile drainage water. As indicated earlier (Fig. 14.4) the loss of N_2O dissolved in tile drainage water could be similar to, or much less than, surface emissions depending on management practices and climatic conditions.

14.4.4 Application of Manure

Animal manures and other organic nutrient sources are increasingly applied as nutrient amendments in potato rotations and as a means of increasing soil organic matter content. These amendments have the potential to increase denitrification and/or N_2O emissions as a result of both increased N and C availability. A study of the form of manure (solid vs. liquid) and the timing of manure application prior to potato production (autumn vs. spring), applied at equivalent rates, demonstrated that neither form of manure nor timing of application influenced N_2O emissions compared with NH_4NO_3 as an N source (Table 14.2). Greater differences were observed between the various phases of the crop rotation (i.e., crop species) than the manure application

14.5 Conclusions

The basic biology of the denitrification process is well understood and the environmental factors that control denitrification have been well studied under controlled conditions. The supply of electron acceptors (O_2, NO_3^-) in combination with the demand for electron acceptors (C availability) determines the extent of denitrification and N_2O molar ratio of the products of denitrification. In the potato production system the formation of a potato ridge results in two distinct environments – a more aerobic ridge and a less aerobic furrow. In Atlantic Canada the banding of fertilizer in the ridge results in increased N_2O emissions from the ridge but reduced emissions from the furrow resulting in lower overall emissions. As a result the N_2O emissions from potato production in Atlantic Canada during the growing season are relatively modest, frequently less than the assumed value for annual emissions of 1% of applied fertilizer N application used by the IPCC (2007). Furthermore, the split application of N fertilizer has the potential to reduce N_2O emissions in years when there is rainfall in the early part of the growing season. The potential for N_2O loss during the non-growing season and/or the loss of N_2O in tile drainage water has yet to be adequately assessed. Annual N_2O emissions are usually related to N application rate but are also influenced by other management factors such as the timing of

Table 14.2 Cumulative N_2O emissions as influenced by location (ridge vs. furrow), nitrogen source and timing of application in (a) 2001-potato; (b) 2002-barley under-seeded to red clover; (c) 2003-red clover; and (d) Spring 2004 (following fall manure application and plow-down of red clover). Note that manure was applied in the fall 2000/spring 2001 and again in fall 2003/spring 2004 (Burton et al. 2005)

(a) Potato 2001

Nitrogen source	Location	Gross emissions (kg N ha^{-1})	Field estimates (kg N ha^{-1})
Solid manure (Fall)	Hill	0.42	0.46
	Furrow	0.50	
Solid manure (Spring)	Hill	0.42	0.42
	Furrow	0.41	
Liquid manure (Spring)	Hill	0.71	0.61
	Furrow	0.51	
Inorganic fertilizer (Spring)	Hill	0.71	0.61
	Furrow	0.51	

(b) Barley 2002

Nitrogen source	Field estimates (kg N ha^{-1})
Solid manure (Fall)	0.13
Solid manure (Spring)	0.15
Liquid manure (Spring)	0.20
Inorganic fertilizer (Spring)	0.12

(c) Red Clover 2003

Nitrogen source	Field estimates (kg N ha^{-1})
Solid manure (Fall)	0.77
Solid manure (Spring)	0.57
Liquid manure (Spring)	0.49
Inorganic fertilizer (Spring)	0.52

(d) Spring 2004 (following plow down)

Nitrogen source	Field estimates (kg N ha^{-1})
Solid manure (Fall)	0.74
Solid manure (Spring)	0.66
Liquid manure (Spring)	0.75
Inorganic fertilizer (Spring)	0.40

N fertilizer application, manure application and the preceding crop. In general, management practices that reduce the accumulation of NO_3^- in the soil will likely result in reduced N_2O emissions. This underscores the need for better tools to manage nitrogen in potato production (Chaps. 13 and 14) and effective monitoring tools (nitrate exposure).

References

Alvarez R, Alvarez DR, Lorenzo G (2001) Carbon dioxide fluxes following tillage from a mollisol in the Argentine rolling pampa. Eur J Soil Biol 37:161–166

Baggs E, Rees R, Smith K, Vinten A (2000) Nitrous oxide emission from soils after incorporating crop residues. Soil Use Manage 16:82–87

Ball B, Scott A, Parker J (1999) Field N_2O, CO_2 and CH_4 fluxes in relation to tillage, compaction and soil quality in Scotland. Soil Tillage Res 53:29–39

Ball B, Crichton I, Horgan G (2008) Dynamics of upward and downward N_2O and CO_2 fluxes in ploughed or no-tilled soils in relation to water-filled pore space, compaction and crop presence. Soil Tillage Res 101:20–30

Bateman E, Baggs E (2005) Contributions of nitrification and denitrification to N_2O emissions from soils at different water-filled pore space. Biol Fertil Soils 41:379–388

Beauchamp EG, Trevors JT, Paul JW (1989) Carbon sources for bacterial denitrification. Adv Soil Sci 10:113–142

Betlach MR, Tiedje JM (1981) Kinetic explanation for accumulation of nitrite, nitric oxide, and nitrous oxide during bacterial denitrification. Appl Environ Microbiol 42:1074–1084

Birch HF (1958) The effect of soil drying on humus decomposition and nitrogen availability. Plant Soil 10:9–31

Blackmer AM, Bremner JM (1978) Inhibitory effect of nitrate on reduction of N_2O to N_2 by soil microorganisms. Soil Biol Biochem 10:187–191

Burton DL (2005) Best management practices for the reduction of greenhouse gas emissions. Final report to Climate Change Funding Initiative in Agriculture, Agriculture and Agri-Food Canada

Burton DL, Beauchamp EG (1994) Profile nitrous oxide and carbon dioxide concentrations in a soil subject to freezing. Soil Sci Soc Am J 58:115–122

Burton DL, Zebarth BJ, Gillam KM, MacLeod JA (2008a) Effect of split application of fertilizer nitrogen on N_2O emissions from potato. Can J Soil Sci 88:229–239

Burton DL, Thangaraj B, Gordon R (2008b) Nitrous oxide emissions and tile drain water quality during spring thaw. Nova Scotia Technology Development Program Final Report. Project DEV21-094

Cho CM, Burton DL, Chang C (1997) Denitrification and fluxes of nitrogenous gases from soil under steady state oxygen distribution. Can J Soil Sci 77:261–269

Christensen S, Tiedje JM (1990) Brief and vigorous N_2O production by soil at spring thaw. J Soil Sci 41:1–4

Crutzen PJ, Ehhalt DH (1977) Effects of nitrogen fertilizers and combustion on the stratospheric ozone layer. Ambio 6:112–117

Dandie CE, Miller MN, Burton DL, Zebarth BJ, Trevors JT, Goyer C (2007) Nitric-oxide reductase-targeted real-time PCR quantification of denitrifier populations in soil. Appl Environ Microbiol 73:4250–4258

Dandie CE, Burton DL, Zebarth BJ, Henderson SL, Trevors JT, Goyer C (2008) Changes in bacterial denitrifier community abundance over time in an agricultural field and their relationship with denitrification activity. Appl Environ Microbiol 74:5997–6005

Dendooven L, Splatt P, Anderson J (1996) Denitrification in permanent pasture soil as affected by different forms of C substrate. Soil Biol Biochem 28:141–149

Environment Canada (2010) National Inventory Report 1990–2008: greenhouse gas sources and sinks in Canada. ISBN 978-1-100-15579-1. http://www.ec.gc.ca/ges-ghg. Accessed 2 May 2011

Firestone MK, Davidson EA (1989) Microbiological basis of NO and N_2O production and consumption in soil. In: Andreae MO, Schimel DS (eds) Exchange of trace gases between terrestrial ecosystems and the atmosphere. John Wiley and Sons, New York

Firestone MK, Smith MS, Firestone RB, Tiedje JM (1979) The influence of nitrate, nitrite, and oxygen on the composition of the gaseous products of denitrification in soil. Soil Sci Soc Am J 43:1140–1144

Flessa H, Ruser R, Schilling R, Loftfield N, Munch J, Kaiser E, Beese F (2002) N_2O and CH_4 fluxes in potato fields: automated measurement, management effects and temporal variation. Geoderma 105:307–325

Focht DD (1992) Diffusional constraints on microbial processes in soil. Soil Sci 154:300–307

Gillam KM, Zebarth BJ, Burton DL (2008) Nitrous oxide emissions from denitrification and the partitioning of gaseous loss as affected by nitrate and carbon addition and soil aeration. Can J Soil Sci 88:133–143

Granli T, Bøckman OC (1994) Nitrous oxide from agriculture. Norw J Agric Sci Suppl 12:128pp
Haile-Mariam S, Collins HP, Higgins SS (2008) Greenhouse gas fluxes from an irrigated sweet corn (*Zea mays* L.) – potato (*Solanum tuberosum* L.) rotation. J Environ Qual 37:759–771
Henderson SL, Dandie CE, Patten CL, Zebarth BJ, Burton DL, Trevors JT, Goyer C (2010) Changes in denitrifier abundance, denitrification gene mRNA levels, nitrous oxide emissions, and denitrification in anoxic soil microcosms amended with glucose and plant residues. Appl Environ Microbiol 76:2155–2164
Hojberg O, Revsbech NP, Tiedje JM (1994) Denitrification in soil aggregates analyzed with microsensors for nitrous oxide and oxygen. Soil Sci Soc Am J 58:1691–1698
Hyatt CR, Venterea RT, Roen CJ, McNearney M, Wilson ML, Dolan MS (2010) Polymer-coated urea maintains potato yields and reduces nitrous oxide emissions in a Minnesota loamy sand. Soil Sci Soc Am J 74:419–428
IPCC (2007) Climate Change 2007: the physical science basis. Contribution of Working Group I to the Fourth Assessment Report of the Intergovernmental Panel on Climate Change (Solomon S, Qin D, Manning M, Chen Z, Marquis M, Averyt KB, Tignor M, Miller HL (eds)). Cambridge University Press, Cambridge, United Kingdom and New York
Matson PA, Parton WJ, Power AG, Swift MJ (1997) Agricultural intensification and ecosystem properties. Science 277:504–509
Miller MN, Zebarth BJ, Dandie CE, Burton DL, Goyer C, Trevors JT (2008) Crop residue influence on denitrification, N_2O emissions and denitrifier community abundance in soil. Soil Biol Biochem 40:2553–2562
Miller MN, Zebarth BJ, Dandie CE, Burton DL, Goyer C, Trevors JT (2009) Influence of liquid manure on soil denitrifier abundance, denitrification, and nitrous oxide emissions. Soil Soc Am J 73:760–768
Mosier A (1998) Soil processes and global change. Biol Fert Soils 27:221–229
Murray RE, Parsons LL, Smith MS (1990) Aerobic and anaerobic growth of rifampin-resistant denitrifying bacteria in soil. Appl Environ Microbiol 56:323–328
Paul JW, Etches V, Zebarth BJ (1997) Increased denitrification below the root zone in the fall following a spring manure application. Can J Soil Sci 77:249–251
Robertson GP, Tiedje JM (1987) Nitrous oxide sources in aerobic soils: nitrification, denitrification and other biological processes. Soil Biol Biochem 19:187–193
Ruser R, Flessa H, Schilling R, Steindl H, Beese F (1998) Effects of soil compaction and fertilization on N_2O and CH_4 fluxes in potato fields. Soil Sci Soc Am J 62:1587–1598
Ruser R, Flessa H, Schilling R, Beese F, Munch JC (2001) Effect of crop-specific field management and N fertilization on N_2O emissions from a fine-loamy soil. Nutr Cycl Agroecosyst 59:177–191
Ruser R, Flessa H, Russow R, Schmidt G, Buegger F, Munch JC (2006) Emission of N_2O, N_2 and CO_2, from soil fertilized with nitrate: effect of compaction, soil moisture and rewetting. Soil Biol Biochem 28:263–274
Ryden JC (1983) Denitrification loss from a grassland soil in the field receiving different rates of nitrogen as ammonium nitrate. J Soil Sci 34:355–365
Saffigna PG, Tanner CB, Keeney DR (1976) Non-uniform infiltration under potato canopies caused by interception, stemflow and hilling. Agron J 68:337–342
Saleh-Lakha S, Shannon KE, Henderson SL, Goyer C, Trevors JT, Zebarth BJ, Burton DL (2009) Effect of nitrate and acetylene on *nirS*, *cnorB* and *nosZ* expression and denitrification activity in *Pseudomonas mandelii*. Appl Envir Micro 75:5082–5087
Sey BK, Manceur AM, Whalen JK, Gregorich EG, Rochette P (2010) Root-derived respiration and nitrous oxide production as affected by crop phenology and nitrogen fertilization. Pant Soil 326:369–379
Shannon KEM, Saleh-Lakha S, Burton DL, Zebarth BJ, Goyer C, Trevors JT (2011) Response of denitrification and nitric oxide reductase (*cnorB*) gene abundance and mRNA levels to additions of nitrate and glucose in anoxic soil microcosms. Antonie van Leeuwenhoek J Microbiol 100:183–195
Shoji S, Delgado J, Mosier A, Miura Y (2001) Use of controlled release fertilizers and nitrification inhibitors to increase nitrogen use efficiency and to conserve air and water quality. Commun Soil Sci Plant Anal 32:1051–1070

Smith KA (1980) A model of the extent of anaerobic zones in aggregated soils, and its potential application to estimates of denitrification. J Soil Sci 31:263–277

Snowden E (2010) Effects of soil nitrate and carbon availability and soil aeration on nitrous oxide emissions in two-year potato rotations. M.Sc. Thesis, Nova Scotia Agricultural College, Truro, NS, Canada, 89pp

Tenuta M, Beauchamp E (1996) Denitrification following herbicide application to a grass sward. Can J Soil Sci 76:15–22

Thomas S, Barlow H, Francis G, Hedderley D (2004) Emission of nitrous oxide from fertilised potatoes. In: Proceedings of the Australia New Zealand SuperSoil 2004 Conference, Sydney, Australia, 6–9 December 2004. http://www.regional.org.au/au/asssi

Tillman D, Cassman KG, Matson PA, Naylor R, Polasky S (2002) Agricultural sustainability and intensive production practices. Nature 418:671–677

Wagner-Riddle C, Thurtell GW (1998) Nitrous oxide emissions from agricultural fields during winter and spring thaw as affected by management practices. Nutr Cycl Agroecosyst 52:151–163

Weier K, Doran J, Power J, Walters D (1993) Denitrification and the dinitrogen to nitrous oxide ratio as affected by soil water, available carbon and nitrate. Soil Sci Soc Am J 57:66–72

Wertz S, Dandie CE, Goyer C, Trevors JT, Patten CL (2009) Diversity of *nirK* denitrifying genes and transcripts in an agricultural soil. Appl Environ Microbiol 75:7365–7377

Wilson ML, Rosen CJ, Moncrief JF (2009) Potato response to polymer-coated urea on an irrigated coarse-textured soil. Agron J 101:897–905

Zebarth BJ, Milburn PH (2003) Spatial and temporal distribution of soil inorganic nitrogen concentration in potato hills. Can J Soil Sci 83:183–195

Zebarth BJ, Rosen CJ (2007) Research perspective on nitrogen BMP development for potato. Am J Potato Res 84:3–18

Zebarth BJ, Leclerc Y, Moreau G, Gareau R, Milburn PH (2003) Soil inorganic nitrogen content in commercial potato fields in New Brunswick. Can J Soil Sci 83:425–429

Zebarth BJ, Rochette P, Burton DL (2008a) Growing season N_2O emissions from spring barley production as influenced by fertilizer nitrogen rate. Can J Soil Sci 88:197–205

Zebarth BJ, Rochette P, Burton DL, Price M (2008b) Effect of fertilizer nitrogen management on N_2O emissions in commercial corn fields. Can J Soil Sci 88:189–195

Zebarth BJ, Drury CF, Tremblay N, Cambouris AN (2009a) Opportunities for improved fertilizer nitrogen management in production of arable crops in eastern Canada: a review. Can J Soil Sci 89:113–132

Zebarth BJ, Scott P, Sharifi M (2009b) Effect of straw and fertilizer nitrogen management for spring barley on soil nitrogen supply to a subsequent potato crop. Am J Potato Res 86:209–217

Zebarth BJ, Arsenault WJ, Moorehead S, Kunelius HT, Sharifi M (2009c) Italian ryegrass management effects on nitrogen supply to a subsequent potato crop. Agron J 101:1573–1580

Ziadi N, Grant C, Samson N, Nyiraneza J, Bélanger G, Parent L-É (2011) Efficiency of controlled-release urea for a potato production system in Quebec, Canada. Agron J 103:60–66

Part V
Sustaining Potato Production in the Cool-Temperate Climate of Tasmania, Australia

Chapter 15
Potato Production in Tasmania, Australia – An Overview of Climate, Soils and Practices

Leigh A. Sparrow and William E. Cotching

Abstract Tasmania is an important potato production area of Australia. Its unique combination of climate and soils help determine the scale of operations and shape the diverse farming system practiced by growers. This chapter briefly describes Tasmania's geography, the farming systems of which potato production is part, and introduces some of the management challenges facing the industry, which are the subjects of further discussion in Chaps. 16–19.

15.1 Tasmania, Australia – Its Geography and Its Potato Industry

The island of Tasmania lies off the south east of mainland Australia (Fig. 15.1), and is its smallest state by both area (7 million ha or 0.9%) and population (0.5 million or 2.3%). At latitude spanning 40 to 42 degrees south, Tasmania lies in the path of westerly wind patterns commonly known as the Roaring Forties, and enjoys a cool-temperate climate with mean annual temperature of 9°C in the central highlands and 13°C nearer the coast. Agriculture, including potato production, is predominantly confined to landscapes below 300 m altitude in the north and central east of the state, where annual rainfall is between 600 and 1,000 mm, snowfall is extremely rare, and frosts are generally confined to the winter months of June to August (Fig. 15.2). Potatoes are usually grown in the period October to March.

L.A. Sparrow (✉)
Tasmanian Institute of Agriculture, University of Tasmania,
PO Box 46, 7249, Kings Meadows, Tasmania, Australia
e-mail: Leigh.Sparrow@utas.edu.au

W.E. Cotching
Tasmanian Institute of Agriculture, University of Tasmania,
PO Box 3523, 7315, Burnie, Tasmania, Australia
e-mail: Bill.Cotching@utas.edu.au

Z. He et al. (eds.), *Sustainable Potato Production: Global Case Studies*,
DOI 10.1007/978-94-007-4104-1_15,

Fig. 15.1 Map of Australia showing its island state of Tasmania

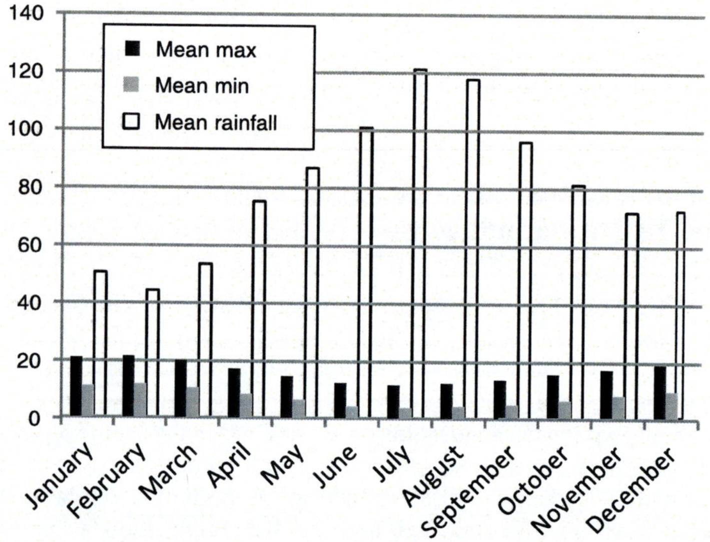

Fig. 15.2 Mean monthly maximum and minimum temperatures (°C, 1966–1996) and rainfall (mm, 1966–2010) for Forthside Research Station in northern Tasmania (lat. 41.2°S; long. 146.3°E)

Potato production in Tasmania is small by global standards, about 300,000 tonnes annually. However, it is the most valuable vegetable crop produced in the State, and constitutes about 25% of Australian potato production (ABS 2010). The crop has been grown ever since European settlement in the nineteenth century and until about 50 years ago comprised mainly fresh potatoes for export to Sydney, Australia's largest city (DPIWE 2003). Since then, French fries have steadily displaced fresh potatoes to the point where the former now constitute about 80% of total production, with only about 10% for the fresh market and 10% for seed (DPIW 2009). Russet Burbank remains the dominant cultivar, grown by almost 80% of producers (Sparrow 2009) and accounting for about 60% of total production. Over the years, the advent of improved practices, particularly irrigation and mechanized planting and harvesting in the 1950s, has seen average processing potato yields increase to 49 Mg/ha in 2008/09 (ABS 2010).

15.2 Pests and Diseases

The relative geographic isolation of Australia means that it enjoys freedom from some potato pests and diseases common in other countries. The Australian potato industry does not currently have to contend with Colorado potato beetle (*Leptinotarsa decemlineata* (Say)), tuber flea beetle (*Epitrix tuberis* Gentner), tomato-potato psyllid (*Bactericera cockerelli* (Sulc)), the A2 mating strain of late blight (*Phytophthora infestans* (Mont.) de Bary), potato wart (*Synchytrium endobioticum* (Schilb.) Percival) or pale potato cyst nematode (*Globodera pallida* (Stone) Behrens). Being an island state of Australia, Tasmania also enjoys freedom from golden potato cyst nematode (*Globodera rostochiensis* (Wollenweber)) and potato virus Y which are found in other Australian states.

As well as being free of certain pests and diseases, in Tasmania the climate and isolation tend to moderate the impact of some of the pests and diseases that are present. This means that, for example, the number of insecticide applications to control insect pests is generally less than for crops in mainland Australia, Europe or North America.

While Tasmanian potato growers face fewer pest and disease challenges than many of their overseas counterparts, soil-borne diseases such as common and powdery scab (caused by the pathogens *Streptomyces scabiei* and *Spongospora subterranea*), as well as black scurf and stem canker (*Rhizoctonia solani*) can be significant problems. Chapters 18 and 19 describe Tasmanian research into the management of these diseases.

15.3 Soils and Land

The major soils used for potato production in Tasmania are known as Red Ferrosols in the Australian Soil Classification (Isbell 2002) or Humic Eutrodox in Soil Taxonomy (Soil Survey Staff 2006). These are deep, well structured, clay loam soils formed on Tertiary basalt flows, with organic carbon contents of 3–6% in the surface horizons. They occur in a rolling landscape where slopes are commonly 5–15% and can be as much as 25%. Water erosion is a significant threat. Other important potato growing soils include Dermosols or Udic Kanhaplustults, which are soils with structured B2 horizons and which lack strong texture contrast between the A and B horizons, Sodosols or Udic Haplustalfs, which are light textured topsoils overlying heavy clay at depths of 300 to 500 mm, and Tenosols or Humic Dystroxerepts, which are deep, windblown sands. These occur on broader plains and river valleys where poor drainage and wind erosion can limit production. The range of potato soils in Tasmania is therefore large, and is reflected in the range of practices required to maintain their physical, chemical and biological fertility. Tasmanian research on the impacts of potato production on soil health, and on the management needed to minimize these impacts, is presented in Chaps. 16 and 17.

15.4 Farmer and Farm Characteristics and Practices

A recent Tasmanian survey of 74 potato growers, about 20% of the Tasmanian potato grower population (Sparrow 2009), showed that most growers are middle-aged, experienced, owner-operators, with 52% between 41 and 60 years old, more than two-thirds having greater than 10 years experience growing potatoes, and over 70% owing all of their potato growing land. Most growers operate on a relatively small scale, with an average annual potato area of 22 ha and two-thirds of farmers growing 6–20 ha of potatoes each year. Tasmanian potato growers are involved in a variety of other agricultural enterprises (Fig. 15.3), with about a third operating five other enterprises. Grazing was one other enterprise for 82% of Tasmanian growers.

Tasmanian potato growers have a significant diversity of crops in their rotation with about 30% growing three or more other crops (Fig. 15.4). This diversity provides breaks for disease management and varied intensity of machinery use. Poppies (*Papaver somniferum* L.) for pharmaceuticals and cereals were the most popular alternate crops, followed by a number of vegetables (Fig. 15.5). The average time between successive potato crops in the same paddock was 6.1 years (Fig. 15.6), with a maximum rest period of 11 years and 90% of growers having 5 or more years

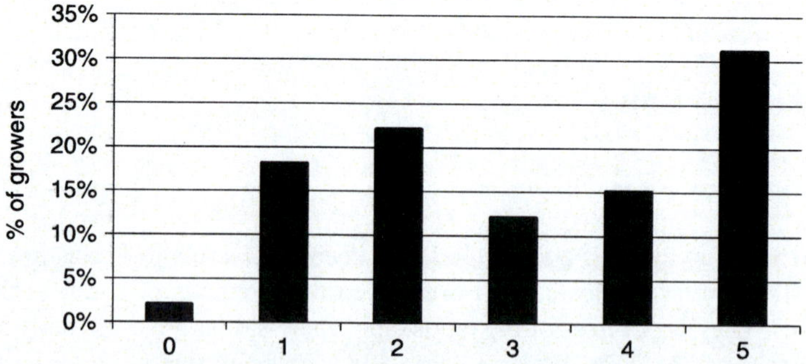

Fig. 15.3 Number of agricultural enterprises other than potatoes operated by Tasmanian growers

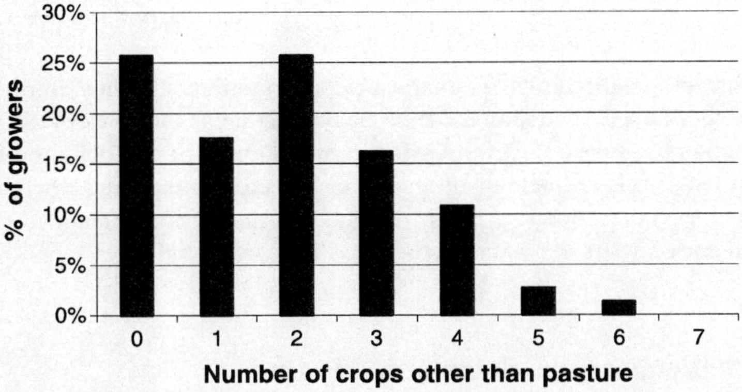

Fig. 15.4 Diversity of crops other than pasture in Tasmanian growers' rotations

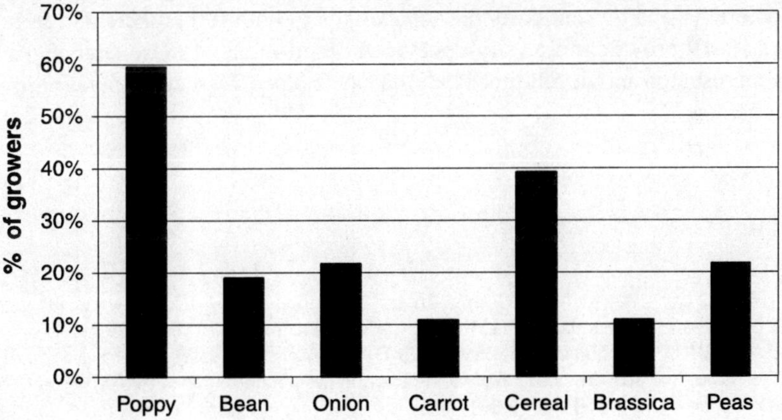

Fig. 15.5 Presence of crops other than potatoes in Tasmanian rotations

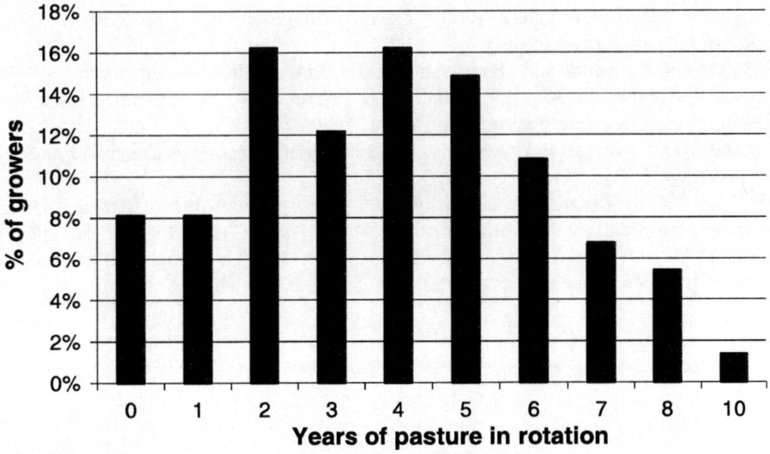

Fig. 15.6 Presence of pasture in Tasmanian potato rotations

between successive plantings. Tasmanian potato rotations are thus considerably longer and more diversified than those in many other areas of the world, including North America (Carter et al. 2009; Griffin et al. 2009; Po et al. 2009) and Europe (Van Loon 1992), where intervals of 3 years or less are common and where cereals predominate as other rotation crops. About two-thirds of Tasmanian growers included at least 3 years of pasture in their rotations (Fig. 15.6).

15.5 Challenges for a Sustainable Industry

The Tasmanian potato industry is somewhat unique globally because, despite its small scale, it comprises a variety of land, soil and rotation combinations. These combinations present particular challenges to both researchers and farmers. Chapters 16–19 provide more fully-developed descriptions of these challenges and how recent research and development efforts have helped Tasmanian potato growers to meet them.

References

ABS (2010) 7121.0 – Agricultural Commodities, Australia, 2008–09 http://www.abs.gov.au/AUSSTATS/abs@.nsf/DetailsPage/7121.02008-09?OpenDocument. Accessed 29 January 2011

Carter MR, Peter RD, Noronha C, Kimpinski J (2009) Influence of 10 years of conservation tillage on some biological properties of a fine sandy loam in the potato phase of two crop rotations in Atlantic Canada. Can J Soil Sci 89:391–402

DPIW (2009) The Tasmanian potato industry. Department of Primary Industries and Water, Tasmania

DPIWE (2003) Tasmania's potato history. In: Taylor T (ed) Department of Primary Industries, Water and Environment, Tasmania

Griffin TS, Larkin RP, Honeycutt CW (2009) Delayed tillage and cover crop effects in potato systems. Am J Potato Res 86:79–87

Isbell RF (2002) The Australian soil classification. CSIRO Publishing, Collingwood

Po EA, Snapp SS, Kravchenko A (2009) Rotational and cover crop determinants of soil structural stability and carbon in a potato system. Agron J 101:175–183

Soil Survey Staff (2006) Keys to soil taxonomy, 10th edn. Natural Resources Conservation Service, USDA, Washington DC

Sparrow L (2009) "Crop Rotations". In: Kirkwood I (ed) Australian Potato Research Program Final Report to Horticulture Australia for project PT04016. Horticulture Australia, Sydney

Van Loon CD (1992) Integrated crop management, the basis for environmentally friendly crop protection of potatoes. Netherlands J Plant Pathol Suppl 2:231–240

Chapter 16
Sustainable Soil Management for Potatoes

William E. Cotching and Leigh A. Sparrow

Abstract This chapter discusses recent work that has assessed soil organic matter decline, soil structure degradation and erosion which are the major factors that impact the sustainable use of soils for potato production in Tasmania, Australia. Soil organic carbon (SOC) has declined by approximately 30% over the past 25 years. Soil compaction has resulted in increased cloddiness, reduced infiltration resulting in increased runoff and soil erosion, reduced trafficability and increased soil strength. Potato tuber yield at paddock scale was found to be significantly correlated with a visual assessment of structure and with penetration resistance. Crop yield decreased and prior soil loss increased with increasing slope. Potato growers in Tasmania are using fewer tillage passes than 5 or 10 years ago to prepare their soil prior to planting. Recommended best practices to prevent further degradation of soil structure and decline in organic carbon content are presented.

16.1 Sustainability

Intensive agriculture, including potato production, has the potential to have dramatic on-site and off-site effects due to the tillage and harvesting practices employed, and the high inputs of fertilizer and agrichemicals. The community as well as farmers are increasingly demanding that agricultural activities do not adversely affect either the natural resources that farmers use, or those of nearby non-agricultural areas.

W.E. Cotching (✉)
Tasmanian Institute of Agriculture, University of Tasmania,
PO Box 3523, 7320 Burnie, Tasmania, Australia
e-mail: Bill.Cotching@utas.edu.au

L.A. Sparrow
Tasmanian Institute of Agriculture, University of Tasmania,
PO Box 46, 7249 Kings Meadows, Tasmania, Australia
e-mail: Leigh.Sparrow@utas.edu.au

Z. He et al. (eds.), *Sustainable Potato Production: Global Case Studies*,
DOI 10.1007/978-94-007-4104-1_16, © Springer Science+Business Media B.V. 2012

In Tasmania, the major factors that impact on the sustainable use of soils for potato production are organic matter decline, soil structure degradation and erosion. Soil degradation associated with growing potatoes arises from deep mixing of profiles, overworking during seedbed preparation, compaction by heavy machinery during harvesting, and harvesting in wet conditions when soils are more susceptible to compaction. This chapter discusses recent work that has assessed these impacts and how they can be minimised. The work is not exclusive to potato production but includes Tasmanian cropping and farm systems of which potatoes are an important part.

16.2 Soil Organic Matter Dynamics

Organic matter is widely regarded as a vital component of a healthy soil (Doran and Parkin 1994; Gregorich et al. 1994). The maintenance of organic matter is critical in the long-term management of soils because of its contribution to soil physical, chemical and biological fertility. Research in Tasmania has shown strong associations between soil carbon and a range of soil physical, chemical and biological properties in all the main soil types used for potato production (Table 16.1).

Table 16.1 Soil properties that have been strongly associated ($P < 0.05$) with soil organic carbon in Tasmanian soils

Soil order	Soil properties correlated to soil carbon
Ferrosol	% of water stable aggregates
	Plastic limit
	Bulk density (↓)
	pH (↓)
	Subsoil field capacity
	Microbial biomass
Sodosol	Cation exchange capacity
	Microbial biomass
	Plastic limit
	Bulk density (↓)
	Porosity
	%Water stable aggregates
	Infiltration rate
	Field capacity (↓)
Dermosol	Microbial biomass
	Plastic limit
	Bulk density (↓)
	Salinity
Tenosol	Bulk density (↓)
	Cation exchange capacity
	Clay
	pH

Data sources: Sparrow et al. 1999; Cotching et al. 2001, 2002a, b
↓ means a decrease with increasing SOC, whilst all other properties showed an increasing trend with increasing SOC

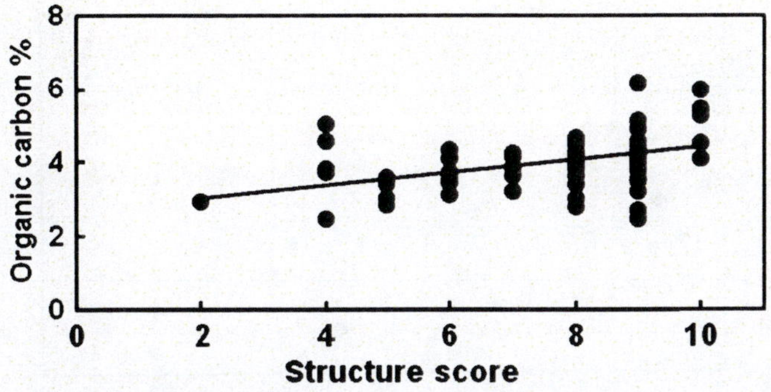

Fig. 16.1 Relationship between soil organic carbon and visible structure score on Ferrosols ($r = 0.42$, $P = 0.01$) (Cotching et al. 2004)

Organic matter was found to have an influence on soil physical properties, such as bulk density with all of the soils studied having significant negative correlations between SOC and bulk density (Table 16.1). A significant correlation between organic carbon and visible structure score was found on Ferrosols (Isbell 2002) (Humic Eutrudox; Soil Survey Staff 2006) (Fig. 16.1). The different pools of SOC are generally considered to turn over at different rates (Baldock and Skjemstad 1999) but these rates have not been fully investigated in Tasmanian soils. Microbial biomass C is considered a sensitive indicator of the effects of management practices over relatively short periods compared with total SOC (Dalal 1998). Significant positive relationships were found between total SOC and other measures of soil organic activity (e.g. microbial biomass C and readily oxidisable carbon) on several soil types in Tasmania. Total SOC appears to be as sensitive an indicator of change in soil health in Tasmania as other measures of the various carbon pools.

Changing from a pasture system to cropping usually results in less carbon being produced. Carbon is removed in harvested produce and is lost by oxidation following tillage and during times of bare fallow resulting in less carbon being returned to the soil (Dalal and Chan 2001). Sparrow et al. (1999) showed that soil carbon concentrations in a selection of Tasmanian Ferrosols including many used for potatoes were less in soils that were more frequently cultivated (Fig. 16.2). However, the data provided no evidence that Tasmanian Ferrosols were approaching a new carbon equilibrium even though current carbon models predict this will happen 10–20 years following land use change (Baldock and Skjemstad 1999). In practice, the almost constant readjustment of crop rotations by farmers in Tasmania means that equilibria may hardly ever be reached. Similar reductions in soil carbon levels due to more frequent cropping have been found in other soil types in Tasmania as shown in Table 16.2 (Cotching et al. 2001, 2002a, b). In most of the soil types studied in Tasmania the greatest change in organic carbon was found in the near surface soil (0–75 mm) with minimal change in the 75–150 mm layer.

Despite the SOC in the surface 150 mm being approximately 30% less in cropping paddocks than pasture paddocks on Ferrosols (Table 16.2), the absolute values in cropping paddocks were of a magnitude which suggests that Tasmanian

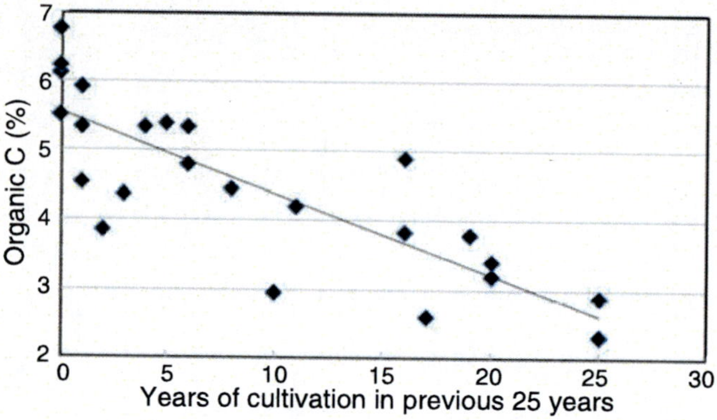

Fig. 16.2 Soil organic carbon change with increasing cropping on Ferrosols in Tasmania (r = −0.78) (Sparrow et al. 1999)

Table 16.2 Organic carbon concentrations (%) in Tasmanian soils subject to different agricultural management (all tests were LECO furnace except for Ferrosols which were Walkley-Black and approximately 90% of that determined by LECO furnace method (Rayment and Higginson 1992))

Soil[a] (depth)	Pasture	Intermittent cropping	Frequent cropping
Ferrosol (0–150 mm)			
Average	6.4	4.9	3.8
Range	4.5–6.8	2.6–5.4	2.3–4.5
Dermosol (0–75 mm)			
Average	7.0	4.3	4.2
Range	3.3–10.6	2.8–7.1	2.0–7.4
Sodosol (0–150 mm)			
Average	2.7	2.3	1.8
Range	1.8–3.3	1.6–2.9	1.4–2.5
Tenosol (0–75 mm)			
Average	2.6	2.1	1.1
Range	2.4–3.0	1.3–2.8	0.4–1.4

Data sources: Sparrow et al. 1999; Cotching et al. 2001, 2002a, b

[a]Ferrosol = Humic Eutrudox; Dermosol = Udic Kanhaplustults; Sodosol = Udic Haplustalfs; Tenosol = Humic Dystroxerepts (Soil Survey Staff 2006)

Ferrosols still have relatively high levels of soil carbon (Bridge and Bell 1994; Cotching et al. 2004; Sparrow et al. 2006). Concentrations of SOC of less than 2% are viewed with concern as these levels have been associated with severe soil structural deterioration and soil-based impediments to plant productivity (Geeves et al. 1995).

16.3 Soil Structure and Related Physical Properties

The use of machinery in modern agriculture is probably the activity with the most potential to degrade soil structure. Tillage and traffic, particularly when soils are wet and have their least bearing strength, can destroy soil structure and result in

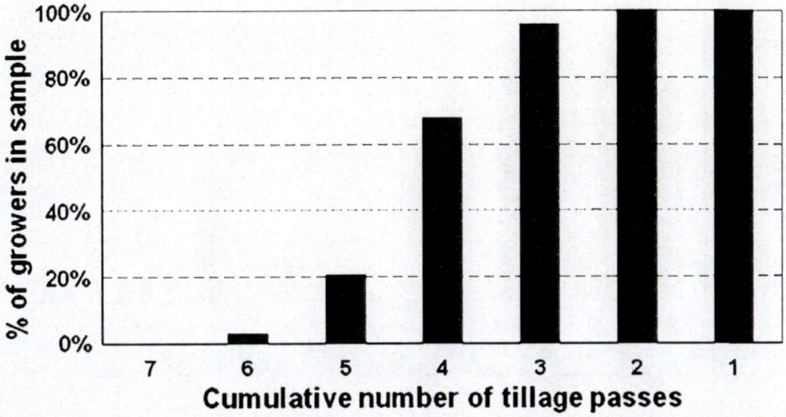

Fig. 16.3 Number of tillage passes used prior to planting potatoes in Tasmania

compaction. Compaction can be identified by a soil's over firmness when dug or cultivated, the presence of clods, or the presence of pans or layers restricting root growth and water movement. Soil physical properties such as structure are notoriously difficult to amend compared to chemical properties. Consequently, it is much more advantageous to prevent soil structure degradation than to remediate it. Excessive tillage can degrade structure by causing the production of fine soil aggregates (0.2–2 mm) or primary particles of sand that are more susceptible to compaction than irregularly shaped coarser (2–10 mm) aggregates. The finer soil particles are also more susceptible to erosion. Tillage also destroys the many fungal hyphae that bind soil particles into aggregates and these hyphae are unable to re-form if tillage is frequent as under annual cropping (Pung et al. 2003).

In a survey of 74 potato growers in Tasmania, Sparrow (2009) found they are using fewer tillage passes than they did 5 or 10 years ago to prepare their soil prior to planting with some using a combination of ripping and rotary cultivation in one pass (Fig. 16.3). The survey also found that most growers still rely on the mouldboard plough in combination with a rotary hoe or power harrow (Fig. 16.4) to cultivate the soil to approximately 0.3 m.

The differences in physical properties between soil orders in Tasmania are pronounced and require correspondingly different tillage, irrigation and drainage management strategies (Cotching and Chilvers 1998). Texture, bulk density, porosity, moisture-holding properties and aggregation vary widely between soil orders, with horizons and horizon boundaries playing important roles in the physical behavior of soils, particularly in relation to water movement. Evidence of the impact of land use (pasture versus intermittent cropping compared to pasture versus frequent cropping) on each of the soil orders used for potato production in Tasmania is presented below.

Many physical attributes of Dermosols showed no effect of land use, not even the presence of potatoes in the rotation, indicating a degree of robustness in these soils (Cotching et al. 2002a). However, cropped paddocks did have larger clods than long

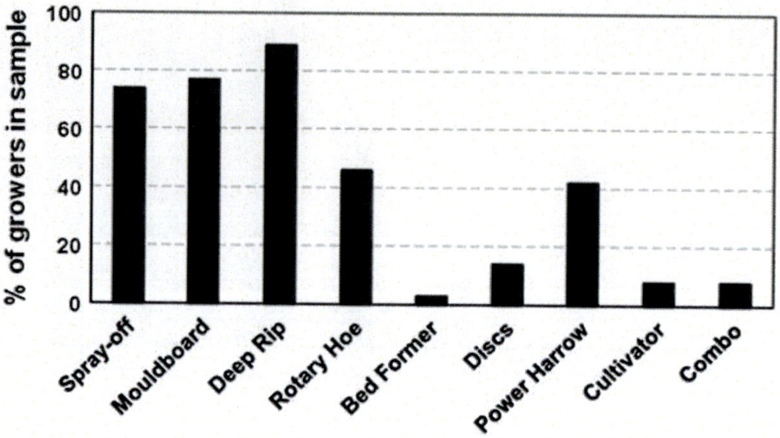

Fig. 16.4 Tillage methods used by potato growers in Tasmania

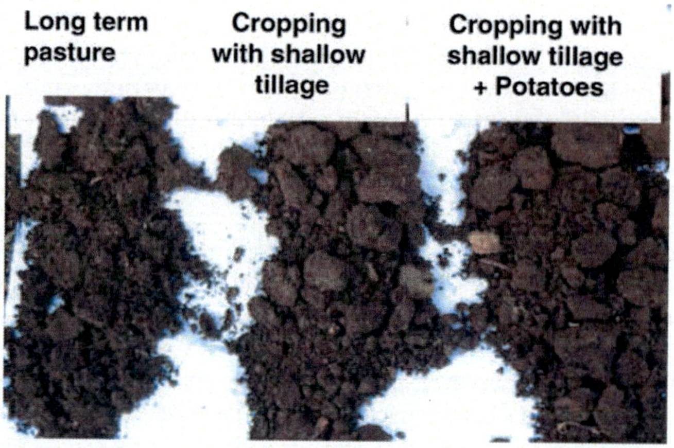

Fig. 16.5 Visible differences in soil structure on Dermosols between long term pasture (left), cropping with shallow tillage (centre) and cropping with shallow tillage plus potatoes (right). The larger clods are 50–100 mm in size

term pasture paddocks (Fig. 16.5) as well as greater surface bulk densities, but they did not have greater penetration resistance or lower infiltration rates, both of which were at levels not likely to impede root growth.

The macroporosity associated with strongly structured Ferrosols is sensitive to compaction both by machinery and livestock. When soils are wet due to seasonal rainfall any working or loading, such as during seedbed preparation or potato harvesting, results in plastic deformation and compaction. The most significant aspect involved in soil structure degradation on Ferrosols appears to be soil compaction.

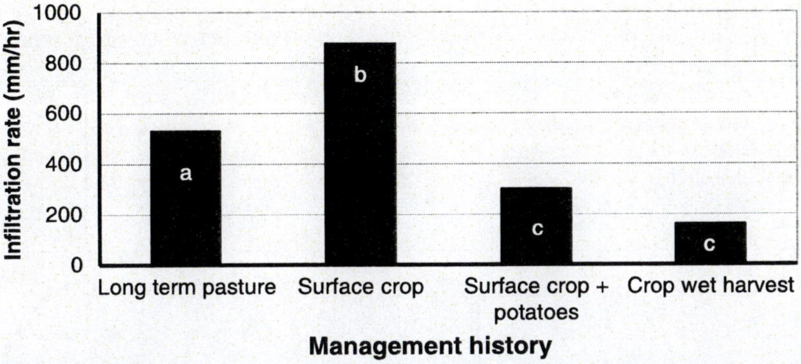

Fig. 16.6 Infiltration rates on Sodosols in Tasmania's Midlands. Values followed by the same letter are not significantly different at $P=0.05$ (Cotching et al. 2001)

Changes associated with, but not necessarily solely due to soil compaction include: increased cloddiness; reduced infiltration resulting in increased runoff and soil erosion; poorer drainage and reduced trafficability; and increased mechanical root impedance through increased soil strength. One of the difficulties in convincing farmers who are managing Ferrosols of the importance of soil structure is that these soils have inherently better physical attributes than many other soils. Even following years of poor husbandry, resulting in significant soil structural decline, degraded Ferrosols can still have better attributes than other soils, and can still be in relatively good condition (Sparrow et al. 1999). Maintaining good soil structure is essential if Ferrosols are to remain under intensive use. Amelioration of increased bulk density and compaction arising from poor cropping practices can only be achieved by mechanical means or active plant growth, as Ferrosols have no natural repair mechanisms, such as shrinking and swelling of clay minerals (Bridge and Bell 1994).

Significant degradation of soil structure was found on Sodosols (sandy topsoils overlying clay subsoil) following cropping with potatoes, particularly after a wet harvest (Cotching et al. 2001). Infiltration rate on Sodosols was greater in paddocks that had grown cereals or crops requiring little tillage than in long term pasture paddocks, but was 43% less in paddocks that had grown potatoes and 70% less after a wet potato harvest (Fig. 16.6). Dry aggregate size showed no change under minimal tillage compared to long term pasture but decreased by 39% in paddocks that had grown potatoes. Aggregate stability in all cropped paddocks was nearly 50% less than in long term pasture paddocks, with intensively tilled potato cropping paddocks having only 35% water stable aggregates which is relatively low. Farmers also reported more unhealthy soil attributes when potatoes were included in their rotation. The lower SOC and poorer physical properties of Sodosols that had grown potatoes (Cotching et al. 2001) indicate that they are less robust than clay textured soils. The choice of crops grown in a cropping rotation and the associated soil tillage and crop harvesting practices are critical for sustainable management of Tasmanian Sodosols.

Growing potatoes on sandy textured Tenosols resulted in greater dry bulk density, reduced total porosity and macroporosity, and subsoil compaction compared to

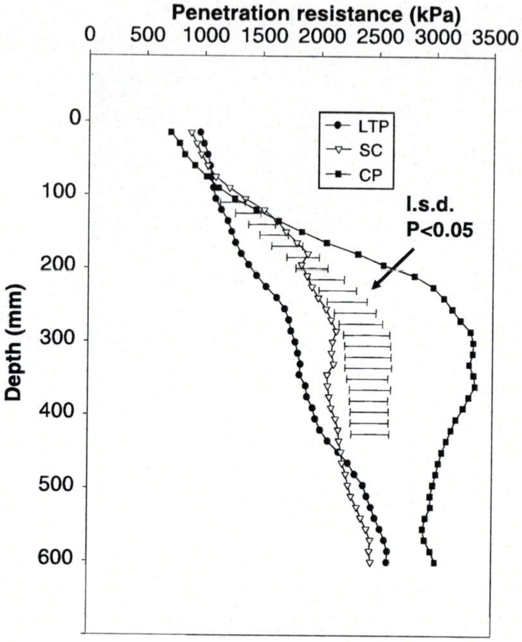

Fig. 16.7 Penetration resistance on Tenosols under long term pasture (LTP), cropping with shallow tillage (SC), and cropping including potatoes (CP) (Cotching et al. 2002b)

other land uses on these soils (Fig. 16.7) (Cotching et al. 2002b). The high penetration resistance (>2000 kPa) indicates that subsoil tillage after potato harvesting may be beneficial to root growth in subsequent crops. Many soil physical properties showed no ill effects due to cropping which may be due to these sandy soils having inherently weak structure even in a healthy state.

The mixing of soils by deep (greater than 0.3 m) tillage, or deeper tillage than is appropriate for the soil type, can bring poorer quality soil nearer to the surface. This is a particular problem in Tasmania's sodic soils in which the clay dominated by sodium occurs beneath the topsoil. Left in their natural horizons, these sodic clays have little impact on plant growth. However, deep tillage that mixes this subsoil clay with topsoil results in dispersive topsoils that are structurally unstable and tend to form surface crusts that inhibit seedling emergence. Once these soils have been mixed, they are difficult to manage and rehabilitate.

16.4 Linking Crop Yield to Soil Structure

Potato tuber yield at paddock scale was found to be significantly correlated with a visual assessment of structure ($r=0.57$, $P<0.05$) and with penetration resistance ($r=0.60$, $P<0.05$) on 28 Ferrosols and 14 Dermosols (Cotching et al. 2004). Average potato yields on 14 Ferrosols (half the sites) with pre-plant visual structure scores of 9 or 10 (i.e. good structure – porous aggregates, with a good range of aggregate sizes and many fibrous roots) were 61.5±11.5 Mg/ha. However, on 6 paddocks (21% of Ferrosol sites) with structure scores of 6 or less (smaller angular clods present, with smooth rather than porous faces, and a narrow size range of natural soil aggregates) average yields were 46.6±11.9 Mg/ha. If 21% of all potato paddocks on Ferrosols suffered the same average yield loss, this would amount to several million dollars of lost production each year due to degraded soil structure.

Sandy loam to sand textured topsoils (Sodosols, Tenosols) are the soils with the least clay and organic matter contents, making their structures inherently weak and very susceptible to physical damage as measured by infiltration rate, porosity and penetration resistance. However, we found no correlation between soil structure measures and potato yields on these sandy soils (Cotching et al. 2004) which may have been due to the inherent variability of soil types within the paddocks sampled.

16.5 Soil Erosion

Water erosion is the main erosion concern in Tasmania due to the winter dominant rainfall pattern (see Chap. 15). Wind erosion can occur on sandy textured soils when the soils have been recently cultivated. These wind erosion events are infrequent but can result in loss of the finer soil fraction, particularly organic matter, and so soil preparation that ensures retention of grass turf and crop stubbles is recommended. Potatoes like many other crops in Tasmania are sown in the spring, which means that tillage is usually required in late winter prior to potato planting to allow for weathering of soil clods, germination of weeds, and preparation of seedbeds by tillage. Sixty percent of potato growers in Tasmania start preparing their soil 1–2 months prior to planting while 21% start 3–5 months prior and 8% are sometimes starting 6 months before planting. In northern Tasmania, where potatoes are grown on Ferrosols with slopes of 10–20%, winter is the period of maximum rainfall and lowest evaporation, which makes the pre-planting period one of high water erosion risk. The risk is further increased by the practice of planting potatoes up and down the slopes rather than on the contour in order to accommodate harvesters (Fig. 16.8). Long term (40+ years) soil loss rates averaging 2.5 – 5.3 Mg/ha/year under intensive cropping rotations have been measured on Ferrosols in northern Tasmania (Richley et al. 1997) and from 10 to 142 Mg/ha/rainfall event (Cotching 1995) with a recurrence interval of less than 2 years. All of these soil loss rates greatly exceed rates of soil formation.

16.5.1 Impacts of Water Erosion

Crop yield and soil health attributes were measured on Ferrosols of varying steepness in paddocks growing processing potatoes to investigate the relationship between slope and past erosion on yield (Cotching et al. 2002c). Crop yield was found to

Fig. 16.8 Potatoes grown on up to 20% slopes in northern Tasmania

decline with increasing slope (Cotching et al. 2002c) and 16–18% crop yield decline was recorded on land of 19–28% slope, which also had the greatest prior soil loss. These yield losses are similar to those reported for corn on steep slopes under fertilised conditions (13–19%) by Pimental et al. (1995) but less than the 50% reported for wheat in an undulating landscape (Papiernik et al. 2005). The amount of erosion to date on slopes up to 18% was not enough to affect crop yield but continued erosion at current rates will inevitably have that effect. Erosion on Ferrosols redistributed soil to concave or toe slope positions resulting in over-thickened topsoils (Cotching et al. 2002c). However, crops grown in concave slope positions, that had accumulated soil, did not compensate with increased yields. There was little apparent difference in topsoil thickness between flat sites and eroded slopes but this was probably due to the uniform depth of cultivation evenly mixing the upper part of the soil profile. There were no significant differences in most soil physical properties between eroded and flat sites which was unexpected because on all sites we did find a greater percentage of large aggregates in the 200–275 mm depth range compared to the surface (0–75 mm depth). Thus it was expected that following erosion of the topsoil, more of the larger aggregates would be incorporated into the topsoil by cultivation.

Water turbidity and calculated volumes of water borne sediment (TSS = total suspended solids) reveal that erosion is an active process at both catchment and paddock scales on the north west coast of Tasmania during frequent rainstorm events (Cotching and Sims 2000). Water quality has been monitored at 173 sites around Tasmania at least once since 1992, which includes monthly sampling from 2003 until June 2009, at 51 of the 56 operational flow monitoring sites around the state as part of the Baseline Water Quality Monitoring Program. This data is available on the Water Information Systems Tasmania (WIST) website (www.water.dpiw.tas.gov.au). However, the Tasmanian State Government has stopped the monitoring program since June 2009. The data gathered have not been used for regulatory purposes and there is no effective legislation relating to the impact of diffuse source pollution on water quality in Tasmania. One gap in the available dataset is that there were no data collected from predominately cropping catchments in Tasmania, despite a significant vegetable cropping industry (Broad et al. 2010). Turbidity readings of more than 200 nephelometric turbidity units (NTUs) have been recorded during rainstorm events with stream turbidity correlated to the percentage of land cropped or fallow in a range of catchments (Cotching and Sims 2003). These very high turbidities recorded in water from farm paddocks and measurements of in-paddock soil erosion rates, lead to the conclusion that soil erosion from cropping paddocks is the major contributor to sediment load in streams in north west Tasmania rather than stream bank erosion as the soils are non-dispersive and stream banks are well vegetated. Soil erosion on Ferrosols was found to result in significantly lower soil carbon content on slopes of 13–18% and 19–28% compared to level or accumulating sites (Cotching et al. 2002c). Steeper slopes (19–28%) had more variable carbon contents (2.1–4.7%) than less steep slopes and this variability was attributed to rill erosion which removes furrows of organic-rich topsoil whilst leaving inter-rill areas unaffected. The loss of soil organic matter through erosion is possibly the single greatest impact of erosion on soil health because of the relationship of organic carbon to other soil properties.

16.6 Soil Management

Tasmanian farmers are now cropping more of the land they farm and have less time under pasture in their rotations, but are not shortening the interval between potato crops (Cotching and Sims 2000). However, there is a constant change in crop rotations driven mainly by profit, but soil degradation, the unavailability of crop contracts and/or their incompatibility with current farming systems are also factors. Farmers need to produce more because the margins are smaller. In addition, there has been continued growth in irrigated cropping in Tasmania, including irrigated potatoes, with much of the expansion in areas dominated by soils with sandy textured topsoils. These soils are the most vulnerable and so greater management skill is required to prevent damage. The more soils are cropped, the more likely it is that timing of operations will be compromised, due to constraints of the weather, finances and contractual obligations. Consequently, there is a greater potential for further soil degradation to occur in the future. Irrigated cropping, that includes potato production, is considered to have some of the greatest impacts on soil physical properties whereas extensive grazing has minimal impact. Physically degraded soils often take a long period of good management to recover, so prevention of soil structure degradation is preferable to amelioration. The use of deep ripper implements is increasing (Fig. 16.4) and there has been a decrease in the depth to which farmers rip as the majority of farmers now rip to less than 300 mm depth (Cotching and Sims 2000). Farmers now use ripping more as a surface soil loosening operation rather than as deep tillage and there is recent trialling of one-pass cultivation to prepare seedbeds for potato planting. Some farmers are able to overcome degraded soil structure through their good management of tillage, irrigation, nutrition, weeds and pathogens.

Recommended best practices to prevent degradation of soil structure include: harvest potatoes before soils become wet (i.e. when drier than the plastic limit); restrict traffic to designated tracks and keep trucks off paddocks; minimise tillage; combine tillage operations in a single pass; and don't use stock for grazing during wet winter days (Chilvers 1996). If signs of degraded soil structure are apparent, growers should seek advice on appropriate soil management practices such as timing and depth of remedial tillage, rotations to include a rejuvenating pasture phase, and growing green manure or cover crops. Formal property management plans are important in the adoption of sustainable land use practices (Curtis et al. 2009). Compared to farmers who do not have a property management plan, farmers who do are more likely to use soil conservation practices on their farm and have a much greater area with soil conservation practices established (Cotching and Sims 2000). The majority of potato growers have 3 or more years of pasture in their rotation and the reasons given for growing pasture are to rest the paddock, improve soil structure, provide fodder and improve organic matter content (chapter 15). Locally, the best known mechanism to rejuvenate degraded soil is to grow a vigorous grass pasture and farmers appreciate that pasture is an important soil management tool that is needed in a good rotation.

The longer a farmer can afford to leave a soil under pasture the better the soil improvement. In Tasmania's climate with a summer dry period, the benefit is enhanced by growing pasture under irrigation in order to ensure high levels of organic matter production. Pasture is probably more effective than cereal crops because pasture grasses (e.g. perennial ryegrass – *Lolium perenne* L.) have a greater proportion of their biomass as roots. Pasture grass roots are fibrous creating a multitude of holes, splitting clods into aggregates and binding soil particles together. Pastures also encourage a higher level of fungal hyphae that bind soil particles into aggregates, and growing perennial pasture encourages proliferation of earthworms as regular cultivation is physically disrupting. Growing green manure crops has increased in popularity with approximately half of the average area cropped per farm under green manure each year (Cotching and Sims 2000). There has been an increase in the number of farmers claiming to be doing something to reduce soil erosion which indicates that farmers are becoming more aware of the issue due to local natural resource management agencies promoting the use of innovative soil erosion control technology (Cotching 2002).

References

Baldock JA, Skjemstad JO (1999) Soil organic carbon/soil organic matter. In: Peverill KI, Sparrow LA, Reuter DJ (eds) Soil analysis: an interpretation manual. CSIRO, Australia

Bridge BJ, Bell MJ (1994) Effect of cropping on the physical fertility of krasnozems. Aust J Soil Res 32:1253–73

Broad ST, Corkrey R, Cotching WE, Coad JR (2010) Estimating nutrient loads and turbidity for Tasmanian catchments. Landscape Logic Technical Report No. 18

Chilvers WJ (1996) Managing Tasmania's Cropping Soils – a practical guide for farmers. Department of Primary Industry and Fisheries, Tasmania

Cotching WE (1995) A review of the challenges for long term management of krasnozems in Australia. Aust J Soil Water Conserv 8:18–27

Cotching WE (2002) Mulched rip lines stop soil erosion under annual cropping on steep slopes in North-West Tasmania. Nat Res Manage 5(2):28–30

Cotching WE, Chilvers WJC (1998) To dig or not to dig. Soil specific cropping management strategies for Tasmania. Nat Resour Manage 1(2):25–27

Cotching WE, Sims CC (2000) Changes in farmer perceptions towards soil conservation and management in North-West Tasmania. Rural Soc 10(3):379–392

Cotching WE, Sims CC (2003) Linking water quality and land use in northwest Tasmania. Nat Resour Manage 6(2):12–20

Cotching WE, Cooper J, Sparrow LA, McCorkell BE, Rowley W (2001) Effects of agricultural management on sodosols in northern Tasmania. Aust J Soil Res 39:711–735

Cotching WE, Cooper J, Sparrow LA, McCorkell BE, Rowley W (2002a) Effects of agricultural management on dermosols in northern Tasmania. Aust J Soil Res 40:65–79

Cotching WE, Cooper J, Sparrow LA, McCorkell BE, Rowley W (2002b) Effects of agricultural management on tenosols in northern Tasmania. Aust J Soil Res 40:4563

Cotching WE, Hawkins K, Sparrow LA, McCorkell BE, Rowley W (2002c) Crop yields and soil properties on eroded slopes of red ferrosols in northwest Tasmania. Aust J Soil Res 40:625–642

Cotching WE, Sparrow LA, Hawkins K, McCorkell BE, Rowley W (2004) Linking Tasmanian potato and poppy yields to selected soil physical and chemical properties. Aust J Exp Agric 44:1241–1249

Curtis A, Race D, Sample R (2009) Management of riparian zones in Tasmanian agricultural catchments Landscape Logic Technical Report No.11. Institute for Land, Water and Society, Charles Sturt University, Albury–Wodonga

Dalal RC (1998) Soil microbial biomass – what do the numbers really mean? Aust J Exp Agric 38:649–665

Dalal RC, Chan KY (2001) Soil organic matter in rainfed cropping systems of the Australian cereal belt. Aust J Soil Res 39:435–464

Doran JW, Parkin TB (1994) Defining and assessing soil quality. In: Doran JW, Coleman DC, Bezdicek DF, Stewart BA (eds) Defining soil quality for a sustainable environment SSSA special publication 35. Soil Science Society of America, Madison

Geeves GW, Cresswell HP, Murphy BW, Gessler PE, Chartres CJ, Little IP, Bowman GM (1995) The physical, chemical and morphological properties of soils in the wheat-belt of southern N.S.W. and Northern Victoria. CSIRO Australia and NSW Department of Conservation and Land Management, Melbourne

Gregorich EG, Carter MR, Angers DA, Monreal CM, Ellert BH (1994) Towards a minimum data set to assess soil organic matter quality in agricultural soils. Can J Soil Sci 74:367–385

Isbell RF (2002) The Australian soil classification Australian soil and land survey series 4. CSIRO Publishing, Melbourne

Papiernik SK, Lindstrom MJ, Schumacher JA, Farenhorst A, Stephens KD, Schumacher TE, Lobb DA (2005) Variation in soil properties and crop yield across an eroded prairie landscape. J Soil Water Conserv 60(6):388–395

Pimental D, Harvey C, Resosudarmo P, Kurz SD, McNair M, Crist S, Shpritz L, Fitton L, Blair SF (1995) Environmental and economic costs of soil erosion and conservation benefits. Science 267:1117–1123

Pung H, Olsen J, Stirling M, Moody P, Pankhurst C, Jackson, S, Hickey M, Cotching WE (2003) A survey approach to investigate the soil factors associated with the productivity and sustainability of vegetable production in Australia. Horticulture Australia Project No. VG99057. Horticulture Australia, Sydney NSW

Rayment GE, Higginson FR (1992) Australian laboratory handbook of soil and water chemical methods Australian soil and land survey handbook vol 3, Inkata, North Ryde NSW
Richley L, Loughran RJ, Elliott GL, Saynor MJ (1997) A national reconnaissance survey of soil erosion, Tasmania a report prepared for the Australian National Landcare Program, Department of Primary Industries and Energy, Canberra
Soil Survey Staff (2006) Keys to soil taxonomy, 10th edn. USDA – Natural Resource Conservation Service, Washington
Sparrow L (2009) "Crop Rotations". In: Kirkwood I (ed) Australian potato research program final report to horticulture Australia for project PT04016. Horticulture Australia, Sydney
Sparrow LA, Cotching WE, Cooper J, Rowley W (1999) Attributes of Tasmanian ferrosols under different agricultural management. Aust J Soil Res 37:603–622
Sparrow LA, Belbin KC, Doyle RB (2006) Organic carbon in coarse and fine particle size fractions as indicators of change in ferrosols and sodosols under mixed farming in Tasmania, Australia. Soil Use Manage 22:219–220

Chapter 17
Potato Nutrient Management in Tasmania, Australia

Leigh A. Sparrow

Abstract This chapter presents data on the nutrient management practices of Tasmanian potato growers and data from Tasmanian and other research on tuber yield and quality responses to fertilizers. Together, this information shows that most growers apply more nitrogen, phosphorus and potassium than is removed in the tubers they harvest and in many cases more than can be justified on the basis of known responses to these elements. This behavior is discussed in relation to the risks of underfertilising on the one hand and the risks of environmental and food safety impacts on the other.

17.1 Introduction

Tasmanian potato growers operate in a global marketplace and therefore are subject to economic pressures that require them to make the most efficient use of inputs including fertiliser, which can constitute as much as 30% of variable costs of production. Foreign imports of processed potato products are increasing (DPIF 2009) which puts downwards pressure on the prices offered to local farmers for their crops, while burgeoning global demand for fertiliser saw its cost more than double in 2008. While fertiliser costs have remained relatively static since, it is difficult to see the general cost–price "squeeze" on growers easing in the immediate future. In addition to increasing economic pressures, Tasmanian potato growers are also subject to growing community expectations that farming will produce safe food and have minimal impact on the natural environment. This chapter summarises the results of Tasmanian research aimed at better understanding the fertiliser needs

---

L.A. Sparrow (✉)
Tasmanian Institute of Agriculture, University of Tasmania,
PO Box 46, 7249, Kings Meadows, Tasmania, Australia
e-mail: Leigh.Sparrow@utas.edu.au

Z. He et al. (eds.), *Sustainable Potato Production: Global Case Studies*,
DOI 10.1007/978-94-007-4104-1_17,

of potatoes, how soil and plant tests can help predict these needs, and also at understanding the barriers Tasmanian farmers face in adopting more efficient management practices and complying with regulatory requirements.

17.2 Fertiliser Use by Tasmanian Potato Growers and Tasmanian Potato Soil Fertility

Potatoes in Tasmania are almost all grown with nitrogen (N), phosphorus (P) and potassium (K) fertiliser. Some micronutrients are occasionally applied even though there is no evidence of responses to their application (see Sect. 17.5). Macronutrients such as sulphur, calcium and magnesium have not been shown to be deficient, and they are not routinely applied in fertiliser. Most of the NPK is banded as a mix of diammonium phosphate and potassium chloride at planting, with some N also applied as urea during the growth of the crop. Potassium sulphate is not used because it is more expensive than potassium chloride and, as noted above, sulphur deficiencies have not been observed in potatoes in Tasmania.

A survey of 54 growers (about 15% of the Tasmanian industry) in 2001 (Cotching and Sparrow unpublished data) showed median N, P and K fertiliser rates in excess of 200 kg N ha^{-1} and 200 kg P ha^{-1} and almost 300 kg ha^{-1} for K (Table 17.1), with little difference in rates across soil types. In a separate Tasmanian study, Lisson et al. (2009) found that seven growers applied 279–442 kg N ha^{-1} to potatoes. The N and K rates above are not atypical of potato production elsewhere (e.g. Maier 1986 in South Australia, Hyatt et al. 2010 in Minnesota, USA), but the rates of P in Tasmania are some of the world's highest (c.f. 40–75 kg P ha^{-1} in Minnesota, USA (Rosen and Bierman 2008), 80–115 kg P ha^{-1} in Prince Edward Island, Canada (Sanderson et al. 2003), and 30–70 kg P ha^{-1} in South Australia (Maier et al. 1989)), mostly because Ferrosols in Tasmania have extremely high P fixation capacity (Moody 1994, Sparrow et al. 1992), a characteristic which is also the main reason why the P fertiliser is banded at planting and not broadcast on these soils beforehand (Sparrow et al. 1992).

Extractable P and K and pH of the surveyed soils (Table 17.1) were higher in Ferrosols than in other soils because Ferrosols in Tasmania have been more intensively cropped (Cotching 1995). Tuber yields were comparable with the Tasmanian average of 49 Mg ha^{-1} reported in Chap. 15.

Growers presumably apply high rates of NPK because on responsive sites the potential benefits are thought to outweigh the potential risks. Based on 2010 fertiliser prices and crop returns (see footnote to Table 17.2), tuber yields have to increase only by about 5 kg per kg of N applied, 12 kg per kg of P and 6 kg per kg of K applied, to pay for the fertiliser application. An examination of published Russet Burbank N, P and K yield responses obtained in Tasmanian field trials (Sparrow and Chapman 2003a; Sparrow et al. 1992; Chapman et al. 1992), and relevant P trials on Ferrosols from Victoria, Australia (Freeman et al. 1998) shows that the break-even point where adding extra N, P or K becomes unprofitable is not reached until crop

Table 17.1 Median rates (and range) of NPK applied by 54 Tasmanian potato growers in 2001, concentrations of extractable P and K and pH of their soils (0–150 mm), and tuber yields

Soil type[a]	Fertiliser N (kg ha^{-1}) at planting	Total fertiliser N (kg ha^{-1})	Fertiliser P (kg ha^{-1}) at planting	Fertiliser K (kg ha^{-1}) at planting
Ferrosol (n=28)	195 (146–245)	265 (146–368)	260 (193–337)	291 (153–380)
Dermosol (n=11)	163 (55–180)	193 (55–280)	195 (68–245)	309 (116–373)
Sodosol (n=10)	162 (110–202)	213 (159–310)	195 (96–225)	283 (203–408)
Tenosol (n=5)	180 (100–213)	253 (238–282)	175 (84–215)	279 (200–318)
Weighted average	186	240	235	288
	Extractable[b] P (mg kg^{-1})	Extractable[b] K (mg kg^{-1})	pH (1:5 H_2O)	Tuber yield (Mg ha^{-1})
Ferrosol (n=28)	127 (38–231)	319 (147–696)	6.5 (6.0–7.0)	54 (26–77)
Dermosol (n=11)	63 (12–125)	186 (55–346)	6.2 (5.8–6.5)	50 (38–66)
Sodosol (n=10)	62 (18–122)	209 (79–337)	6.2 (5.8–6.4)	51 (39–62)
Tenosol (n=5)	95 (43–107)	178 (96–344)	6.1 (5.9–6.4)	53 (49–62)

[a]Australian Soil Classification (Isbell 2002); Ferrosol=Humic Eutrodox; Dermosol=Udic Kanhaplustults; Sodosol=Udic Haplustalfs, Tenosol=Humic Dystroxerepts (Soil Survey Staff 2006)
[b]1:100, 0.5M $NaHCO_3$ at pH 8.5 (Colwell 1963)

Table 17.2 Examples of break-even profit points* for N, P and K fertilisation of potatoes expressed as % of maximum yield and as kg element ha^{-1}

Element	% Ymax at break-even	kg element ha^{-1} at break-even
Nitrogen	99.3%, 99.0%, 99.3%, 99.4%, 98.9%	278, 362, 272, 305, 344
Phosphorus	99.7%, 99.8%, 99.7%, 99.0%, 99.9%, 98.6%, 97.0%, 98.5%, 98.0%, 99.0%, 98.7%, 97.1%	172, 248, 125, 236, 154, 156, 296, 189, 180, 101, 161, 212
Potassium	98.7%, 98.8%, 98.7%,99.4%	603, 480, 508, 199

*Based on published marketable tuber yield responses and 2010 Tasmanian fertiliser prices ($1.26 kg^{-1} N; $3.25 kg^{-1} P; $1.54 kg^{-1} K) and crop returns ($270 Mg^{-1})

yield is very close to maximum (Table 17.2). Corresponding N, P and K fertiliser rates are also shown, and largely vindicate the grower rates shown in Table 17.1. Furthermore, break-even points are relatively insensitive to changes in potato price. For example, decreases in price of 10% only drop the yield targets by about 0.1% (data not shown).

17.3 Assessing the Need for Fertiliser

The targets shown in Table 17.2 only apply to *responsive* sites, and there is evidence in Tasmania for N and K that many sites may be *non-responsive*. For example, 4 of the 13 N trials described in Sparrow and Chapman (2003a) and 4 of the 8 K trials of Chapman et al. (1992) were non-responsive. How can growers tell if their site is likely to respond to fertiliser? For N, an immediate history of grass/clover (usually

Lolium spp. / *Trifolium* spp.) pasture was associated with a lower potato crop N requirement (48 kg ha^{-1}) compared to that of continuously cropped paddocks (193 kg ha^{-1}) (Sparrow and Chapman 2003a), because pasture residues provided more mineral N than did residues of crops (Sparrow and Chapman 2003b). Compared with grass species, legumes as short term cover crops have been shown to provide more N to potatoes, but do not totally replace the need for N fertiliser (Porter and Sisson 1991; Honeycutt et al. 1996; Sincik et al. 2008). In Tasmania, many grass-legume pastures are grown for several years before being ploughed in (Chap. 15), and so should accumulate more N than a cover crop docs. Potato crops grown after pasture may therefore need only modest rates of N, for example 50–100 kg ha^{-1} or sometimes none at all. The time of ploughing is important to maximise the N benefit by ensuring mineralised N is available when the potatoes need it (Griffin and Hestermann 1991, Sanderson and MacLeod 1994).

Petiole testing for nitrate-N is widely used to diagnose the N-status of potato crops (Williams and Maier 1990; Westcott et al. 1991) and has been promoted on the basis that deficient crops will benefit from supplementary N, and also that a better match between N supply and demand will improve N use efficiency. In the 2001 survey mentioned previously Tasmanian potato growers applied an average of 56 kg supplementary N ha^{-1} (data not shown). In Tasmania, Sparrow and Chapman (2003a) conducted 13 N experiments on Russet Burbank which confirmed the trend of decreasing petiole nitrate concentrations with crop age found elsewhere (Williams and Maier 1990, Westcott et al. 1991). However, in only one of 11 crops which responded to N at planting were yield responses to supplementary N recorded (Sparrow and Chapman 2003a). Thus, while petiole N testing may be a good diagnostic tool, overcoming a diagnosed deficiency may not be straightforward. Sparrow and Chapman (2003a) concluded that the poor response to topdressed N may be because potatoes absorb most of their N early in the life of the crop, and that later on the plant shows a greater emphasis on internal relocation of N from shoots to tubers. While the evidence for this is equivocal, there are very few published instances of split N applications outyielding equivalent rates of N applied at planting (see Sparrow and Chapman 2003a for discussion of this point).

For K, a measure of soil sodium bicarbonate-extractable K (Colwell 1963) was found to be a reliable predictor of potato responsiveness (Chapman et al. 1992), with a critical value for 95% maximum yield on Ferrosols of 400 mg K kg^{-1} for Russet Burbank and 350 mg kg^{-1} for Kennebec. For Sodosols and Tenosols the critical value of 200 mg K kg^{-1} derived by Maier (1986) on similar soils in South Australia is used in Tasmania.

For P, the critical extractable P value (Colwell 1963) derived by Maier et al. (1989) of 34 mg P kg^{-1} is used for light-textured soils in Tasmania, but the lack of a good soil calibration for P on Ferrosols, the state's major potato growing soil, remains a limitation to good P management. The best such calibration is that of Freeman et al. (1998), who predicted a critical Olsen P (Olsen et al. 1954) value of 46 mg kg^{-1} for 90% maximum Russet Burbank yield on Victorian Ferrosols. However, this predicted value was beyond the range of their data. In Tasmania, 3 of 4 trials on Ferrosols high in extractable P (Colwell P 112–210 mg kg^{-1}) did not reach maximum yield even at 240 kg P ha^{-1} (Sparrow et al. 1992). The results from this study and that of Freeman et al. (1998) suggest that Ferrosols may require significant P loading for potatoes not to respond to additional fertiliser P. More research on such heavily loaded soils is needed to define a calibration.

17.4 Potato Grower Fertiliser Practices

Most Tasmanian potato growers are reluctant to cut back on fertiliser because they know how responsive their crops can be, and they are unwilling to risk losing yield. Figures 17.1 and 17.2 show P and K fertiliser and soil test data and NPK balances of the Tasmanian potato crops surveyed in 2001 (Sparrow et al. 2003). Extractable K and P concentrations from individual paddocks on different soil types are plotted against grower K and P fertiliser rates for the same paddocks (Fig. 17.1). In each case, there was no relationship between soil test values and fertiliser rates, either within a soil type or across all soils. It is particularly disturbing to see that most

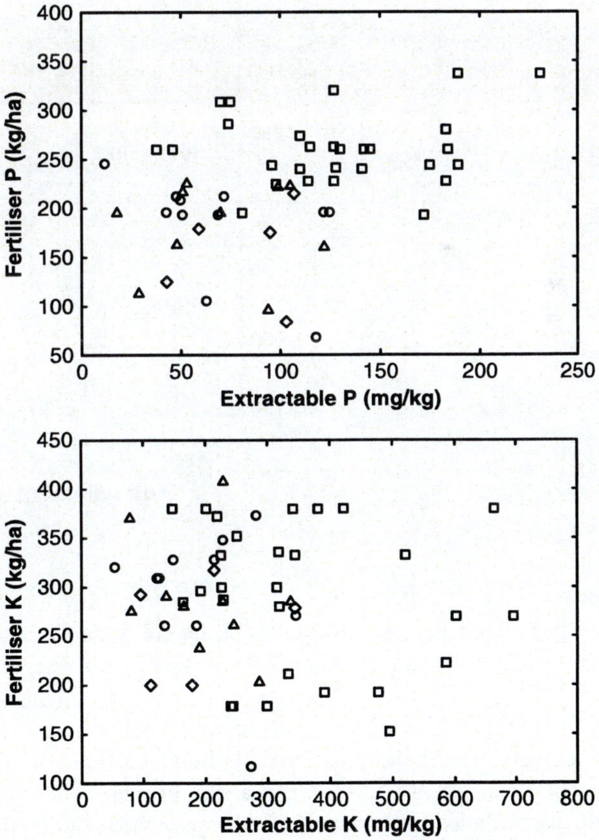

Fig. 17.1 Rates of fertiliser P (top) and K (bottom) applied to 54 Tasmanian potato crops in 2001 in relation to extractable P and K (Colwell 1963) in the soil (0–150 mm) (squares = Ferrosols; circles = Dermosols; triangles = Sodosols; diamonds = Tenosols)

Fig. 17.2 NPK nutrient balances (kg element applied ha^{-1} in fertiliser – kg element removed ha^{-1} in tubers) of 54 Tasmanian potato fields surveyed in 2001 (squares = Ferrosols; circles = Dermosols; triangles = Sodosols; diamonds = Tenosols). Crop yields and unpublished average Tasmanian fresh tuber nutrient concentrations of 0.3% N, 0.042% P and 0.4% K were used to calculate rates of removal

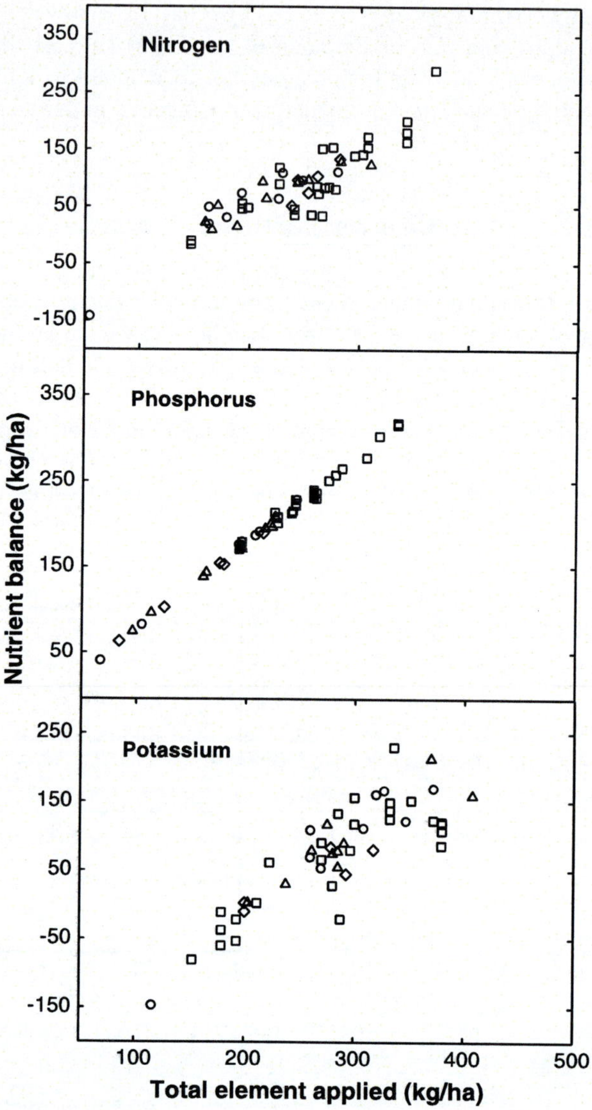

growers on Sodosols and Tenosols appear to follow the P and K practices of their counterparts on Ferrosols despite (1) the calibrated soil tests at their disposal, (2) successful demonstrations that such high rates are not needed on these soils (Sparrow et al. 1993c) and (3) 90% of growers using soil tests (Sparrow 2009). Most growers applied more N, P and K than was removed in the tubers (Fig. 17.2), and there were obvious positive linear relationships between rates of applied NPK and nutrient balance. Mistrust of soil tests and a lack of market or regulatory signals regarding fertiliser use may be barriers to the adoption of better fertiliser practices.

Table 17.3 Trends in extractable P and K and soil pH of Tasmanian Ferrosol topsoils used for cropping

	1960s	1980s	2001	ANOVA
Soil test	16 samples	83 samples	60 samples	F-prob
*Colwell P	40	109	141	<0.001
*Colwell K	295	312	370	0.037
pH in water	5.5	6.1	6.5	<0.001

Source: Wright and Temple-Smith, unpublished data, Sparrow, unpublished data, Sparrow et al. 2003.
*Colwell (1963)

It seems likely that Tasmanian vegetable growers apply more than maintenance fertiliser rates to many of their crops, not just potatoes. This is supported by a comparison of 3 populations of Tasmanian Ferrosol cropping soil test data since the 1960s (Table 17.3), which indicates a steady increase in extractable K and P. In addition, Sparrow et al. (submitted) monitored bicarbonate-extractable P (Colwell 1963) in the topsoil of 15 other cropped Tasmanian Ferrosols between 1997 and 2010 and found an average increase during this period of 69 mg kg^{-1}. If these trends continue, Tasmanian Ferrosols may yet reach a non-responsive P status. Research to define this status will be needed. Soil pH has also increased (Table 17.3) due to the consistent use of lime for other acid-sensitive rotation crops, notably onions and poppies (e.g. Temple-Smith et al. 1983), rather than on potatoes where lime may increase the incidence and severity of common scab.

17.5 Micronutrients

While micronutrient fertilisers are not routinely used on Tasmanian potatoes, they are frequently marketed as "silver bullets" or potential missing links in the production chain. Copper (Cu), zinc (Zn) boron (B) and molybdenum (Mo) are sometimes applied. Tasmanian soils are very high in iron (Fe) and manganese (Mn), so these micronutrients are not normally applied. In the 1990s, the potato and fertiliser industries and government invested in micronutrient field trials for a range of Tasmanian crops including potatoes (Salardini and Sparrow 2001). Eleven trials on potatoes were conducted, seven with foliar-applied micronutrients (Cu, Zn and B), three with soil-applied micronutrients (Cu, Zn, B, Mo), and one with a combination of soil and foliar Zn application. The soil at the selected sites had low relative concentrations of one or more micronutrients as determined by either DTPA extraction for Zn and Cu, or hot water extraction for B. Micronutrients were applied as both inorganic and EDTA salts. Rates for soil and foliar application were consistent with those used in published research elsewhere.

There were no statistically significant tuber yield increases in any of the trials, and one instance of B toxicity when 4 kg B ha^{-1} was added to a very sandy soil. Salardini and Sparrow (2001) concluded that the micronutrient status of Tasmanian potato soils is generally adequate for potatoes, and this was borne out by petiole Zn,

Cu and B concentrations surveyed as part of the study: the vast majority of these lay within the ranges considered adequate for potatoes (Salardini and Sparrow 2001). In addition, responses to some elements e.g. Zn would be expected to be small or absent because these elements are or were previously applied regularly at low rates in many fungicides.

17.6 Effects of Fertiliser on Tuber Quality

In addition to the effect of nutrition on tuber yield, much of the research referred to in this chapter included assessment of effects on tuber quality, including tuber size and shape, tuber number, specific gravity, hollow heart, crisp colour and bruising susceptibility. The quality responses observed included increased tuber number with applied P (Sparrow et al. 1992) and decreased specific gravity and increased yields of misshapen tubers with increasing rates of applied N (Sparrow and Chapman 2003a). However, such responses were far from universal and sometimes they depended on whether the site was yield-responsive or not. For example, Chapman et al. (1992) found that on highly K-deficient sites, specific gravity and crisp colour increased with low rates of fertiliser K, but across all sites the trend was the opposite. A general observation arising from all of this work was that applying enough fertiliser for optimum yield usually gave good tuber quality.

17.7 Cadmium in Potatoes

One aspect of tuber quality that has received particular attention in Tasmania and elsewhere in Australia in recent times is tuber cadmium (Cd) concentration. Cadmium is a natural impurity in rock-phosphate and therefore of P fertiliser, and accumulates both in soils to which P fertiliser is added, and in crops grown on those soils. The Australian potato and fertiliser industries supported a significant Cd research program in the 1990s. This was in response to data (Stenhouse 1991) indicating that, because of their propensity for Cd accumulation and their popularity amongst consumers, potatoes were potentially a significant source of dietary Cd in Australia, even though there has never been any evidence of Cd-related health impacts in the country. The Tasmanian research on Cd in potatoes conducted as part of the national program has a number of implications for the management of soil fertility.

Russet Burbank tuber Cd concentrations increased with rate of P fertiliser (and hence Cd in fertiliser), but only when those rates of P also increased tuber yield (Sparrow et al. 1992, 1993a). Fertiliser P rates in excess of crop needs did not further increase tuber (or petiole) Cd concentrations although they did increase petiole P concentrations (Sparrow 2000). It appears that Cd uptake at least in Russet

Burbank may be related to its root growth or function more than to Cd supply. This view is supported by other work which showed that, at equivalent P rates, increasing the Cd-content of P fertiliser by 6-fold did not increase tuber Cd concentration (Sparrow et al. 1993b). On the basis of immediate Cd risk, there thus appears to be little incentive to use P fertiliser wisely, and the need to minimise soil Cd load in the longer term does not seem to be a strong influence on grower behaviour. However, the Australian fertiliser industry did respond to this need by sourcing phosphate rocks with relatively low Cd content compared with Australia's traditional sources supplied by Christmas Island and Nauru.

The form of K fertiliser can affect tuber Cd uptake (Sparrow et al. 1994), with potassium chloride increasing uptake compared with potassium sulphate applied at the same K rate. This is because the chloride ion forms soluble chloro-cadmium complexes in the soil solution, increasing Cd availability (O'Connor et al. 1984). Switching to potassium sulphate is a short term management option that growers can use to decrease tuber Cd uptake. In 2010, potassium sulphate was about 70% more expensive than potassium chloride, so this switch would come at significant additional cost.

Cd availability is considered to decrease with increasing soil pH (Page 1981), yet field studies in Tasmania (Sparrow et al. 1993a) and South Australia (Maier et al. 1997) failed to show that lime in the short term decreased tuber Cd concentrations. However, in repeat crops grown 2 or 3 years later on the Tasmanian sites (Sparrow and Salardini 1997), tuber Cd concentrations were decreased by about 30% in the residual lime treatments. This medium term effect was attributed to the more even and deeper mixing of the lime with the soil by the harvest of the initial crop of potatoes. Lime is therefore a suitable medium to long term Cd control measure, and the pH trends in Table 17.3 indicate that use of lime by Tasmanian farmers is probably helping to control Cd uptake. However, soil pH is only one control factor, and Sparrow and Salardini (1997) found it explained only 22% of the variation in tuber Cd concentrations under controlled conditions. Moreover, at one site they also found that the previous liming increased the severity of common scab symptoms to the point of tubers being unmarketable.

A number of other strategies for minimising tuber Cd uptake were identified in other Australian research (McLaughlin 1999), including choice of cultivar, chloride content of irrigation water, addition of zinc and maintenance of soil organic matter. Fortunately, Russet Burbank, Tasmania's most popular cultivar, is a relatively low Cd accumulator (McLaughlin et al. 1994), while Tasmania's irrigation water is mostly low in chloride salts, and its soils relatively high in organic matter (Chapter 16).

By far the most significant outcome of the national program was that its data formed the foundation of a submission by the potato industry to the Australian government which allowed a more informed dietary risk assessment to be conducted and which resulted in the maximum allowed level for Cd in potatoes to be revised from 0.05 to 0.10 mg kg^{-1} (FSANZ 2010). The revised level currently accommodates virtually all Australian potatoes, including those from Tasmania. Whether that continues to be the case will depend on how well growers manage their Cd inputs and their crops.

17.8 Environmental Impacts of Fertiliser

While potatoes could be considered to have a potentially high environmental impact because of their high inputs, heavy cultivation and soil disturbance during harvesting, quantitative data on this impact in Tasmania is scarce. Here, recent Tasmanian research on the modelling of potential impacts of potato production on N leaching and nitrous oxide emissions is summarised.

Lisson et al. (2009) modelled nitrogen leaching and crop yield on seven cropping farms in northern Tasmania using the Agricultural Production Systems Simulator (APSIM, Keating et al. 2003). Information about farm practices (including rotation, fertiliser use and irrigation management) was collected by interviewing the farmers involved. Potatoes were part of the rotation of six of the seven farms. Other rotation crops included poppies, cereals, beans, peas and broccoli, and pasture was also grown in each rotation. The modelling approach allowed for year-to-year carry-over effects (i.e. nitrogen remaining in the soil after harvest that was available for use by the following crop or lost via leaching) and the effects of seasonal climate variability to be fully assessed.

The average N fertiliser rate for potatoes was 397 kg N ha^{-1}, with a range of 279–480 kg N ha^{-1} used across the 6 farms growing potatoes. These rates were the highest of any crop and higher than the median rates found in 2001 (Table 17.1), and not surprisingly potatoes had the highest estimated N leaching over the 25 year simulation (29 kg N ha^{-1} $year^{-1}$), four times more than estimates for other crops. Across all farms and years there was an average excess of N supplied as fertiliser over N uptake by potatoes of 89 kg ha^{-1}. This did not include an estimate of N mineralisation from soil organic matter, which Sparrow and Chapman (2003b) measured in 13 potato paddocks from just before planting to the time of early tuber growth. They measured average mineralisation rates of 3.5 kg N ha^{-1} day^{-1} in ex-pasture paddocks and 1.2 kg N ha^{-1} day^{-1} in previously cropped paddocks. Clearly, mineralisation will add substantially to the pool of excess mineral N and the risk of N leaching.

However, while the modelling indicated that N fertiliser rate did affect N leaching loss, changes to irrigation management were much more influential. By scheduling irrigation according to crop demand (only one farmer monitored soil water content, the rest used a fixed irrigation schedule interrupted only by heavy rain), the model predicted that N leaching under potatoes on the farm with the highest loss could be cut from 53 to 6 kg N ha^{-1}. However, reducing the N fertiliser rate from 480 to 284 kg N ha^{-1} only cut N leaching by a further 3 kg N ha^{-1}. Significantly, neither changes to irrigation nor N fertiliser rates affected modelled crop yields, indicating that growers have little to lose in adopting these changes. The maximum seasonal N loss ranged up to 135 kg N ha^{-1}, illustrating the variability in N leaching.

In their survey of Tasmanian potato grower practices, Sparrow (2009) found that only about a third of growers used some form of soil moisture monitoring. Almost half assessed soil moisture by visual inspection. Thus the savings available to the farmers through better irrigation scheduling in the modelling study would seem to be much more widely available in Tasmania. The average of the break-even N rates shown in Table 17.2, 312 kg N ha^{-1}, is consistent with the reduced rate modelled by

Lisson et al. (2009), further supporting the notion that the growers in that study were using too much N. There is a need to raise the awareness of Tasmanian potato growers to the potential for more efficient water and N use as this appears to be a win-win situation, benefiting both grower returns and environmental quality.

Nitrogen fertiliser was also found to be one the main sources of greenhouse gases (GHG) in a recent assessment of the on-farm environmental footprint of the Australian potato industry (Norton et al. 2008). This work comprised life cycle analyses (LCA, e.g. Grant and Beer 2008) of a number of regional case studies including three for processed potato production in Tasmania. On-farm practices from seed production to the delivery of the harvested crop to the factory gate were assessed. Off-site losses through leaching were included. Rate of N fertiliser use was linearly related to emissions of GHG, with estimates of its contribution ranging from 25–56% of total emissions (8–25 kg N_2O ha^{-1}) across the case studies. Other emissions came from diesel use (26–40%), agri-chemical use (4–9%), infrastructure (11–16%) and electricity (0–19%). Estimated average GHG emissions from Tasmanian potato production were 14% higher than those for onions and 95% than those for broccoli, partly because the latter crops received only 180 and 91 kg N ha^{-1} respectively. At a regional scale, the study found that potato production could account for about 2.5% of total agricultural GHG emissions. The rates of N fertiliser used in the LCA were those volunteered by industry experts in the case study regions. The average rate for the three Tasmanian potato case studies was 413 kg N ha^{-1}, much higher than those for the other three potato case studies in South Australia (200 kg N ha^{-1}), Queensland (164 kg N ha^{-1}) and New Zealand (288 kg N ha^{-1}). This adds further weight to the call for a re-examination of N fertiliser rates on potatoes in Tasmania.

17.9 Conclusion

The Tasmanian potato industry is a large user of N, P and K fertiliser. There is no evidence that other nutrients are lacking in Tasmanian potato soils. For N and P, there is a continuing tension between maximising economic returns and minimising the risk of environmental impact. The tension is driven by 1) the relatively low cost of fertiliser compared to the potential returns which encourages high application rates and 2) the current weak signals to growers about water and food quality impacts. When these signals intensify, growers may have to fertilise to a target yield closer to 95% of maximum. This would reduce fertiliser rates considerably but still allow for profitable production.

There is adequate evidence that the rates of N and K currently used are often excessive, and for N, may be leading to adverse environmental impacts. A better soil test calibration for P on Ferrosols, the main soil for potato production in Tasmania (see Chap. 15), would provide growers with a soil P target, but getting growers to adhere to such a target will require a substantial change to the current culture of high P fertiliser use.

Acknowledgments The author would like to thank the anonymous reviewers for their constructive comments on the manuscript.

References

Chapman KSR, Sparrow LA, Hardman PR, Wright DN, Thorp JRA (1992) Potassium nutrition of Kennebec and Russet Burbank potatoes in Tasmania: effect of soil and fertiliser potassium on yield, petiole and tuber potassium concentrations and tuber quality. Aust J Exp Agric 32:521–527

Colwell JD (1963) The estimation of the phosphorus fertilizer requirements of wheat in southern New South Wales by soil analysis. Aust J Exp Agric Anim Husbandry 3:190–197

Cotching WE (1995) A review of the challenges for long term management of krasnozems in Australia. Aust J Soil Water Conserv 8:18–27

DPIF (2009) The Tasmanian Potato Industry. Department of Primary Industries and Water, Tasmania

Freeman KL, Franz PR, de Jong RW (1998) Effect of phosphorus on the yield, quality and petiolar phosphorus concentrations of potatoes (cvv. Russet Burbank and Kennebec) grown in the krasnozem and duplex soils of Victoria. Aust J Exp Agric 38:83–93

FSANZ (2010) Australia New Zealand Food Standards Code http://www.foodstandards.gov.au/foodstandards/foodstandardscode/. Accessed 29 January 2011

Grant T, Beer T (2008) Life cycle analysis of greenhouse gas emissions from irrigated maize and their significance in the value chain. J Exp Agric 48:375–381

Griffin TS, Hestermann OB (1991) Potato response to legume and fertilizer nitrogen sources. Agron J 83:1004–1012

Honeycutt CW, Clapham WM, Leach SS (1996) Crop rotation and N fertilization effects on growth, yield and disease incidence in potato. Am Potato J 73:45–61

Hyatt CR, Venterea RT, Rosen CJ, McNearney M, Wilson ML, Dolan MS (2010) Polymer-coated urea maintains potato yields and reduces nitrous oxide emissions in a Minnesota loamy sand. Soil Sci Soc Am J 74:419–428

Isbell RF (2002) The Australian soil classification. Revised edition. CSIRO, Collingwood

Keating BA, Carberry PC, Hammer GL, Probert ME, Robertson MJ, Holzworth D, Huth NI, Hargreaves JNG, Meinke H, Hochman Z, McLean G, Verburg K, Snow V, Dimes JP, Silburn M, Wang E, Brown S, Bristow KL, Asseng S, Chapman S, McCown RL, Freebairn DM, Smith CJ (2003) An overview of APSIM, a model designed for farming systems simulation. Eur J Agron 18:267–288

Lisson S, Cotching B, Rawnsley R (2009) Assessing the risk of nutrient loss under intensive crop production in the Panatana Catchment. Final report for CERF Landscape Logic Project 4, Australian Department of Environment and Heritage. CSIRO, Hobart

Maier NA (1986) Potassium nutrition of irrigated potatoes in South Australia. 1. Effect on tuber yield and the prediction of tuber yield response by soil analysis. Aust J Exp Agric 26:717–725

Maier NA, Potock-Pacay KA, Jacka JM, Williams CMJ (1989) Effect of phosphorus fertiliser on the yield of potato tubers (Solanum tuberosum L.) and the prediction of tuber yield response by soil analysis. Aust J Exp Agric 29:419–432

Maier NA, McLaughlin MJ, Heap M, Butt M, Smart MK (1997) Effect of current season applications of calcitic lime on pH, yield and cadmium concentration of potato (*Solanum tuberosum* L.) tubers. Nutr Cycl Agroecosyst 47:1–12

McLaughlin MJ (1999) A national strategy to reduce cadmium in potatoes. Final Report, Project PT423, Horticultural Research and Development Corporation. CSIRO Land and Water, Glen Osmond

McLaughlin MJ, Williams CMJ, McKay A, Gunton J, Jackson K, Dowling B, Kirkham R, Partington D, Smart MK, Tiller KG (1994) Effect of potato variety on cadmium accumulation in potato tubers. Aust J Agric Res 45:1483–1495

Moody PW (1994) Chemical fertility of krasnozems: a review. Aust J Soil Res 32:1015–1041

Norton N, Hay F, Lambert S, Sparrow L, Kirkwood I (2008) Enhancing environmental sustainability in the processing potato industry in Australia Final Report Project PT07060. Horticulture Australia, Sydney

O'Connor GA, O'Connor C, Cline GR (1984) Sorption of cadmium by calcareous soils: influence of solution composition. Soil Sci Soc Am J 48:1244–1247

Olsen SR, Cole CV, Watanabe FS, Dean LA (1954) Estimation of available phosphorus in soils by extraction of sodium bicarbonate. USDA Circular 939. Washington, DC

Page AL (1981) Cadmium in soils and its accumulation by crops. In: Proceedings 3rd international conference on heavy metals in the environment, Amsterdam, September 1981. CEP Consultants, Edinburgh, pp 206–213

Porter GA, Sisson JA (1991) Response of Russet Burbank and Shepody potatoes to nitrogen fertilizer in two cropping systems. Am Potato J 68:425–443

Rosen CJ, Bierman PM (2008) Potato yield and tuber set as affected by phosphorus fertilization. Am J Potato Res 85:110–120

Salardini A, Sparrow L (2001) The trace element requirements of vegetables and poppies in Tasmania. Final Report of Project PT320 to Horticulture Australia Limited. Horticulture Australia, Sydney

Sanderson JB, MacLeod JA (1994) Soil nitrate profile and response of potatoes to fertilizer N in relation to time of incorporation of lupin (*Lupinus albus*). Can J Soil Sci 74:241–246

Sanderson JB, MacLeod JA, Douglas C, Coffin R, Bruulsema T (2003) Phosphorus research on potato in PEI. Acta Hortic 619:409–417

Sincik M, Turan ZM, Göksoy AT (2008) Response of potato (*Solanum tuberosum* L.) to green manure cover crops and nitrogen fertilization rates. Am J Potato Res 85:150–158

Soil Survey Staff (2006) Keys to soil taxonomy, 10th edition. Natural Resources Conservation Service, USDA. Washington, DC. Available at: http://www.statlab.iastate.edu/soils/keytax/

Sparrow LA (2000) Excess P fertiliser does not increase tuber cadmium concentrations in potato. In: Adams JA, Metherell AK (eds) Soil 2000: new horizons for a new century proceedings of the Australian and New Zealand Second Joint Soils Conference 3–8 December 2000, Lincoln University, NZ, vol 2 oral papers. NZ Society of Soil Science, Lincoln

Sparrow L (2009) Crop rotations. In: Kirkwood I (ed) Australian Potato Research Program Final Report to Horticulture Australia for project PT04016. Horticulture Australia, Sydney

Sparrow LA, Chapman KSR (2003a) Effects of nitrogen fertiliser on potato (*Solanum tuberosum L.*, cv. Russet Burbank) in Tasmania. 1. Yield and quality. Aust J Exp Agric 43:631–641

Sparrow LA, Chapman KSR (2003b) Effects of nitrogen fertiliser on potato (*Solanum tuberosum L.*, cv. Russet Burbank) in Tasmania. 2. Petiole and soil analysis. Aust J Exp Agric 43: 643–650

Sparrow LA, Salardini AA (1997) Effects of residues of lime and phosphorus fertilizer on cadmium uptake and yield of potatoes and carrots. J Plant Nutr 20:1333–1349

Sparrow LA, Chapman KSR, Parsley D, Hardman PR, Cullen B (1992) Response of potatoes (*Solanum tuberosum* cv. Russet Burbank) to band-placed and broadcast high cadmium phosphorus fertiliser on heavily cropped krasnozems in north-western Tasmania. Aust J Exp Agric 32:113–119

Sparrow LA, Salardini AA, Bishop AC (1993a) Field studies of cadmium in potatoes (*Solanum tuberosum* L.). I. Effect of lime and phosphorus on cv. Russet Burbank. Aust J Agric Res 44:845–853

Sparrow LA, Salardini AA, Bishop AC (1993b) Field studies of cadmium in potatoes (*Solanum tuberosum* L.) II. Response of cvv. Russet Burbank and Kennebec to two double superphosphates of different cadmium concentration. Aust J Agric Res 44:855–861

Sparrow LA, Walker AR, Johnstone J, Chapman KSR (1993c) New fertiliser recommendations for potato growers on sandy soils in Tasmania's northern midlands. In: DR Baigent (ed) Proc First ASPAC National Workshop on Soil and Plant Analysis, Ballarat, 2–4 March 1993. ASPAC, Tatura

Sparrow LA, Salardini AA, Johnstone J (1994) Field studies of cadmium in potatoes (*Solanum tuberosum* L.). III. Response of cv. Russet Burbank to sources of banded potassium. Aust J Agric Res 45:234–249

Sparrow LA, Cotching WE, Hawkins K (2003) Fertiliser practices of Tasmanian potato and poppy grows: do they show any relation to pre-plant soil test information? In: LD Currie and JA Hanly (eds) Tools for Nutrient and Pollutant Management: Applications to Agriculture and

Environmental Quality Occasional Report No. 17 Fertilizer and Lime Research Centre, Massey University, Palmerston North, New Zealand. Massey University, Palmerston North

Stenhouse F (1991) The 1990 Australian Market Basket Survey Report. Australian Government Publishing Service, Canberra, Australia

Temple-Smith MJ, Wright DN, Laughlin JC, Hoare BJ (1983) Field responses of poppies (*Papaver somniferum* L) to lime application on acid krasnozems in Australia. J Agric Sci Cambridge 100:485–492

Westcott MP, Stewart VR, Lund D (1991) Critical petiole nitrate levels in potato. Agron J 83:844–850

Williams CMJ, Maier NA (1990) Determination of the nitrogen status of irrigated potato crops. I. Critical nutrient ranges for nitrate-nitrogen in petioles. J Plant Nutr 13:971–984

Chapter 18
Managing and Monitoring Viral and Soil-Borne Pathogens in Tasmanian Potato Crops

Leigh A. Sparrow and Calum R. Wilson

Abstract The characteristics of Tasmanian potato production, including its geographic isolation, seed certification scheme, and long rotations have helped to minimize the incidence of important potato viruses. However, many soil-borne pathogens have steadily built-up in Tasmanian potato soils. The main influence on pathogen concentrations, especially for the powdery scab pathogen, seems to be the presence of the host crop in the rotation. A useful predictive relationship between pathogen DNA and powdery scab severity has emerged from work to date but needs testing across a range of potato cultivars.

18.1 Virus Incidence in Tasmanian Potatoes

Seven viruses have been recorded infecting potatoes in Tasmania (Table 18.1; Sampson and Walker 1982). Most are common pathogens of potato present in the majority of potato growing regions of the world. The incidences of major yield debilitating viruses of potato are historically very low in Tasmanian production (Table 18.2). Most viruses which induce obvious visual symptoms have been successfully controlled through reduction in virus inoculum.

Of critical importance has been the operation of the Tasmanian potato seed certification scheme. First established in 1930, the scheme is currently based on mini-tuber production from pathogen-tested, tissue-cultured stocks which ensures

L.A. Sparrow (✉)
Tasmanian Institute of Agriculture, 46, Kings Meadows, Tasmania 7249, Australia
e-mail: Leigh.Sparrow@utas.edu.au

C.R. Wilson
Tasmanian Institute of Agriculture, 13 St Johns Avenue, New Town, Tasmania 7008, Australia
e-mail: Calum.Wilson@utas.edu.au

Z. He et al. (eds.), *Sustainable Potato Production: Global Case Studies*,
DOI 10.1007/978-94-007-4104-1_18, © Springer Science+Business Media B.V. 2012

Table 18.1 Viruses recorded in Tasmanian potato crops

Virus		Frequency
Potato leafroll *luteovirus*	PLRV	Moderately rare
Potato virus Y *potyvirus*	PVY	Rare
Potato virus X *potexvirus*	PVX	Common
Potato virus S *carlavirus*	PVS	Very common
Potato virus A *potyvirus*	PVA	Rare
Tomato spotted wilt *tospovirus*	TSWV	Rare
Alfalfa mosaic *alfamovirus*	AMV	Very rare

Information from Sampson and Walker 1982, personal observations and recent survey data

planting material is free of all viral and soil-borne pathogens prior to field bulking. Rigorous visual inspections and rogueing of symptomatic plants, particularly in the early years of establishment of the seed certification scheme, has greatly reduced incidence of important pathogens such as PLRV and PVY. Furthermore, the relative remoteness of Australia from other major potato regions and the physical isolation of the Tasmanian potato production region from the rest of Australia with subsequent severe restrictions on potato imports into the state has limited the risk of introduction of exotic virus pathogens (such as Potato mop virus and Tobacco rattle virus) and new strains of important viral pathogens (such as the N strains of PVY which are present in some parts of mainland Australia).

Also, the long and diverse rotations employed in potato production regions of Tasmania (see Chap. 15) have contributed to inoculum reduction. Typical rotations of 5–7 years between potato crops have increased the capacity of growers to manage volunteer potatoes and thus reduce these as a major source of virus inoculum.

The very low inoculum levels of the serious aphid-borne viruses (PLRV, PVY and PVA), and the absence of certain exotic insect pests (such as the Colorado potato beetle, and the Tomato and potato psyllid) has meant seed growers have little or no need for insecticides in seed tuber production. This is rather unique as most potato seed production regions in the world employ routine insecticide treatments to reduce PLRV and in particular PVY spread. Insecticide use in the Tasmanian industry (particularly targeting aphids) is extremely low. This has allowed natural predators of aphids to establish and aphid populations within potato crops are modest and seldom if ever cause damage through feeding (Paul Horne; IPM Technologies Pty Ltd, Pers. Comm.).

Virus pathogens that are less well controlled in the Tasmanian production system include Tomato spotted wilt virus (TSWV) and the contact transmitted viruses Potato virus S (PVS) and Potato virus X (PVX) which commonly occur as symptomless infections in potato.

18.1.1 Tomato Spotted Wilt Virus

TSWV is a relatively uncommon pathogen of potatoes. Whilst this virus commonly infects a wide range of crop and plant species world-wide, its occurrence as a pathogen of significance within potato production is very rare in regions outside of

Table 18.2 Prevalence (percentage of crops recording infections) and incidence (mean percentage of infected plants within crops) of potato virus infections in Tasmanian seed potato crops over 9 years

	PLRV		PVY		PVS		PVX		TSWV	
Season	Prevalence	Incidence	Prevalence	Incidence	Prevalence	Incidence	Prevalence	Incidence	Prevalence	Incidence
2001/02	7	2.6	0	0	34.6	15.9	7.5	1.4	0	0
2002/03	5.4	0	5.5	0	69.4	12.6	14.2	0.3	0.9	0
2003/04	0	0	3.7	0	43.8	8.7	5.6	1.0	0	0
2004/05	0	0	0	0	52.4	13.0	7.3	1.9	0	0
2005/06	0	0	0	0	67.8	14.8	11.9	2.0	0	0
2006/07	0	0	0	0	43.2	43.0	2.7	0.4	0	0
2007/08	0	0	0	0	37.2	13.2	2.2	0.3	0	0
2008/09	0	0	0	0	39.0	5.3	5.5	2.3	0	0
2009/10	0	0	0	0	34.7	5.4	0.6	0	0	0

Australia (Wilson 2001). Symptoms associated with TSWV infection are atypical of viral infections (necrosis sometimes leading to plant death) and may be confused with fungal infections (such as early blight caused by infections with *Alternaria solani*; Stevenson et al. 2001). TSWV epidemics within potato have been noted in the USA (Abad et al. 2005) and the Middle East (Al-Shahwan et al. 1997).

TSWV occurs sporadically in potato crops in Australia but can be devastating when epidemics occur (Wilson 2001). In the 1996/97 season 26 seed crops in Tasmania were affected with TSWV with incidence varying from trace levels to 28%. One thousand tonnes of seed failed certification in this season due to TSWV infections (Wilson 2001). Most of the TSWV infections recorded in this epidemic were situated in the low lying non-traditional Southern seed production regions selected for low levels of soil-borne pathogen inoculum. The 1996/97 epidemic led to a major reduction in seed production for these regions. One further complication was, as growers looked to attempt to control TSWV, they initiated insecticide applications targeting the thrips vector of this virus. These insecticides had an adverse effect on the natural enemies of aphids and led to sporadic aphid population blooms on these crops. This in turn led to some small spikes in occurrence of PLRV in these regions (e.g. season 2001/02; Table 18.2).

Control of TSWV in potato is problematic, as unlike the other major potato viruses, the major sources of inoculum come from outside of the crop (weed and/or alternate crop species). Thus seed certification and volunteer control has limited effect on reduction of TSWV inoculum. Currently the Australian potato industry is examining opportunities for enhancing resistance to TSWV and avoidance of growing regions where epidemics of TSWV occur (presumably due to high incidence of external TSWV inoculum).

18.1.2 Potato Virus S and Potato Virus X

PVS and PVX seldom induce obvious foliar symptoms in potato. As such visual observation and rogueing is ineffective in their control. Pathogen tested tissue culture stocks are free from infection with both pathogens, but recent surveys have shown that re-infection of early generation crops, especially with PVS, commonly occurs (Table 18.2).

PVS may be aphid transmitted, and it is possible that early re-infection may be a result of spread by aphids from other potato crops and/or volunteer plants in the environment. Both PVS and PVX are commonly transmitted by plant contact. In Tasmania most seed crops are planted with cut seed (large seed tubers cut into 50 g setts for planting). Seed cutting is commonly managed by a few commercial seed cutting enterprises. Thus the majority of Tasmanian seed passes through a small number of seed cutting machines. This provides the potential for infection of early generation seed with contact transmitted viruses if contaminated (later generation) seed lots have previously passed through the cutters. Experimental evidence of substantial PVX and PVS transmission in this manner has been ambiguous (Lambert 2008).

PVS (and often PVX) are sometimes disregarded as important pathogens in potato due to the perceived lack of impact on yield and tuber quality. However, Tasmanian studies have shown that individually or in co-infection yield losses of 10–15% frequently occur (Lambert 2008). Recent efforts to decrease the incidence of PVS infections through removal of highly infected early generation seed stocks appear to have had some success. Whilst the prevalence of infected crops has remained relatively high, at 69–35%, the mean incidence of PVS infection has decreased in recent years from 43% to 5% (Table 18.2).

18.2 Other Factors Influencing Success of Virus Control

Certain locally popular varieties grown for the Tasmanian domestic market have been propagated outside of the potato certification scheme. When tested these varieties often have very high incidence of common viral pathogens including PVX, PVS and PLRV. Increasing popularity of some of these varieties probably led to recent minor epidemics of PLRV in certified crops particularly in Southern production regions where these are mostly cropped. However, most of these local boutique varieties have now been introduced into the seed scheme following selection of virus free material.

18.3 Soil-Borne Potato Pathogens in Tasmania

Soil-borne potato diseases such as common and powdery scab, caused by *Streptomyces scabiei* and *Spongospora subterrranea* respectively, and black scurf and Rhizoctonia canker, caused by *Rhizoctonia solani*, can be significant problems in Tasmania. While estimates of losses from these diseases in Tasmania are not available, across Australia they are thought to cost the potato industry up to AUD $50 m each year (Australian Processing Potato Industry, unpublished data). In a 2005 survey (Sparrow 2009), just over half of Tasmanian potato growers said that they actively managed at least one of these diseases. In this chapter, some recent Tasmanian research on the potential of tests of the DNA of the pathogens in soil to monitor soil pathogen load and to indicate disease risk is reported. DNA probes for a number of soil-borne potato pathogens were developed and evaluated as a key part of the 2004–2009 Australian Potato Research Program (Horticulture Australia 2009). These real-time PCR assays are part of a high-throughput DNA extraction and analysis system available at the South Australian Research and Development Institute (Ophel-Keller et al. 2008). Assays are available for *Sp. subterranea*, *St. scabiei* and the anastomosis groups of *R. solani* most relevant to potatoes in Australia, i.e. AG2.1, AG2.2, AG3 and AG4. While the primary application of the probes is intended to be the prediction of disease risk, they are also a potentially useful tool to show the effects of time and land use on pathogen populations.

18.4 Monitoring Pathogen DNA in Soil

In Tasmania, the DNA probes were used to annually monitor concentrations of pathogen DNA in soil in 28 farmer paddocks for 4 years from 2005 to 2008 (Sparrow 2009). The aim was to see whether pathogen DNA concentrations changed and if the changes could be explained by seasonal conditions or land use. All paddocks were planted to potatoes in the 2005–06 season. Within a few days of commercial potato planting, a 20 m strip of a single potato crop row was marked in each paddock by reference to pegs on boundary fences and by global positioning system, so that the strips could reliably be relocated. The farmers' seed was then removed from the strip and replaced by surface sterilized seed of cv. Shepody. At the same time, forty cores of soil 150 mm deep were taken along each strip to form a composite sample, and this sampling was repeated in 2006, 2007 and 2008. Land use in each paddock at the time of soil sampling was noted and land owners were subsequently asked to confirm or correct land use details for the period 2000/01 to 2008/09 so that land use effects prior to the start of the sampling could also be assessed. Soil samples were dried at 40°C, ground <2 mm, and analysed for pathogen DNA.

Land use is shown in Table 18.3. The frequent presence of pasture or ryegrass in Tasmania is consistent with the results from the growers survey (see Chap. 15), as is the frequent occurrence of crops including poppies, various cereals and other vegetables in the rotation. The number of sites where each pathogen was found and average pathogen DNA concentrations (across all sites) for each region and year are shown in Table 18.4. Only *R. solani* AG2.1 and *Sp. subterranea* were widespread in all years, while *R. solani* AG3 was more prevalent after 2005 (Table 18.4), suggesting its introduction at some sites via potato seed. There was a low occurrence of *R. solani* AG2.2 and *Streptomyces spp.* in all years. *R. solani* AG4 was found at about a quarter of the sites (Table 18.4).

There were statistically significant temporal differences in pathogen DNA for *Sp. subterranea*, and for *R. solani* AG2.1, and AG3 (Table 18.4). The most notable changes were the increases from 2005 to 2006 of *R. solani* AG3 and *Sp. subterranea* DNA. Presumably these were at least partly because the known host, potato, was grown at all sites in the intervening season. This conclusion is supported by the observation that average concentrations of *Sp. subterranea* DNA were significantly less in 2007 and 2008 than in 2006, as the time since potatoes increased (Tables 18.4 and 18.5). However, *Rhizoctonia* AG3 DNA concentrations remained high in these later years (Table 18.4), which supports other studies showing that it is able to readily survive by infecting or associating with a range of other crop and weed plants (Carling et al. 1986, Sturz et al. 1995, Bains et al. 2002).

Pathogen DNA has also been monitored in a replicated field trial near Devonport in northern Tasmania where the effects of different green manures on soil condition and crop production are being measured (Sparrow 2009). Like the monitoring sites described above, the presence of a potato crop in this controlled experiment was associated with a significant subsequent increase in DNA of *Sp. subterranea* and *R. solani* AG3 (Table 18.5), but unlike the monitoring sites, in the trial *R. solani* AG3 DNA concentrations subsequently decreased while those for *Sp. subterranea* remained steady (Table 18.5). *R. solani* AG2.1 and AG4 DNA also decreased after June 2007.

Table 18.3 Land use at the sites

Site	2008/09	2007/08	2006/07	2005/06	2004/05	2003/04	2002/03	2001/02	2000/01
TAS 05 001	Lucerne	Lucerne	Oats	Potato	Lucerne	Lucerne	Lucerne	Lucerne	Lucerne
TAS 05 002	Pasture	Poppies	Barley	Potato	Wheat	Poppies	Lucerne	Wheat	Wheat
TAS 05 003	Oats	Fallow	Cereal	Potato	Oats	Pasture	Pasture	Pasture	Pasture
TAS 05 004	Pasture	Pasture	Pasture	Potato	Pasture	Pasture	Pasture	Pasture	Pasture
TAS 05 005	Pasture	Pasture	Pasture	Potato	Pasture	Pasture	Pasture	Pasture	Pasture
TAS 05 006	Pasture	Pasture	Fallow	Potato	Pasture	Pasture	Pasture	Poppies	Beans
TAS 05 007	Poppies	Wheat	Peas	Potato	Onions	Peas	Poppies	Beans	Potato
TAS 05 008	Oats	Barley	Carrots	Potato	Pasture	Pasture	Pasture	Pasture	Pasture
TAS 05 009	Pasture	Oats	Carrots	Potato	Peas	Poppies	Pasture	Pasture	Pasture
TAS 05 010	Pasture	Pasture	Wheat	Potato	Pasture	Pasture	Pasture	Pasture	Pasture
TAS 05 011	Ryegrass	Oats	Poppies	Potato	Pyrethrum	Pyrethrum	Pyrethrum	Pyrethrum	Pyrethrum
TAS 05 012	Ryegrass	Wheat	Poppies	Potato	Wheat	Poppies	Beans	Ryegrass	Poppies
TAS 05 013	Pasture	Fallow	Ryegrass	Potato	Poppies	Beans	Pasture	Pasture	Pasture
TAS 05 014	Wheat	Peas	Poppies	Potato	Onions	Broccoli	Poppies	Ryegrass	Potato
TAS 05 015	Pasture	Pasture	Poppies	Potato	Pasture	Pasture	Pasture	Pasture	Pasture
TAS 05 016	Lucerne	Lucerne	Triticale	Potato	Ryegrass	Pasture	Triticale	Triticale	Potato
TAS 05 017	Pyrethrum	Pyrethrum	Wheat	Potato	Poppies	Pasture	Pasture	Pasture	Pasture
TAS 05 018	Pyrethrum	Carrots	Peas	Potato	Ryegrass	Poppies	Swedes	Carrots	Peas
TAS 05 019	Pasture	Pasture	Peas	Potato	Pasture	Pasture	Pasture	Pasture	Pasture
TAS 05 020	Pasture	Pasture	Fallow	Potato	Ryegrass	Ryegrass	Poppies	Wheat	Not known
TAS 05 022	Pasture	Pasture	Pasture	Potato	Pasture	Pasture	Pasture	Poppies	Pasture
TAS 05 023	Onions	Ryegrass	Ryegrass	Potato	Ryegrass	Ryegrass	Ryegrass	Broccoli	Poppies
TAS 05 024	Poppies	Onions	Barley	Potato	Ryegrass	Pyrethrum	Pyrethrum	Pyrethrum	Pyrethrum
TAS 05 025	Ryegrass	Wheat	Wheat	Potato	Lucerne	Lucerne	Lucerne	Lucerne	Lucerne
TAS 05 026	Pasture	Pasture	Pasture	Potato	Ryegrass	Poppies	Potato	Pasture	Pasture
TAS 05 027	Pyrethrum	Pyrethrum	Pyrethrum	Potato	Poppies	Ryegrass	Carrots	Poppies	Ryegrass
TAS 05 029	Wheat	Ryegrass	Ryegrass	Potato	Cauliflower	Ryegrass	Poppies	Potato	Not known
TAS 05 030	Potato	Ryegrass	Poppies	Potato	Pasture	Pasture	Pasture	Beans	Poppies

Table 18.4 Average pathogen DNA concentrations (pg/g soil, $\log_{10}$ transformed), and number of monitoring sites* where pathogens were detected in each year of sampling

	2005		2006		2007		2008	
Pathogen	DNA	#sites	DNA	#sites	DNA	#sites	DNA	#sites
Rhizo. AG2.1	1.75ab	24	1.97b	27	1.77b	25	1.38a	24
Rhizo. AG2.2	0.09a	3	0.09a	3	0.11a	2	0.08a	1
Rhizo. AG3	0.42a	7	0.88b	17	1.49c	22	1.04b	17
Rhizo. AG4	0.41a	5	0.79a	7	0.95a	9	0.78a	7
Streptomyces	0.29a	3	0.42a	5	0.44a	7	0.38a	6
Spongospora	1.86a	27	3.62c	28	3.22b	28	3.10b	27

Within each row, means followed by the same letter are not significantly different ($P=0.05$)
*Only 27 sites were sampled in 2008

Table 18.5 Mean pathogen DNA (pg/g soil, $\log_{10}$ transformed) at five dates in a single field trial in Tasmania

Pathogen	Sept 2006	June 2007	Nov 2007	Aug 2008	Dec 2008	F-prob
Rhizo. AG2.1	2.19bc	2.40c	1.69a	1.48a	1.83ab	0.002
Rhizo. AG2.2	Not	detected	at	any	date	
Rhizo. AG3	0.98a	2.25c	1.61b	1.09a	1.01a	<0.001
Rhizo. AG4	2.07bc	2.83c	1.93b	0.80a	0.82a	<0.001
Streptomyces	0.00a	0.13a	0.00a	0.13a	0.00a	0.564
Spongospora	1.42a	2.27b	2.30b	2.24b	2.29b	<0.001

For each pathogen, means followed by the same letter are not significantly different at the 5% level. Potatoes, cv. Russet Burbank, were grown from October 2006 to March 2007

18.5 Influence of Prior Land Use on Changes in Pathogen DNA in Soil

ANOVA was conducted to determine if the absolute and relative temporal changes in pathogen DNA at the monitoring sites could be related to the immediate prior land use. Zero data were excluded so that changes in pathogen concentrations were only investigated in paddocks with measurable pathogen DNA. The analyses showed a significant effect of immediate prior land use only for *Sp. subterranea* (F-Prob<0.001) and *R. solani* AG3 (F-Prob$=0.01$). For *Sp. subterranea* (Table 18.6), the effect of prior land use was driven largely by increased concentrations of DNA after potatoes, compared to decreases for most other land uses. This may therefore simply reflect the general increase in *Sp. subterranea* DNA under favourable conditions in 2005/2006 when all the sites were in potatoes, and the subsequent decrease in inoculum when they were not, as reflected by the changes in average annual concentrations (Table 18.4).

The changes in *R. solani* AG3 DNA (Table 18.7) appear to be less strongly tied to potatoes, which again suggests this organism readily survives on a range of plant material, and which may also explain the absence of any prior land use effect for the other anastomosis groups of *R. solani*. The main absolute differences in AG3 were

Table 18.6 Absolute (Δ DNA) and relative (Log DNA before/Log DNA after) changes in *Spongospora subterranea* DNA (pg/g soil) averaged for particular land uses in the prior season

Prior land use	Triticale	Poppy	Carrot	Pea	Barley	Oat	Pasture	Lucerne	Wheat	Fallow	Pyrethrum	Ryegrass	Cereal	Onion	Potato
# of obs.	1	6	3	4	3	3	12	2	6	4	3	6	1	1	28
Δ DNA	−6,798	−5,623	−4,850	−4,353	−3,698	−2,684	−2,550	−1,561	−1,462	−1,476	−1,455	−662	−126	43	7,244
LogDNA before LogDNA after	−1.29	−0.39	−0.30	−0.30	−0.33	−0.31	−0.18	−0.21	−0.43	0.13	−0.22	−0.08	−0.30	0.05	1.86

Sites with no measurable DNA in any year excluded from the analysis

Table 18.7 Absolute (Δ DNA) and relative (Log DNA before/Log DNA after) changes in *Rhizoctonia solani* AG3 DNA (pg/g soil) averaged for particular land uses in the prior season

Prior land use	Oat	Wheat	Pasture	Pyrethrum	Triticale	Barley	Lucerne	Potato	Poppy	Pea	Carrot	Ryegrass	Fallow
# of obs.	2	5	11	3	1	3	1	23	6	3	2	4	3
Δ DNA	−521	−387	−243	−2	8	8	12	69	163	179	682	868	1,027
LogDNA before / LogDNA after	−0.99	−0.91	−0.48	−0.31	0.42	0.59	0.29	0.56	0.42	1.34	0.25	1.04	1.51

Sites with no measurable DNA in any year excluded from the analysis

Table 18.8 Average concentrations ($\log_{10}$ transformed) of DNA in 2005 in soil at four sites where potatoes were grown in the previous 5 years, compared to average concentrations at all sites in 2005

Pathogen	Rh. AG2.1	Rh. AG2.2	Rh. AG3	Rh. AG4	Spong.	Strept.
Prior potatoes	1.9	0.0	0.3	0.0	2.0	0.0
All sites	1.8	0.1	0.4	0.4	1.9	0.3

between fallow, carrot and ryegrass, which tended to increase DNA, and oats, wheat and pasture, which were associated with decreases. Relative changes following fallow and ryegrass were also positive, and negative for oats and wheat. Peas were also associated with a relative increase in DNA (Table 18.7).

Regressions of pathogen DNA against Tasmanian land use over the entire 9 years of records (Table 18.3) did not identify any additional effects of land use. There was no evidence that the four sites (TAS 05 007, 014, 016, 029 see Table 18.3) where potatoes were grown between 2000 and 2005 had higher than average 2005 pathogen DNA concentrations (Table 18.8). However, further monitoring since 2008 (Sparrow unpublished data) has shown that at the six sites where potatoes have again been grown, concentrations of *Sp. subterranea* DNA have increased further, although the corresponding incidence and severity of powdery scab in the potatoes at these sites was low. Increases in DNA without increased symptoms of powdery scab may occur if *Sp. subterranea* proliferates as galls on the potato roots rather than as lesions on the tubers (Falloon et al. 2003). Root galls were observed at many of these sites (Fig. 18.1).

18.6 Poppy Versus Pasture as Influences on Soil-Borne Diseases of Subsequent Potato Crops

In the Tasmanian potato grower survey (Chap. 15), pasture was the most common other element of potato rotations and poppies were the most common other crop. We therefore decided to investigate whether the presence of these land uses immediately prior to potatoes would generate different soil-borne disease outcomes in the potatoes that followed.

Thirteen commercial potato fields were selected in 2007/08 and nine in 2008/09. The fields were chosen such that they had been in either (a) grass pasture or grass/legume pasture (hereafter referred to as pasture) or (b) poppies in the year immediately prior to the potatoes. All fields were on loam to clay loam soils in northern Tasmania.

After the fields had been commercially planted with potatoes, a single 10 m length of planted row was marked and the commercial potato seed excavated and replaced by surface-sterilised, cut and cured pieces of certified seed of cv. Desiree. This was done to remove the influence of seed-borne inoculum and to introduce a cultivar more susceptible to soil-borne diseases than Russet Burbank, the usual commercial variety. Prior to excavating the commercial seed at each site, 40 soil

Fig. 18.1 *Spongospora subterannea* galls on the roots of potato, cv. Desiree, at one of the monitoring sites where potatoes were grown for a second time during the study period (the largest tuber is about 100 mm long)

cores 150 mm deep were taken from along the row. The cores were bulked, air-dried, ground to <2 mm and analysed for pathogen DNA.

The strips were otherwise managed by the growers as part of the commercial crop. Four plants were dug from each strip 8–10 weeks after planting to inspect stems, stolons and roots for symptoms of *R. solani* canker and *Sp. subterranea* root galls. Canker was scored on a scale of 1 to 9, with 1 denoting severe canker and stem girdling and 9 denoting no symptoms. The remaining plants were harvested at maturity and total tuber yield recorded. A 50-tuber subsample was washed and scored for symptoms of soil-borne diseases according to standard protocols (Crump 2005). These scores included disease incidence (% of tubers showing any symptoms) and a severity index (the incidence weighted by a score reflecting the tuber surface area affected by symptoms). Symptoms of powdery and common scab were distinguished by checking by microscope for the presence of *Sp. subterranea* spore balls in scrapings from lesions, and by use of Bioreba Sss test strips (www.bioreba.com).

Results were analysed by one-way ANOVA using Genstat. Associations between variables were explored using simple correlation and by fitting standard curves. Soil DNA data were normalised by log-transformation prior to statistical analysis.

The only significant effect of immediate paddock history was for the pre-plant concentration of *Sp. subterranea* DNA (Table 18.9). Paddocks with an immediate

Table 18.9 Mean pre-plant soil DNA (pg/g soil) and tuber and root disease scores for poppy vs pasture sites in 2007–08 and 2008–09

	Log_{10} *Spongospora* DNA	Powdery scab incidence	Powdery scab disease index	Log_{10} *Streptomyces* DNA	Common scab incidence	Common scab disease index	Log_{10} *Rhizoctonia* AG2.1 DNA	Log_{10} *Rhizoctonia* AG2.2 DNA	Log_{10} *Rhizoctonia* AG3 DNA	Log_{10} *Rhizoctonia* AG4 DNA	Stem canker score	Black scurf incidence	Black scurf disease index
Mean 07–08 Poppy	2.15	19%	4.8	0.00	63%	17.2	1.88	0.00	0.90	0.54	8.5	1%	0.3
Mean 07–08 Pasture	3.65	47%	12.2	0.00	50%	27.2	1.26	0.00	0.52	0.30	8.5	1%	0.3
Mean 08–09 Poppy	2.90	23%	9.8	0.00	32%	19.8	1.14	0.00	1.10	0.44	8.9	13%	5.8
Mean 08–09 Pasture	3.89	41%	15.5	0.52	55%	25.9	2.46	0.00	0.15	0.00	8.1	0%	0.0
Overall mean poppy	2.42	20%	6.6	0.00	52%	18.2	1.61	0.00	0.97	0.51	8.7	5%	2.3
Overall mean pasture	3.76	44%	13.7	0.23	52%	26.6	1.80	0.00	0.35	0.16	8.3	1%	0.1
Overall mean 07–08	2.8	32%	8.2	0.0	57%	21.9	1.6	0.0	0.7	0.4	8.5	1%	0.3
Overall mean 08–09	3.4	33%	13.0	0.3	45%	23.2	1.9	0.0	0.6	0.2	8.5	6%	2.6
Anova FProb history, years combined	0.022	NS	NS	NS	NS	NS	NS	NS	NS	NS	NS	NS	NS
Anova FProb year, histories combined	NS	NS	NS	NS	NS	NS	NS	NS	NS	NS	NS	NS	NS
Anova FProb year x history	NS	NS	NS	NS	NS	NS	0.013	NS	NS	NS	NS	NS	NS

NS not statistically significant, $P>0.05$

history of poppies had a lower average concentration than paddocks with an immediate history of grass or grass/clover pasture. The difference was more than one order of magnitude. While this difference was not reflected in significant differences between the incidence or severity of powdery scab at these sites, mean incidence and severity values in poppy paddocks were about half those of pasture paddocks (Table 18.9).

There were no significant effects of year, but there was a significant interaction between year and history for pre-plant DNA of *R. solani* AG2.1 (Table 18.9), with pasture sites having more DNA than poppy sites in 2008 but not in 2007. Although the lower mean stem canker score (i.e. more canker) for pasture (8.1) than poppy (8.9) sites in 2008 was consistent with the DNA difference, the stem canker difference was not statistically significant (Table 18.9), and there were no significant associations between any *R. solani* DNA measure and either stem canker or black scurf. These two diseases were not severe with only one site having a black scurf severity index greater than 20 and two sites scoring stem canker lower than 8. There were no year or history effects on the other Rhizoctonia anastomosis groups or on the incidence or severity of black scurf (Table 18.9).

Other observations of note were the lack of measurable *R. solani* AG2.2 DNA in soil at any site, and the very low level of measured *St. scabiei* DNA (recorded at only 2 out of the 22 sites) despite an average incidence of common scab of about 50% and an average severity of about 20% (Table 18.9).

18.7 DNA Testing of Soil as a Predictor of Disease

While an understanding of temporal changes in pathogen DNA concentrations in soil is important for medium to long term management of pathogen loads, farmers also want to know the threshold concentrations above which disease risk increases substantially (Ophel-Keller et al. 2008). In the Tasmanian DNA monitoring work described earlier in this chapter, the only significant relationship between pathogen load and disease was for *Sp. subterranea* and powdery scab in cv. Shepody grown in 2005–06 (Fig. 18.2). For a disease index of 20, above which processors have previously applied price penalties, the DNA threshold for cv. Shepody is about 10 pg/g.

However, in our work comparing the disease impacts of prior growth of either poppies or pasture, the powdery scab – *Spongospora* DNA relationship, fitted across all sites in both years of this study, was exponential not linear (Fig. 18.3) and the threshold for cv. Desiree in these years was much higher, about 1,000 pg/g. This may reflect the different potato cultivar, because both Falloon et al. (2003) and Slater (2009) found from cultivar screening trials that Desiree was more resistant than Shepody to powdery scab, and may also reflect the wetter growing season in 2005–06 which was highly favourable for powdery scab.

The relationship in Fig. 18.3 was much stronger than the corresponding relationships between powdery scab disease severity and either years since the last potato crop (Fig. 18.4) or the number of potato crops in the last 20 years (Fig. 18.5) at these sites. The latter two measures do not account for the degree of pathogen introduction on seed or pathogen build-up from previous potato crops.

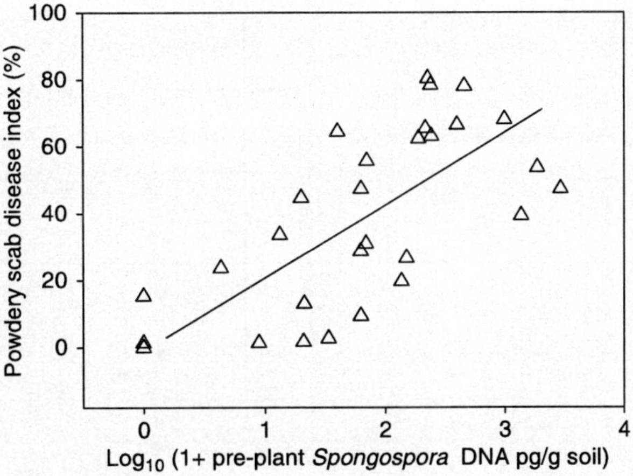

Fig. 18.2 Powdery scab disease severity index in Shepody potatoes vs soil DNA at 29 monitoring sites in Tasmania in 2005–06 ($r^2 = 0.46$)

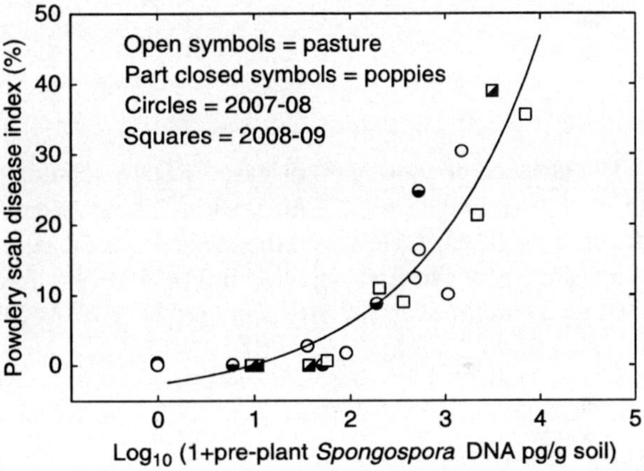

Fig. 18.3 Powdery scab disease severity index in Desiree potatoes vs soil DNA in 22 Tasmanian field sites (13 in 2007–08 and 9 in 2009–10). *The equation of the line is* $y = -4.06 + 0.801 \times 2.249^x$ *(*$P < 0.001$*, 83% of variance explained)*

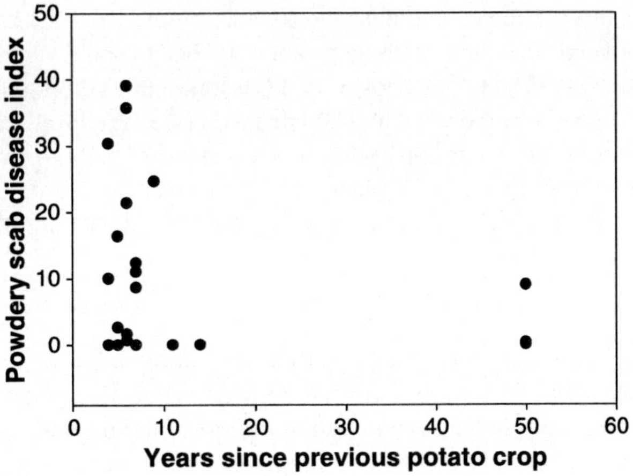

Fig. 18.4 Relationship between powdery scab disease index and number of years since potatoes were last grown at each site

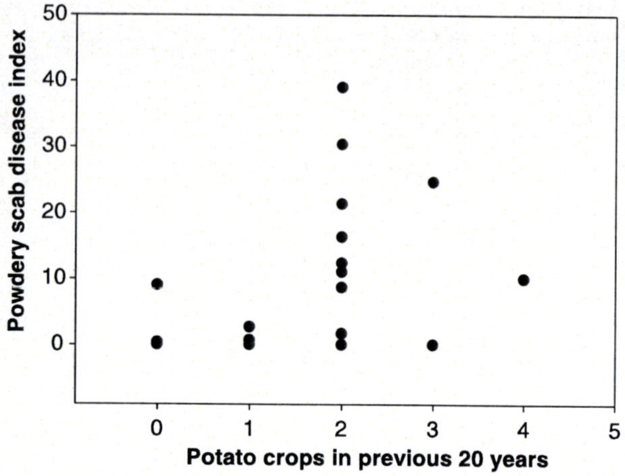

Fig. 18.5 Relationship between powdery scab disease index and number of potato crops in the previous 20 years at each site

The absence of significant disease vs DNA relationships for other pathogens in our work was, in the case of *Rhizoctonia*, largely because we found almost no stem canker or black scurf. However, the reverse was the case in some instances for common scab; e.g. at the sites depicted in Fig. 18.3, we found almost no *Streptomyces* DNA yet common scab severity averaged over 20% (Table 18.9).

18.8 Conclusions

Isolation, long rotations, seed certification and rigorous rogueing of symptomatic plants has led to a potato production system in Tasmania that has very low incidence of important viral pathogens such as PLRV and PVY common in other regions of the world. This has allowed seed production without the need for insecticide application for aphid control. Reliance on visual growing season inspections and a centralised seed cutting system for most seed stocks prior to planting has however allowed increased incidence of symptomless contact-transmitted viruses such as PVS and PVX. Recent efforts to reduce incidence of PVS within seed crops appear to have had some success.

Many soil-borne potato pathogens are well established in Tasmania but their impact is variable in space and time. Not surprisingly, the presence of potatoes appears to have the biggest effect on pathogen DNA concentrations in the soil. The lack of evidence from Tasmanian work to date of effects of other land uses may be at least partly because in this work most land uses were replicated at relatively few sites. It is likely that other crop management practices e.g. tillage, cultivar, irrigation management and seed quality, exert equal or greater influence on disease. Our *Sp. subterranea* DNA threshold for powdery scab in cv. Desiree is consistent with results from South Australia (Ophel-Keller et al. 2009) and gives hope that this test has utility as an indicator of disease risk. However it also appears that thresholds may vary for different cultivars and this warrants further testing for other common Tasmanian processing cultivars.

Acknowledgments This work was supported in part by Horticulture Australia Limited (HAL) in partnership with the Potato Processing Association of Australia and was funded by the potato levy. The Australian Government provides matched funding for all HAL's R&D activities.

References

Abad JA, Moyer JW, Kennedy GG, Holmes GA, Cubeta MA (2005) *Tomato spotted wilt virus* on potato in Eastern North Carolina. Am J Potato Res 82:255–261

Al-Shahwan IM, Abdalla OA, Al-Saleh MA (1997) Viruses in the northern potato-producing regions of Saudi Arabia. Plant Pathol 46:91–94

Bains PS, Bennypaul HS, Lynch DR, Kawchuk LM, Schaupmeyer CA (2002) Rhizoctonia disease of potatoes (*Rhizoctonia solani*): fungicidal efficacy and cultivar susceptibility. Am J Potato Res 79:99–106

Carling DE, Kebler KM, Leiner RH (1986) Interactions between *Rhizoctonia solani* AG-3 and 27 plant species. Plant Disease 70:577–578

Crump N (2005) Assessing diseases in potatoes for the processing potato group. Department of Primary Industries, Victoria

Falloon RE, Genet RA, Wallace AR, Butler RC (2003) Susceptibility of potato (*Solanum tuberosum*) cultivars to powdery scab (caused by *Spongospora subterranea* f. sp. *subterranea*), and relationships between tuber and root infection. Aust Plant Pathol 32:377–385

Horticulture Australia (2009) Australian potato research program final report to Horticulture Australia for project PT04016 (I. Kirkwood ed). Horticulture Australia, Sydney

Lambert S (2008) Epidemiology of potato virus S and potato virus X in seed potato in Tasmania, Australia. Dissertation, University of Tasmania

Ophel-Keller K, McKay A, Hartley D, Herdina, Curran J (2008) Development of a routine DNA-based testing service for soilborne diseases in Australia. Australas Plant Pathol 37:243–253

Ophel-Keller K, McKay A, Harding R, Hall B, Wicks T, Wiechel T, Crump N, Shah F, Marshall J, Falloon R (2009) DNA monitoring tools for soil-borne pathogens of potato. In: Kirkwood I (ed) Australian potato research program final report to Horticulture Australia for project PT04016. Horticulture Australia, Sydney

Sampson PJ, Walker J (1982) An annotated list of plant diseases in Tasmania. Department of Agriculture, Tasmania

Slater T (2009) Disease resistance screening (common scab, powdery scab and TSWV). In: Kirkwood I (ed) Australian potato research program final report to Horticulture Australia for project PT04016. Horticulture Australia, Sydney

Sparrow L (2009) Crop Rotations. In: Kirkwood I (ed) Australian potato research program final report to Horticulture Australia for project PT04016. Horticulture Australia, Sydney

Stevenson WR, Loria R, Franc GD, Weingartner DP (2001) Compendium of potato diseases, 2nd edn. APS Press, St Paul

Sturz AV, Johnston HW, MacWilliams CK (1995) Weed hosts of *Rhizoctonia solani* in Prince Edward Island. Can J Plant Pathol 17:346–352

Wilson CR (2001) Resistance to infection and translocation of *Tomato spotted wilt virus* in potato. Plant Pathol 50:402–410

Chapter 19
In Vitro Cell Selection Techniques for Enhancing Disease Resistance – Case Study: Common Scab Resistance in Russet Burbank

Calum R. Wilson and Robert S. Tegg

Abstract Here we demonstrate the value of *in vitro* cell selection techniques for potato cultivar improvement using enhancement of common scab resistance as a case study. Common scab is an important disease of potato world-wide. A diversity of *Streptomyces* species are associated with disease, however all pathogenic species and strains produce the phytotoxin, thaxtomin A. This toxin is an essential pathogenicity factor and provides a convenient positive selection agent for *in vitro* cell selection to obtain disease resistant potato variants. Using such techniques we obtained many variants of Russet Burbank with enhanced resistance to common scab, interestingly some of which did not express toxin tolerance. Agronomic assessments showed many variants had impaired yield, however several variants had equivalent or even superior yield to the unselected parent cultivar and these have been progressed toward commercial exploitation. Associated studies showed many of the common scab resistant variants showed resistance to a second unrelated disease, powdery scab. Preliminary studies suggest altered suberisation within lenticels may be a possible mechanism for broad spectrum disease resistance to tuber-invading pathogens.

19.1 *In Vitro* Cell Selection Versus Traditional Breeding Approaches

Management of soil-borne diseases of potato is often difficult and not particularly effective. There are limited management tools available. Fungicides or other agrichemical options may be ineffective due to the difficulty in maintaining sufficient concentrations of the active materials in the soil during the time of pathogen invasion (often many weeks after planting).

C.R. Wilson (✉) • R.S. Tegg
Tasmanian Institute of Agriculture, University of Tasmania, 13 St Johns Avenue, New Town, TAS 7008, Australia
e-mail: Calum.Wilson@utas.edu.au

Z. He et al. (eds.), *Sustainable Potato Production: Global Case Studies*,
DOI 10.1007/978-94-007-4104-1_19,

Host plant resistance is a valuable tool to combat debilitating diseases. Most potato breeding programs around the world strive to enhance resistance to locally important soil-borne diseases, however, selecting effective disease resistance traits whilst maintaining other essential agronomic and quality features is often extremely difficult. This is particularly problematic in French fry production as strict demands are placed on tuber quality and cooking characteristics as well as yield.

An alternate approach to conventional breeding to obtain genetic disease resistance (or other traits of value) is through the use of *in vitro* cell selection to produce somaclonal variants (Shepard et al. 1980; Larkin and Scowcroft 1981; van Harten 1998). Potato is readily amenable to tissue culture and may be regenerated from undifferentiated callus with relative ease (Shepard and Totten 1977). Variants of a parent potato cultivar may be generated during tissue cell culture with or without the use of chemical mutagens. They can then be screened for specific traits of interest. Selection of somaclonal variants with traits such as enhanced disease resistance has a significant advantage over traditional breeding in that it avoids the major genetic re-assortment associated with sexual crossing. Thus the likelihood that derived variants retain most or all desirable traits of the parent cultivars is greater (Shepard et al. 1980; Larkin and Scowcroft 1981). Furthermore, the process for selection of elite cultivars is generally faster and cheaper than conventional breeding approaches.

In vitro cell selection procedures have been used to obtain or refine traits of interest in potato. These include enhanced disease resistance to pathogens such as *Phytopthora infestans, Verticillium dahliae, Alternaria solani, Fusarium oxysporum* and *Streptomyces scabiei* (Matern et al. 1978; Behnke 1980; Gunn et al. 1985; Taylor et al. 1993; Sebastiani et al. 1994).

19.2 Case Study: Selection of Resistance to Common Scab Disease in Commercial Potato Cultivars

Common scab is regarded as one of the most economically important diseases of potato in Tasmania reflecting its prevalence and disease severity. Common scab is induced following infection of developing tubers with pathogenic *Streptomyces* spp. Yield is seldom affected, but reduction in tuber quality and consequent value of fresh produce, and losses to seed tuber producers through failure of certification standards can be substantial. Deep-pitted lesions frequently occur in common scab affected tubers in Australia (Wilson et al. 1999). These affect processing quality where the normal steam-peeling processes do not remove the lesions adding substantial costs to the processing sector.

Recent industry figures estimate common scab costs the Tasmanian French fry processing industry A$3.7 M per annum (p.a.) or approximately 4% of the annual industry value (P. Hardman, Simplot Australia, personal communication). Seed producers face the largest losses. Crops may be rejected if infection levels exceed the national certification guidelines of ≤4% tubers possessing lesions. To cope with anticipated loss of certified seed, the processing companies routinely contract 20% more seed than required. The cost to the seed sector is estimated at A$2.3 M p.a. (31% of this sector's value). Growers of processing crops face loss of tuber quality and of tuber yield (as severely affected crops are harvested early) resulting in an estimated A$0.4 M p.a loss. In addition, rejection of crops at the factory (an average of 1% of crops are rejected) due to unacceptable levels of common scab disease result in an additional A$1.0 M p.a. loss. In the factory, sophisticated processing lines can eliminate much of the lesioned tissues that escape the peeling process. Infected potatoes also will be processed according to the quality standards of the customer (the worst affected lines going to cheaper generic brand products). Reject fries and post-peel potato waste (including some lesioned materials) may be processed into hash browns or potato flour.

Management options for the control of common scab in potato are limited and generally not very effective (Loria et al. 2006). Host plant resistance is regarded as an important control strategy for durable disease management. No commercial potato cultivars are immune to common scab disease, but there is considerable variation in cultivar susceptibility (McKee 1958). Following infection the potato tuber is stimulated to produce suberised cork layers at the site of attack that limit further penetration by the pathogen (Fischl 1990). However, it is these defensive cork layers which disfigure the tubers.

Potato breeding programs around the world include enhancing resistance to common scab as one of their key priorities. This objective, however, has proved very difficult to achieve, due at least partially to the polygenic nature of known heritable resistance (Haynes et al. 1997). Furthermore, there are logistical difficulties in accurately and efficiently screening for common scab resistance within the population of seedlings from breeding crosses. Also, as mentioned above, the French fry processing industry has very strict requirements for acceptable yield, tuber size, shape, reducing sugars, solids etc. which severely limits suitable varieties.

Russet Burbank is one of the most widely grown potato cultivars in the world (comprising *c.* 40% of all potatoes grown in the USA and Australia) primarily because it possesses tuber characteristics and cooking qualities that make it ideal for French fries. Russet Burbank was first selected over 130 years ago and despite the concerted efforts of breeding programs to obtain alternative varieties with equivalent or better traits it remains the dominant cultivar for French fry production around the world. The variety is male sterile which reduces its value as a parent in conventional breeding and has meant incorporation of its genetic qualities into new varieties has not been widely attempted.

Russet Burbank possesses moderate resistance to common scab disease (Lambert et al. 2006), however, it frequently suffers severe disease under conducive conditions (Sparrow and Salardini 1997; Wilson et al. 1999; Wanner and Haynes 2009). There are several different clones of Russet Burbank characterised (Coleman et al. 2003). Small but significant variability in the levels of resistance to common scab (incidence and severity) have been shown (Wilson 2001).

19.3 Pathogen Characterisation from Tasmanian Soils

S. scabiei is the predominant species associated with this disease, although a diversity of distinct pathogenic *Streptomyces* species and strains have been identified in various studies around the world (Tashiro et al. 1990; Doering-Saad et al. 1992; Faucher et al. 1992; Bouchek-Mechiche et al. 2000; Park et al. 2003). A common feature of all pathogenic strains and species of *Streptomyces* is their ability to synthesise the thaxtomin group of phytotoxins (of which thaxtomin A is the predominant compound), whilst non-pathogenic strains fail to produce the phytotoxins (King et al. 1989, 1991; Babcock et al. 1993; Loria et al. 2006).

A selection of pathogenic isolates obtained from diseased Tasmanian potatoes have been analysed for morphological and biochemical traits (Kurster 1972), partial 16s cDNA sequencing, thaxtomin A production by thin layer chromatography, and the presence of the *nec1* gene (commonly linked with thaxtomin synthesis genes) using a PCR assay with nested primers (Cullen et al. 2000).

Examination of the 16s data showed most of these isolates fell within a single clade clustering with sequences from *S. scabiei* isolates from various parts of the world (Fig. 19.1). Morphological characteristics of these isolates varied to some extent from those of the type strain of *S scabiei* (ATCC 49173) including greater tolerance of acid growth conditions (down to pH 4.0). One diverse isolate was identified. It clustered most closely to *S. turgidiscabies* (ATCC 700248, AB026221) of the known pathogenic species (Fig. 19.1), but again possesses characters which differ from the type strain of this species. *S. turgidiscabies* was described from Japan, and has also been reported associated with common scab in Korea (Park et al. 2003) and other parts of the world. This Tasmanian isolate did not produce a melanin pigment, favoured acid growth conditions (down to pH 3.5), did not possess the *nec1* gene but did produce thaxtomin A. It is interesting to note the presence of an acid-tolerant strain from Tasmanian soils, as the soil environment is only mildly acidic and well buffered, not necessarily suggestive of a major evolutionary advantage toward acid tolerance.

19.4 *In Vitro* Selection of Disease Resistant Russet Burbank Somaclones

Thaxtomin A is thought to inhibit cellulose biosynthesis and to trigger a programmed cell death response in plants (Fry and Loria 2002; Duval et al. 2005; Bischoff et al. 2009). Experimental evidence has confirmed thaxtomin A has an essential role in the induction of common scab (Goyer et al. 1998; Healy et al. 2000; Kers et al. 2005). This requirement of thaxtomin A for disease induction has provided a specific target for resistance studies. Selecting for tolerance to thaxtomin A may be an effective way of obtaining or identifying resistance. Indeed the use of purified phytotoxin to rapidly screen new lines generated in breeding programs for

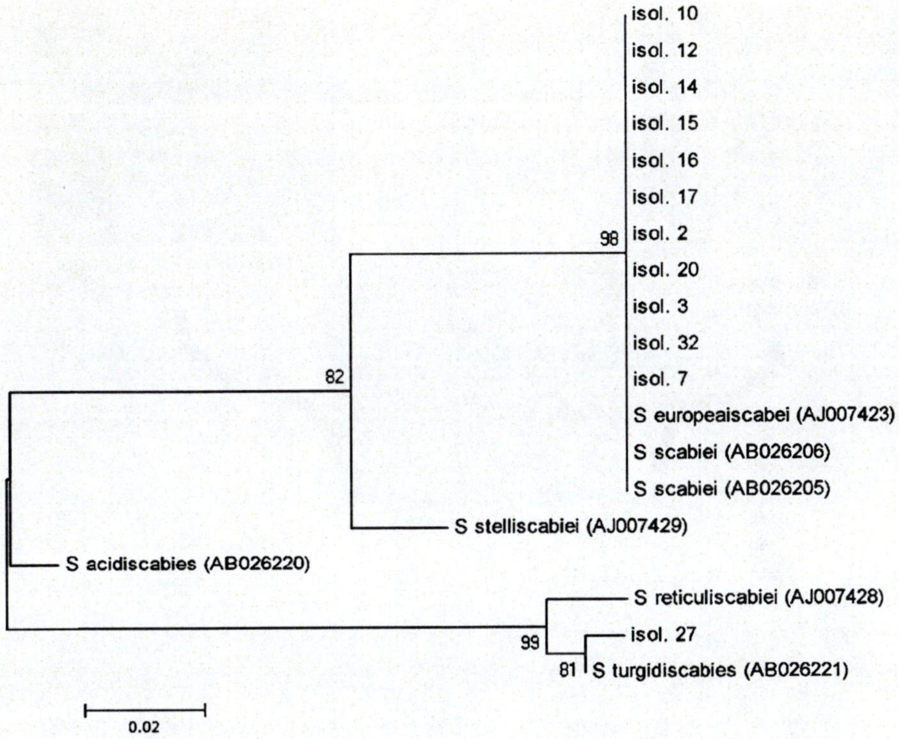

Fig. 19.1 Phylogenetic analysis of partial 16s cDNA sequences for 12 Tasmanian pathogenic *Streptomyces* isolates (compared to typed pathogenic species from GENBANK). The maximum likelihood tree was constructed using MEGA4 with 1,000 bootstrap replicates and the pairwise deletion option. Bootstrap values greater than 50% are shown

probable common scab resistance has been evaluated (Hiltunen et al. 2006). The phytotoxin also provides a convenient selection tool for *in vitro* cell selection studies to obtain toxin tolerant (and disease resistant) variants of existing commercial cultivars.

We utilised tissue culture technologies to select for thaxtomin A tolerant potato clones of current commercial varieties (Wilson et al. 2009, 2010b). This has the advantage of seeking enhanced resistance phenotypes whilst retaining important commercial characteristics of the parent varieties. The approach used thaxtomin A as a cell selection agent against cultured potato cells (callus). Incorporation of thaxtomin A into the culture medium was used to impose a toxic stress on a large population of plant cells from which rare variants tolerating the toxin treatment were recovered.

The procedure we used for *in vitro* cell selection of disease resistant potato variants is outlined in Fig. 19.2. In brief, we obtained highly purified thaxtomin A by solvent extraction and column chromatography from pathogen cultures in liquid medium (Wilson et al. 2009). Potato callus cultures were initiated from stem

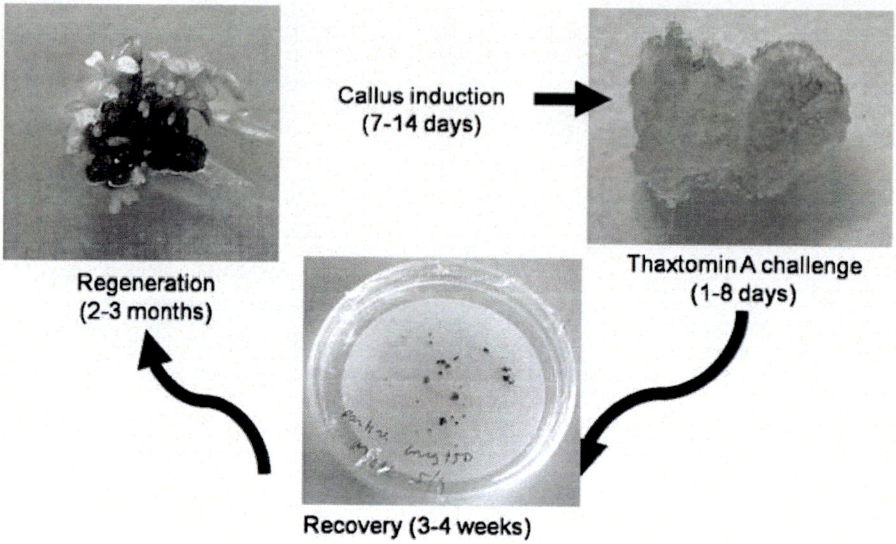

Fig. 19.2 Schematic of the *in vitro* cell selection procedure used to obtain common scab resistant potato variants

internodes of tissue culture plants of Russet Burbank. Friable callus cells were isolated from calloused stems and suspended in liquid medium containing toxic quantities of thaxtomin A (4–7 μM) and incubated for 1–8 days.

Recovery of the rare surviving potato cell colonies required amelioration of the inhibitory effect of the majority of dead and decaying cells. This was achieved by plating toxin-treated cells onto a sterile filter paper placed on top of a nurse culture of tobacco cells (*Nicotiana plumbaginifolia)* grown on a standard callus inducing medium (Conner and Meredith 1985). This was further aided by incorporation of a coconut water additive within the medium (Wilson et al. 2009, 2010b).

After 3–4 weeks, surviving potato cell colonies were identifiable. Following transferral to a regeneration medium shoots were obtained after a further 2–3 months, multiplied in tissue culture and used for subsequent assays. From 29 toxin-challenge selection trials, a total of 253 variants (from 212 cell colonies) of Russet Burbank were obtained.

19.5 Assessment for Toxin Tolerance and Disease Resistance

Variants obtained following selection were screened for thaxtomin A tolerance in comparison to the unselected parent using both detached leaflet and tuber slice bioassays (Wilson et al. 2010b). In these bioassays potato tissues were incubated with lethal concentrations (7 μM) of thaxtomin A and resulting necrosis scored.

In the leaflet bioassay necrosis scores following thaxtomin A treatment of the variants tested ranged from only 10% up to 250% of the necrosis shown by the parent with a mean result equivalent to the parent. The tuber slice bioassay gave similar although more conservative results. Between 16% and 22% of all variants tested showed enhanced thaxtomin A tolerance (with 5% showing a surprising increased sensitivity to the toxin in the leaflet bioassay; Table 19.1). Variants were consistent in their response to toxin treatment either showing enhanced thaxtomin A tolerance (e.g. A260) or no apparent change in thaxtomin A sensitivity (e.g. A380; TC-RB8; NZ-22c; Fig. 19.3).

Variants along with the unselected parent cultivar were then screened for common scab resistance in glasshouse trials. Those lines showing significantly less disease than the unselected parent cultivar were repeatedly screened in further glasshouse and subsequent field trials. Only those lines consistently showing reduced disease across all trials were considered as possessing enhanced disease resistance. In glasshouse trials inoculum (vermiculite colonised with known strains of pathogenic *S. scabiei*) was introduced to the soil medium at planting, whereas field assessment relied on natural soil inoculum. Growth conditions were adjusted to favour disease (limited irrigation and addition of lime at 1 kg/ha equivalent to the potting media or field soils). Harvested tubers were scored for disease incidence (measured by determining the proportion of tubers with visible lesions), and severity (measured by two assessments. First the tuber surface cover score was estimated visually, and second, where lesions were present on a tuber, the depth of the deepest lesion was scored; Wilson et al. 1999).

A total of 18–20% of Russet Burbank variants consistently demonstrated enhanced resistance to common scab in both incidence and severity assessments (Table 19.1) under varying disease pressure in both glasshouse and field assessment. Pooling data for all variants tested showed a mean disease reduction of 68% (incidence) and 65% (severity – tuber surface area covered with lesions) compared to the parent cultivar. The most disease resistant variants (e.g. A260 and NZ-22c) had a mean reduction in disease incidence of 97–100% and in disease severity of 98–100% (Table 19.2; Fig. 19.3). Visual symptoms were rare in the most resistant variants, while tubers of the unselected parent frequently had severe disease symptoms (Fig. 19.4).

19.6 Agronomic Assessment of Yield and Tuber Quality

In selection of somaclonal variants for valuable characteristics such as disease resistance, undesirable traits may be co-selected. These may be a direct result of the genetic change to achieve the desirable trait, or more commonly, the result of additional independent genetic changes (van Harten 1998). In order to determine the commercial merit of the disease resistant variants we undertook agronomic assessment of plant growth, yield and tuber quality (Wilson et al. 2010a).

Variants that expressed reduced mean common scab severity in one or more trials (70% or less than expressed by the parent cultivar) and four susceptible clones were

Table 19.1 Summary of toxin tolerance, disease resistance, agronomic performance and tuber quality for Russet Burbank somaclonal variants obtained by *in vitro* cell selection relative to the unselected parent cultivar

		Mean variant score relative to parent (%)	Range of variant scores relative to parent (%)	Variants significantly greater than parent (%)	Variants significantly less than parent (%)
Thaxtomin A tolerance	Leaflet bioassay	100	10–250	5	16
	Tuber slice bioassay	90	50–110	0	22
Disease resistance	Incidence (tubers with lesions)	32	0–156	0	20
	Severity (lesion surface coverage)	35	0–176	0	20
	Severity (lesion depth)	90	38–253	0	18
Tuber yield	Tuber number	36	2–121	0	9
	Tuber weight (total)	66	2–124	3	68
	Tuber weight (fry grade)	64	2–127	3	72
Tuber quality	Specific gravity	100	99–100	2	8
	Dry matter content	94	87–107	2	8
	Flesh colour	115	100–200	6	0
Cooking quality	Dark end (defect)	69	0–420	2	0
	Best fry colour (c000)	86	0–320	3	0
	Unacceptable fry colour (c1–2)	82	0–1,350	6	0
Tuber defects	Misshapen tubers	104	0–515	3	0
	Cracked tubers	57	0–790	5	0
	Hollow heart	61	0–820	2	0
	Internal browning	55	0–350	2	0

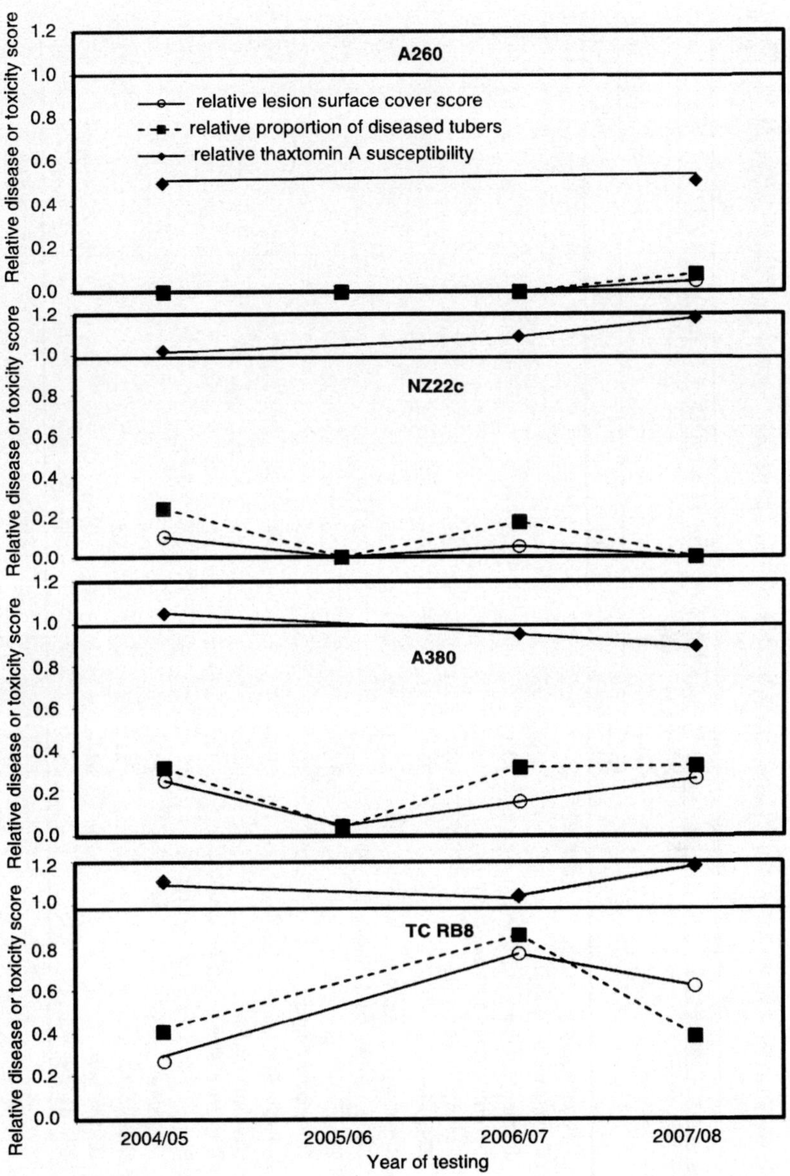

Fig. 19.3 Consistency of thaxtomin A response, and enhanced disease resistance (relative to parent cultivar (1.0)) within four somaclonal variants

then evaluated for relative yield, tuber quality and processing performance in a series of field trials. Tubers harvested from each plot were assessed for tuber yield, quality and cooking characteristics

Harvested tubers were graded into size classes and weighed. Total tuber yields per plant were calculated (tuber number and tuber mass) and compared to that

Table 19.2 Performance data for a selection of common scab disease-resistant somaclonal variants showing distinct traits relative to unselected parent cultivar

Phenotype		Increased yield	Equivalent yield	Very poor yield	
		Equivalent toxin sensitivity			Toxin tolerance
Clone name		TC-RB8	A380	NZ-22c	A260
Thaxtomin A tolerance	Leaflet bioassay	–	–	–	–
	Tuber slice bioassay	110	100	100	52
Disease resistance	Incidence (tubers with lesions)	70	18	9	3
	Severity (lesion surface coverage)	52	14	5	2
	Severity (lesion depth)	74	68	80	87
Tuber yield	Tuber number	113	112	49	76
	Tuber weight (total)	124	85	14	31
	Tuber weight (fry grade)	127	84	7	25
Tuber quality	Specific gravity	101	100	–	–
	Dry matter content	105	95	–	–
	Flesh colour	100	110	–	–
Cooking quality	Dark end (defect)	110	32	–	–
	Best fry colour (c000)	37	10	–	–
	Unacceptable fry colour (c1–2)	0	0	–	–
Tuber defects	Misshapen tubers	100	12	–	–
	Cracked tubers	21	29	–	–
	Hollow heart	824	0	–	–
	Internal browning	281	23	–	–

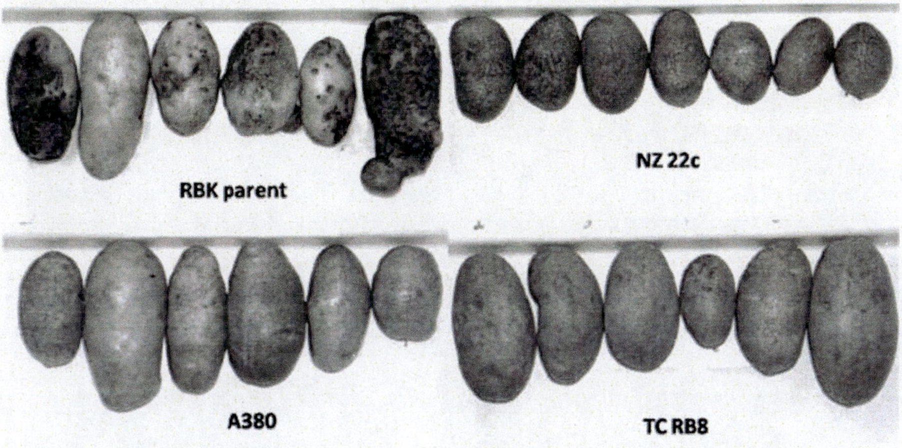

Fig. 19.4 Typical severity of common scab on unselected Russet Burbank parent cultivar and three disease resistant somaclonal variants

obtained from the parent cultivar within the same trial to give a relative yield score. Tubers with external defects (misshapen or cracked tubers) and small tubers (<75 g) were then removed and tubers reassessed to give a relative commercial (fry grade) yield assessment. Tubers were then assessed for internal defects (hollow heart and internal browning), tuber flesh colour (against a standard chart), specific gravity and dry matter content. Samples were cooked following industry standard protocols. Chip colour (using standard colour charts) and presence of dark end defects (representing sugar accumulation and subsequent caramelisation after cooking) were assessed.

Yield performance of the variants varied markedly, from very poor to equivalent or in some cases (3% of variants) better than the parent Russet Burbank cultivar (Table 19.1; Fig. 19.5). The majority of variants evaluated showed significant reductions in yield compared to the parent (68% of variants for total tuber yield, and 72% for fry grade yield) indicating co-selection of detrimental traits with disease resistance (Table 19.1; Fig. 19.5). We found a weak negative correlation between tuber yield (as assessed by weight of tubers per plant) and relative disease resistance within selected variants tested (Fig. 19.6). On average variants yielded 66% (total tuber weight) and 64% (fry grade) of the tuber mass of the parent. Tuber number was often suppressed as well (average of 36% of the parent).

However, we did identify a reasonable number of disease-resistant variants with equivalent yields to the parent cultivar, and furthermore, a few disease-resistant variants (e.g. TC-RB8; Table 19.2; Fig. 19.5) consistently yielded more tuber mass than the parent.

A few (2–6%) variants showed significantly greater tuber or cooking defects than the parent, however, the majority had equivalent or better tuber quality characteristics and cooking qualities than the parent cultivar (Table 19.1). Independent testing

Fig. 19.5 Typical tuber yield (per plot) of Russet Burbank parent cultivar and three disease resistant somaclonal variants

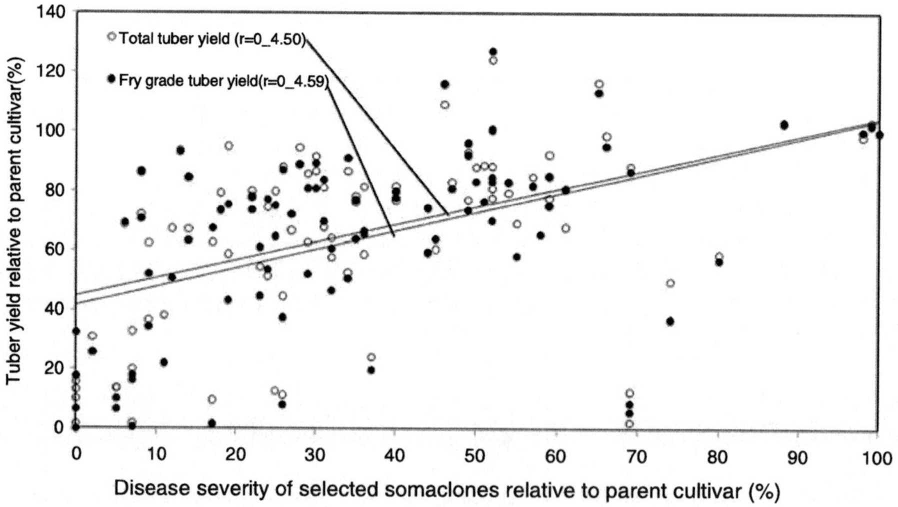

Fig. 19.6 Correlation between mean relative disease severity (tuber lesion coverage) and mean relative tuber yields (total and commercial grade weight) for the Russet Burbank variants

of a selection of the variants by a commercial French fry processing company confirmed these quality characteristics. High yielding disease resistant variants (TC-RB8 and A380; Table 19.2) have been progressed through to Plant Breeders Rights registration and are under assessment for commercial exploitation.

19.7 Possible Novel Resistance to Broad Range of Tuber-Invading Pathogens

During field assessment of the common scab resistant variants, we found that many appeared to exhibit enhanced resistance to a second important soil-borne disease, powdery scab (Fig. 19.7). This was interesting because although powdery scab produces symptoms that somewhat resemble common scab, this disease is caused by infection with a very different pathogen, the Plasmodiophoromycete *Spongospora subterranea* (de Boer 1991). The genetics, pathogenicity factors and environmental conditions favoured by this pathogen are completely different to those for the common scab pathogen. The only obvious linkage is that both pathogens penetrate developing potato tubers through immature lenticels on the tuber skin (Diriwachter and Parbery 1991; Adams and Lapwood 1978).

Lenticels are natural openings in the tuber periderm (outer skin layer) used for gas exchange (Wiggington 1973). They are recognised as important sites for pathogen invasion for many important diseases of potato. Common scab (Adams and

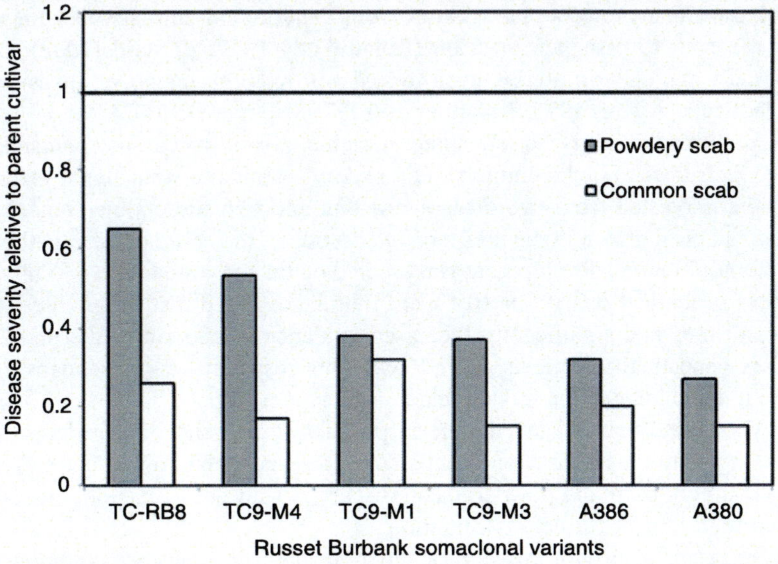

Fig. 19.7 Severity of powdery and common scab relative to the parent cultivar (1.0) for a selection of Russet Burbank somaclones (mean of five field trials)

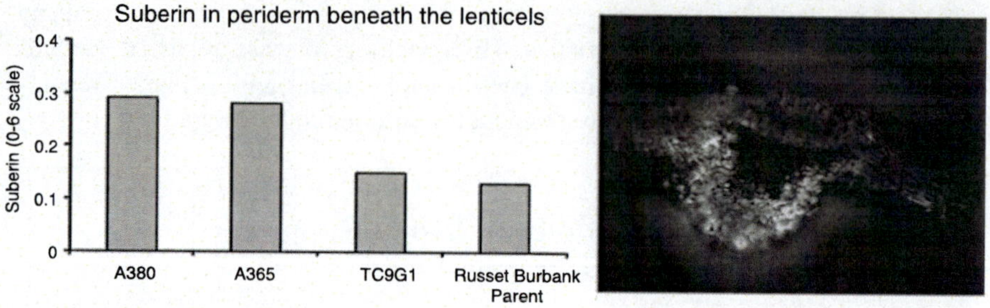

Fig. 19.8 Accumulated suberin in lenticel periderm of disease resistant (A380 and A365) and susceptible (TC9G and Russet Burbank parent) somaclones. Transverse lenticel section (A380) showing fluorescent suberin

Lapwood 1978), powdery scab (de Boer 1991; Diriwachter and Parbery 1991), tuber soft rot (Fox et al. 1971; Scott et al. 1996), and the tuber rot form of late blight (Adams 1975; Miller et al. 2002) are all examples of major tuber-infecting potato diseases associated with lenticel invasion. Potato cultivars resistant to certain tuber-invading pathogens have been characterised with low lenticel density, and/or more cell layers in the periderm and intensive suberisation and cuticularisation in lenticel tissues (e.g. Weber and Bartel 1986). For many such diseases, susceptibility of tubers to infection is also known to decrease as tubers mature. As tubers develop, lenticels mature, become filled with specialised "packing cells" and suberin and cuticular waxes are deposited within the lenticel periderm. Suberin is a cell wall lipid polyester that protects plant tissues from dehydration and is well known to provide a barrier to pathogen invasion (Kolattukudy 1977; Bernards 2002). Both packing cells and suberin are believed to assist in reducing pathogen invasion of mature lenticels (Adams 1975; Kolattukudy 1977; Miller et al. 2002).

In a preliminary study to elucidate possible physiological mechanisms of enhanced resistance to both common and powdery scab in our somaclonal variants, tubers were harvested from two disease-resistant and two susceptible somaclonal lines that had been grown in the presence or absence of the common scab pathogen (Khatri 2009). Lenticel development was studied on the harvested tubers. Whilst no differences in lenticel density or size were found, we noted that the two resistant somaclonal lines had significantly increased evidence of suberin in the periderm cell layers beneath the lenticels (Fig. 19.8). This response was dependent upon exposure to the pathogen (an inducible defence response).

We hypothesise that enhanced suberin production associated with developing lenticels on potato tubers may represent a common mechanism for enhanced resistance to common scab and powdery scab disease within our somaclonal variants. This is currently under further investigation.

Diseases associated with invasion of tubers through lenticels are common and highly damaging to potato production worldwide (Stevenson et al. 2001). The ability to use *in vitro* cell selection techniques to select for broad-scale resistance against a suite of tuber-infecting diseases (possibly through reduced susceptibility of tuber lenticels), whilst retaining important characters of the parent cultivars, would be highly significant and desirous.

Acknowledgements This work was partially funded by Horticulture Australia Limited (HAL) in partnership with the Potato Processing Association of Australia. The Australian Government provides matched funding for all HAL's R&D activities.

References

Adams MJ (1975) Potato tuber lenticels – susceptibility to infection by *Erwinia caratovora* var. *atroseptica* and *Phytopthora infestans*. Ann Appl Biol 79:275–282

Adams MJ, Lapwood DH (1978) Studies on the lenticel development, surface microflora and infection by common scab (*Streptomyces scabies*) of potato tubers growing in wet and dry soils. Ann Appl Biol 90:335–337

Babcock MJ, Eckwall EC, Schottel JL (1993) Production and regulation of potato-scab-inducing phytotoxins by *Streptomyces scabies*. J Gen Microbiol 139:1579–1586

Behnke M (1980) Selections of dihaploid potato callus for resistance to the culture filtrates of *Fusarium oxysporum*. Z Pflanzenzucht 85:254–258

Bernards MA (2002) Demystifying suberin. Can J Bot 80:227–240

Bischoff V, Cookson SJ, Wu S, Scheible WR (2009) Thaxtomin A affects CESA-complex density, expression of cell wall genes, cell wall composition, and causes ectopic lignifications in *Arabidopisis thaliana* seedlings. J Exp Bot 60:955–969

Bouchek-Mechiche K, Gardan L, Normand P, Jouan B (2000) DNA relatedness among strains of *Streptomyces* pathogenic to potato in France: *S. europaeiscabiei* sp. nov. and *S. stelliscabiei* sp. nov. associated with common scab, and *S. reticuliscabiei* sp. nov. associated with netted scab. Int J Syst Evol Microbiol 50:91–99

Coleman WK, Lil XQ, Tremblay-Deveau E, Coleman S (2003) Chemical maturation and storage performance of eleven Russet Burbank clones. Can J Plant Sci 83:893–902

Conner AJ, Meredith CP (1985) Strategies for the selection and characterisation of aluminium-resistant variants from cell cultures of *Nicotiana plumbaginifolia*. Planta 111:466–473

Cullen DW, Lees AK, Toth IK, Bell KS, Duncan JM (2000) Detection and quantification of fungal and bacterial potato pathogens in plants and soil. Bull OEPP 30:485–488

DeBoer RF (1991) Evaluation of potato cultivars in the greenhouse and field for resistance to powdery scab. Aust J Exp Agric 31:699–703

Diriwachter G, Parbery DG (1991) Infection of potato by *Spongospora subterranea*. Mycol Res 95:762–764

Doering-Saad C, Kampfer P, Manulis KG, Schneider J, Zakrezewska-Czerwinska J, Schrempf H, Barash I (1992) Diversity among *Streptomyces* strains causing potato scab. Appl Environ Microbiol 58:3932–3940

Duval I, Brochu V, Simard M, Beaulieu C, Beaudoin N (2005) Thaxtomin A induces programmed cell death in *Arabidopsis thaliana* suspension-cultured cells. Planta 222:1432–2048

Faucher E, Savard T, Beaulieu C (1992) Characterisation of actinomycetes isolated from common scab lesions on potato tubers. Can J Plant Pathol 14:197–202

Fischl G (1990) Biological characterisation and pathohistory of host parasite relationship related to potato scab. Novenytermeles 39:400–406

Fox RTV, Manners JG, Myers A (1971) Ultrastructure of entry and spread of *Erwinia caratovora* var. *atroseptica* into potato tubers. Potato Res 14:61–73

Fry BA, Loria R (2002) Thaxtomin A: evidence for a plant cell wall target. Physiol Mol Plant Pathol 60:1–8

Goyer C, Vachon J, Beaulieu C (1998) Pathogenicity of *Streptomyces scabies* mutants altered in thaxtomin A production. Phytopathology 88:442–445

Gunn RE, Jellis GJ, Starling NC (1985) Improved resistance to common scab (*Streptomyces scabies*) in protoplast-derived potato somaclones previously selected for high yield. Ann Appl Biol 106((suppl) 6):162–163

Haynes KG, Goth RW, Young RJ (1997) Genotype x environment interactions for resistance to common scab in tetraploid potato. Crop Sci 37:1163–1167

Healy FG, Wach M, Krasnoff SB, Gibson DM, Loria R (2000) The *txtAB* genes of the plant pathogen *Streptomyces acidiscabies* encode a peptide synthetase required for phytotoxin thaxtomin A production and pathogenicity. Mol Microbiol 38:794–804

Hiltunen LH, Laakso I, Chobot V, Hakala KS, Weckman A, Valkonen JPT (2006) Influence of thaxtomins in different combinations and concentrations on growth of micropropagated potato shoot cultures. J Agric Food Chem 54:3372–3379

Kers JA, Cameron KD, Joshi MV, Bukhalid RA, Morello JE, Wach MJ, Gibson DM, Loria R (2005) A large mobile pathogenicity island confers plant pathogenicity on *Streptomyces* species. Mol Microbiol 55:1025–1033

Khatri BB (2009) Potato tuber anatomy and susceptibility to common scab (*Streptomyces scabiei*). Dissertation, University of Tasmania

King RR, Lawrence CH, Clark MC, Calhoun LA (1989) Isolation and characterisation of phytotoxins associated with *Streptomyces scabies*. J Chem Soc Chem Commun 13:849–850

King RR, Lawrence CH, Clark MC (1991) Correlation of phytotoxin production with pathogenicity of *Streptomyces scabies* isolated from scab infected potato tubers. Am Potato J 68:675–680

Kolattukudy P (1977) Lipid polymers and associated phenols, their chemistry, biosynthesis and role in pathogenesis. Recent Adv Phytochem 77:185–246

Kurster E (1972) Simple working key for the classification and identification of natural taxa included in the international *Streptomyces* project. Int J Syst Bacteriol 22:139–148

Lambert DH, Reeves AF, Goth RW, Grounds GS, Giggie EA (2006) Relative susceptibility of potato varieties to *Streptomyces scabiei* and *S. acidiscabies*. Am J Potato Res 83:67–70

Larkin PJ, Scowcroft WR (1981) Somaclonal variation – a novel source of variability from cell-cultures for plant improvement. Theor Appl Genet 60:197–214

Loria R, Kers J, Joshi M (2006) Evolution of plant pathogenicity in *Streptomyces*. Ann Rev Phytopathol 4:469–487

Matern U, Styrobel G, Shepard J (1978) Reaction to phytotoxins in a potato population derived from mesophyll protoplasts. Proc Natl Acad Sci U S A 75:4935–4939

McKee RK (1958) Assessment of the resistance of potato varieties to common scab. Eur Potato J 1:65–80

Miller JS, Cummings TF, Mikitzel LJ, Johnson DA (2002) Influence of timing of harvest in relation to haulm killing and planting date on potato tuber rot caused by *Phytopthora infestans*. Plant Dis 86:264–268

Park DH, Cho JM, Hur JH, Lim CK, Kim JS (2003) Characterization of Streptomycetes causing potato scab in Korea. Plant Dis 87:1290–1296

Scott RI, Chard JM, Hocart MJ, Lennard JH, Graham DC (1996) Penetration of potato tuber lenticels by bacteria in relation to biological control of black leg disease. Potato Res 39:333–344

Sebastiani L, Lenzi A, Pugliesi C, Fambrini M (1994) Somaclonal variation for resistance to *Verticillium dahliae* in potato (*Solanum tuberosum* L.) plants regenerated from callus. Euphytica 80:5–11

Shepard JF, Totten RE (1977) Mesophyll cell protoplasts of potato – isolation, proliferation, and plant regeneration. Plant Physiol 60:313–316

Shepard JF, Bidney D, Shahin E (1980) Potato protoplasts in crop improvement. Science 208:17–24

Sparrow LA, Salardini AA (1997) Effects of residues of lime and phosphorus fertilizer on cadmium uptake and yield of potatoes and carrots. J Plant Nutr 20:1333–1349

Stevenson WR, Loria R, Franc GD, Weingarter DP (2001) Compendium of potato diseases, 2nd edn. APS Press, St Paul

Tashiro N, Miyashita K, Suzui T (1990) Taxonomic studies on the *Streptomyces* species, isolated as causal organisms of potato common scab. Ann Phytopathol Soc Japan 56:73–82

Taylor RJ, Secor GA, Ruby CL, Orr PH (1993) Tuber yield, soft rot resistance, bruising resistance and processing quality in a population of potato (cv. Crystal) somaclones. Am Potato J 70: 117–130

van Harten AM (1998) Mutation breeding – theory and practical applications. Cambridge University Press, Cambridge

Wanner LA, Haynes KG (2009) Aggressiveness of *Streptomyces* on four potato cultivars and implications for common scab resistance. Am J Potato Res 86:335–346

Weber J, Bartel W (1986) Potato lenticels – starting point and point of entry for wet rots. Kartoff Aktuell 65:45–54

Wiggington MJ (1973) Diffusion of oxygen through lenticels in potato tuber. Potato Res 16:85–87

Wilson CR (2001) Variability within clones of potato cv. Russet Burbank to infection and severity of common scab disease of potato. J Phytopathol 149:625–628

Wilson CR, Ransom LM, Pemberton BM (1999) The relative importance of seed-borne inoculum to common scab disease of potato and the efficacy of seed tuber and soil treatments for disease control. J Phytopathol 147:13–18

Wilson CR, Luckman GA, Tegg RS, Yuan ZQ, Wilson AJ, Eyles A, Conner AJ (2009) Enhanced resistance to common scab through somatic cell selection in cv. Iwa with the phytotoxin thaxtomin A. Plant Pathol 58:137–144

Wilson CR, Tegg RS, Hingston LH (2010a) Yield and cooking qualities of somaclonal variants of cv. Russet Burbank selected for resistance to common scab disease of potato. Ann Appl Biol 157:283–297

Wilson CR, Tegg RS, Wilson AJ, Luckman GA, Eyles A, Yuan ZQ, Hingston LH, Conner AJ (2010b) Stable and extreme resistance to common scab of potato obtained through somatic cell selection. Phytopathology 100:460–467

Part VI
Water-Saving Managements of Potato Production in the Semi-Arid Areas of Northern China

Chapter 20
Potato Evapotranspiration and Productivity as Affected by Drip Irrigation Frequency and Soil Matric Potential

Fengxin Wang and Zhongqi He

Abstract Drip irrigation has been shown to be an effective method for achieving high potato yields. Soil matric potential (SMP) and irrigation frequencies are two important factors in optimizing potato production and tuber quality. This chapter reviews and discusses a case study of potato evapotranspiration (ET) and productivity in drip irrigated potato systems in the North China Plain in 2001 and 2002, as affected by SMP and irrigation frequency. The experiment in this case study included five treatments for SMP: F1 (–15 kPa), F2 (–25 kPa), F3 (–35 kPa), F4 (–45 kPa) and F5 (–55 kPa) and six treatments for irrigation frequency: N1 (daily irrigation), N2 (2 day intervals), N3 (3 day intervals), N4 (4 day intervals), N6 (6 day intervals) and N8 (8 day intervals). In general, ET was reduced at lower SMP, as F5 had the lowest ET (150 mm) among the five treatments, and F2 had the highest (208 mm). F1 experienced somewhat waterlogged conditions. Potato ET losses declined as SMP dropped from –25 to –55 kPa. Higher frequency of irrigation enhanced both potato tuber growth and water use efficiency (WUE). Reducing irrigation frequency from N1 to N8 resulted in significant yield reductions by 33.4% and 29.1% in the 2001 and 2002 growing seasons, respectively. Based on the results, the authors of this case study suggested a SMP threshold of –25 kPa, and an irrigation frequency of once a day for the as a target for drip irrigation management for potato production in the North China Plain.

F. Wang
Center for Agricultural Water Research in China, China Agricultural University, Beijing 100083, China
e-mail: fxinwang@yahoo.cn

Z. He (✉)
USDA-ARS, Southern Regional Research Center, New Orleans, LA 70124, USA
e-mail: Zhongqi.He@ars.usda.gov

Z. He et al. (eds.), *Sustainable Potato Production: Global Case Studies*,
DOI 10.1007/978-94-007-4104-1_20,

20.1 Introduction

Arid and semi-arid areas account for one third of the land area of the Earth's surface. Semi-arid regions are defined as transition zones between arid and sub-humid belts, where precipitation is less than the potential evaporation (ET) (Huang et al. 2010). One of these regions is the Northern Area of China. As China is a country not only with a shortage of water resources, but also supporting a large population with great food needs, the development of sustainable and economically viable food production techniques is important in order to achieve security of both food and water resources. Research on water-saving and high-efficiency agriculture in China has been primarily focused on five aspects: (1) biological water-saving technologies such as identification and creation of water-efficient and draught-tolerant plant species, (2) development of technologies that can recycle and utilize inferior water resources such as brackish and reclaimed irrigation water, (3) creation of highly efficient water using technologies for dry farming, (4) development of water saving irrigation technology and equipment, and (5) creation of comprehensive technologies that meet regional needs for high-efficiency water-saving agriculture (Wu 2010). There has been a significant body of research conducted towards these efforts. For example, the effects of different planting pattern on water use and yield performance of winter wheat were examined in the Huang-Huai-Hai plain of China, and the furrow irrigated raised bed-planting was recommended as a sound opportunity for sustainable farming in that region (Zhang et al. 2007). The technique of micro-rainwater harvesting with ridges and furrows for potato production in semi-arid areas of China was evaluated (Wang et al. 2008). In addition to this chapter, Chap. 22 reviews the effect of plastic mulch on reduction of water loss and soil temperature regulation for potato growth. Chapter 23 reports the development of the transgenic potato plants with improved tolerance to drought and salinity stresses.

Efficient water delivery systems can contribute towards increasing higher crop yields and improved water and fertilizer use efficiency (WUE) (Badr et al. 2010; Jensen et al. 2010). Drip irrigation has been shown to be an effective method for achieving high potato yields (Eldredge et al. 2003; Yuan et al. 2003). However, precise irrigation management is essential to ensure the most efficient attainment of optimal yield and quality, as water storage under drip irrigation conditions is generally less than that for surface and sprinkler irrigation techniques, and most roots are concentrated in the wetted soil volume near each emitter or along each lateral line (Kang et al. 2004). Soil matric potential (SMP) and irrigation frequencies are two important factors in optimizing the potato production and quality (Wilson et al. 2001; Shock and Wang 2011). Based on published data (Kang et al. 2004; Wang et al. 2006, 2007), this chapter reviews and discusses a case study of potato ET and productivity in drip irrigated potato systems in the North China Plain affected by the SMP and irrigation frequency.

20.2 Irrigation Management and Precipitation

Field experiments were conducted at Luancheng (Hebei Province) Agro-ecosystem Station (LAES), Chinese Academy of Sciences, during the 2001 and 2002 growing seasons (Kang et al. 2004; Wang et al. 2006, 2007). Annual precipitation in the area is about 480 mm. and is normally concentrated between July and September, as precipitation is very rare in the spring and early summer. The dominant soil type is loam, with an average bulk density of 1.53 $g\,cm^{-3}$ for the upper 30 cm soil layer. The mineral content of the groundwater is less than 0.5 $g\,L^{-1}$, with the water table about 28 m below the surface. The drip irrigation system was installed after the experimental field was ploughed and bedded. Thin-wall drip tapes with a flow rate of 3.72 $L\,m^{-1}\,h^{-1}$ at 0.042 MPa were placed in the center of the raised beds. Dripper spacing was 30 and 20 cm, respectively, in the 2001 and 2002 growing seasons.

The six treatments for irrigation frequency were: daily (N1), and at intervals of every 2 days (N2), 3 days (N3), 4 days (N4), 6 days (N6), and 8 days (N8). In 2001, due to the lack of information on potato ET under drip irrigation the N1 treatment was irrigated according to SMP readings, weather broadcasts, and visual observation of potato growth. In 2002, the irrigation depth of N1 was determined from the previous day's ET, as measured by a lysimeter on the SMP (−25 kPa) treatment conducted at the same site and time. Irrigation quantities for N1 and precipitation after potato emergence are listed in Fig. 20.1. The total amount of irrigation was 192 and 142 mm

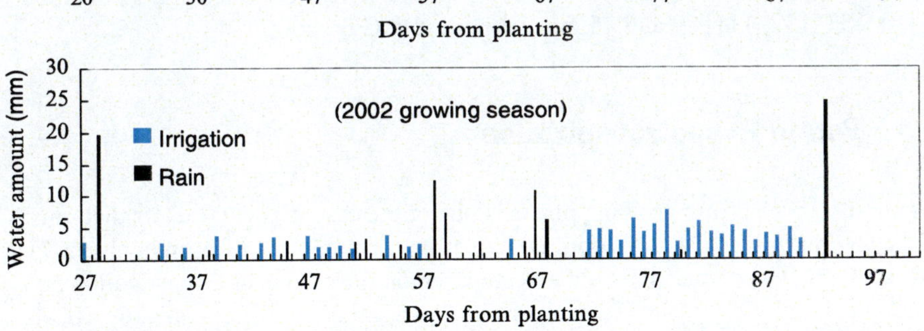

Fig. 20.1 Total amount of water received from daily irrigation (*N1*) and precipitation following potato emergence (Wang et al. 2006)

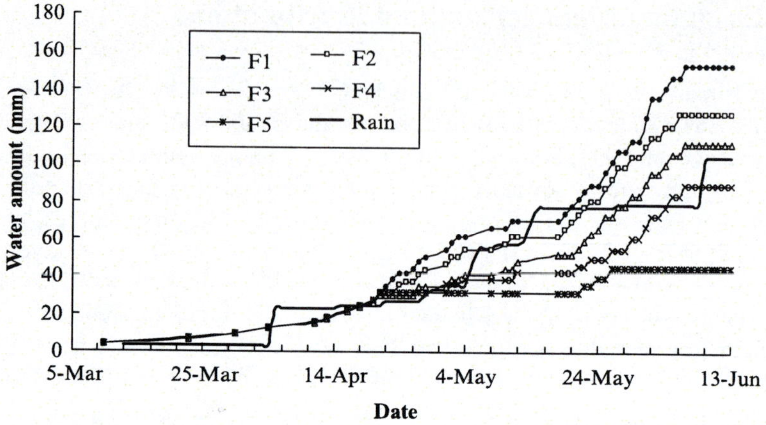

Fig. 20.2 Cumulative precipitation and irrigation water for potatoes in the 2002 growing season under different soil matric potentials: –15 kPa (F1), –25 kPa (F2), –35 kPa (F3), –45 kPa (F4) and –55 kPa (F5) with rain precipitation (Kang et al. 2004)

during the 2001 and 2002 growing seasons, respectively. Total rainfall was 77 and 104 mm, respectively, for the two seasons (Wang et al. 2006). Irrigation water in the other treatments was taken as the cumulative value of N1 treatment. When it rained, effective rainwater was subtracted from irrigation application. Before the potatoes sprouted, all treatments plots were well irrigated with the same quantity of water, at the same frequency, in order to ensure a uniform germination rate. After that, each plot was irrigated according to prescribed frequency treatments.

SMP was measured at 0.2 m, immediately under the drip emitter, and the 5 SMP treatments were: –15 kPa (F1), –25 kPa (F2), –35 kPa (F3), –45 kPa (F4) and –55 kPa (F5). Irrigation was applied only when the SMP reached the targeted values for F1, F2, F3, F4 and F5. The depth of water for each irrigation event of all the SMP treatments (the constant) was constantly changing during the growing season and varied from about 3 to 6 mm. Seasonal total applied irrigation for F1, F2, F3, F4 and F5 was 153, 132, 111, 89 and 45 mm, respectively (Fig. 20.2).

20.3 Potato Evapotranspiration

Evapotranspiration from growing plants is one of the essential parameters that must be assessed in order to achieve optimal irrigation management (Timlin et al. 2007). Badr et al. (2010) conducted a field investigation on the effects of drip irrigation and ET on yield and yield components, using four irrigation levels representing 100%, 80%, 60% and 40% of potato ET. Marutani and Cruz (1989) found that 3 to 5 mm of water per day is necessary to fulfill ET requirements and maintain optimal soil water potential (–50 to –10 kPa) for growing potatoes in the tropics. For a sub-humid region in India under furrow irrigation conditions, Kashyap and Panda (2001, 2003) evaluated ET estimation methods, and developed the crop-coefficients for potato. They found that the maximum and average daily ET of potatoes was 4.24 and 2.49 mm, respectively. Under hot dry conditions in northeastern Portugal, peak ET rates reached 12 to 13 mm per day on days immediately following irrigation, but then declined logarithmically to about 3 mm per day within 5 days following sprinkler irrigation (Ferreira and Carr 2002).

In this case study, the ET varied from 275.6 mm (N4) to 293.0 mm (N3) in the 2001 season, and from 192.1 (N8) to 214.5 mm (N1), among the six irrigation frequency treatments. The differences between the two seasons revealed a 74.4 mm higher average ET of the six irrigation frequency treatments in 2001 than that in 2002. However, it should be pointed out that the differences in ET between treatments in both growing seasons were statistically insignificant at $P=0.05$ (Wang et al. 2006).

In the 2002 growing season, the highest ET value for the SMP treatments was 63.4 mm greater than that for the lowest SMP, representing a decrease of 32.1%. A drip-irrigation frequency of 4 days or less had little effect on potato ET; however, irrigation frequencies of 6 and 8 days resulted in ET values were lower than in the other frequency treatments. The highest ET value was 36.7 mm (19.2%) more than the lowest value. As potato ET is affected by SMP and irrigation frequency, Kang et al. (2004) proposed that potatoes would sustain severe water stress and reduced crop ET when the SMP was below –45 kPa. Further analysis indicated that total ET was a function of SMP and irrigation frequency. The relationships are $ET\ (mm)=0.0216P^2-0.1217P+205.7$ (P, –KPa; $R^2=0.987$) and $ET\ (mm)=0.6076D^2+0.1945D+187.81$ (D, days; $R^2=0.843$), for the ET and SMP, respectively. These relationships indicate that total ET decreases significantly as SMP and irrigation frequency decrease.

Based on the observations in the work, a SMP threshold of –25 kPa and an irrigation frequency of once a day were suggested for potato drip-irrigation scheduling in the North China Plain (Kang et al. 2004). This suggestion was further fine tuned by Wang et al. (2007). In the later analysis, F5 had the lowest ET among the five treatments (150 mm), while F2 had the highest ET (208 mm). The difference was 48 mm (30%) higher with F2 than with F5. Although F1 had the highest irrigation frequency and greatest depth percolation (14.7 mm), the lower ET of F1 than F2 suggested that the F1 experienced somewhat water-logged conditions. This observation implied that plant performance and water uptake could have been constrained by leaching of nutrients from the rooting zone, and a decrease in root activity due to poor soil aeration. ET values declined as SMP dropped from –25 to –55 kPa, with the sharpest reduction (22 mm) occurring when the SMP dropped from –45 to –55 kPa. This suggests that potatoes sustain notable water stress at SMP values below –45 kPa. In other words, an SMP of –25 kPa (F2) was the most favorable criterion for potato production and water use efficiency (WUE), whereas –15 kPa (F1) was too wet and –45 kPa (F4) too dry, leading to severe water stress.

20.4 Potato Root Distribution

Based on the root length density (RLD) and root weight density (RWD) data (Tables 20.1), researchers concluded that potato roots were concentrated in the top 40 cm of the soil layer for all treatments. The highest root density appeared between 0 and 10 cm, where about 26–41% and 63–82% of the total root density was concentrated based on RLD and RWD, respectively. For the entire soil profile (0–60 cm) the data revealed the trend of increasing RLD but decreasing RWD with the more irrigation frequency. This inconsistency between RLD and RWD has been reported previously for irrigated cotton plants (Plaut et al. 1996). For this reason, some researchers use RLD only to characterize root systems (Coelho and Or 1999).

In the SMP experiment Wang et al. (2007) found that RWD had a tendency to increase gradually as SMP dropped from −15 to −35 kPa in the 0–30 cm soil layer, reaching a maximum at F3 (−35 kPa) and then declining as SMP decreased from −35 to −55 kPa (Fig. 20.3). One exception was observed with the F4 (−45 kPa) treatment at the depth of 0–10 cm, and could perhaps be due to sampling errors. It was clear that most potato roots grew in the upper 0 to 40 cm of the soil, with only a few roots growing deeper than 40 cm. The affect of SMP on RWD in the horizontal direction of root growth (Fig. 20.3b) was similar to that vertically (Fig. 20.3a), with few roots extending more than 30 cm from the potato plant. Previously, Lahlou and Ledent (2005) reported that the average root length of four potato cultivars whose root dry mass had been reduced by drought were all below 38.5 cm, under either well irrigated or water-stressed conditions. It was recently proposed that partial rootzone drying (PRD) is one technique that offers potential for reducing the

Table 20.1 Effects of potato root distribution under different irrigation frequencies in the 2001 growing season. Irrigation frequency treatments were: once every day (N1), once every 2 days (N2), once every 3 days (N3), once every 4 days (N4), once every 6 days (N6), and once every 8 days (N8) (Wang et al. 2006)

	Soil layer					
Treatment	0–10 (cm)	10–20 (cm)	20–30 (cm)	30–40 (cm)	40–60 (cm)	0–60 (cm)
(a) Root length density (cm cm^{-3})						
N1	0.227	0.153	0.160	0.172	0.083	0.146
N2	0.193	0.139	0.137	0.136	0.065	0.122
N3	0.276	0.102	0.111	0.108	0.042	0.113
N4	0.255	0.107	0.111	0.115	0.040	0.112
N6	0.268	0.110	0.096	0.097	0.044	0.110
N8	0.246	0.107	0.109	0.103	0.048	0.110
(b) Root weight density (mg cm^{-3})						
N1	0.883	0.178	0.109	0.154	0.043	0.235
N2	1.030	0.108	0.099	0.087	0.081	0.248
N3	1.486	0.237	0.186	0.150	0.072	0.367
N4	1.614	0.132	0.097	0.122	0.051	0.345
N6	1.924	0.166	0.127	0.074	0.029	0.392
N8	2.175	0.221	0.125	0.106	0.090	0.468

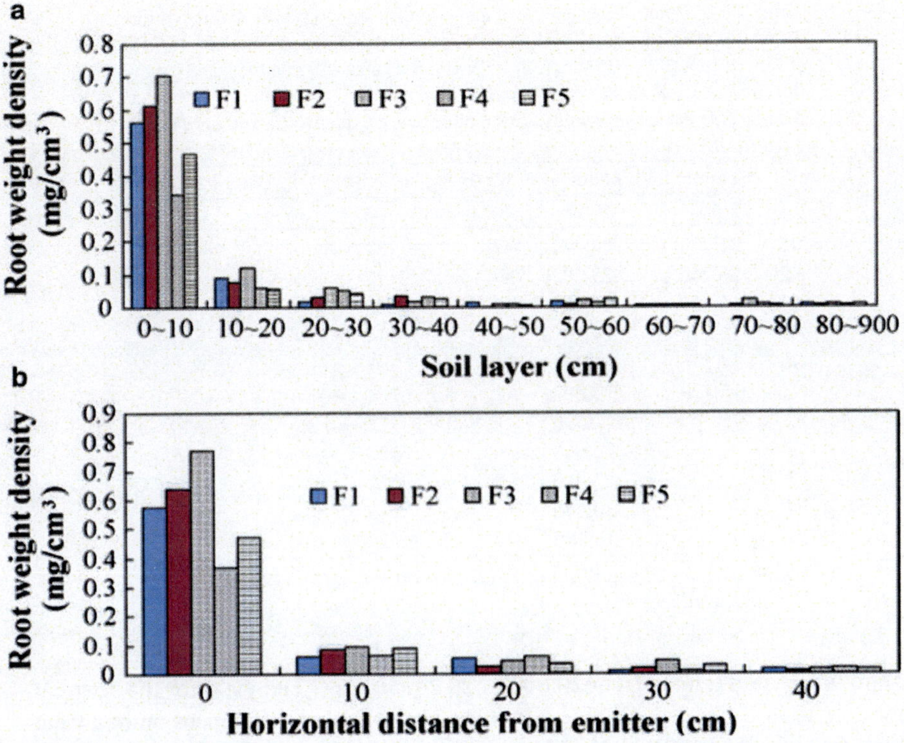

Fig. 20.3 Root weight density at (**a**) different soil depths and (**b**) different horizontal distances from crop base. The five soil matric potential treatments were: –15 kPa (F1), –25 kPa (F2), –35 kPa (F3), –45 kPa (F4) and –55 kPa (F5) (Wang et al. 2007)

use of irrigation water (Saeed et al. 2008). PRD may have potential in achieving sustainable potato production in the North China Plain. However, more field research on the impacts of drip irrigation parameters on root growth is needed prior to its widespread implementation, in order to ensure conservation of irrigation water with minimal loss of potato yield.

20.5 Potato Yields and Water-Use Efficiency (WUE)

Although the absolute tuber yield values were different in 2001 and 2002, tuber yield decreased as irrigation frequency declined (Table 20.2). In both season, the yields with irrigation frequency of N2, N3, and N4 were very close, whereas there was a sharp yield reduction with N6 and N8. In the 2001 season, the yield of potato tubers ranged from 14,139 to 21,234 kg ha^{-1}. The highest yield was with N1, and was 50.2% higher than the lowest yield produced (N8); however, differences among treatments were not significant according to F-test ($P>0.05$). In the 2002 season,

Table 20.2 Potato tuber yields, water use efficiency (WUE) and irrigation water use efficiency (I_{WUE}) with soil matric potential treatments of −15 kPa (F1), −25 kPa (F2), −35 kPa (F3), −45 kPa (F4) and −55 kPa (F5), and irrigation frequency of once every day (N1), once every 2 days (N2), once every 3 days (N3), once every 4 days (N4), once every 6 days (N6), and once every 8 days (N8)

	Tuber yield (kg ha^{-1})		WUE (kg ha^{-1} mm)		I_{WUE} (kg ha^{-1} mm)	
	2001	2002	2001	2002	2001	2002
N1	21,234	28,241	77	132	–[a]	–
N2	16,750	25,307	59	125	–	–
N3	16,889	25,405	58	120	–	–
N4	16,872	24,109	61	114	–	–
N6	14,176	21,157	50	103	–	–
N8	14,139	20,000	51	104	–	–
F1	–	22,590	–	115	–	−194
F2	–	26,660	–	128	–	155
F3	–	23,410	–	122	–	126
F4	–	20,640	–	113	–	38
F5	–	18,980	–	119	–	–

Data are adapted from Kang et al. (2004) and Wang et al. (2006, 2007)
[a]No data available

potato tuber yield ranged from 20,000 to 28,241 kg ha^{-1}. The yield for the six treatments were in order of N1 > N3 > N2 > N4 > N6 > N8, and differences among treatments was significant ($P<0.05$). Tuber yield was significantly higher with N1 than with N6, and N1 was 41.2% higher than N8. The yield difference between any other two treatments was insignificant ($P=0.05$) (Wang et al. 2006). The SMP data for the 2002 season revealed that total tuber yield for the different SMP treatments followed the order of F2 > F3 > F1 > F4 > F5 (Table 20.2). It is assumed that tuber growth in treatments F4 and F5 were restrained to some extent by a soil water deficit, while tuber growth at F1 was restrained by soil water excess (Wang et al. 2007). Statistical analysis ($P=0.05$) indicated that treatment F2 resulted in a significantly higher yield than treatments at F1, F4 and F5. F3 also had a significantly higher yield than F5 (Wang et al. 2007).

The cumulative irrigation for each drip irrigation treatment was 192 and 142 mm during 2001 and 2002 growing seasons, respectively. Cumulative rainfall was 77 and 104 mm, respectively, for the two seasons (Wang et al. 2006). In 2001, the change in soil water content that occurred between planting and harvesting of each treatment was negative, indicating that the soil became drier at the end of the growing season. In 2002, the change was positive, suggesting that the soil became wetter. Perhaps due to this difference in soil moisture content, the values of WUE were nearly twice as high in 2002 than 2001: however, despite higher WUE values in 2002, the effects of irrigation frequency on WUE and tuber yield for both seasons displayed the same general pattern.

Wang et al. (2007) noticed that WUE at the F1 SMP treatment was higher than that at F4, implying that WUE was not a good criterion for evaluating the effectiveness of irrigation in that particular experiment; therefore, an improved parameter, irrigation water use efficiency (I_{WUE}) was introduced as an additional factor. This parameter is defined as:

$$I_{WUE_i} = \frac{Y_i - Y_{i-1}}{I_i - I_{i-1}}$$

where Y_i and Y_{i-1} are yield at irrigation levels I_i and I_{i-1}, respectively.

Evaluation of this factor revealed that the F2 SMP treatment had the highest I_{WUE}, and suggested that a SMP of −25 kPa should be an irrigation target (Table 12.2). F1 had a negative I_{WUE} value consequent with lower tuber yield than F2. Furthermore, F3 had a higher I_{WUE} than F4, implying that the irrigation increase for F3 was more worthwhile than that for F4. Based on both the WUE and I_{WUE} analysis, the authors concluded that both F2 and F3 are good SMP thresholds for favorable potato production, with F2 superior due to both higher WUE and I_{WUE} values.

20.6 Potato Tuber Quality

In the 2001 season, the total tuber number per plant varied between 4.0 and 6.9 among the drip irrigation treatments (Table 20.3). Treatments N1, N2, N3, N4 and N6 resulted in more tubers per plant than did N8; however, the difference among the six treatments were not significant ($P>0.05$) (Wang et al. 2006). In contrast, there were significant differences in the number of marketable tubers among these treatments, with N2 producing the most, N1 producing the second, and N8 producing the lowest quantity of tubers that were over 50 g in weight. Overall, irrigation frequency increased the number of marketable tubers, as N1, N2, N3, N4 and N6 all had significantly higher marketable tuber yield than N8. In 2002, the effects of irrigation frequency on tuber quality was more significant than in 2001, with significant differences found in total tuber number per plant among the treatments. The difference in marketable tuber number among the treatments was also highly significant ($P<0.05$), and N1 produced significantly more marketable tubers than the other treatments. While the marketable tuber yield of N2 and N3 were very similar, irrigation frequency did increase production, and N3 was significantly higher than the N6 treatment. In summary, data for both the 2001 and 2002 growing seasons suggest that potato tuber production was optimized under the N1 irrigation frequency (daily irrigation), and that potatoes grown under the N8 (once every 8 days) irrigation regime had lower total tuber and marketable tuber yield than potatoes grown with higher irrigation (Wang et al. 2006).

In 2002, while the total number of tubers produced per plant was not affected by SMP, both the number of marketable tubers and the marketable tuber weight were significantly affected by SMP (Table 20.3). Treatments of SMP at F2 and F3 produced more marketable tubers per plant than potatoes grown at F4 and F5. The marketable tuber weight per plant was also significantly affected by SMP,

Table 20.3 Potato tuber quality impacted by soil matric potential treatments of −15 kPa (F1), −25 kPa (F2), −35 kPa (F3), −45 kPa (F4) and −55 kPa (F5), and irrigation frequency of once every day (N1), once every 2 days (N2), once every 3 days (N3), once every 4 days (N4), once every 6 days (N6), and once every 8 days (N8)

	Total tubers (plant^{-1})		Marketable tubers (plant^{-1})*		Marketable tuber weight (g plant^{-1})	
	2001	2002	2001	2002	2001	2002
N1	6.9a**	7.3a	3.3ac	3.9a	–***	–
N2	6.7a	5.2b	3.5a	3.3ab	–	–
N3	6.0a	6.1ab	2.7ac	3.3ab	–	–
N4	6.1a	5.4b	2.7ac	2.9b	–	–
N6	6.6a	4.5b	2.6ac	2.5b	–	–
N8	4.0a	4.4b	1.7b	2.6b	–	–
F1	–	5.2a	–	3.0ab	–	311.3ab
F2	–	6.4a	–	3.5ac	–	385.9a
F3	–	6.4a	–	3.7a	–	355.5ab
F4	–	5.3a	–	2.6b	–	271.2b
F5	–	4.5a	–	2.3b	–	225.6b

Data are adapted from Wang et al. (2006, 2007)
*Tuber weight not less than 50 g each
**For each treatment, values within a column followed by the same letter are not significantly different (P = 0.05)
***No data available

with F2 producing a much greater mass than the F4 and F5 treatments. Wilson et al. (2001) suggested that management of soil moisture deficit during tuber initiation and early development was important for the optimization of net and marketable tuber yields, as they found marketable tuber yield was better at −25 and −35 kPa than at −55 and −75 kPa. Lynch and Tai (1989) showed that marketable potato yield decreased as soil water potential dropped from −30 to −120 kPa. The observations in the field studies of Wang et al. (2006, 2007) for potato production in the North China Plain are consistent with the two previous studies.

20.7 Conclusions

In Northern China, potatoes are generally planted in raised beds and furrow-irrigated. Due to regional water shortages, farmers in this region are currently encouraged to adopt drip irrigation techniques to maximize water conservation. This chapter reviewed a field experiment that compared the effects of different drip irrigation frequencies and SMP thresholds on potato ET and productivity in a loam soil in the North China Plain. Potato ET increased as irrigation frequency and SMP increased, with the highest ET 63.4 mm (32.1%) more than the lowest ET value. Analysis of potato growth and productivity data revealed that a SMP of −25 kPa is most favorable for potato production, while a SMP of −15 kPa is too high and a

SMP of –45 kPa or more leads to notable water stress. Based on irrigation frequency treatments, the highest ET was 36.7 mm (19.2%) more than the lowest value. Drip irrigation frequency also affected root growth, as higher irrigation frequency resulted in a higher root length density. Increasing the irrigation frequency enhanced potato tuber growth, yield of marketable tubers, and water use efficiency. Based on these results these researchers recommend that a SMP threshold of –25 kPa and an irrigation frequency of once a day, could be used as a target reference for drip irrigation management for potato production in this region.

References

Badr MA, Abou Hussein SD, El-Tohamy WA, Gruda N (2010) Efficiency of subsurface drip irrigation for potato production under different dry stress conditions. Gesunde Pflanzen 62:63–70

Coelho EF, Or D (1999) Root distribution and water uptake patterns of corn under surface and subsurface drip irrigation. Plant Soil 206:123–136

Eldredge EP, Shock CC, Saunders LD (2003) Early and late harvest potato cultivar response to drip irrigation. In: Yada RY (ed) Potatoes – healthy food for humanity. International Society for Horticultural Science, Madison, pp 233–239

Ferreira TC, Carr MKV (2002) Responses of potatoes (*Solanum tuberosum* L.) to irrigation and nitrogen in a hot, dry climate – I. Water use. Field Crop Res 78:51–64

Huang J et al (2010) Dust aerosol effect on semi-arid climate over Northwest China detected from A-Train satellite measurements. Atmos Chem Phys 10:6863–6872

Jensen CR et al (2010) Deficit irrigation based on drought tolerance and root signalling in potatoes and tomatoes. Agric Water Manage 98:403–413

Kang YH, Wang FX, Liu HJ, Yuan BZ (2004) Potato evapotranspiration and yield under different drip irrigation regime. Irrig Sci 23:133–143

Kashyap PS, Panda RK (2001) Evaluation of evapotranspiration estimation methods and development of crop-coefficients for potato crop in a sub-humid region. Agric Water Manage 50:9–25

Kashyap PS, Panda RK (2003) Effect of irrigation scheduling on potato crop parameters under water stressed conditions. Agric Water Manage 59:49–66

Lahlou O, Ledent JF (2005) Root mass and depth, stolons and roots formed on stolons in four cultivars of potato under water stress. Eur J Agron 22:159–173

Lynch DR, Tai GCC (1989) Yield and yield component response of 8 potato genotypes to water-stress. Crop Sci 29:1207–1211

Marutani M, Cruz F (1989) Influence of supplemental irrigation on development of potatoes in the tropics. Hortscience 24:920–923

Plaut Z, Carmi A, Grava A (1996) Cotton root and shoot responses to subsurface drip irrigation and partial wetting of the upper soil profile. Irrig Sci 16:103–117

Saeed H, Grove IG, Kettlewell PS, Hall NW (2008) Potential of partial rootzone drying as an alternative irrigation technique for potatoes (*Solanum tuberosum*). Ann Appl Biol 152:71–80

Shock CC, Wang FX (2011) Soil water tension, a powerful measurement for productivity and stewardship. Hortscience 46:178–185

Timlin D, Fleisher D, Kim SH, Reddy V, Baker J (2007) Evapotranspiration measurement in controlled environment chambers: a comparison between time domain reflectometry and accumulation of condensate from cooling coils. Agron J 99:166–173

Wang FX, Kang YH, Liu SP (2006) Effects of drip irrigation frequency on soil wetting pattern and potato growth in North China Plain. Agric Water Manage 79:248–264

Wang FX, Kang YH, Liu SP, Hou XY (2007) Effects of soil matric potential on potato growth under drip irrigation in the North China Plain. Agric Water Manage 88:34–42
Wang Q, Zhang E, Li F, Li F (2008) Runoff efficiency and the technique of microwater harvesting with ridges and furrows for potato production in semi-arid areas. Waste Resour Manage 22:1431–1443
Wilson CR, Pemberton BM, Ransom LM (2001) The effect of irrigation strategies during tuber initiation on marketable yield and development of common scab disease of potato in Russet Burbank in Tasmania. Potato Res 44:243–251
Wu P-T (2010) Strategic considerations for Chinese agricultural water and food safety: issues, challenges and suggestions. Afr J Biotechnol 9:5251–5261
Yuan BZ, Nishiyama S, Kang YH (2003) Effects of different irrigation regimes on the growth and yield of drip-irrigated potato. Agric Water Manage 63:153–167
Zhang J, Sun J, Duan A, Wang J, Shen X, Liu X (2007) Effects of different planting pattern on water use and yield performance of winter wheat in the Huang-Huai-Hai plain of China. Agric Water Manage 92:41–47

Chapter 21
Effects of Plastic Mulch on Potato Growth

Fengxin Wang and Zhongqi He

Abstract Northern China is a major potato production region, and water-saving measures that can enhance both potato yield and quality play an important role in this region due to general water shortages. Plastic mulch has been used as an effective water-saving measure for potato cultivation in China. This chapter reviews the case studies on the effects of plastic mulching on potato growth, conducted at two areas of North China. Data from these experiments indicated that plastic mulching could save irrigation water and reduce evapotranspiration in most cases. Daily mean soil temperature under mulch was 2–9°C higher than that without mulch, especially during the early growth stage. However, as the plant canopy enlarged, the soil temperature difference between mulched and non-mulched plots became smaller. Plastic mulch could restrain or enhance potato plant growth during the early growth, dependent on the microenviromental air and soil temperatures. The possible negative effects of plastic mulching included a lower emergence, lower potato tuber yield, and poorer tubers quality, which may be attributable to the poorer soil aeration and detrimentally high soil temperature associated with plastic mulch when the air temperature is high. As mulch duration is an influential factor, data from these case studies suggested that 60 days of mulching duration was most favorable for potato production in the tested areas. Mulch removal after 60 days was proposed to avoid subsequent negative effects. To complement the current knowledge on the plastic mulching research, future research should be focused on the hydrothermal dynamics and its effect on potato growth with different drip irrigation regimes under plastic mulching conditions.

F. Wang (✉)
Center for Agricultural Water Research in China, China Agricultural University, Beijing 100083, China
e-mail: fxinwang@yahoo.com.cn

Z. He (✉)
Southern Regional Research Center, USDA-ARS, New Orleans, 70124, LA, USA
e-mail: zhongqi.he@ars.usda.gov

Z. He et al. (eds.), *Sustainable Potato Production: Global Case Studies*,
DOI 10.1007/978-94-007-4104-1_21,

21.1 Introduction

China is currently the world's leading potato-growing country, and potatoes are widely cultivated across China (Wei 2005). Northern China is a major potato production region and potatoes are generally planted in raised beds and furrow-irrigated. However, frequent droughts, general water shortages and poor irrigation management often lead to low yields and poor tuber quality. Therefore, water-saving measures that can enhance both potato yield and quality play an important role in this region (Wang et al. 2006).

Soil mulching with plastic film, which results in reduced water loss and more regulated soil temperature, has been widely used in agriculture (Li et al. 2004a). Additional benefits of plastic mulching can include decreased nitrate leaching (Schmidt and Worthington 1998; Bowen and Frey 2002; Romic et al. 2003), increased crop yields (Romic et al. 2003; Tiwari et al. 2003; Xie et al. 2005; Ramakrishna et al. 2006), increased death of pathogens (Vos et al. 1995; Triki et al. 2001), suppressed weeds (Ramakrishna et al. 2006; Ghosh et al. 2006), lessened soil bulk density (Anikwe et al. 2007), and enhanced use of rainwater (Tian et al. 2003). At the same time, the negative effects of mulching are rather notable. Tiquia et al. (2002) and Li et al. (2004b) found that the CO_2 content in the soil under mulch was much higher than in non-mulched areas, which means poor soil aeration and a detriment to potato growth (Phene and Sanders 1976). There are controversial reports on potato yield being impacted when plastic mulch is used, with yield reduction reported in some studies (Liang et al. 1998; Baghour et al. 2002), but yield increases in other studies (Cheng and Zhang 2000; Sun and Li 2004). As this yield variability might be attributed to differences in climatic conditions (Doring et al. 2005), information is needed on how potato growth would be affected by plastic mulch in Northern China. Therefore, researchers at China Agricultural University have conducted field research on the effects of plastic mulch on potato growth, began in 2001 and 2006 in North China Plain, and continued from 2006 to 2010 in Northwestern China (Wang et al. 2009, 2011; Hou et al. 2010). This chapter reviews these research activities with particular focus on the effects of mulching-drip irrigation regimes on potato growth.

21.2 Irrigation Water and Evapotranspiration

The total irrigation amounts varied with the year and irrigation criteria (Tables 21.1 and 21.2). At the Luancheng (Hebei Province) Agro-ecosystem Station, Chinese Academy of Sciences in 2001, mulching duration was for 60 days after planting with six drip irrigation frequencies with the same amount of irrigation water for all the mulched treatments. The non-mulched treatment (M0F1) had 20 mm more irrigation than the mulching treatments. Precipitation was 77 mm, which was normal for the local climate. Potato evapotranspiration (ET, the sum of soil surface evaporation and crop's transpiration) among the mulching treatments were in a range from 236 to 249 mm, showing little impact by irrigation frequency (Table 21.1).

Table 21.1 Seasonal rainfall precipitation, irrigation amount, and evapotranspiration in mm for each irrigation treatment at two sites of North China Plain (Wang et al. 2009)

Year	Treatment	Precipitation	Irrigation	Evapotranspiration
2001	M60F1	77	172	249
	M60F2	77	172	236
	M60F3	77	172	247
	M60F4	77	172	246
	M60F6	77	172	243
	M60F8	77	172	248
	M0F1	77	192	275
2006	M50	81	–[a]	–
	M70	81	–	–
	Mw	81	–	–
	M0	81	–	–

Treatment codes are defined as M: with plastic mulch of 0, 50, 60, and 70 days after planting, and the entire growing season (w), respectively. F: drip irrigation with once every (F1), 2 (F2), 3 (F3), 4 (F4), 6 (F6) and 8 (F8) days

[a]No data available

However, ET for the non-mulched treatment (M0F1) was 26–39 mm more than the mulched treatments. At the Tongzhou (Beijing) Farmland Water Cycle and Modern Water-saving Irrigation Experimental Station, Institute of Geographic Sciences and Natural Resources Research, Chinese Academy of Sciences in 2006, precipitation was 81 mm, which was very close to that at the Luancheng (Hebei Province) Agro-ecosystem Station as both sites are in the North China Plain. Unfortunately, the data of irrigation and soil water changed were unavailable (Wang et al. 2009).

At the Xuebai Experimental Station of Minqin Agricultural Extension Center of Gansu, precipitation in 2006 was 56 mm, which was less than normal in the region. In 2007, however, precipitation was 170 mm, nearly doubled the normal amount. The average ET of all treatments in 2006 was 121 mm less than that in 2007. In both the growing seasons, treatment M0 without plastic mulching received the most irrigation water and had the highest ET. Similarly, higher irrigation amounts and potato evapotranspirations without plastic mulch were observed in the 2008, 2009, and 2010 seasons at Shiyanghe Experimental Station for Water-Saving in Agriculture and Ecology, China Agricultural University, Wuwei, Gansu province in the same arid region of Northwest China (Table 21.2). These data confirmed the effectiveness of plastic mulching in reduction of irrigation water as this technique reduced the severity of the dry and windy climate and the strong soil surface evaporation under non-mulched conditions, especially early in the growing season (Chakraborty et al. 2008; Wang et al. 2003, 2009).

As shown in Table 21.2, irrigation ranged from 303 to 359 mm in 2008, from 127 to 200 mm in 2009, and from 177 to 312 mm in 2010. Irrigation during 2009 and 2010 was less than in 2008 due to more abundant and more effective rainfall and the selective reduction of the SMP criteria during parts of the growing season. In 2009, treatments MS2, MS3 and MS4 had less irrigation than MS1, implying that the lower soil matric potential (SMP) criterion, –35 kPa, could save irrigation water no

Table 21.2 Seasonal rainfall precipitation, irrigation amount, and evapotranspiration in mm for each irrigation treatment during five potato growing seasons at two sites in Gansu Province (Hou et al. 2010; Wang et al. 2011)

Year	Treatment	Precipitation	Irrigation	Evapotranspiration
2006	M0	56	330	428
	M60	56	250	361
	M75	56	260	357
	M90	56	250	371
	M105	56	260	391
	Mw	56	290	385
2007	M0	170	284	527
	M40	170	284	512
	M60	170	257	491
	M90	170	265	522
	Mw	170	245	564
2008	F1	54	359	465
	MF1	54	359	476
	MF2	54	339	448
	MF4	54	303	424
	MF8	54	338	447
2009	S1	97	200	336
	MS1	97	190	331
	MS2[a]	97	175	298
	MS3	97	147	289
	MS4	97	127	262
2010	S1	63	312	343
	MS1	63	296	314
	MS2	63	305	335
	MS3	63	259	258
	MS4	63	177	243

M: with plastic mulch of 0, 40, 60, 75, 90, and 105 days after planting, and the entire growing season (w), respectively. F: drip irrigation with once every 1 (F1), 2 (F2), 3 (F3), 4 (F4), and 8 (F8) days. S1: irrigation threshold at −25 kPa soil matric potential (*SMP*) at the three growth stages; S2: irrigation threshold at −25 kPa *SMP* at potato initiation and bulking stages and −35 kPa *SMP* at maturing stage; S3: irrigation threshold at −25 kPa *SMP* at potato initiation and maturing stages and −35 kPa *SMP* at bulking stage; S4: irrigation threshold at −35 kPa *SMP* at the three growth stages
[a]Irrigation amount of MS2 was unexpectedly higher than MS1. This should be experimental error caused by spatial variability of soil water (Wang et al. 2011)

matter at which potato development stages the criterion was used (Wang et al. 2011). In 2008, the mulched treatments irrigated once every 2, 4, and 8 days received less water than the mulched and non-mulched treatments with daily irrigation due to irrigation delays caused by rain.

Potato ET ranged from 424 to 476 mm in 2008, 262 to 336 mm in 2009, and 243 to 343 mm in 2010 season (Table 21.2). Since more irrigation water was applied in 2008, the ET values in 2008 were higher than that in 2009 and 2010. In 2008, the highest ET occurred in the treatments with the highest irrigation frequency; water use from plants under mulch was comparable to the non-mulched treatment, which is similar to the results of Kang et al. (2004). The highest ET (476 mm) was found in treatment MF1, slightly higher than F1. Tolk et al. (1999) and Xie et al. (2005) also reported similar results. They both attributed the greater water use to the higher leaf area index where the crop was grown with plastic mulch.

21.3 Soil Temperature Changes with Plastic Mulching

In an experiment carried out in 2006 at the Xuebai Experimental Station at the Minqin Agricultural Extension Center in Gansu Province, a typical arid region of Northwest China, Wang et al. (2009) observed that daily mean soil temperatures under mulching conditions was 2–9°C higher than non-mulching conditions, especially during the early growing period (Fig. 21.1). The effects were more apparent in the top 5 and 10 cm soil layers. As the plant canopy enlarged, more soil surface was shaded and the soil temperature difference between mulching and non-mulching decreased. During mid-July, the difference at 25 cm soil depth even became negative. Finally, as the plant stems and leaves senesced, the difference became positive again. The soil temperature variations under plastic mulching conditions were consistent with previous observation for potato (Ma and Li 1996; Cheng and Zhang 2000) and further confirmed in 2007 between mulched and non-mulch potato beds by the same research group (Hou et al. 2010).

Kar and Kumar (2007) suggested that a maximum temperature above 30°C could inhibit tuber growth. Thus, Wang et al. (2009) recorded daily soil highest temperature at 2:00 pm. The daily soil highest temperature at 5 and 10 cm depths were above 30°C during the period from early May to late June, which was when tuber initiation and tuber bulking occurred. A maximum temperature of 42°C was reached at 5 cm

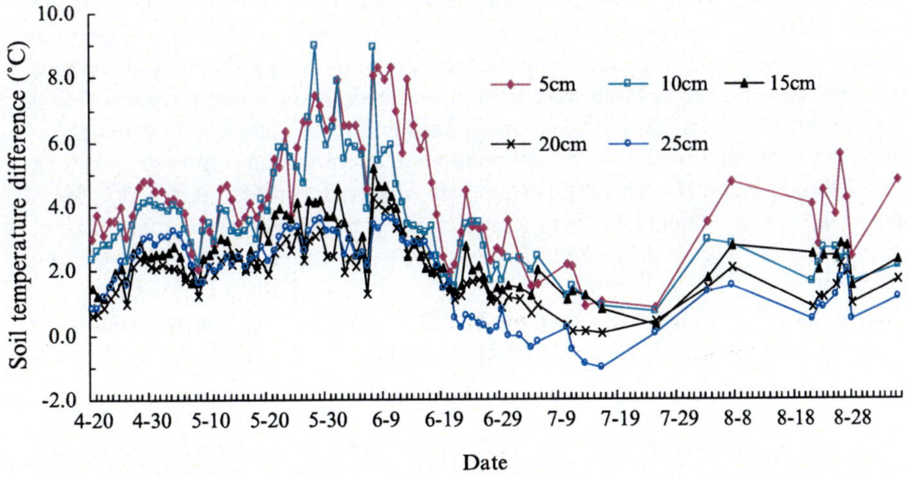

Fig. 21.1 Difference in daily mean soil temperature at different soil depths between potato beds with and without plastic mulch (Wang et al. 2009)

depth during these days. After that, highest daily soil temperature gradually fell within a range from 30.2°C to 20.0°C as potato canopy continued to enlarge. In the last phase (throughout August), soil temperature rose again regardless the falling of air temperature. The higher soil temperatures under plastic mulch during the middle of the day could have been detrimental to potato seed piece health and spouting, explaining the observation on a lower potato emergence rate with plastic mulch than that under no-mulch conditions (Wang et al. 2009).

21.4 Effects of Plastic Mulch on Potato Plant Height

Inconsistent effects of plastic mulching on plant height were observed in the early experiments conducted in the North China Plain (Wang et al. 2009). At the Luancheng Agro-ecosystem Station, in 2001, potato plant height without plastic mulch was higher than with mulch during the early growing period. After removing the mulch, the difference of crop height became smaller with little difference during the late growing period. In another experiment conducted at the Tongzhou (Beijing) Farmland Water Cycle and Modern Water-saving Irrigation Experimental Station, in 2006, however, plant heights without mulch were lower than those with mulch during the early growing period, but became similar during the late growing stage. Wang et al. (2009) noticed that the maximum air temperature during the early growing period in the 2001 experiment was typically higher than 30°C at Luancheng area, and therefore the soil temperature under mulch conditions was probably too high to detrimental to potato plant development (Kooman et al. 1996); however, soil temperature data was not recorded.

The results of the three experiments conducted in the arid Gansu Province of Northwestern China are more consistent, showing the promotion of plant height by plastic mulching (Wang et al. 2009; Hou et al. 2010). In both the 2006 and 2007 growing seasons, early season plant heights without plastic mulch were lower than those with mulch (Fig. 21.2). During the later growing period, the air and soil temperatures were higher as there were 51 and 41 days mostly in June–August (Wang et al. 2009; Hou et al. 2010), the further raising soil temperatures by mulching, which could be detrimental to potato plant elongation and tuber growth (Kooman et al. 1996; Kar and Kumar 2007). Thus, with the earliest mulch removal in 2006, plant height of treatment M60 was significantly greater than other mulched treatments ($P<0.05$). Other mulched treatments had very similar plant heights and the difference among these treatments was not significant. This suggested that later mulch-removal had little effect on plant height. In 2007, however, it was found that mulching duration had significantly ($P<0.01$) affected plant height during the early growing period (From June 6 to June 26): M0<M40<M60<M90≈Mw. This height difference could be explained by the mild higher soil temperature under mulch, which enhanced stem elongation according to Marinus and Bodlaender (1975). In June, the average highest air temperatures were 29.3°C and 28.5°C in 2006 and 2007, respectively (Hou et al. 2010). Later in 2008, 2009, and 2010 seasons

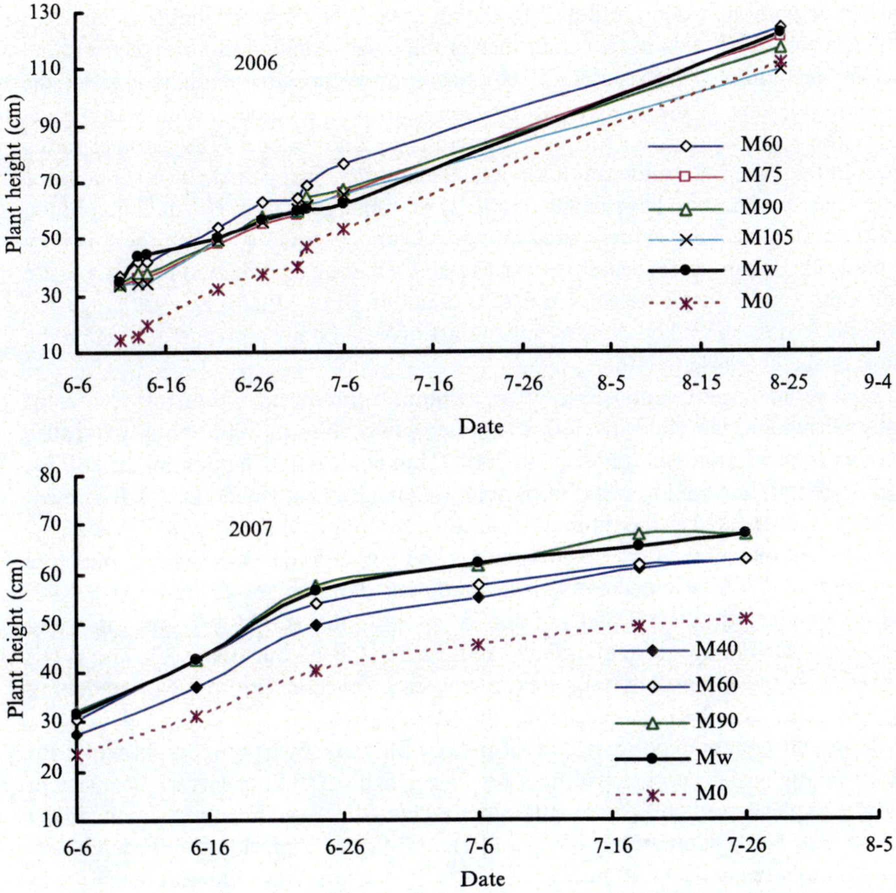

Fig. 21.2 Potato plant heights affected by the plastic mulching durations in the 2006 and 2007 growing seasons. Mulching durations are M0, M40, M60, M75, M90, M105, and Mw for 0 (control), 40, 60, 75, 90, and 105 days after planting, and the entire growing season , respectively (Hou et al. 2010)

with experiments conducted at a nearby experiment station, the same researchers (Wang et al. 2011) consistently observed that plant heights without plastic mulch were lower than those with mulch under various experimental conditions.

21.5 Tuber Yield and Water Use Efficiency (WUE)

In the two experiments conducted at two sites in North China Plain, the tuber yields in the plots without plastic mulch were 21.2 and 26.0 Mg ha^{-1}, respectively (Wang et al. 2009). The average tuber yields in the plots under different plastic mulching conditions in the two sites were 14.6 and 24.1 Mg ha^{-1}, respectively, both lower than the yields without plastic mulch. More than that, none of tuber yields in any plot with plastic mulch was higher than that in the corresponding control plat without mulch. The authors (Wang et al. 2009) proposed two possible explanations for the observation. One is the higher soil temperature under mulch, which was almost above the optimum (18°C) for tuber growth (Hay and Allen 1978); another is the lower emergence rates under mulch. Thus, the authors (Wang et al. 2009) concluded that plastic mulch may have negative effects on tuber growth in North China Plain. Plastic mulch has been widely used in North China, especially where there is very limited rain resources for water saving. Based on these experimental results, plastic mulching should not be used for potato production in this Plain. But we suggested more field trials should be done to get an unequivocal conclusion and determine the cause of the decreased yield.

At Xuebai Experimental Station of Minqin Agricultural Extension Center of Gansu Province, the yield in 2007 was much lower than in 2006, mainly because there were many more cloudy days in 2007 (Hou et al. 2010). Mulching the soil for 60 days (M60) had the highest (2006) or among the highest yields (2007). However, tuber yield decreased as the mulching duration got longer than 60 days (Table 21.3). In the meantime, all the mulched treatments had higher WUE than the non-mulched treatment and WUE also decreased as the mulching duration got longer in 2006. M40 had lower yield than M60 and regression suggests that slightly longer mulching durations than 40 days from planting were beneficial. Both the tuber yield and WUE suggest that the early plastic mulching was beneficial, and 60 days of mulching duration seemed to be most favorable for potato growth.

In the subsequent experiments at the nearby Shiyanghe Experimental Station for Water-saving in Agriculture and Ecology, Wang et al. (2011) observed inconsistent impacts of plastic mulching on potato tuber yields. Similar to the observations in the two sites of North China Plain (Wang et al. 2009), the control without mulch, F1, had the highest potato yield in 2008 (Table 21.3) as the yield in treatment F1 was 9.9%, 14.8%, 15.0% and 16.2% higher than the mulched treatments MF1, MF2, MF4 and MF8, respectively. The highest potato tuber yield was also observed in the treatment without mulch in 2009. Yields of treatments in MS1, MS2, MS3 and MS4 were 23.3%, 19.7%, 8.5% and 27.0% lower, respectively, than that in S1. In 2010, there is an exception as the highest potato tuber yield was obtained in treatment MS2. This was due to the fact that this treatment inadvertently received the most irrigation water (Wang et al. 2011). Potato tuber yield in the treatment without mulch (S1) was higher than any other mulched treatments in 2010.

WUE values ranged from 136 to 186 kg ha^{-1} mm^{-1} in 2006, from 59 to 86 kg ha^{-1} mm^{-1} in 2007, 68to 77 kg ha^{-1} mm^{-1} in 2008, 80 to 109 kg ha^{-1} mm^{-1} in 2009, and 96 to 115 kg ha^{-1} mm^{-1} in 2010 (Table 21.3). In 2009, WUE in S1 was 27.5% higher than the corresponding mulched treatment MS1, a similar result to the difference between F1 and MF1 in 2008. The results in 2008 and 2009 are consistent with the results in 2006 and 2007 that plastic mulch may negatively affect WUE if it is not removed in a timely manner (Hou et al. 2010).

The effect of irrigation frequency on WUE in F1 was 13% higher than the corresponding mulched treatment (MF1) in 2008. Similar to the earlier observation (Wang et al. 2009), the effect of irrigation frequency on potato WUE with plastic

Table 21.3 Potato tuber yield and water use efficiency during five potato growing seasons at two sites in Gansu Province (Hou et al. 2010; Wang et al. 2011)

Year	Treatment	Yield (Mg ha −1)	WUE (kg ha −1 mm −1)
2006	M0	58.2	136
	M60	66.3	184
	M75	58.8	165
	M90	56.3	152
	M105	55.9	143
	Mw	54.2	141
2007	M0	30.9	59
	M40	37.0	72
	M60	42.0	86
	M90	41.2	79
	Mw	39.4	85
2008	F1	35.7	77
	MF1	32.5	68
	MF2	31.1	69
	MF4	31.0	73
	MF8	30.7	69
2009	S1	34.4	102
	MS1	26.4	80
	MS2	27.6	93
	MS3	31.5	109
	MS4	25.1	96
2010	S1	35.3	103
	MS1	34.5	110
	MS2	38.6	115
	MS3	29.7	115
	MS4	23.3	96

M: with plastic mulch of 0, 40, 60, 75, 90, and 105 days after planting, and the entire growing season (w), respectively. F: drip irrigation with once every 1 (F1), 2 (F2), 3 (F3), 4 (F4), and 8 (F8) days. S1: irrigation threshold at −25 kPa *SMP* at the three growth stages; S2: irrigation threshold at −25 kPa *SMP* at potato initiation and bulking stages and −35 kPa *SMP* at maturing stage; S3: irrigation threshold at −25 kPa *SMP* at potato initiation and maturing stages and −35 kPa *SMP* at bulking stage; S4: irrigation threshold at −35 kPa *SMP* at the three growth stages

mulch was not clear. This observation is in contradiction with the effects of irrigation frequency on potato WUE without mulch as high frequency irrigation enhanced potato tuber growth and WUE (Wang et al. 2006).

21.6 Potato Tuber Quality

Data from tuber grading analysis indicated that the number of marketable tuber (W ≥ 50 g) was affected by mulching (Table 21.4). Mulching reduced marketable tubers per plant in 2006, but increased marketable tubers per plant in 2007 season.

Table 21.4 Potato tuber grades impacted by plastic mulch treatments

	Total tubers (plant^{-1})	Marketable tubers (plant^{-1})[a]	Total tuber weight (g plant^{-1})	Marketable tuber weight (g plant^{-1})
2006				
M0	–[b]	8.1	1,420	1,347
M60	–	6.7	1,603	1,551
M75	–	5.8	1,386	1,326
M90	–	6.7	1,382	1,286
M105	–	6.1	1,382	1,310
Mw	–	4.9	1,441	1,392
2007				
M0	6.8	4.3	629	572
M40	7	4.5	595	538
M60	7.7	4.9	733	667
M90	8.6	4.8	675	606
Mw	8.9	5.4	735	659

Mulching durations are M0, M40, M60, M75, M90, M105, and Mw for 0 (control), 40, 60, 75, 90, and 105 days after planting, and the entire growing season, respectively. Data are adapted from Hou et al. (2010)

[a]Tuber weight not less than 50 g each

[b]No data available

Consistently, mulching duration affected tuber quality as treatment M60 had the heaviest marketable tubers in both seasons. The yields of the marketable tubers were then reduced with mulching duration either shorter or longer than 60 days. The pattern of the yields of the marketable tubers (Table 21.4) is in the same trend of the total tuber yields impacted by mulching (Table 21.3). It is apparent that plastic mulch was beneficial to potato growth during the early growth stage shown by mean early emergence and plant height (thus aboveground biomass) because the air temperature was relatively low and mulching could increase temperature to an extent not restraining potato growth. As air temperature went high, the negative effects of plastic mulch became dominant: higher soil temperature and poorer soil aeration were both detrimental to tuber growth.

Wang et al. (2011) further measured several quality parameters for the potato harvested in 2008, 2009, and 2010 growing seasons (Table 21.5). No substantial differences were found in tuber firmness between the non-mulched and the corresponding mulched treatments in the 3 years, suggesting that this quality index was not affected by mulch. However, potatoes grown without plastic mulch (F1 and S1) always had higher tuber specific gravity, starch content, and vitamin C content than those grown with plastic mulch (MF1 and MS1). This observation suggests that plastic mulching negatively affected the three quality parameters. Lower tuber quality could be due to the poorer soil aeration and higher temperature associated with plastic mulch as such circumstances were not conducive to the accumulation of biomass and nutrients (Mendoza and Estarda 1979; Ge and Zhang 2003; Li et al. 2004b; Guo et al. 2008).

Further analysis of those mulching data revealed that potato tuber nutrient content increased with the increased irrigation frequency, consistent with published

Table 21.5 Potato tuber quality for the different irrigation treatments with plastic mulching

Year	Treatment	Specific gravity	Starch content (%)	Vitamin C content (mg·100 g^{-1})	Tuber firmness (MPa)
2008	F1	1.080 a*	13.91 a*	25.96 NS	1.986 NS
	MF1	1.074 ab	12.68 ab	23.38	2.028
	MF2	1.071 ab	12.04 ab	22.00	1.999
	MF4	1.071 ab	11.92 b	21.08	1.964
	MF8	1.066 b	10.96 b	19.71	2.038
2009	S1	1.080 NS	13.98 NS	29.00 NS	1.861 NS
	MS1	1.069	11.77	24.00	1.768
	MS2	1.077	13.20	21.00	1.814
	MS3	1.075	12.82	20.33	1.786
	MS4	1.076	13.16	27.67	1.773
2010	S1	1.080 a	14.00 a	28.50 a	1.100 b
	MS1	1.066 b	10.83 b	24.80 ab	1.130 b
	MS2	1.072 ab	12.29 ab	22.54 ab	1.170 a
	MS3	1.078 a	13.41 a	21.24 b	1.090 b
	MS4	1.073 ab	12.32 ab	22.55 ab	1.120 b

M: with plastic mulch. F: drip irrigation with once every (F1), 2 (F2), 3 (F3), 4 (F4), and 8 (F8) days. S1: irrigation threshold at –25 kPa soil matric potential (*SMP*) at the three growth stages; S2: irrigation threshold at –25 kPa *SMP* at potato initiation and bulking stages and –35 kPa *SMP* at maturing stage; S3: irrigation threshold at –25 kPa *SMP* at potato initiation and maturing stages and –35 kPa *SMP* at bulking stage; S4: irrigation threshold at –35 kPa *SMP* at the three growth stages. Data are from Wang et al. (2011)

NS difference among different treatments is not significant, F-test ($P > 0.05$)

* Difference among different treatments is significant by F-test ($P < 0.05$). Values in a column with the same letter are statistically similar according to Duncan's multiple range test ($P < 0.05$)

data (Zhou et al. 2004). Wang et al. (2011) assumed that high-frequency irrigation created a more humid soil environmental conditions where the potato tuber grew, favoring nutrient uptake by tuber roots. In 2009, no significant difference was found in any of the four quality indexes among the SMP treatments. The tuber specific gravity and starch content in treatment MS2 were slightly higher than those in the other mulched treatments. In 2010, however, treatment MS3 rather than MS2 had the highest tuber specific gravity and starch content among the mulched treatments. Tuber specific gravity and starch content for treatment MS3 were significantly higher than MS1. These findings were different from the results in 2009. For vitamin C content, no significant differences were found between any of the mulched treatments, implying that the irrigation threshold change by SMP had little effect on vitamin C content. Tuber firmness, was greatest in treatment MS2 among all the mulched treatments and was significantly greater than any other treatment. Günel and Karadoĝan (1998) reported significant increases in specific gravity, dry matter, starch content, chip yield and significant decreases in protein content and oil absorption rate of chips by frequent irrigations at early planting-stolen initiation and stolen initiation-tuber bulking growth stages. However, they found frequent irrigations at the final tuber bulking growth stage had deleterious effects on specific gravity, dry matter, starch content and chip yield especially when irrigation continued until maturity. Thus, further research should be focused on more sophistic experiments of drip irrigation schemes to elucidate accurate impacts of plastic mulching on potato quality in the arid region of Northwest China.

21.7 Conclusion

This chapter examines the case studies on the effectiveness of plastic mulching for potato production under drip irrigation in two typical regions of Northern China. Results suggest that daily mean soil temperature under mulch was 2–9°C higher than without mulch. The mulch effect on soil temperature was greatest during the early growth and became less as the plant canopy increased. Mulch reduced irrigation water and evapotranspiration. Reduction of soil matric potential irrigation criterion from −25 kPa to −35 kPa during tuber bulking and maturing stage could enhance water use efficiency, but the constant irrigation threshold of −35 kPa could lead to lower water use efficiency. However, inconsistent results were obtained on the effects of potato growth and tuber yield with reduction of this irrigation criterion to −35 kPa during tuber bulking stage.

Both tuber yield and water use efficiency demonstrated benefits from early plastic mulching. The possible negative effects of mulching included a lower emergence, lower potato tuber yield, and fewer marketable tubers per plant. Plastic mulch can also negatively affect tuber specific gravity, starch content, and vitamin C content. These quality parameters were less affected by plastic mulching with higher irrigation frequency. During the later part of the growing season, the negative effects (i.e. poor soil aeration and potentially higher soil temperature higher than the optimum for potato growth) may outweigh water saving effect of plastic mulch. These case studies suggested that 60 days of mulching duration was most favorable for potato production in the experimental regions. Mulch removal after 60 days is apparently necessary in order to avoid subsequent negative effects.

References

Anikwe MAN, Mbah CN, Ezeaku PI, Onyia VN (2007) Tillage and plastic mulch effects on soil properties and growth and yield of cocoyam (*Colocasia esculenta*) on an ultisol in southeastern Nigeria. Soil Till Res 93:264–272

Baghour M, Moreno DA, Hernández J, Castilla N, Romero L (2002) Influence of root temperature on uptake and accumulation of Ni and Co in potato. J Plant Physiol 159:1113–1122

Bowen P, Frey B (2002) Response of plasticultured bell pepper to staking, irrigation frequency, and fertigated nitrogen rate. Hort Sci 37:95–100

Chakraborty D, Nagarajan S, Aggarwal P, Gupta VK, Tomar RK, Garg RN, Sahoo RN, Sarkar A, Chopra UK, Sarma KS, Sundara KN (2008) Effect of mulching on soil and plant water status, and the growth and yield of wheat (*Triticum aestivum* L.) in a semi-arid environment. Agric Water Manage 95:1323–1334

Cheng JS, Zhang XX (2000) Study on the ecological effects of plastic mulching on potato in cool and humid highland of Wei'nan county in Gansu province. Chin Potato 14:83–84

Doring TF, Brandt M, Heß J, Finckh MR, Saucke H (2005) Effects of straw mulch on soil nitrate dynamics, weeds, yield, and soil erosion in organically grown potatoes. Field Crops Res 94:238–249

Ge Z, Zhang B (2003) Effect of mulch-removal on potato growth under mulched condition. Chin Potato 17(4):244–246

Ghosh PK, Dayal D, Bandyopadhyay KK, Mohanty M (2006) Evaluation of straw and polythene mulch for enhancing productivity of irrigated summer groundnut. Field Crops Res 99:76–86

Günel E, Karadoĝan T (1998) Effect of irrigation applied at different growth stages and length of irrigation period on quality characters of potato tubers. Potato Res 41:9–19

Guo ZM, Sun ZP, Liu YL (2008) Effect of different rhizospheric ventilation treatments on yield and quality of potatoes. Crops 3:59–61

Hay RKM, Allen EJ (1978) Tuber initiation and bulking in the potato (Solanum tuberosum) under tropical conditions: the importance of soil and air temperature. Trop Agric Trinidad 55:289–295

Hou XY, Wang FX, Han JJ, Kang SZ, Feng SY (2010) Duration of plastic mulch for potato growth under drip irrigation in an arid region of Northwest China. Agric For Meteorol 150:115–121

Kang YH, Wang FX, Liu HJ, Yuan BZ (2004) Potato evapotranspiration and yield under different drip irrigation regimes. Irrig Sci 23:133–143

Kar G, Kumar A (2007) Effects of irrigation and straw mulch on water use and tuber yield of potato in eastern India. Agric Water Manage 94:109–116

Kooman PL, Fahem M, Tegera P, Haverkort AJ (1996) Effects of climate on different potato genotypes 2. Dry matter allocation and duration of the growth cycle. Eur J Agron 5:207–217

Li FM, Wang J, Xu JZ, Xu HL (2004a) Productivity and soil response to plastic film mulching durations for spring wheat on entisols in the semiarid Loess Plateau of China. Soil Till Res 78:9–20

Li FM, Song QH, Jjemba PK, Shi YC (2004b) Dynamics of soil microbial biomass C and soil fertility in cropland mulched with plastic film in a semiarid agro-ecosystem. Soil Biol Biochem 36:1893–1902

Liang DC, Li WG, Gong XF (1998) Effects of plastic mulching on potato production. Inn Mong Agric Sci Technol (Chinese journal) (Suppl.):67–68

Ma SM, Li SB (1996) Effectiveness and technique of plastic mulching for potato. Rain Fed Crops (Chinese journal) 6:30–32

Marinus J, Bodlaender KBA (1975) Response of some potato varieties to temperature. Potato Res 18:189–204

Mendoza HA, Estarda RN (1979) Breeding potatoes for tolerance to stress: Heat and Frost. In: Mussell H, Staples RC (eds) Stress Physiology in Crop Plants. Willey, New York, pp 227–262

Phene CJ, Sanders DC (1976) High-frequency trickle irrigation and row spacing effects on yield and quality of potatoes. Agron J 68:602–607

Ramakrishna A, Tam HM, Wani SP, Long TD (2006) Effect of mulch on soil temperature, moisture, weed infestation and yield of groundnut in northern Vietnam. Field Crops Res 95:115–125

Romic D, Romic M, Borosic J, Poljak M (2003) Mulching decreases nitrate leaching in bell pepper cultivation. Agric Water Manage 60:87–97

Schmidt JR, Worthington JW (1998) Modifying heat unit accumulation with contrasting colors of polyethylene mulch. Hort Sci 33:210–214

Sun SP, Li SZ (2004) Climatic ecological effect to plastic-mulched potato planting in high latitude and low temperature area. J Shanxi Agric Sci 32(1):26–28

Tian Y, Su D, Li F, Li X (2003) Effect of rainwater harvesting with ridge and furrow on yield of potato in semiarid areas. Field Crops Res 84:385–391

Tiquia SM, Lloyd J, Herms DA, Hoitink HAJ, Michel FC Jr (2002) Effects of mulching and fertilization on soil nutrients, microbial activity and rhizosphere bacterial community structure determined by analysis of TRFLPs of PCR-amplified 16S rRNA genes. Appl Soil Ecol 21:31–48

Tiwari KN, Singh A, Mal PK (2003) Effect of drip irrigation on yield of cabbage (Brassica oleracea L.Var. capitata) under mulch and non-mulch conditions. Agric Water Manage 58:19–28

Tolk JA, Howell TA, Evett SR (1999) Effect of mulch, irrigation, and soil type on water use and yield of maize. Soil Till Res 50:137–147

Triki MA, Priou S, Mahjoub ME (2001) Effects of soil solarization on soil-borne populations of *Pythium aphanidermatum* and *Fusarium solani* and on the potato crop in Tunisia. Potato Res 44:271–279

Vos JGM, Uhan TS, Sutarya R (1995) Integrated crop management of hot pepper under tropical lowland conditions: effects of rice straw and plastic mulches on crop health. Crop Prot 14:445–452

Wang FX, Kang YH, Liu SP (2003) Plastic mulching effects on potato under drip irrigation and furrow irrigation. Chin J Eco-Agric (Chinese journal) 11(4):99–102

Wang FX, Kang YH, Liu SP (2006) Effects of drip irrigation frequency on soil wetting pattern and potato growth in North China Plain. Agric Water Manage 79:248–264

Wang FX, Feng SY, Hou XY, Kang SZ, Han JJ (2009) Potato growth with and without plastic mulch in two typical regions of Northern China. Field Crops Res 110:123–129

Wang FX, Wu XX, Shock CC, Chu LY, Gu XX, Xue X (2011) Effects of drip irrigation regimes on potato tuber yield and quality under plastic mulch in arid Northwestern China. Field Crop Res 122:78–84

Wei YA (2005) Status quo and characteristics of potato industry development in the world. World Agric (Chinese journal) 3:29–32

Xie ZK, Wang YJ, Li FM (2005) Effect of plastic mulching on soil water use and spring wheat yield in arid region of northwest China. Agric Water Manage 75:71–83

Zhou NN, Zhang XJ, Qin YB, Xu Q (2004) Effect on different quantities of drip irrigation and nitrogen fertilization for yield and quality of potato. Soils Fertil (Chinese journal) 6:11–16

Chapter 22
Drought and Salinity Tolerance in Transgenic Potato

Huaijun Si, Ning Zhang, and Di Wang

Abstract Drought and salinity are the most important environmental stress factors that limit agricultural production worldwide. Complex responses to drought and salinity stresses in plants are quantitative traits, thus involve cooperative functions of many genes and biochemical-molecular mechanisms. It is generally accepted that drought and salinity tolerance could be increased through transgenic approaches by incorporating genes involved in stress protection into plants that lack them. Potato is regarded as a moderately salt-sensitive and drought-sensitive crop. Transgenic potato plants with improved tolerance to drought and salinity stresses have been produced using various genes. This chapter presented the case study of enhanced drought and salinity tolerance of transgenic potato plants with a betaine aldehyde dehydrogenase (BADH) gene from spinach under the control of the constitutive expression promoter CaMV 35S and the stress-inducible expression promoter rd29A, respectively. The recent advance was summarized in improving drought and salinity tolerance through transgenic approaches in potato. The role of transgenic potato in sustainable production and its biosafety was also discussed. It is concluded that the transgenic approach is one of the powerful tools to improve potato crop for sustainable production and food supply in response to the coming increase of world population in the future.

22.1 Introduction

Abiotic stresses, such as drought, salinity, extreme temperatures, chemical toxicity and oxidative stresses, are serious threats to agriculture besides their deteriorative impact to the environment. Drought and salinity are the most important environmental stress

---

H. Si (✉) • N. Zhang • D. Wang
Gansu Provincial Key Laboratory of Aridland Crop Science; Gansu Key Laboratory of Crop Genetic and Germplasm Enhancement; College of Life Science and Technology, Gansu Agricultural University, Lanzhou 730070, People's Republic of China
e-mail: hjsi@gsau.edu.cn

Z. He et al. (eds.), *Sustainable Potato Production: Global Case Studies*,
DOI 10.1007/978-94-007-4104-1_22,

factors that limit food production worldwide, and may cause a serious salinization on more than 50% of all arable lands by the year 2050 (Wang et al. 2003). In China, almost half of the land is arid or semi-arid and the crop production is strongly affected by drought and salinity seasonally even if in the irrigated farm land.

Potato is regarded as a moderately salt-sensitive (Ahmad and Abdullah 1979), and drought-sensitive crop compared with other crops from the field experiments (Salter and Goode 1967; van Loon 1981). Drought always influences development and growth of stem, root and tuber (Ojala et al. 1990), and reduces number of tubers and yield in potato (Cavagnaro et al. 1971).

Complex reaction of plant in response to drought and salinity stresses is a quantitative character which genetic control includes the functions of many genes and biochemical-molecular mechanisms. These responses lead to a wide variety of biochemical and physiological changes such as the accumulation of various organic compounds of low-molecular weight, collectively known as compatible solutes or osmolytes, synthesis of late-embryogenesis-abundant (LEA) proteins, and activation of several detoxification enzymes (Bajaj et al. 1999).

Using transgenic approaches to enhance drought and salinity tolerance has been thoroughly reviewed lately in plants (Apse and Blumwald 2002; Rontein et al. 2002; Wang et al. 2003; Chen and Murata 2008; Kolodyazhnaya et al. 2009) and potato (Byun et al. 2007). It is generally accepted that drought and salinity tolerance could be increased through transgenic approaches by incorporating genes involved in stress protection into plants that lack them. Transgenic potato plants with improved tolerance to drought and salinity stress have been produced using various genes which have been summarized in Table 22.1 and briefly discussed below.

Trehalose is a non-reducing disaccharide of glucose. A plant that produces trehalose is often highly tolerant to desiccation stress. Goddijn et al. (1997) engineered trehalose biosynthesis in potato by introducing the *otsA* and *otsB* genes from *Escherichia coli*, which encode trehalose-6-phosphate synthase and trehalose-6-phosphate phosphatase, respectively. Jeong et al. (2001) introduced *GPD* gene for glyceraldehydes-3-phosphate dehydrogenase from the oyster mushroom (*Pleurotus sajor-caju*) into potato and obtained transgenic potato plants with enhanced salinity tolerance. Ambard-Bretteville et al. (2003) suppressed *FDH* gene encoding for formate dehydrogenase in transgenic potato plants which formate levels are increased. Thus, the suppression resulted in accumulation of praline in response to osmotic stress. Turhan (2005) developed transgenic potato plants with higher salinity tolerance by expressing *oxo* gene which enhanced synthesis of oxalate oxidase for catabolizing oxalic acid. The dehydration-responsive element (DRE) is essential for regulating dehydration-responsive gene expression (Yamaguchi-Shinozaki and Shinozaki 1994). The transformation of plants using regulatory genes is an attractive approach for producing dehydration-stress tolerant plants. The overexpression of *DREB1A* gene for DRE-binding protein from *Arabidopsis* in transgenic potato showed that the tolerance to salt-stress was increased in proportion to its copy number of the gene in tetrasomic tetraploid potato (Behnam et al. 2006). Oxidative stress is a major damaging factor for plants exposed to environmental stresses. Tang et al. (2006) obtained transgenic potato plants with increased tolerance to multiple environmental

Table 22.1 Genes overexpressed in transgenic potato plants for drought and salinity tolerance

Gene	Gene product	Performance of transgenic plant	Reference
otsA	Trehalose-6-phosphate synthase	Trehalose accumulation	Goddijn et al. (1997)
otsB	Trehalose-6-phosphate phosphatase	Trehalose accumulation	Goddijn et al. (1997)
OLP	Osmotin-like protein	Salt resistance	Evers et al. (1999)
TPS1	Trehalose-6-phosphate synthase	Increased tolerance to drought	Yeo et al. (2000)
GPD	Glyceraldehydes-3-phosphate dehydrogenase	Improvement of salt tolerance	Jeong et al. (2001)
FDH	Formate dehydrogenase	Accumulate proline rapidly to resist drought	Ambard-Bretteville et al. (2003)
oxo	Oxalate oxidase	Higher salinity tolerance	Turhan (2005)
DREB1A	Dehydration-responsive element (DRE)-binding protein	Tolerance to salt stress	Celebi-Toprak et al. (2005)
DREB1A	Dehydration-responsive element (DRE)-binding protein	Highly tolerant to salinity	Behnam et al. (2006)
SOD and APX	Cu/Zn superoxide dismutase and ascorbate peroxidase	Multiple stresses including drought, salinity, oxidative stress and high temperature	Tang et al. (2006)
SST/FFT	Fructan	Proline accumulation	Knipp and Honermeier (2006)
StEREBP1	Ethylene responsive element binding protein 1	Tolerance to NaCl stress	Lee et al. (2007)
AtNDPK2	Nucleoside diphosphate kinase	Enhanced tolerance to salt	Tang et al. (2008)
codA	Choline oxidase	Enhanced tolerance to oxidative, salt, and drought stresses	Ahmad et al. (2008)
GLOase	L-gulono-c-lactone oxidase	Enhanced tolerance to various abiotic stresses like oxidative, salt and drought stresses	Hemavathi et al. (2010)
BADH	Betaine aldehyde dehydrogenase	Enhanced drought and salinity tolerance	Zhang et al. (2011)

stress due to the overexpressed both superoxide dismutase (SOD) and ascorbate peroxidase (APX) in chloroplasts. Glycine betaine (GB) is a common compatible solute in many different organisms including higher plants. Many plant species can accumulate GB in response to drought and salinity. Ahmad et al. (2008) and Zhang et al. (2011) showed that the transgenic potato plants were more tolerant to drought and salinity stress because of overexpressing *codA* and betaine aldehyde dehydrogenase (BADH) genes for GB synthesis.

This chapter, based on our research, presented the case study of enhancing drought and salinity tolerance in transgenic potato plants expressing *BADH* gene from spinach. Roles of transgenic potato in sustainable crop production and its biosafety concerns were also addressed.

22.2 Enhancement of Drought and Salinity Tolerance in Transgenic Potato by Expressing *BADH* Gene

To ensure their own survival and prosperity of their offspring, plants have evolved a range of strategies to cope with various abiotic stresses. One common mechanism is the accumulations of compatible solutes including certain polyols, sugars, amino acids, betains and related compounds. Glycine betaine (GB) is one of the most important of osmolytes (Chen and Murata 2008). Many plant species accumulate betaine in response to drought and salinity, thus adapt arid and saline areas (Rhodes and Hanson 1993). In higher plants, GB is synthesized by conversion from choline to GB through a two-step oxidation via the intermadiate betaine aldehyde (Hanson and Scoff 1980). The relevant enzymes are choline monooxygenase (CMO) and betaine aldehyde dehydrogenase (BADH) (Sakamoto and Murata 2000). Transgenic plants of various species have been produced, which tolerance to drought and salinity has been enhanced because they have elevated levels of GB by expressing CMO or BADH transformed with the corresponding genes (Sakamoto and Murata 2000).

The fact that many important crops, such as rice, potato and tomato, are betaine-deficient has inevitably led to the proposal that it might be possible to increase drought and salinity tolerance by genetic engineering of GB synthesis (McCue and Hanson 1990). In this chapter, the spinach BADH gene was transformed into potato under the control of the constitutive expression promoter CaMV 35S and stress-inducible expression promoter rd29A respectively, and the resultant transgenic potato plants gained the ability resistant to drought and salinity stresses (Zhang et al. 2009, 2011).

22.2.1 Isolation of BADH Gene and Plasmid Construction

The 1,556 bp cDNA of *BADH* gene (GenBank accession AY156694) was isolated from spinach using reverse transcription-polymerase chain reaction (RT-PCR) method (Zhang et al. 2004). The sequence analysis showed that the *BADH* cDNA contains 1,494 bp open reading frame (ORF) encoding a protein of 497 amino acids. The nucleotide sequence of *BADH* cDNA shared 99.87% identity with *BADH* gene which was previously cloned from spinach (GenBank accession M31480) (Weretilnyk and Hanson 1990). The nucleotide sequence at position of 781–813 nt

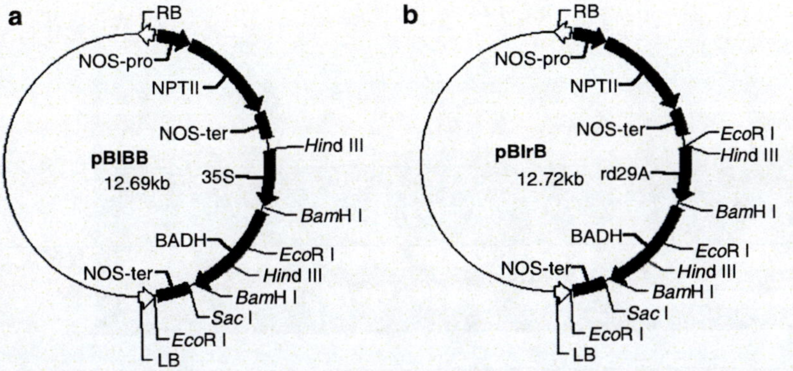

Fig. 22.1 Schematic diagram of the expression vectors pBIBB (**a**) and pBIrB (**b**). *RB* right border, *LB* left border, *NOS-pro* nopaline synthase promoter, *NOS-ter* nopaline synthase terminator, *NPTII* neomycin phosphotransferase II gene, *35 S* the constitutive expression promoter CaMV 35S, *rd29A* the stress-inducible expression promoter rd29A, *BADH* betaine aldehyde dehydrogenase gene. *Hind III, BamH I, EcoR I, Sac I* restriction endonuclease recognition sites

encodes the deca-peptide Val-Thr-Leu-Glu-Leu-Gly-Gly-Lys-Ser-Pro and at position of 992–994 nt encodes cysteine (Cys) related to the function of enzyme activity, and the deca-peptide and Cys are highly conserved among general dehydrogenase (Weretilnyk and Hanson 1989).

The 824 bp of rd29A promoter was amplified from *Arabidopsis thaliana* genome by the PCR technique (Zhang et al. 2005). Sequence analysis showed that the cloned fragment shared 99.39% identity with reported rd29A promoter (GenBank accession D13044) (Yamaguchi-Shinozaki and Shinozaki 1993) and contained several cis-acting elements including dehydration responsive element (DRE) and ascorbic acid (ABA) responsive element (ABRE) (Shinozaki and Yamaguchi-Shinozaki 1997). The result from transgenic potato plants showed that the expression of β-glucuronidase (GUS) gene under control of the rd29A promoter was induced by drought, salinity, low temperature and ABA (Zhang et al. 2005).

The plant expression vectors pBIBB and pBIrB were constructed by fusing *BADH* gene with the constitutive expression promoter CaMV 35S in plasmid pBI121 and the stress-inducible expression promoter rd29A in plasmid pBIrd (Fig. 22.1) (Si et al. 2007; Zhang et al. 2005). The expression vectors were then introduced into *Agrobacterium tumefaciens* strain LBA4404 by freeze-thaw method (Hofgen and Willmitzer 1988) and proved by the enzyme digestion and PCR amplification.

22.2.2 Potato Transformation and Molecular Analysis

Microtubers of potato cultivar Gannongshu 2 were used as the receptor for *Agrobacterium*-mediated transformation performed as described previously (Si et al. 2003). Green shoots were produced directly from surface of the transformed

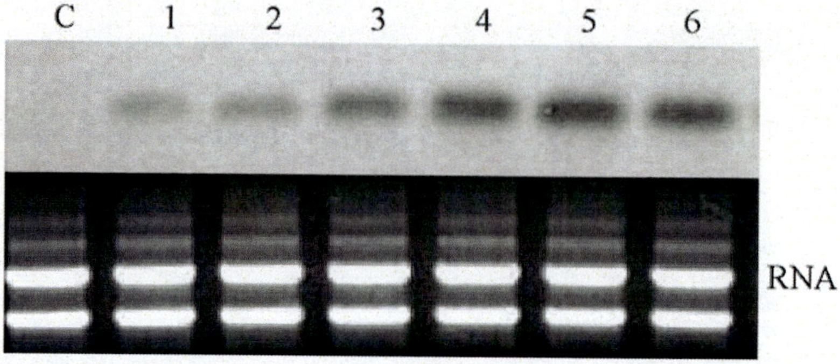

Fig. 22.2 Northern blotting analysis of *BADH* gene expression in the transgenic potato transformed with *BADH* gene under the control of the stress-inducible expression promoter rd29A. Plants were grown in vermiculite in 3 L pots in a greenhouse under natural light at 25 °C, and watered and fertilized weekly with a complete nutrient solution. When the plants reached 10 cm in height, they were treated with NaCl and polyethylene glycol (PEG, MW 6,000), respectively. NaCl treatment was begun at a concentration of 50 mM and increased stepwise by 50 mM every day until the final concentration, 500 mM, was reached. PEG treatment was conducted with 15% PEG solution once a day for 10 days. *Lane C* untransformed potato plant, *Lane 1 and 2* untreated transgenic potato plants, *Lane 3 and 4* transgenic potato plants subjected to NaCl and PEG treatments for 5 days, respectively, *Lane 5 and 6* transgenic potato plants 3 days after being subjected to NaCl and PEG treatments for 10 days, respectively. Each lane in electrophoresis contained the similar 30 μg RNA sample stained with ethidium bromide. This figure was reproduced from Zhang et al. (2011)

microtuber slices after 4 weeks cultured in MS medium (Murashige and Skoog 1962) containing 1 mg/L indole-3-acetic acid (IAA), 0.2 mg/L gibberellic acid (GA_3), 0.5 mg/L 6-benzyladenine (BAP) and 2 mg/L zeatin riboside (ZR) supplemented with 75 mg/L kanamycin and 400 mg/L carbenicillin. Roots were formed in about 10 days when green shoots transferred to MS medium supplemented with 50 mg/L kanamycin and 200 mg/L carbenicillin. The plantlets with well-developed roots were propagated for further molecular analysis. PCR and Southern blot analysis showed that *BADH* gene has been integrated into genome of potato (data not shown). Northern hybridization analysis demonstrated that expression of *BADH* gene was induced by drought and NaCl stress in the transgenic potato plants transformed with *BADH* gene driven by the promoter rd29A (Fig. 22.2), while was not induced in the transgenic plants driven by the promoter CaMV 35S as shown in Fig. 22.3 (Zhang et al. 2009; Zhang et al. 2011).

22.2.3 BADH Activities and Relative Electrical Conductivities of the Transgenic Plants

The analysis of BADH activity demonstrated that it could be detected in the transgenic potato plants transformed with *BADH* gene driven by the promoter CaMV 35S, but the enzyme activity could not be detected in the untransformed

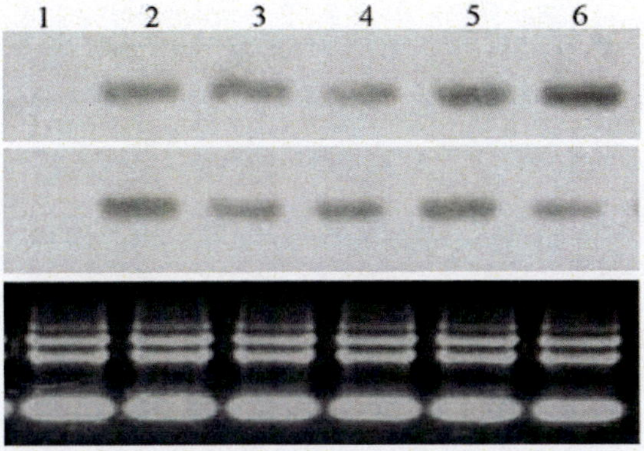

Fig. 22.3 Northern blotting analysis of transgenic potato plants transformed with *BADH* gene under the constitutive expression promoter CaMV 35S. Treatment was same as Fig. 22.2. *Lane 1* untransformed potato plant, *Lane 2–6* transgenic potato plants 3 days after being subjected to NaCl and PEG treatments for 10 days, respectively. Each lane in electrophoresis contained the similar 30 μg RNA sample stained with ethidium bromide. This figure was reproduced from Zhang et al. (2009)

control plants. The activity of BADH varied from 2.1 to 11.5 U among different transgenic individuals. The BADH activity and relative electrical conductivity on the transgenic potato leaves were highly negatively related ($y = -3.7738x + 57.083$, $r = 0.989^{**}$) (Zhang et al. 2009).

The activities of BADH in the transgenic potato plants transformed with *BADH* gene driven by the promoter rd29A were rather low when they were not stressed, but increased greatly 3 days after the treatment with NaCl and PEG had applied. The BADH activities varied between 10.8 and 11.7 U and varied a little among the different transgenic plant lines. The relative electrical conductivities among the transgenic plants were 17.4–19.6% under NaCl and PEG stress, much less than those among the control plants (45.6%). The low conductivities showed that the cell membranes of the transgenic plants were less injured than those of the control plants under NaCl and PEG stress. A significant negative linear relationship between the relative electrical conductivity (y) and BADH activity (x) was observed, which could be represented by a function of $y = -2.2083x + 43.329$ ($r = 0.9495$), revealing that BADH activity was positively related to protection of cell membrane permeability (Zhang et al. 2011).

22.2.4 Drought and Salinity Tolerance in Transgenic Potato Plants

The growth of the transgenic potato plants *in vitro* was normal and better than the untransformed plants under NaCl and PEG stresses. Plant height increased 0.41–1.0 cm and fresh weight per plant increased 10–35% for the transgenic potato plants transformed with *BADH* gene under the control of the promoter CaMV 35S

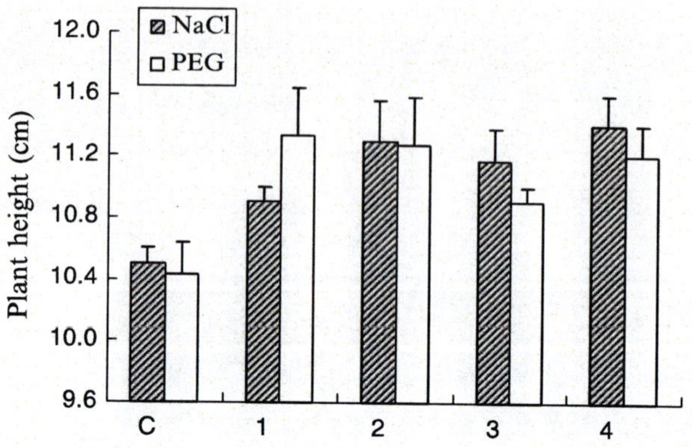

Fig. 22.4 Height of transgenic potato plants 3 days after 10 days of NaCl and PEG treatment had been completed, respectively. Plants were grown in vermiculite in 3 L pots in a greenhouse under natural light. Treatment was same as Fig. 22.2. The data are the mean ± standard error (*SE*) from three replicates. *C* nontransgenic potato plant, *1–4* transgenic potato plant lines. This figure was reproduced from Zhang et al. (2011)

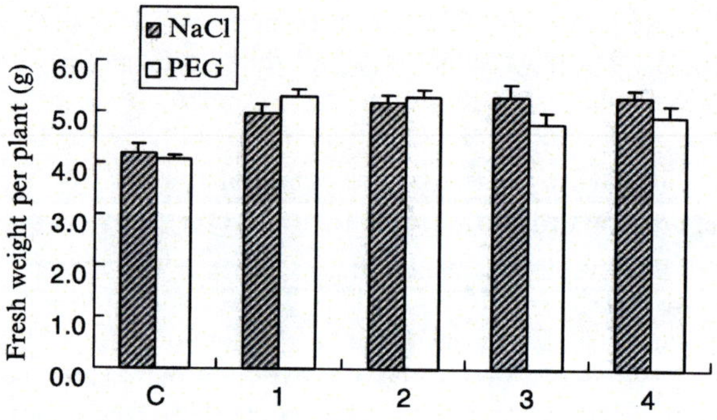

Fig. 22.5 Fresh weight per plant of transgenic potato plants 3 days after 10 days of NaCl and PEG treatment had been completed, respectively. Treatment was same as Fig. 22.2. Plants were grown in vermiculite in 3 L pots in a greenhouse under natural light. The data are the mean ± standard error (*SE*) from three replicates. *C* nontransgenic potato plant, *1–4* transgenic potato plant lines. This figure was reproduced from Zhang et al. (2011)

compared with the control potato plants (Zhang et al. 2009). For the transgenic potato plants with the promoter rd29A, plant height of the transgenic plants increased 0.4–0.9 cm and fresh weight per plant increased 17–29% compared with the control potato plants as shown in Fig. 22.4 and 22.5 (Zhang et al. 2011).

When exposed to various degree of NaCl stress (0%, 0.3% and 0.6%) for 2 months, the leaves of transgenic potato plant still stayed green, while the leaves of nontransgenic potato plant became yellow and wilting under 0.3% NaCl stress. A better growth performance was still observed in transgenic plants when they grew under the condition supplemented with 0.6% NaCl in comparison to nontransgenic control (Fig. 22.6), demonstrating that the transgenic plants with *BADH* gene acquired higher tolerance to NaCl stress than that of nontransgenic ones (Li et al. 2007).

Fig. 22.6 Comparison on growth of the transgenic potato plants transformed with *BADH* gene and nontransgenic potato plants under stress of NaCl concentration of 0%, 0.3% and 0.6% for 2 months respectively. **(a–c)** nontransgenic potato plants under stress of NaCl concentration of 0%, 0.3% and 0.6% respectively, **(d–f)** transgenic potato plant line GN-4 under stress of NaCl concentration of 0%, 0.3% and 0.6% respectively. This figure was reproduced from Li et al. (2007)

The result suggested that the *BADH* gene can be used to improve the drought and salinity tolerance of important crops that are betaine-deficient through genetic engineering.

22.3 Role of Transgenic Potato in Sustainable Potato Production

With increasing population, higher demand for economic development and rapid urbanization, the global potentials for food will continue to grow; furthermore, climate change is leading to production uncertainties. Against this backcloth, a range of transgenic crops, that are herbicide-tolerant, insect-resistant, drought and other forms of stress-tolerant, or higher yielding, have been developed, and a few are now being grown in many parts of the world. The area of transgenic crops grown globally has increased from 2 to 134 million hectares in 2009 since wide-scale planting started in 1996 (Park et al. 2011). The principal transgenic crops are soya bean, maize, cotton and canola, which are modified for agronomic input traits such as herbicide tolerance and/or insect resistance (*Bacillus thuringiensis*-Bt). Recent work has focused on the use of biotechnology to produce abiotic stress-tolerant and nutritionally enhanced food and feed with a range of new events being predicted by 2015 (Newell-McGloughlin 2008).

Potato is an easy-to-grow plant and can provide more nutritious food faster and on less land than any other food crop, and in almost any habitat. A shift towards the use of potato in convenience foods, such as potato chips and French fries, has been recorded in developed nations, but in developing economies the majority of the potato crop is still used for direct consumption. Today, potatoes are grown worldwide and more than a billion people consume them on a daily basis. To support this demand, a lot of varieties have been developed (Mullins et al. 2006).

It generally requires 10–15 years to develop a single potato cultivar through traditional breeding (Byun et al. 2007). It also relies on utilizing existing genetic stocks, whose quantities are limited (Bajaj et al. 1999). In addition, the cultivated species *Solanum tuberosum* is autotetraploid which has a highly complicated quantitative inheritance pattern, and is difficult to hybridize sexually with the related relative wild species *Solanum*. The most obvious advantage of using transgenic approaches for crop improvement is that genes from any organism can be utilized, whereas in conventional breeding the gene pool is restricted to closely related plant species. Thus, a large variety of genes are available for transfer into plants to confer a specific desirable effect (Bajaj et al. 1999).

Park et al. (2011) deemed that transgenic crops are a potential ‘tool’ giving options for ongoing sustainable development if the growing world population is to be adequately fed, both in terms of quantity and quality, without further compromising the environmental services that the planet provides, and elucidated the considerable contribution of transgenic crops in relation to the three traditional pillars of sustainability, i.e. economically, environmentally and socially (Park et al. 2011 and references therein).

22.4 Biosafety of Transgenic Potato

Despite the growth and use of transgenic crops in many areas of the world, some governments, organizations and individuals still hesitate to acknowledge that transgenic crops provide economic and environmental benefits that are unobtainable in a timely manner via non-transgenic advances in plant breeding. Hall and Moran (2006) described some of the organizations that believe that there are unacceptable risks associated with the release of transgenic crops. Conner et al. (2003) and Nap et al. (2003) summarized the current status of environmental release of genetically modified (GM) crops around the globe. They provided an overview of the approaches used for regulating GM crop release into the environment and presented a detailed description of risk assessments and how they are performed, followed by a discussion of the perceived risks associated with the release of GM crops. Craig et al. (2008) summarized general features of risk assessments of GM crops, which provided an introduction to some of the main considerations made in the compilation and evaluation of risk assessments.

In order to alleviate some of the public concerns over the deployment of GM crops in agriculture, Conner and colleagues developed a novel transformation vector (intragenic vector) system (Conner et al. 2007; Barrell et al. 2010). Intragenic vector system is a gene transfer system composed of only DNA that originates from that host plant species (or related species to which it can be hybridised). Gene transfer using intragenic vectors will facilitate the well-defined genetic improvement of plants with all transferred DNA originating from within the gene pool already available to plant breeders. In this manner, genes can be introgressed into elite cultivars without the incorporation of any foreign DNA. The resulting plants are non-transgenic, although they are derived using the tools of molecular biology and plant transformation. With gene transfer using intragenic vectors, there is no longer a clear biological distinction between traditional plant breeding approaches and development of GM crops (Conner et al. 2007). This opens up the possibility of using an intragenic vector system to create non-transgenic GM crops.

Many nontheological ethical objections to genetic engineering are associated with interfering with nature or natural evolution or the natural order of life. The intragenic vector system will help alleviate some of the ethical issues associated with transferring DNA across wide taxonomic boundaries, and will provide a socially acceptable and responsible way forward for the development of GM crops (Barrell et al. 2010).

For potato crop, genetic modification using intragenic vectors can therefore provide a valuable breeding tool. There are several groups improving potato traits, such as disease resistance, abiotic stress tolerance, pigmentation and processing quality, using intragenic/cisgenic approaches (Barrell et al. 2010). The intragenic vector system offers opportunities to accelerate the efficiency and extent of further potato improvement. Ongoing improvements can be expected in characteristics such as resistant to pests, diseases, herbicides, and environmental stress, as well as quality traits such as improved post-harvest storage, flavor, nutrition, shape and colour. Genetic manipulation of these characteristics will allow breeders to respond much more quickly to the market need for new and improved potato cultivars. The anticipated result is higher quality, blemish-free tubers with reduced chemical residues as demanded by the processors and consumers (Conner 2004).

The adoption rate of GM potato in agriculture will be become higher along with higher satisfaction of growers and benefits for the whole production chain result from best characters of GM potato. GM potato will be proved to be a promising solution for sustainable potato production, which could serve as one of powerful tools in combating famine and malnutrition in developing countries with increasing of world population in the future.

22.5 Conclusion

In the twenty-first century, food security has potential crisis in the world. Potato crop is grown worldwide and plays very important role in food security in developing countries. Since potato crop is introduced to China in the seventeenth century, it is steadily spread throughout the country. In order to overcome drought climate and backward production conditions in north China, the potato makes a greater contribution to solve food problem since it can be grown to yield relatively better production while other food crops is not in the arid and semi-arid areas.

Improvement of cultivated potato by traditional breeding method is slow and unpredictable due to its tetraploid genetics and the quantitative nature of inheritance. Genetic engineering provides a faster and more reliable means for potato crop improvement and these techniques are especially applicable to development of resistance to abiotic and biotic stresses such as drought, salt, cold, and pathogens (Byun et al. 2007).

In the chapter, the transgenic potato plants that are resistant to drought and salinity stresses are developed by transforming into potato with *BADH* gene from spinach. However, the transgenic potato plants have yet to be grown commercially since they must be performed assessment on field performance and biosafety. For practical applications, the useful stress-tolerant transgenic potato plants must give higher yield under stress and field conditions compared with the control plants. There is still much more work ahead in assessment of the transgenic potato plants under field performance.

Food insecurity is one of the most important social issues faced today. Strategies to address food insecurity must aim to increase agricultural productivity in order to tackle poverty, and must provide long-term improvements in crop yields to keep up with demands as the world's population grows. Genetically enhanced plants provide one route to sustainable higher yields, either by increasing the intrinsic yield capability of the crop or by protecting them from biotic and abiotic constraints (Christou and Twyman 2004). In conclusion, genetic engineering has the potential to help increase production and productivity in agriculture. While there is no easy solution to improve potato crop for sustainable production and food supply, transgenic approach is an alternative way to meet the aim and contribute to safe food supply.

Acknowledgments We would like to thank Junlian Zhang, Dongkui Li, Yikai Wen, Liang Li, Tao Yang, Honghui Du, Bailin Liu, and Chunfeng Zhang for performing valuable work on the research. This work was supported in part by the National High-Technology (863) Program of China (2006AA100107 and 2009AA10Z103), Research Fund for the Doctoral Program of Higher Education of China (20050733003), and the Research Grant from Agri-biotechnology and Development Program of Gansu Province of China (GNSW-2006-01).

References

Ahmad R, Abdullah Z (1979) Salinity induced changes in the growth and chemical composition of potato. Pak J Bot 11:103–112

Ahmad R, Kim MD, Back KH, Kim HS, Lee HS, Kwon SY, Murata N, Chung WI, Kwak SS (2008) Stress-induced expression of choline oxidase in potato plant chloroplasts confers enhanced tolerance to oxidative, salt, and drought stresses. Plant Cell Rep 27:687–698

Ambard-Bretteville F, Sorin C, Rébeillé F, Hourton-Cabassa C, des Francs-Small CC (2003) Repression of formate dehydrogenase in *Solanum tuberosum* increases steady-state levels of formate and accelerates the accumulation of praline in response to osmotic stress. Plant Mol Biol 52:1153–1168

Apse MP, Blumwald E (2002) Engineering salt tolerance in plants. Curr Opin Biotech 13:146–150

Bajaj S, Targolli J, Liu LF, Ho THD, Wu R (1999) Transgenic approaches to increase dehydration-stress tolerance in plants. Mol Breed 5:493–503

Barrell PJ, Jacobs JME, Baldwin SJ, Conner AJ (2010) Intragenic vectors for plant transformation within gene pools. CAB reviews: perspectives in agriculture, veterinary science, nutrition and natural resources 5, No. 010. doi: 10.1079/PAVSNNR20105010. http://www.cabi.org/cabreviews

Behnam B, Kikuchi A, Celebi-Toprak F, Yamanaka S, Kasuga M, Yamaguchi-Shinozaki K, Watanabe KN (2006) The *Arabidopsis DREB1A* gene driven by the stress-inducible rd29A promoter increases salt-stress tolerance in proportion to its copy number in tetrasomic tetraploid potato (*Solanum tuberosum*). Plant Biotechnol 23:169–177

Byun MO, Kwon HB, Park SC (2007) Recent advances in genetic engineering of potato crops for drought and saline stress tolerance. In: Jenks MA, Hasegawa PM, Jain SM (eds) Advances in molecular breeding toward drought and salt tolerant crops. Springer, Dordrecht, pp 713–737

Cavagnaro JB, De Lis BR, Tizio RM (1971) Drought hardening of the potato plant as an after-effect of soil drought conditions at planting. Potato Res 14:181–192

Celebi-Toprak F, Behnam B, Serrano G, Kasuga M, Yamaguchi-Shinozaki K, Naka H, Watanabe AJ, Yamanaka S, Watanabe KN (2005) Tolerance to salt stress of the transgenic tetrasomic tetraploid potato, *Solanum tuberosum* cv. Desiree appears to be induced by the *DREB1A* gene and rd29A promoter of *Arabidopsis thaliana*. Breeding Sci 55:311–319

Chen THH, Murata N (2008) Glycinebetaine: an effective protectant against abiotic stress in plants. Trends Plant Sci 13:499–505

Christou P, Twyman RM (2004) The potential of genetically enhanced plants to address food insecurity. Nutr Res Rev 17:23–42

Conner AJ (2004) Global environmental and food safety issues of GM potatoes. Abstracts of the proceedings of the fifth world potato congress, Kunming, China, p 11

Conner AJ, Barrell PJ, Baldwin SJ, Lokerse AS, Cooper PA, Erasmuson AK, Nap JP, Jacobs JME (2007) Intragenic vectors for gene transfer without foreign DNA. Euphytica 154:341–353

Conner AJ, Glare TR, Nap JP (2003) The release of genetically modified crops into the environment. II. Overview of ecological risk assessment. Plant J 33:19–46

Craig W, Tepfer M, Degrassi G, Ripandelli D (2008) An overview of general features of risk assessments of genetically modified crops. Euphytica 164:853–880

Evers D, Overney S, Simon P, Greppin H, Hausman JF (1999) Salt tolerance of *Solanum tuberosum* L. overexpressing an heterologous osmotin-like protein. Biol Plantarum 42:105–112

Goddijn O, Verwoerd T, Voogd E, Krutwagen R, Graaf P, Poels J, Dun K, Ponstein A, Damm B, Pen J (1997) Inhibition of trehalase activity enhances trehalose accumulation in transgenic plants. Plant Physiol 113:181–190

Hall C, Moran D (2006) Investigating GM risk perceptions: a survey of anti-GM and environmental campaign group members. J Rural Stud 22:29–37

Hanson AD, Scoff NA (1980) Betaine synthesis from radioactive precursor in attached water stressed barley leaves. Plant Physiol 66:342–348

Hemavathi UCP, Akula N, Young KE, Chun SC, Kim DH, Park SW (2010) Enhanced ascorbic acid accumulation in transgenic potato confers tolerance to various abiotic stresses. Biotechnol Lett 32:321–330

Hofgen R, Willmitzer L (1988) Storage of competent cells for *Agrobacterium* transformation. Nucleic Acid Res 16:9877

Jeong MJ, Park SC, Byun MO (2001) Improvement of salt tolerance in transgenic potato plants by glyceraldehydes-3-phosphate dehydrogenase gene transfer. Mol Cells 12:185–189

Knipp G, Honermeier B (2006) Effect of water stress on proline accumulation of genetically modified potatoes (*Solanum tuberosum* L.) generating fructans. J Plant Physiol 163:392–397

Kolodyazhnaya YS, Kutsokon NK, Levenko BA, Syutikova OS, Rakhmetov DB, Kochetov AV (2009) Transgenic plants tolerant to abiotic stresses. Cytol Genet 43:132–149

Lee HE, Shin D, Park SR, Han SE, Jeong MJ, Kwon TR, Lee SK, Park SC, Yi BY, Kwon HB, Byun MO (2007) Ethylene responsive element binding protein 1 (StEREBP1) from *Solanum tuberosum* increases tolerances to abiotic stress in transgenic potato plants. Biochem Biophys Res Commun 353:863–868

Li DK, Wand D, Zhang JL, Si HJ, Liu N (2007) Determination of salt-tolerance of the second asexual multiplication generation of transgenic potato (*Solanum tuberosum* L.) with *BADH* gene. Plant Physiol Commun 43:873–875

McCue KF, Hanson AD (1990) Drought and salt tolerance: towards understanding and application. Trends Biotechnol 8:358–362

Mullins E, Milbourne D, Petti C, Doyle-Prestwich BM, Meade C (2006) Potato in the age of biotechnology. Trends Plant Sci 11:254–260

Murashige T, Skoog F (1962) A revised medium for rapid growth and bioassays with tobacco tissues cultures. Physiol Planta 15:473–479

Nap JP, Metz PLJ, Escaler M, Conner AJ (2003) The release of genetically modified crops into the environment. Part I. Overview of current status and regulations. Plant J 33:1–18

Newell-McGloughlin M (2008) Nutritionally improved agricultural crops. Plant Physiol 147:939–953

Ojala JC, Stark JC, Kleinkopf GE (1990) Influence of irrigation and nitrogen management on potato yield and quality. Am J Potato Res 67:29–43

Park JR, McFarlane I, Phipps RH, Ceddia G (2011) The role of transgenic crops in sustainable development. Plant Biotechnol J 9:2–21

Rhodes D, Hanson AD (1993) Quaternary ammonium and tertiary sulfonium compounds in higher plants. Annu Rev Plant Physiol Plant Mol Biol 44:357–384

Rontein D, Basset G, Hanson AD (2002) Metabolic engineering of osmoprotectant accumulation in plants. Metab Eng 4:49–56

Sakamoto A, Murata N (2000) Genetic engineering of glycinebetaine synthesis in plants: current status and implications for enhancement of stress tolerance. J Exp Bot 51:81–88

Salter PJ, Goode JE (1967) Crop responses to water at different stages of growth. Res Rev Commonwealth Bureau Hortic, East Malling 2:93–97

Shinozaki K, Yamaguchi-Shinozaki K (1997) Gene expression and signal transduction in water-stress response. Plant Physiol 115:327–334

Si HJ, Xie CH, Liu J (2003) An efficient protocol for *Agrobacterium*-mediated transformation of microtuber and the introduction of an antisense class I patatin gene into potato. Acta Agron Sin 29:801–805

Si HJ, Zhang N, Wang D (2007) Enhancement of drought and salt resistances in tobacco by transformation of betaine aldehyde dehydrogenase gene. Acta Agron Sin 33:1335–1339

Tang L, Kim MD, Yang KS, Kwon SY, Kim SH, Kim JS, Yun DJ, Kwak S, Lee HS (2008) Enhanced tolerance of transgenic potato plants overexpressing nucleoside diphosphate kinase 2 against multiple environmental stresses. Transgenic Res 17:705–715

Tang L, Kwon SY, Kim SH, Kim JS, Choi JS, Cho KY, Sung CK, Kwak SS, Lee HS (2006) Enhanced tolerance of transgenic potato plants expressing both superoxide dismutase and ascorbate peroxidase in chloroplasts against oxidative stress and high temperature. Plant Cell Rep 25:1380–1386

Turhan H (2005) Salinity response of transgenic potato genotypes expressing the oxalate oxidase gene. Turk J Agric For 29:187–195

van Loon CD (1981) The effect of water stress on potato growth, development, and yield. Am J Potato Res 58:51–69

Wang WX, Vinocur B, Altman A (2003) Plant responses to drought, salinity and extreme temperatures: towards genetic engineering for stress tolerance. Planta 218:1–14

Weretilnyk EA, Hanson AD (1989) Betaine aldehyde dehydrogenase from spinach leaves: purification, *in vitro* translation of the mRNA and regulation by salinity. Arch Biochem Biophys 271:55–63

Weretilnyk EA, Hanson AD (1990) Molecular cloning of a plant betaine aldehyde dehydrogenase, an enzyme implicated in adaptation to salinity and drought. Proc Natl Acad Sci USA 87:2745–2749

Yamaguchi-Shinozaki K, Shinozaki K (1993) Characterization of the expression of a desiccation-responsive *rd29A* gene of *Arabidopsis thaliana* and analysis of its promoter in transgenic plants. Mol Gen Genet 236:331–340

Yamaguchi-Shinozaki K, Shinozaki K (1994) A novel *cis*-acting element in an *Arabidopsis* gene is involved in responsiveness to drought, low-temperature, or high-salt stress. Plant Cell 6:251–264

Yeo ET, Kwon HB, Han SE, Lee JT, Ryu JC, Byu MO (2000) Genetic engineering of drought resistant potato plants by introduction of the trehalose-6-phosphate synthase (TPS1) gene from *Saccharomyces cerevisiae*. Mol Cell 10:263–268

Zhang N, Si HJ, Li L, Yang T, Zhang CF, Wang D (2009) Drought and salinity tolerance in transgenic potato expressing the betaine aldehyde dehydrogenase gene. Acta Agron Sin 35:1146–1150

Zhang N, Si HJ, Wang D (2005) Cloning of *rd29A* gene promoter from *Arabidopsis thaliana* and its application in stress-resistance transgenic potato. Acta Agron Sin 31:159–164

Zhang N, Si HJ, Wen G, Du HH, Liu BL, Wang D (2011) Enhanced drought and salinity tolerance in transgenic potato plants with a *BADH* gene from spinach. Plant Biotechnol Rep 5:71–77

Zhang N, Wang D, Si HJ (2004) Isolation and expression of betaine aldehyde dehydrogenase gene from spinach. J Agric Biotech 12:612–613

Part VII
Increasing Sustainability of Potato Production Systems in Brazil and Peru

Chapter 23
Statistical Models in Plant Diagnosis and Calculating Recommended Nitrogen Rates

Paulo Cezar Rezende Fontes, Marcelo Cleón de Castro Silva, and Glauco Vieira Miranda

Abstract The potato, *Solanum tuberosum* L., is an important crop in Brazil and in the world. In addition to other factors, the potato plant needs an adequate nutrient supply. Best N management in potato plants is aimed not only to improve tuber yield and quality but also to increase N fertilizer use efficiency and to reduce environmental risk. It is common to use mathematical models in establishing relationships between N rate and crop yield and plant N content. Those relationships are essential to best N management in potato plant at diagnosis and recommendation phases. In establishing a plant index it is necessary to adjust the data using some mathematical model. Therefore, either in the assessment of plant index or in the rate recommendation it is necessary to select a model. In the text will be discussed the relationship between potato yield and nitrogen rates obtained by different mathematical models and how the model chose affects plant nitrogen indices under Brazilian conditions.

23.1 Introduction

The potato, *Solanum tuberosum* L., is an important crop in Brazil and throughout the world. In addition to other factors, the potato plant needs an adequate nutrient supply. Nitrogen (N) is one of the nutrients of greatest impact on crop productivity. The N effect on potato tuber yield has been well documented worldwide (Meyer and Marcum 1998; Bélanger et al. 2000; Rodrigues et al. 2005; Silva et al. 2007). N has a marked effect on the vegetative and reproductive plant compartments. It is essential for the fast cycle and high growth rate of the potato plant. A higher N

P.C.R. Fontes (✉) • M.C. de Castro Silva • G.V. Miranda
Departamento de Fitotecnia, Universidade Federal de Viçosa,
s/n. 36570-000 Estado de Minas Gerais, Brazil
e-mail: pacerefo@ufv.br

Z. He et al. (eds.), *Sustainable Potato Production: Global Case Studies*,
DOI 10.1007/978-94-007-4104-1_23,

availability has positive effects on stem and leaf growth and intercepted radiation, which generally leads to high tuber yield and N accumulation (Nunes et al. 2006).

Low N supply will not only result in lower yield but will also reduce tuber size due to reduced leaf area and early defoliation. On the other hand, excess of N leads to dry matter yield in other parts of the plant than the tubers (Goffart et al. 2008). When superfluous, N promotes excessive stolon and leaf growth, delays both leaf maturation and tuber differentiation, and reduces the length of the tuber bulking period, tuber solid contents and yield. Additionally, the excess of N remains available in the soil to be leached.

Best N management in potato plants is aimed not only to improve tuber yield and quality but also to increase N fertilizer use efficiency and to reduce environmental risk. It is common to use mathematical models in establishing relationships between N rate and crop yield and plant N content. Those relationships are essential to best N management in potato plant at diagnosis and recommendation phases. All nitrogen recommendation approaches, *ex post*, *ex-ante* and "when necessary" imply uncertainty about which model should be chosen and usually local and temporal variations are neglected.

23.2 Nitrogen Rate Recommendation

In an optimistic view, the N rate recommended by the extension service (*ex ante*) is based on the yield potential, soil organic matter and N use efficiency (NUE). Usually the *ex ante* recommendation is based on the yield potential and the NUE (50%). In a broad sense, under Southeastern Brazilian conditions it is necessary to apply 1.0 kg of N fertilizer for the expected yield of 190 ± 40 kg potato (Fontes 1997).

Alternatively, the recommendation may be based on a mathematical function previously obtained in experiments where the effects of N fertilizer rates on crop yield were evaluated. The relationship between fertilizer application and crop yield is generally represented by a mathematical function which seeks to estimate the optimal rate or the maximum economic rate of N (Zimmermann and Conagin 1986; Seefeldt et al. 1995). As the mathematical function is *ex-post* chosen, this is an *ex-post* recommendation.

Different models, *ex-post* chosen, provide differing values for the estimated optimum fertilizer rate (Nelson et al. 1985; Fontes and Ronchi 2002; Berzsenyi and Dang 2006) and the model chosen largely determines the maximum economic N rate or economically optimal fertilization rate (ENR) affecting the crop profitability and may cause adverse impact on the environment. To illustrate the impact of model choice on estimating N rate, Olness et al. (1998) compared the relative accuracy of three models on 48 corn data sets. In about one-third of the cases all models performed about equally well. The ENR depended on several factors (soil texture, tillage, hybrid, climatic zone) including the model. This renders broad generalizations of ENR quite misleading.

Another approach is applying an intentionally small N rate before planting and to decide, in real time, the supplemental N needs of the crop (Fontes and Araujo 2006). This decision should be based on a plant nitrogen index at the appropriate time. Schepers et al. (1992) called "fertilization when necessary" the decision of topdressing fertilizer based on plant N status. To answer the question how much N should be applied it is necessary to construct an algorithm, a finite sequence of operations including a plant N index.

In theoretical terms, the N rate should be dictated by demand-capacity-efficiency factors combination based on the complex soil-plant-environment. The factors combination may lead to an algorithm. In the algorithm construction process at least three types of information are utilized: (1) plant N demand; (2) soil capacity to provide N; and (3) fertilizer N use efficiency. Ideally, the N rate to be applied as fertilizer should be estimated using the assessment of the integrated system in both providing (soil, water and incorporated organic residues) and demanding N (yield potential of the cultivar in a given production system) modulated by the processes efficiency (Fontes and Araujo 2007). By combining strategies it is possible to reduce the potential for nitrate leaching in the potato crop, but the progress of research has not allowed a quick and ready solution to prevent the potential for nitrate leaching to groundwater in certain regions of the world (Shrestha et al. 2010).

There are several models available in the literature, with different qualitative and quantitative approaches, especially for the estimation of the demand-capacity terms. Normally, variable effects are known qualitatively but the interactions among them are unknown, making the model empirical. The deterministic mathematical models, with one or more variables, are hardly compatible with a set of data showing high variability and low correlations. Still, the mechanistic and deterministic models are useful in knowledge and information systematization and organization. But there is always the challenge of how to parameterize the various processes occurring in biological systems.

Some models attempt to quantify the various factors that affect the demand-capacity processes in an attempt to infer the efficiency of the combination. Normally, those factors are reflected in a specific N index evaluated at the appropriate time in the plant. This index is simultaneously utilized in the N diagnosis and prescription processes, integrants of an algorithm.

In practical terms, the N plant index should enable answering the following questions: (a) is it necessary to fertilize (especially valid question when the N rate applied was the recommended and not an intentionally small rate at planting) and (b) how much N should be applied? In precision agriculture, these questions should be answered in real time. In most Brazilian conditions, up to 25 days after potato plant emergence (time of side dress application) is necessary to determine answers to the two above questions.

In establishing the plant N index it is necessary to adjust the data using some mathematical model. Therefore, either in the assessment of plant index or in the rate recommendation it is necessary to select a model.

23.3 Selection of Mathematical Model: The N Rate

There are several classifications for mathematical models, for example, static, dynamic, linear, deterministic, stochastic, empirical, mechanistic, and others. In part, the reliability of the information obtained with the aid of a model depends on the fitness of the model to the experimental data. The term model is also adopted for the representation of a system with flow diagrams where several mathematical submodels and algorithm may be involved in implementing the several factors that explain the system.

In the present paper there is interest in explaining the potato yield as a function of the N rates. This relationship can be described by a mathematical model using the quantitative variable N rate as independent variable and yield as dependent variable. Selecting the most appropriate model to describe the relationship between crop yield and fertilizer rate is not an obvious decision (Bock and Sikora 1990; Angus et al. 1993). There seems no possibility of standardizing a specific model to describe the plant response to N rates. The main reason is that the type of curve needed is inherently dependent on the variation in soil N availability reflecting the added rates.

Generally, in research reports there is little description of how the model was chosen. There is no statistical basis for selecting one functional form over another across all sites and years (Rajsic and Weersink 2008). The choice of a statistical model should be based on some criteria, such as a biological explanation of the phenomenon; the significance of the regression mean square; F-statistic significance or lack of fit; high coefficient of determination (R^2) and the significance of the regression parameters. Besides these criteria, it would be recommended to consider the maximization of productivity and profit. These issues will be addressed in the following example.

The relationship between potato yield and N rates obtained by different mathematical models will be described. This relationship was obtained from research conducted in a Red-Yellow Podzolic Cambic soil where five N rates (0, 50, 100, 200 and 300 kg ha^{-1}), as ammonium sulfate, furrow applied, were evaluated (Silva et al. 2007). Potato 'Monalisa' was cultivated under irrigated conditions and 114-day-growth cycle. After natural canopy drying, the tubers were harvested, remaining in the field around an hour and then weighed. Tuber yield data were submitted to analysis of variance procedures and to linear and nonlinear regression analysis and curve fitting using the SAS and SAEG programs. Six mathematical models were selected to relate yield and N rate: linear plateau, quadratic plateau, Mitscherlich, sigmoidal, square root and quadratic. The first four are nonlinear models and the two last are linear. For each considered model four variables were estimated: (a) the maximum N rate; (b) the maximum physical yield of tubers; (c) money spent on N fertilizer; (d) money left over after selling the potato and paying the N fertilizer. The economic optimum N fertilization rate (ENR) was also estimated for quadratic model at unfavorable and favorable potato price conditions in Brazil (Fig. 23.1).

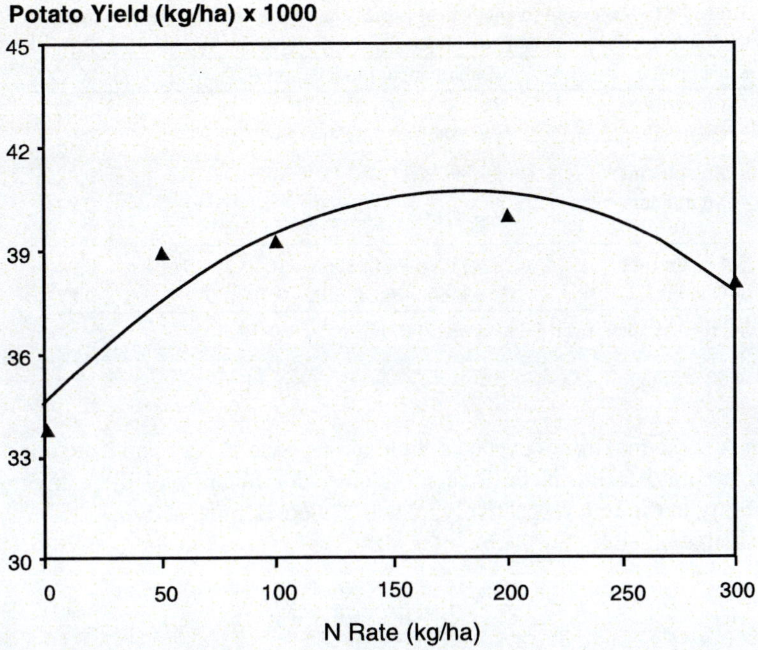

Fig. 23.1 Relationship between potato yield and nitrogen rate as described by the quadratic model

The mathematical expressions of the models are below.

1. Linear plateau, defined by Eqs. 23.1 and 23.2:

$$Y = a + bX, \text{ if } X < C \tag{23.1}$$

$$Y = P, \quad \text{if } X \geq C \tag{23.2}$$

Where Y is the tuber yield (kg ha^{-1}); a and b are intercept and linear coefficient, respectively; X is the N rate (kg ha^{-1}); the constant C is the intersection point of the linear model with the plateau; P is the potato yield when it reaches the plateau.

2. Quadratic plateau, defined by Eqs. 23.3 and 23.4:

$$Y = a + bX + cX^2 \text{ if } X < C \tag{23.3}$$

$$Y = P, \quad \text{if } X \geq C \tag{23.4}$$

Where c is the quadratic coefficient and the others terms were defined in Model 1.

3. Mitscherlich, defined by Eq. 23.5:

$$Y = A / \left(1 - e^{-c(X+b)}\right) \tag{23.5}$$

Table 23.1 Name and mathematical expression of the adjusted models to the relationship between potato yield (kg ha^{-1}) and nitrogen rate (kg ha^{-1})

Model name	Mathematical model expression	R^2
1- Linear plateau	$Y = -83.5986 + 0.0057X$	0.09
2- Quadratic plateau	$Y = 33813 + 178.1X - 1.4937X^2$	0.92
3- Mitscherlich	$Y = 39090/(1 - e^{-1(X+2.0025)})$	0.92
4- Sigmoidal	$Y = 37850 + \frac{200}{1 + e^{-[(X-Xo)/1]}}$	0.92
5- Square root	$Y = 33790 - 44.9876X + 1039.7792X^{1/2}$	0.97
6- Quadratic	$Y = 34608 + 68.8565X - 0.1938X^2$	0.87

Source: Adapted from Silva et al. (2007)

Where A is the maximum expected yield in response to N; c and b are constants and correspond to the N fertilizer efficiency coefficient and the estimated N availability in the soil, respectively; X and Y were defined above.

4. Sigmoidal, defined by Eq. 23.6

$$Y = Yo + \frac{a}{1 + e^{-[(X-Xo)/b]}} \tag{23.6}$$

Where Yo is the yield obtained with the initial rate (g/plant); a and b are non-linear regression model parameters; Xo is the initial N rate or 0 kg ha^{-1}; X and Y were defined above.

5. Square root, defined by Eq. 23.7:

$$Y = a + bX + cX^{1/2} \tag{23.7}$$

Where Y, b, c and X have been defined previously.

6. Quadratic, defined by Eq. 23.8:

$$Y = a + bX + cX^2 \tag{23.8}$$

Where Y, b, c and X have been defined previously.

The models fitted to experimental data (Table 23.1) were evaluated by the following criteria: the significance of the regression mean square (QMRr); significance of the F-statistic or lack of fit (FA); high coefficient of determination (R^2); significance of the regression parameters using the t-test at 1, 5 and 10% and F at 1 and 5% probability (T′ and T″); fidelity to the observed data (FTR). The results are shown (Table 23.2).

In linear plateau, quadratic plateau, Mitscherlich and sigmoidal models the biological explanation of the phenomenon is dependent on the actual rate-yield curve, mainly at the highest N rate. So they were classified as unfaithful to the observed data – FTR – (Table 23.2). Smaller QMRr indicates model better fit the data. The T values showed differences between the N rates evaluated. The coefficient of determination (R^2) is the measure of correlation between N rate and tuber yield. Several models (linear plateau, quadratic plateau, quadratic, square root and

Table 23.2 Regression mean square (QMRr), lack of fit (FA), coefficient of determination (R^2), significance of the regression parameters (T′ and T″) and fidelity to the observed data in the six models

	Criteria					
Models	QMRr	FA	R^2	T′	T″	FTR
1- Linear plateau	17676	NC	0.09	NC	NC	No
2- Quadratic plateau	952641	NC	0.92	NC	NC	No
3- Mitscherlich	639648	NC	0.92	NC	NC	No
4- Sigmoidal	11679632	NC	0.92	NC	NC	No
5- Square root	246196	ns	0.97	6.6**	8.5*	Yes
6- Quadratic	1583441	ns	0.87	3.5***	3.1***	Yes

NC not considered, *ns* not significant by F test
*, **, and *** significant by t test at 1, 5 and 10% probability, respectively
Source: Adapted from Silva et al. (2007)

Table 23.3 Estimated maximum nitrogen rate (DMN), maximum tuber yield (PMFT) and the cost of the nitrogen fertilizer (GAN) in the six models

Models	DMN (kg ha^{-1})	PMFT (kg ha^{-1})	GAN (U$ ha^{-1})
1-Linear plateau	50.00	33,493	97
2-Quadratic plateau	59.64	39,125	116
3-Mitscherlich	65.70	39,125	128
4-Sigmoidal	65.70	39,125	128
5-Square root	133.53	39,797	259
6-Quadratic	177.57	40,720	345
7- Without fertilizing with N	0	33,813	0,00

Source: Silva et al. (2007)

Mitscherlich) were also evaluated by Cerrato and Blackmer (1990) to describe the corn yield response to N rates. The authors obtained R^2 values ranging from 79 to 84 but it was not a reliable criterion for the model selection and the economically optimal N rate identification.

In the present example, with each model but Mitscherlich and Sigmoidal the maximum N rate (DMN), the maximum tuber yield (PMFT), and the cost of the N fertilizer (GAN) were estimated. With asymptotic models it is not possible to calculate the maximum, so it was utilized at 90% of the estimated maximum for Mitscherlich and Sigmoidal models. The results are shown in Table 23.3. Models with distant R^2 values can estimate close values for the estimated DMN which ranged from 50 to 178 kg ha^{-1} depending on the model (Table 23.3). With the quadratic model, the estimated maximum N rate was 178 kg ha^{-1} leading to a maximum tuber yield of 40.7 Mg ha^{-1}.

With the quadratic model, the maximum economic N rate or economically optimal fertilization rate (ENR) was calculated, which was defined as the rate of N application where U$1 of additional N fertilizer returned U$1 of potatoes, and it describes the minimum rate of N application required to maximize economic return (Colwel 1994). ENR was the point where the last increment of N returns a yield

Table 23.4 The lowest and the highest relative price of nitrogen to potato price in two nitrogen fertilizer, from January to October 2010 in Brazil

	Relative price of N/potato price	
Nitrogen fertilizer	Lowest	Highest
Ammonium sulfate	2.22	6.90
Urea	1.76	5.63

Table 23.5 Value receipt from the potato sale (RCVB) and money left over after selling potatoes and paying the nitrogen fertilizer (SAPN) under favorable and unfavorable scenario of potato price in the six models

	Unfavorable scenario		Favourable scenario	
	RCVB	SAPN	RCVB	SAPN
Models	(U$ ha^{-1})			
1- Linear plateau	11,821	11,765	23,642	23,586
2- Quadratc plateau	13,809	13,693	27,618	27,502
3- Mitscherlich	13,809	13,681	27,618	27,502
4- Sigmoidal	13,809	13,681	27,618	27,502
5- Square root	14,046	13,786	28,092	27,832
6- Quadratic	14,372	14,027	28,743	28,399
7- No fertilizer N	11,934	11,934	23,868	23,868

Source: Adapted from Silva et al. (2007)

large enough to pay for the additional N. ENR was calculated by setting the first derivative of the N response curve equal to the ratio between the cost of fertilizer and the price of potatoes. The resulting equation was solved for the ENR. Price ratio was the ratio of N fertilizer price to potato tuber price (U$/kg ÷ U$/kg), in two potato price scenarios, favorable and unfavorable. For the calculations, N price was U$ 1.94/kg. Potato prices were $ 0.35/kg (unfavorable scenario) and U $ 0.71/kg (favorable scenario). The estimated ENR value was 163 or 171 kg ha^{-1} in unfavorable or favorable potato price scenarios, respectively. For reference, the lowest and the highest relative price of N to potato price in two fertilizer sources in Brazilian conditions are shown (Table 23.4).

The value received from the potato sale (RCVB) was calculated by multiplying PMFT by the potato price. Also evaluated was the amount of money left over after selling potatoes and paying the N fertilizer (SAPN), which was obtained from DMN and the corresponding tuber yield. Fertilizer application costs were considered equal at all N rates and any yield variation does not imply extra costs. Results are shown in Table 23.5. The SAPN would be highest with the quadratic model (Table 23.5). Moreover, a higher amount of N estimated by the quadratic model in relation to the square root could be an insurance against possible losses of N. This probably did not occur due to several conditions, among them the N source, ammonium sulfate, applied in the furrow, a loamy soil and the drought period, only 255 mm of rainfall during the growing period supplemented by irrigation.

Three statistical models (quadratic, square root and exponential-Mitscherlich) were compared to describe the potato yield response to N rates at planting, in Canada

(Bélanger et al. 2000). High values of R^2 for the three models were found and the highest optimal N rate was estimated by the quadratic model, followed by the square root and the exponential model. That is, the estimated N rate for the potato crop depended on the mathematical model as also mentioned by Neeteson and Wadman (1987) in Netherlands.

A quadratic model was used to describe the yield response of potato cultivars to N fertilizer rates (0 to 300 kg ha^{-1}) under Brazilian conditions (Fontes et al. 2010). For the maximum marketable tuber yields, the optimum fertilization rates were 168, 212, 175, and 193 kg ha^{-1} of N for Ágata, Asterix, Atlantic, and Monalisa, respectively. For these cultivars at the optimum N fertilization rate the predicted marketable yields were 33.1, 32.3, 33.3, and 25.9 Mg ha^{-1}, respectively. The economic optimum N fertilization rates ranged from 147 to 201 kg ha^{-1} depending upon cultivar and relative prices of N and potato tubers. Depending on the cultivar, under favorable price conditions (low N price and high tuber price), the economic optimum N fertilization rates to be applied by potato growers were 92–95% of the estimated N fertilization rate for obtaining the maximum potato yield. Under unfavorable conditions (high N price and low potato tuber price) the economic optimum N fertilization rates to be applied should be decreased to 86–92% of the rates for maximum yield. Usually, with crops of high value, as potato, the fertilizer price has less impact on the most economic rate than for crops with lower value.

In another potato crop study under Brazilian conditions, among the regression models tested (linear, quadratic and square-root), the quadratic model was more appropriate to describe the relationship between N rates (0 to 400 kg ha^{-1}), as urea, and yield of potato cultivars (Coelho et al. 2010). Nitrogen rates at 297 and 250 kg ha^{-1} provided the highest commercial potato yield of Agata (45.1 Mg ha^{-1}) and Asterix (46.5 Mg ha^{-1}).

The quadratic model is not always the best choice to represent the relationship between N rate and yields, as was found with corn plants (Cerrato and Blackmer 1990; Bullock and Bullock 1994). They found that the quadratic plateau was the most appropriate model.

Depending on many factors such as pre-crops, tuber yield, rainfall, soil type, cultural practices, season year, spacing, source, and cultivar involved in the experiments, the N fertilizer use efficiency in our conditions has been 190 ± 40. That is, for each kg of N fertilizer added the yield has been 190 ± 40 kg of potato. So, to produce 30 Mg ha^{-1} of potato it will be necessary to use from 143 up to 200 kg ha^{-1} of N. In Minas Gerais State, the recommendation has been 190 kg ha^{-1} of N for the 30 Mg ha^{-1} target yield (Fontes 1999).

23.4 Model Selection: Plant N Status

The potato plant N status (ENP) can be monitored by several direct and indirect methods. The main ones are the analysis of N content in the leaf dry matter, the petiole sap nitrate content, and leaf chlorophyll content with several studies trying to use leaf spectral reflectance indices determined by a spectroradiometer or a digital camera (Wu et al. 2007; Goffart et al. 2008; Zebarth et al. 2009; Cohen et al. 2010; Busato et al. 2010; Fontes 2011).

Almost all tests to assess the ENP employ a reference or critical value to assist in making the decision to side dress N in the potato crop. Several factors affect the critical value among them the mathematical procedures employed to calculate its value (Fontes 2001). This was also shown by Fontes and Ronchi (2002) in a study that aimed to establish critical values for several plant N indices. Plant indices assessed included chlorophyll meter (SPAD) readings, petiole sap nitrate content (PSNC), and N contents in the leaf dry matter (ORNL) under different soil and nutrient solution conditions, and determined by three different statistical procedures.

In the procedure designated as 'one', linear, quadratic, square root, potential, exponential, hyperbolic, logarithmic and cubic root models were fitted to statistically significant data using N level as the independent variable. The best fitting model with biological explanation of the phenomenon was used to estimate the maximum shoot dry weight (SDW) obtained by equating the first derivatives of the best fitting model to zero, solving for X, substituting the X values into the model and solving for Y. To estimate SPAD, PSNC, and ORNL critical values (CV) in both experiments, N rate associated with maximum shoot dry weight (CV100) was introduced into the best fit model previously determined, which correlates SPAD, PSNC, and ORNL to N rate. The model also was used to determine the SPAD, PSNC, and ORNL critical values associated with 99.9, 99, 95, and 90% of the maximum SDW.

In the procedure designated 'two', the initial steps were the same as in 'one', but the best fitting model was chosen among only linear, quadratic and cubic models. In the procedure designated 'three', all models listed in procedure one were fitted to SPAD, PSNC and ORNL as independent variables (X) and the SDW as the dependent variable (Y).

In each experiment, the best fitting model within the range of observed X values was used to estimate SPAD, PSNC and ORNL critical values at CV100, CV99.9, CV99, CV95, and CV90. There were considerable disagreement among the statistical procedures, substrates and yield levels selected to estimate critical plant N indices, indicating a need to emphasize them when setting critical values. As expected, all critical N indices in tomato plants grown in soil and nutrient solution were higher when 100% maximum shoot dry weight was selected compared to lower percentage of the maximum shoot dry weight.

Selecting higher maximum values for the critical value imply higher N rate. Using a lower optimum N rate prevents over-fertilization but highest yields can not be assured. As the price of N fertilizer is relatively low in relation to potato, a high percentage of the maximum yield should be chosen. But using enough N fertilization to reach 100% of the maximum yield is usually not economically and ecologically optimal. The impact of uncertainty on the optimum N fertilization rate and agronomic, ecological and economic factors was discussed by Henke et al. (2007).

23.5 Conclusion

The complex relationships between N rate and crop yield and plant N content can be explaining by a model. Models are simplifications to facilitate understanding, organizing, reasoning and eventually allow the prediction of certain complex relationships. Those relationships are essential to best N management in potato plant at diagnosis and recommendation phases. Therefore professionals are involved in the selection of more appropriate models either in the assessment of plant index or in the rate recommendation.

Acknowledgments To CNPq and FAPEMIG by awarding research productivity grants and for financial resources. To UFV for the physical and professional infrastructures.

References

Angus JF, Bowden JW, Keating BA (1993) Modeling nutrient responses in the field. Plant Soil 155(156):57–66

Bélanger G, Walsh JR, Richards JE, Milburn PH, Ziadi N (2000) Comparison of three statistical models describing potato yield response to nitrogen fertilizer. Agron J 92(5): 902–908

Berzsenyi Z, Dang QL (2006) Use of various functions to analyze the fertilizer responses of maize (Zea mays L.) hybrids in long-term experiments. Acta Agron Hung 54(1):1–14

Bock BR, Sikora FJ (1990) Modified-quadratic/plateau model for describing plant responses to fertilizer. Soil Sci Soc Am J 54(6):1784–1789

Bullock DG, Bullock DS (1994) Quadratic and quadratic-plus-plateau models for predicting optimal nitrogen rate of corn: a comparison. Agron J 86(1):191–195

Busato C, Fontes PCR, Braun H, Cecon P (2010) Seasonal variation and threshold values for chlorophyll meter readings on leaves of potato cultivars. J Plant Nutr 33(14):2148–2156

Cerrato ME, Blackmer AM (1990) Comparison of models for describing corn yield response to nitrogen fertilizer. Agron J 82(1):138–143

Coelho FS, Fontes PCR, Puiatti M, Neves JCL, Silva MCC (2010) Dose de nitrogênio associada à máxima produtividade econômica de tubérculos de cultivares de batata e índices do estado de nitrogênio da planta. Rev Bras Ciênc Solo 34(4):1175–1183

Cohen Y et al (2010) Leaf nitrogen estimation in potato based on spectral data and on simulated bands of the VENμS satellite. Precis Agric 11(5):520–537

Colwel JD (1994) Estimating fertilizer requirements: a quantitative approach. CAB International, Wallingford

Fontes PCR (1997) Preparo do solo, nutrição mineral e adubação da batata. Universidade Federal de Viçosa, Viçosa

Fontes PCRB (1999) In: Ribeiro AC et al (eds) Recomendações para o uso de corretivos e fertilizantes em Minas Gerais. 5ª Aproximação. CFSEMG, Viçosa

Fontes PCR (2001) Diagnóstico do estado nutricional das plantas. Editora UFV, Viçosa

Fontes PCR (2011) Nutrição mineral de plantas: avaliação e diagnose. Arka Editora, Viçosa

Fontes PCR, Araujo C (2006) Use of a chlorophyll meter and plant visual aspect for nitrogen management in tomato fertigation. J Appl Hortic 8(1):8–11

Fontes PCR, Araujo C (2007) Adubação nitrogenada de hortaliças: princípios e práticas com o tomateiro. Editora UFV, Viçosa

Fontes PCR, Ronchi CP (2002) Critical values of nitrogen indices in tomato plants grown in soil and nutrient solution determined by different statistical procedures. Pesqui Agropec Bras 37(10):1421–1429

Fontes PCR, Braun H, Busato C (2010) Economic optimum nitrogen fertilization rates and nitrogen fertilization rate effects on tuber characteristics of potato cultivars. Potato Res 53(2):1–13

Goffart JP, Olivier M, Frankinet M (2008) Potato crop nitrogen status assessment to improve N fertilization management and efficiency: past-present-future. Potato Res 51(3–4):355–383

Henke L, Breustedt G, Sieling K, Kage H (2007) Impact of uncertainty on the optimum nitrogen fertilization rate and agronomic, ecological and economic factors in an oilseed rape based crop rotation. J Agric Sci 145(5):455–468

Meyer RD, Marcum DB (1998) Potato yield, petiole nitrogen, and soil nitrogen response to water and nitrogen. Agron J 90(3):420–429

Neeteson JJ, Wadman WP (1987) Assessment of economically optimum application rates of fertilizer N on the basis of response curves. Fertil Res 12:37–52

Nelson LA, Voss RD, Pesek J (1985) Agronomic and statistical evaluation of fertilizer response. In: Engelstad OP (ed) Fertilizer technology and use, 3rd edn. S.S.S.A, Madison

Nunes JCS, Fontes PCR, Araújo EF (2006) Potato plant growth and macronutrients uptake as affected by soil tillage and irrigation systems. Pesqui Agropec Bras 41(12):1787–1792

Olness A, Evans SD, Alderfer R (1998) Calculation of optimal fertilizer rates: a comparison of three response models. J Agron Crop Sci 180(4):215–222

Rajsic P, Weersink A (2008) Do farmers waste fertilizer? A comparison of ex post optimal nitrogen rates and ex ante recommendations by model, site and year. Agric Syst 97(1–2):56–67

Rodrigues MA, Coutinho J, Martins F, Arrobas M (2005) Quantitative side dress nitrogen recommendations for potatoes based upon crop nutritional indices. Eur J Agron 23(1):79–88

Schepers JS, Francis DD, Vigil M, Below FE (1992) Comparison of corn leaf nitrogen concentration and chlorophyll meter readings. Commun Soil Sci Plant Anal 23(17/20):2173–2187

Seefeldt SS, Jensen JE, Fuerst EP (1995) Log-logistic analysis of herbicide dose–response relationships. Weed Technol 9(2):218–227

Shrestha RK, Cooperband LR, Macguidwin AE (2010) Strategies to reduce nitrate leaching into groundwater in potato grown in sandy soils: case study from North Central USA. Am J Potato Res 87(3):229–244

Silva MCC, Fontes PCR, Miranda GV (2007) Modelos estatísticos para descrever a produtividade de batata em função da adubação nitrogenada. Hortic Bras 25(3):360–364

Wu J, Wang D, Rosen CJ, Bauer ME (2007) Comparison of petiole nitrate concentrations, SPAD chlorophyll readings, and QuickBird satellite imagery in detecting nitrogen status of potato canopies. Field Crop Res 101(1):96–103

Zebarth BJ, Drury CF, Tremblay N, Cambouris AN (2009) Opportunities for improved fertilizer nitrogen management in production of arable crops in eastern Canada: a review. Can J Plant Sci 89(2):113–132

Zimmermann FJP, Conagin A (1986) Ajuste de modelos polinomiais do 2° grau em pesquisas com fertilizantes. Pesqui Agrop Bras 21(9):971–978

Chapter 24
Effect of Soil Compaction Alleviation on Quality and Yield of Potato

Carlos Francisco Ragassi, Carlos Alberto Lopes, and Ítalo Moraes Rocha Guedes

Abstract Soils of tropical environments are exposed to the process of degradation when the natural ecosystems are converted to unsustainable agro-ecosystems. Soil compaction, which is part of that degradation process, is widely found in Brazilian cultivated areas, especially in the potato fields in which the intensive soil revolving operations promote the disruption of aggregates. Furthermore, potato is very sensitive to the problems associated to the soil compaction, such as the low development of roots and the diseases that are favored by high humidity in the tuber. In this way, a tillage system based on deep alleviation of soil compaction for the potato crop is presented in this chapter. This system, named "deep soil loosening" was evaluated in two Brazilian producing regions in Bahia and São Paulo States and provided higher yields, which were associated to the improvement of soil physical attributes. Thus, the "deep soil loosening" system is suggested as an alternative to improve potato production in compacted areas and, also, as a tool for promote the recuperation of soils damaged by compaction.

24.1 Introduction

Viewed from a broad sense, not only in the potato production system, but in agricultural systems in general, land degradation is evident in Brazil and the damages occur from the moment a natural ecosystem is converted to an unsustainable agro-ecosystem. For example, the great cycles of sugarcane and coffee that occurred during the colonial period depended on the rainforest natural soil fertility and on the migration to new areas once such natural fertility was depleted (Lopes and Guilherme 2007). During this period, the coffee production system in Brazil

C.F. Ragassi (✉) • C.A. Lopes • Í.M.R. Guedes
Research, Development and Innovation Division, Embrapa Hortaliças, Caixa Postal 218, CEP 70359-970 Brasília, DF, Brazil
e-mail: cragassi@gmail.com; clopes@cnph.embrapa.br; italo@cnph.embrapa.br

Z. He et al. (eds.), *Sustainable Potato Production: Global Case Studies*, DOI 10.1007/978-94-007-4104-1_24, included deforestation, burning of wood and cultivation on deforested areas. The natural soil fertility could support the production system for 20 or 30 years, before the productivity declined below the interest. Afterwards, those areas were generally abandoned and left to extensive farming (Lopes and Guilherme 2007).

Soil degradation after conversion of native ecosystems into agricultural lands is directly linked to the loss of organic matter (Mielniczuk 2008). In tropical intact ecosystems, the organic matter decomposition rate is higher than in ecosystems of temperate regions. However, tropical and temperate forest soils present similar organic matter content due to the larger production and addition of residues in the tropics (Sanchez 1976). Consequently, the organic matter content of tropical soils subjected to unsustainable agricultural use declines much more rapidly than in temperate regions. In cold regions, half the original stock of organic matter is lost within 50 and 100 years of cultivation, whereas in regions with hot and humid climate, similar losses occur in periods of merely 10–15 years (Bayer and Mielniczuk 2008).

Thus, the amount of crop residues needed to maintain stocks of soil organic matter in tropical regions is much higher than that of temperate regions (Bayer and Mielniczuk 2008) and, as the constant supply of organic matter to soil is essential to feed the dynamic processes that maintain its aggregation, low supply of organic matter results in the loss of aggregation (Silva and Mendonça 2007), making the soil more susceptible to compaction, a process that damages soil structure, resulting in higher density with smaller pore space (Thornton et al. 2008).

By this way, maintaining and increasing the soil organic matter content are the basic conditions to avoid the loss of soil fertility and the degradation of its structure, and, consequently, to provide a productive agro-ecosystem (Lopes and Guilherme 2007), especially in humid tropical conditions, where the cycling of organic matter is faster.

Nevertheless, soil management in vegetable crops in Brazil often provides insufficient organic matter to soils. The organic matter additions usually consist of the own vegetable crop residues and animal manure, materials that are easy to decompose and present a transient effect on soil structure. Furthermore, conservation systems for vegetable crops are not common in Brazil, especially for crops that require periodic soil tillage. The impact of equipments used in tilling operations damages the natural soil structure, which consists of clusters of particles, and then the disaggregated particles are deposited in the sub-surface, forming a dense layer just below the volume of plowed soil (Carvalho Júnior 1995). In Brazil, this layer is commonly known as "pé-de-grade" (pressure packing action of discs) and occurs widely in agricultural areas.

Besides the processes involved in soil tillage, traffic also contributes to the compaction of cultivated soils. The tire air pressure is normally set to 1.70 atm (25 psi), whereas only 0.27 atm (4 psi) would be enough to compact the soil (Thornton et al. 2008).

The increase in soil density due to application of pressure, such as the machine traffic pressure (wheel-track packing), depends on both nature and conditions of soil at the time that the pressure is exerted. In soils with low water contents, part of the force applied is consumed by friction among the soil particles. In wetter soils, however, the arrangement between particles during compaction is facilitated by the presence of water. Thus, the pressure exerted by the equipment is even more damaging when the mechanized operations are carried out on excessively wet soil, which often occurs in areas where potato is cultivated in Brazil, mainly due to the difficulty to manage, in large areas, the tillage operations for planting at the beginning of the rainy season.

Superficial or sub-superficial compaction negatively affects water infiltration into the soil due to the reduction of the total porosity and, especially, macroporosity. When the rainfall intensity exceeds the water infiltration capacity into the soil profile, the excess of water can run superficially and accumulate in depressions. The runoff becomes more severe with a higher terrain slope, because the high velocity of water carries mineral and organic particles of soil out of the agro-ecosystem. For example, a study carried out in Paraná State, southern Brazil, between 1985 and 1987 in a clayey Inceptsol, revealed a loss of over 300 Mg ha^{-1} in potato cultivated on a downhill landscape (Farias and Nazareno 2009).

In addition to the productivity losses caused by erosion, which affects mainly the superficial layers and, therefore, more fertile soil, considerable environmental degradation occurs as a consequence of eutrophication and siltation of water bodies, affecting an area larger than the compacted agricultural field.

In flatter areas, which should be preferred for potato cultivation, the increased density of the compacted soil leads to accumulation of water in low areas and, consequently, to poor soil aeration in those sites. Besides compromising the respiration of the root system, excessive water can facilitate the reduction of certain metallic chemical elements such as iron and manganese, which, depending on the levels in the soil and the sensitivity of the cultivar, can cause toxicity symptoms. Some soil diseases may also be favored by the flooding caused by compaction.

The degradation of soil structure in the topsoil, however, varies according to the characteristics of each soil class. Oxisols rich in kaolinite are more susceptible to degradation of the structure by mechanical stress than Oxisols rich in gibbsite, for example. In some potato-producing areas, cultivation on Inceptsols is common, especially on the sandy ones, which are preferred due to the more favorable texture. The presence of sand in higher proportion may delay or prevent soil compaction at some degrees, but the inadequate soil management can even overcome it. Besides the already cited problems from the increased soil density, the position in which Inceptisols commonly occur naturally favors higher erosion rates, which may be exacerbated by compaction. Soils with argillic B horizon (more clay-rich), such as Ultisols and Alfisols, have a natural tendency to present increased density. In such cases, the management to avoid compaction should be even more rigorous.

Although, in principle, more sandy soils, such as the Entisols, are less prone to compaction, they may be more prone to erosion and loss of nutrients. For both cases, the appropriate management of soil organic matter is probably the most recommended measure, because it can improve both the soil structure and the retention of nutrients.

24.2 Soil Compaction in the Potato Crop Fields

24.2.1 Damages to the Potato Crop

Potato (*Solanum tuberosum* L.) is the most cultivated vegetable in Brazil, with an area of 145,000 ha, and generates the biggest number of jobs in the vegetable business (1.304 million). Nevertheless, in the past, the area planted to this crop was bigger. In 2003, for example, it reached 152,000 ha, with a production of 3.089 million tons. In 2008, the total harvest amounted to 3.677 million tons, from 145,000 ha. In 2009, the area receded to 141,000 ha and the production volumes also kept pace with the smaller area, and reached 3.434 million tons (Corrêa 2010).

Potato is cultivated from northeast to south of Brazil (Fig. 24.1), which represents a high diversity of soil and climate conditions. In Brazil, the cultivation of potato is a highly soil-disturbing activity, because intensive soil revolving is adopted

Fig. 24.1 Potato producing regions in Brazil; the map comprises latitudes between approximately 13 and 34°S (Adapted from Wrege et al. 2004)

Fig. 24.2 Crooked taproot and superficial root growth as a consequence of soil compaction

in order to provide the production of tubers with impeccable appearance of skin, which is required by the Brazilian market (Farias and Nazareno 2009). Furthermore, the soil tillage for potato includes the hilling operation, which consists of adding a layer of soil to the base of the plant in order to stimulate the formation of tubers and to protect them against the greening caused by direct light incidence and insect attack; it also helps to eliminate weeds and to incorporate the nitrogen topdressing fertilization.

The southern Minas Gerais State in the Southeast region of Brazil is one of the most traditional potato-producer in the Country. Mesquita et al. (1999) observed during the last decade of the past century a reduction in the potato yields per unity of inputs applied. The authors attributed this reduction to the intensive utilization of the areas, specifically to the damage of soil attributes, such as compaction, nutritional disbalance and chemical residues. For this region, Filgueira (1999) was also emphatically not recommending potato cultivation on hill-slope areas, enhancing the difficulties faced by the local farmers. In the same way, the author highlighted the advantages of the utilization of the Brazilian savannah plane deep soils, even though these soils have to suffer acidity correction and receive heavy fertilization. Nevertheless, a few more than 10 years after the recommendation of Filgueira (1999), Minas Gerais is, nowadays, the greatest potato-producer State in Brazil and it is due to the expansion of potato cultivation on the region known as "Triângulo Mineiro" (Triangle of Minas Gerais State), an hill-slope area, with consequent problems to that State.

Soil compaction affects the growth of potato in many ways. As the movement of water and air is restricted in compacted soils, its availability to the plant is limited. The roots do not grow well and do not penetrate the soil and, thus, remain located superficially (Fig. 24.2). In addition, the plant needs to use more energy for the growth of roots and tubers, which reduces the energy available for growing other organs. The shallow root system limits the ability to obtain nutrients and resistance to periods of drought stress and, consequently, the plants become less vigorous and the quality and yield of tubers are negatively affected (Thornton et al. 2008).

Observed in a series of experiments in England (Stalham et al. 2007), soil compaction delayed emergence, reduced the rate of leaf area expansion, reduced the cycle of potato and restricted light interception, factors, which combined, significantly reduced the tuber yield. Root density and maximum rooting depth were reduced, particularly where compaction was present in shallower layers. In the intensively revolved furrows, the penetration rate of potato roots was 2 cm day^{-1} and halved in places where resistance to penetration was 1.5 MPa, measured using a penetrometer. In a sample of 602 commercial fields in that country, two thirds of the fields showed resistance to penetration higher than 3 MPa and root growth rates inferior to 0.2 cm day^{-1} in the first 55-cm soil layer. Thus, the rooting depth of potato, due to physical restriction, was less than ideal, leading to inefficient use of water and nutrients (Stalham et al. 2007).

By this way, it appears that the root system of potato is fragile, and therefore requires compaction-free soils. The presence of soil clods and, especially, compacted soil layers, cause tuber deformation, increase the frequency of flooding events – and its associated rots – and, also, complicate or even make it impossible to harvest mechanically.

Soil compaction has direct and indirect effects on the potato plant. Among the indirect effects, the principal is reduced oxygenation of the soil that, coupled with a high temperature, causes collapse and death of roots. Moreover, deficient oxygenation within tubers, either by low levels of oxygen in soil or by improper post-harvest stacking of tubers, results in the formation of potato black heart. This disorder occurs by increased respiration at high temperatures accompanied by abnormal enzymatic reactions: the interior of the tuber receive insufficient amount of oxygen, which causes cell death, with the formation of melanin, a brown pigment characteristic of this disorder (Wale et al. 2008; Thornton et al. 2008).

The most striking effect of the excess of water in soil, induced by compaction, is the occurrence of diseases caused by pathogens with high dependence of soil moisture. Except common scab (*Streptomyces* spp.), which is favored by relatively dry soil conditions during the beginning of the tuberization phase, potato diseases are favored by high humidity. Under low temperature, besides late blight (*Phytophthora infestans*), which attacks mainly the shoot but can also cause injury to tubers, reports of white mold (*Sclerotinia* spp.) epidemics are also common (Gudemestad 2008). The most serious, however, is the combination of high temperatures with high humidity. Crops cultivated in the hot and rainy periods are subject to significant losses caused by bacterial wilt (*Ralstonia solanacearum*), nematodes (*Meloidogyne* spp.) and blackleg or soft rot (*Pectobacterium* spp. and *Dickeya* spp.) (Wale et al. 2008).

Water availability in soil is one of the most important factors for the spread, multiplication and colonization of *R. solanacearum*. Actually, water availability determines the incidence of bacterial wilt in the field. This finding is old, in the early nineteenth century, producers in India mentioned the disease as "rasa", or "disease of moisture". Thus, well-drained and properly-irrigated soils result in less disease attack.

Although soil compaction is often associated with a higher intensity of diseases caused by soil borne pathogens, there are few scientific studies that address the effect of drainage on expression of potato diseases. Recently, Copas et al. (2008) studied the influence of soil compaction and subsoiling in the occurrence of "pink eye" in central Wisconsin, USA. The "pink eye" is a potato disorder which causes the formation of pink pigmentation on the buds (eyes) of tubers. Although this disorder, by itself, does not cause much damage, it is the forerunner of many other diseases that damage the tubers. The "pink eye", first attributed to infection by *Pseudomonas fluorescens*, does not have a clearly identified cause. It is known, however, that this disorder is associated to high soil moisture. According to Secor and Gudmestad (2001), the most favorable condition for the appearance of "pink eye" is the presence of free water or high soil moisture followed by environmental conditions that favor soil heating. Furthermore, some studies showed that subsoiling 75–100 cm significantly reduced the soil temperature and the incidence of "pink eye".

In the 1990s, Leach et al. (1993) studied the effect of deep tillage with moldboard and chisel plow in the incidence and severity of *Rhizoctonia* and in the population of the pathogen (*Rhizoctonia solani*) in soil. Chisel plowing presented the greatest effect in reducing the disease. The authors concluded that, for this pathosystem and in the growing conditions of Maine, USA, deep plowing with moldboard plow may even have positive effect in structuring the soil to the depth of development of the potato plant. However, this tillage strategy facilitates the occurrence of a soil compacted layer and revolves propagules of inoculum from the soil surface, placing them closely to the plant organs which are susceptible to infection (potato-seed, sprouts, roots, stolons and tubers). In contrast, the chisel plow does not invert the soil profile and, so, crop residue remains on top. Chisel plow also promotes the mixture of a part of these residues in the lower layers of soil, which favors its decomposition.

Another classic direct association between soil moisture and potato diseases is the complex blackleg and soft rot, caused by pectolytic bacteria, especially of the genera *Pectobacterium* and *Dickeya*. These bacteria occur in all soil types and are considered secondary pathogens. For the pathogen to multiply in the affected tissues, such as tubers, and then trigger the disease, it needs to overcome barriers that hinder the process of infection and colonization, such as availability of water and nutrients and plant resistance. Especially under high temperatures, the tuber is exposed to an anaerobic condition when a water film formed on it prevents the renewal of the oxygen consumed by respiration. This anaerobic condition hinders processes which are associated to the reaction of plants to pathogens, such as formation of substances (phytoalexins, phenols and free radicals) that confer natural resistance to pathogens. In addition, anaerobic condition inhibits suberisation and lignification of tissues, which would form barriers that suppress the development of lesions (Pérombelon 2002). The anaerobic condition also causes lenticelose, which is an exacerbated growth of the lenticels that turns them into ports of entry for pathogens.

24.2.2 *Results of the Deep Soil Loosening for the Potato Crop*

Research results show positive effects of several techniques that remove and/or avoid compaction on the potato crop. Examples of these techniques are zero traffic (Young et al. 1993), precision tillage (Bishop and Grimes 1978; Sojka et al. 1993), soil physical correction up to 80 cm depth for loosening deep compacted layers (Labuschagne and Joubert 2006) and the use of tropical grasses as green manures highly effective for promoting soil aggregation (Silva and Mielniczuk 1997a, b). In this context, it is important to consider that crop rotation, which can be used to reduce the incidence of diseases (Larkin et al. 2011), can also be used to reduce the soil compaction as well (Silva and Mielniczuk 1997a, b).

Zero traffic is a promising strategy to minimize soil compaction. This system restricts all traffic to certain streets, eliminating traffic on the planting rows (Young et al. 1993). It is obtained through the preparation of larger seedbeds, or experimentally by changing the gauge of conventional tractors to 2.8 m (Young et al. 1993). Through this system, yield gains of cereals, potatoes and grasses were obtained (Dickson et al. 1992; Douglas et al. 1992).

With the use of zero traffic, potato yield was 18% higher than in conventional systems due to higher soil porosity in the rainy season and lower soil penetration resistance in the dry season (Dickson et al. 1992). During harvest, the conventional system produced 34% more soil clods, complicating the mechanic harvest. Leaf area, interception of solar radiation and dry matter of leaves, stems and tubers were higher in zero traffic (Young et al. 1993), which explained the increased yield.

Another important benefit of zero traffic is the higher number of days which soil allows mechanized operations. According to Young et al. (1993), the soil physical conditions in the zero traffic system allowed the beginning of mechanized activities 5 days before the conventional system, even though the soil submitted to zero traffic was more humid. This is of considerable importance in the field, in which mechanized operations are frequently carried out under inadequate moisture due to the small interval of dry days during the rainy season.

The "precision tillage" (Carter and Tavernetti 1968) is also a promising technique to reduce compaction in potato crop and consists of chiseling soil 50–60 cm depth exactly below the plant row. The results of this application are greater than those obtained with deep tillage performed randomly in the field (Bishop and Grimes 1978).

By this way, in order to provide adequate soil physical, biological and chemical conditions for growing potatoes, MAFES, a Brazilian private company of agricultural technology in partnership with the University of São Paulo, Brazil, developed the concept named "deep soil loosening", which was studied in two potato growing areas in Brazil. The "deep soil loosening" system comprises the use of deep-action implements, zero traffic, precision tillage and physical correction of soil to 80 cm depth.

In Brazil, a study carried out in a commercial field at 13°17'S, 41°24'W and 1,100 m height in the county of Ibicoara, a region also known as "Chapada Diamantina", Bahia State, Mitsuiki (2006) observed that the cultivation of potatoes with "deep soil loosening" provided positive results on soil physical properties such as penetration resistance and water infiltration, increasing yield and reducing damage to tubers.

In that research, the control treatment was prepared conventionally with two disking operations to reduce crop residues to small pieces and to incorporate lime. After 2 months, when the plant residues were dry, two plowing and two subsoiling operations were carried out to 20 cm depth. In the deep loosening treatments, the operations started with a chopper, which crushed the crop residues followed by two mechanic rotary hoes operation, which revolved soil and incorporated plant residues up to 45 cm depth (Mitsuiki 2006).

Finally, a subsoiling operation was carried out precisely in the center of the seedbed to a depth of 80 cm, removing the compacted layers up to this depth. The results obtained by Mitsuiki (2006) demonstrated higher yield of tubers with deep soil loosening (50.6 Mg ha^{-1}) in comparison to conventional tillage (43.0 Mg ha^{-1}). The author related this gain to a larger volume of soil that deep tillage provided for the root system development.

However, despite the positive effects of the mechanic soil loosening, there is evidence that these effects present short duration, since the reconsolidation of the loosened soil increases with the accumulated volume of rainfall (Busscher et al. 1995, 2002). Thus, in a subsequent study, the association between mechanical soil loosening and the cultivation of green manures with high potential of producing roots was evaluated (Ragassi et al. 2009).

That study was proposed because roots and rhizosphere provide different effects on soil aggregation. Roots intertwine with soil particles and release exsudates that result in physical, chemical and biological modifications that influence soil aggregation. The stability of aggregates is higher in the rhizosphere and suffers the effect of quantity, distribution and rates of deposition and renewal of roots. The rhizosphere also hosts a large population of micro and macro-organisms, contributing to the increase in organic carbon and soil aggregation. Furthermore, the mucilaginous compounds produced by roots, such as polygalacturonic acid, can stabilize the aggregates by increasing the strength of the connection between the soil particles and by reducing the wetting speed of the aggregates, which promotes their preservation (Bronick and Lal 2005).

Other authors also consider the role of roots in the formation of stable soil aggregates through the supply of organic residues for decomposition (Oades 1978), exsudation of organic substances, approximation of particles and microaggregates due to dehydration caused by water absorption (Bradfield 1937; Oades 1978) and physical involvement of microaggregates in soil, especially in the case of grasses (Tisdall and Oades 1979).

The stability of soil aggregates is highly correlated to the level of organic carbon in the soil in areas under forest and under cultivation (Silva and Mielniczuk 1997a, b). However, in areas with perennial grasses, the organic carbon content was not sufficient to explain the high number and stability of aggregates, which indicates that another factor, besides the organic matter content, is responsible for the efficient aggregation found in soils cultivated with these grasses.

Other studies emphasize the action of plant roots in the formation and stabilization of soil aggregates (Bradfield 1937; Tisdall and Oades 1979). However, many researchers have emphasized that within the universe of plants, perennial grasses have shown greater benefits (Bradfield 1937; Carpenedo and Mielniczuk 1990; Paladini and Mielniczuk 1991; Tisdall and Oades 1979; Rizzo 2000). These benefic effects are attributed mainly to the high density of roots (Haynes and Beare 1997), which promotes the approach of particles by the constant absorption of water from the soil profile, the periodic renewal of the root system and the uniform distribution of exsudates through the soil, which stimulates microbial activity, whose products act in the formation and stabilization of aggregates. Moreover, the high productive potential of plants with C_4 metabolism should take into consideration in tropical climates (Webster and Wilson 1980), which certainly improves their capacity to increase the level of soil organic matter.

Thus, Ragassi et al. (2009) evaluated, in Piracicaba, São Paulo State, southeastern Brazil, at 22°42'S, 47°38'W and 569 m height, deep loosening associated to the cultivation of three different grasses, *Panicum maximum* cv. Tanzania, *Brachiaria brizantha* cv. Marandu and corn hybrid 'Dekalb 191', comparing them to conventional tillage (heavy disking to 20 cm depth) associated to the cultivation of the same corn hybrid.

The research was carried out in a clayey Alfisol, which, in water content corresponding to the field capacity, penetration resistance exceeded 2.0 MPa, which is restrictive for the potato root development (Stalham et al. 2007). The experiment consisted of four treatments: the succession maize (*Zea mays* 'Dekalb 191') – potato (*Solanum tuberosum* cv. Atlantic) in shallow soil tillage (control), whose soil profile can be seen in Fig. 24.3. In the other three, the deep soil loosening system was used (Fig. 24.4) with different crop sequences: (1) *Panicum maximum* cv. Tanzania – potato, (2) *Brachiaria brizantha* cv. Marandu – potato, and (3) maize – potato, being the latter, different from the control only by the tillage system.

The soil preparation before sowing the grasses started with a chopper, which crushed the crop residues, followed by two mechanic rotary hoe operations in the deep soil loosening treatments. The first rotary hoe operation revolved soil and incorporated plant residues to 20 cm depth and demarcated the double seedbed with 1.80 m width. The second operation revolved soil and incorporated the plant residues to 40 cm depth. These operations were carried out in all treatments, except the control, in which the second rotary hoe operation, to 40 cm depth, was not carried out. In the control treatment, between the operation of the chopper and the rotary hoe, a heavy disking was carried out to 20 cm depth. Afterwards, in the deep soil loosening treatments, a subsoiling operation in the center of the double seedbed was carried to 70 cm depth, dividing the bed into two rows, and, then, the seeding of grasses was performed. Four months later, the grasses were chopped and let on the soil surface for 45–60 days. Finally, the same tillage operations were carried out and the semi-mechanized planting of potatoes was done.

The soil penetration resistance (SPR) was evaluated 60 days after the potato planting, through impact penetrometer, model IAA / Planalsucar-Stolf (Stolf et al. 1983). In the center of each single row, SPR showed an increase with the depth

Fig. 24.3 Potato crop in a soil prepared by heavy disking to 20 cm depth (shallow tillage system)

Fig. 24.4 Soil profile in a deep soil loosening system

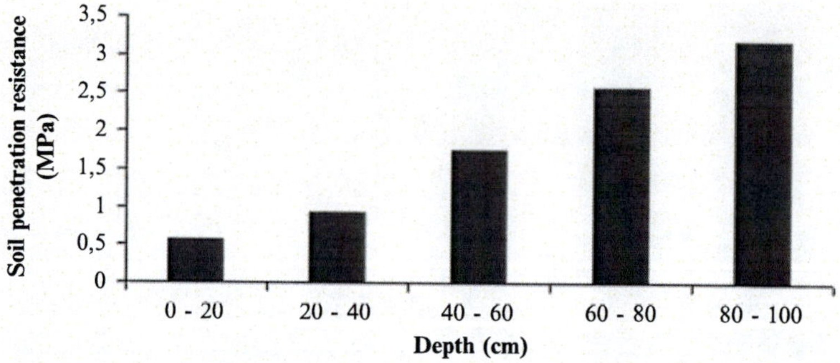

Fig. 24.5 Soil penetration resistance below potato crop ridge line (average among treatments). CV = 11.8% (Adapted from Ragassi et al. 2009)

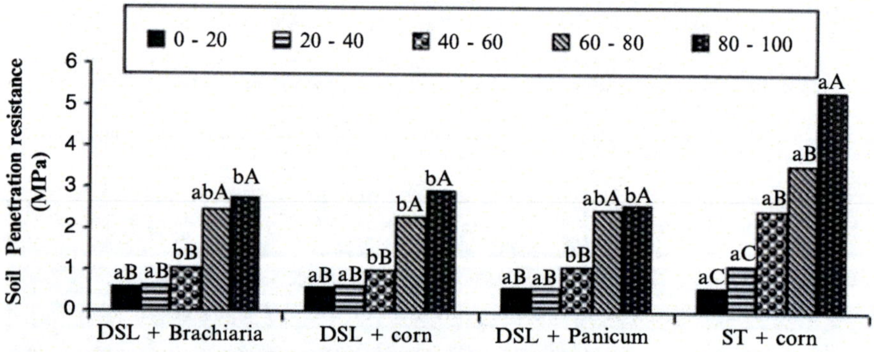

Fig. 24.6 Soil penetration resistance in the subsoiled region at the potato bed center, deep soil loosening (DSL) and shallow tillage (ST) were the assessed soil management methods, followed by the name of the crop cultivated prior to potato. Different small letters indicate statistical difference (Duncan $p < 0.05$) among treatments at the same soil layer (cm) and different capital letters, among soil layers at the same treatment. CV = 18.3% (Adapted from Ragassi et al. 2009)

(Fig. 24.5) which was not influenced by the treatments. A tendency of increasing with depth was also observed for the (SPR) in the center of the double seedbed (between two rows, where the subsoiling operation to 70 cm depth was carried out). However, significant increases occurred only from the 40–60 cm layer in the shallow tillage and from the 60–80 cm layer in the deep soil loosening treatments (Fig. 24.6).

Considering 1.5 MPa (Stalham et al. 2007) as critical for the potato root development, in the row, up to the 20–40 cm layer, no limitation for root growth was observed in any treatment. In the following layer (40–60 cm), all treatments showed higher values than the critical, which can be attributed, in part, to the natural attributes of that Alfisol. In the center of the seedbed (subsoiled in the deep soil loosening treatments), the shallow tillage provided limitation to root growth from the 40–60 cm layer, whereas in deep soil loosening treatments, this limitation only occurred from the 60–80 cm layer. In the planting rows, the penetration resistance in the 20–40 cm layer did not differ among treatments, although the shallow tillage was done only up to 20 cm depth. This suggests that the pressure exerted by the tractor wheel in the lateral area to the bed during the mechanized operations potentiated the increase of density of the revolved soil. It suggests that, for greatest benefit of the deep soil loosening, this operations should be done as far as possible from the traffic line, which should be fixed in the area, according to the zero traffic method (Dickson et al. 1992; Young et al. 1993).

Benefits to the potato crop provided by improving soil physical conditions in the planting row have been previously reported, such as the increase of productivity of 10% (from 53.9 to 59.8 Mg ha^{-1}) (Bishop and Grimes 1978) or 14% (Dickson et al. 1992) and, although in this study the treatments did not influence RSP in the row, deep soil loosening increased the total yield of tubers (from 28.3 Mg ha^{-1} in the shallow tillage with the cultivation of corn to 32.9 Mg ha^{-1} in the deep soil loosening, also with corn).

The higher yield achieved with deep soil loosening is probably related to lower levels of RSP in the central region of the seedbed, especially in the 40–60 cm layer. Another study also demonstrated the ability of potato to explore a compaction-free region of soil located outside the row (Young et al. 1993). In this case, the space between wheels of each axis (gauge) of the tractor was modified to 2.8 m, so that, the seedbed could fit three rows, spaced at 0.81 m. These authors found no difference between plants from the sidelines (adjacent to traffic lines) and those evaluated on the central line, which was not influenced by the wheel compaction. According to the author, this was due to the exploitation of the soil in the center of the seedbed by roots of plants placed in the sidelines (Young et al. 1993).

The proportion (by mass) of tubers in the "Lower" class (diameter inferior to 4 cm) was statistically different between shallow tillage (5.1%) and deep soil loosening with maize (2.9%) and *Brachiaria* (2.2%). The proportion of tubers in the class "Commercial" (4–10 cm diameter) and "Superior" (diameter superior to 10 cm) showed no difference between treatments. The average values for the "Commercial" class ranged from 91.4% in deep soil loosening with *Brachiaria* to 83.4% in the shallow tillage with maize. For the "Superior" class, the values ranged from 11.4% in the shallow tillage to 5.3% in deep soil loosening with *Panicum*.

The potential of deep soil loosening for improving not only yield, but also the quality of the tubers was, thus, verified, which corroborated the results of other works, such as an increase of 4.6% (3.8 Mg ha^{-1}) on yield of "Superior" tubers, without increasing the overall productivity or an increase of 5.6 Mg ha^{-1} on the yield of "Superior" tubers, with an overall yield increase of only 4.2 Mg ha^{-1} (Sojka et al. 1993).

Finally, the research of Ragassi et al. (2009) showed that, especially with respect to the conditions of the studied soil (type and presence of compaction), the deep soil loosening system reduced soil penetration resistance in the center of the seedbed, especially in the 40–60 cm layer, which was associated to higher yields of potato, regardless the grass species used as green manure.

The deep soil loosening, however, only should be recommended after the evaluation of the soil profile, which can be carried out using a penetrometer, to identify where the soil layers of high penetration resistance are located. The depth of soil loosening, in this way, should be set to achieve those hard layers. This is necessary especially in the soils that have argillic B horizon. Also, the cost/benefit should be considered before recommending such mechanized operation, which tends to be a high investment.

24.3 Conclusion

Based on what is presented in this chapter, we suggest that, as soil compaction occurs frequently in Brazilian soils and probably in all tropical countries, the deep soil loosening technique may provide positive results in a wide range of soils and climate conditions in which potato is cultivated. In addition to possible economic gains due to the increase of yield and the higher proportion of "Superior" tubers, the environmental gain of recovering soils degraded by compaction must be considered. The recovery of the soil production capacity would reduce the crop migration and the necessity of deforesting new areas for cultivation. Furthermore, considering the relationship between the soil compaction and the incidence of diseases in potato, further researches aiming to determine the potential of the deep soil loosening technique as an auxiliary measure for the management of potato diseases is suggested.

Acknowledgments MAFES agricultural technology (www.mafes.net) and Agronomy School of the São Paulo University, ESALQ (www.esalq.usp.br), which, in association, developed and evaluated the deep soil loosening system presented in this chapter; CAPES (Coordenação de Aperfeiçoamento de Pessoal de Nível Superior) for scholarship from February, 2007 to March, 2008 and FAPESP (Fundação de Amparo à Pesquisa do Estado de São Paulo – process number 2007/05562-6) for scholarship and financial support from April to December, 2008.

References

Bayer C, Mielniczuk J (2008) Dinâmica e função da matéria orgânica. In: Santos GA, Silva LS, Canellas LP, Camargo FAO (eds) Fundamentos da matéria orgânica do solo: ecossistemas tropicais e subtropicais, 1st edn. Metrópole, Porto Alegre

Bishop JC, Grimes DW (1978) Precision tillage effects on potato root and tuber production. Am Potato J 55:65–71

Bradfield R (1937) Soil conservation from the view point of soil physics. Agron J 29:85–92

Bronick CJ, Lal R (2005) Soil structure and management: a review. Geoderma 124:3–22. doi:10.1016/j.geoderma.2004.03.005

Busscher WJ, Edwards JH, Vepraskas MJ, Karlen DL (1995) Residual effects of slit tillage and subsoiling in a hardpan soil. Soil Tillage Res 35:115–123

Busscher WJ, Bauer PJ, Frederick JR (2002) Recompaction of a coastal loamy sand after deep tillage as a function of subsequent cumulative rainfall. Soil Tillage Res 68:49–57

Carpenedo V, Mielniczuk J (1990) Estado de agregação e qualidade de agregados de latossolos roxos submetidos a diferentes sistemas de manejo. Rev Bras Ciênc Solo 14:99–105

Carter LM, Tavernetti JR (1968) Influence of precision tillage and soil compaction on cotton yields. Am Soc Agric Eng Trans 11:65–67

Carvalho Júnior I (1995) Estimativas de parâmetros sedimentológicos para estudo de camadas compactadas e/ou adensadas em latossolo de textura média, sob diferentes usos. Dissertation, Federal University of Viçosa, Viçosa, Minas Gerais State, Brazil

Copas ME, Bussan AJ, Drilias MJ, Charkoski A (2008) Influence of compaction and subsoil tillage on soil conditions and pink eye. Am J Potato Res 85:342–352. doi:10.1007/s12230-008-9038-y

Corrêa S (2010) Anuário brasileiro de hortaliças. Editora Gazeta Santa Cruz, Santa Cruz do Sul

Dickson JW, Campbell DJ, Ritchie RM (1992) Zero and conventional traffic systems for potatoes in Scotland, 1987–1989. Soil Tillage Res 24:397–419

Douglas JT, Campbell NJ, Crawford CE (1992) Soil and crop responses to conventional, reduced ground pressure and zero traffic systems for grass silage production. Soil Tillage Res 24:421–439

Farias GS, Nazareno NRX (2009) Considerações gerais sobre sistemas de preparo do solo e tratos culturais. In: Nazareno NRX (ed) Produção orgânica de batata: potencialização e desafios, 1st edn. Instituto Agronômico do Paraná, Londrina

Filgueira FAR (1999) Práticas culturais adequadas em bataticultura. Informe Agropecuário 20:34–41

Gudemestad NC (2008) Potato health from sprouting to harvest. In: Johnson DA (ed) Potato health management, 2nd edn. APS Press, Saint Paul

Haynes RJ, Beare MH (1997) Influence of six crop species on aggregate stability and some labile organic matter fractions. Soil Biol Biochem 29:1647–1653

Labuschagne N, Joubert D (2006) Profile modification as a means of soil improvement: promoting root health through deep tillage. In: Uphoff N, Ball SA, Palm C (eds) Biological approaches to sustainable soil systems, 1st edn. CRC Press, Boca Raton

Larkin RP, Honeycutt CW, Griffin TS, Olanya OM, Halloran JM, He ZQ (2011) Effects of different potato cropping system approaches and water management on soilborne diseases and soil microbial communities. Phytopathology 101:58–67

Leach SS, Porter GA, Rourke RV, Clapham WM (1993) Effects of moldboard plowing, chisel plowing and rotation crops on the Rhizoctonia disease of white potato. Am Potato J 70:329–337

Lopes AS, Guilherme LRG (2007) Fertilidade do solo e produtividade agrícola. In: Novais RF, Alvarez VH, Barros NF, Fontes RLF, Cantarutti RB, Neves JCL (eds) Fertilidade do solo, 1st edn. Sociedade Brasileira de Ciência do Solo, Viçosa

Mesquita HA, Paula MB, Alvarenga MIN, Regina SM (1999) O cultivo da batata e a preservação ambiental. Informe Agropecuário 20:20–23

Mielniczuk J (2008) Matéria orgânica e a sustentabilidade de sistemas agrícolas. In: Santos GA, Silva LS, Canellas LP, Camargo FAO (eds) Fundamentos da matéria orgânica do solo: ecossistemas tropicais e subtropicais, 1st edn. Metrópole, Porto Alegre

Mitsuiki C (2006) Efeito de sistemas de preparo de solo e do uso de microrganismos eficazes nas propriedades físicas do solo, produtividade e qualidade da batata. Dissertation, University of São Paulo, Piracicaba, São Paulo State, Brazil

Oades JM (1978) Mucilages at the root surface. Eur J Soil Sci 29:1–16

Paladini FLS, Mielniczuk J (1991) Distribuição de tamanho de agregados de um solo Podzólico Vermelho-Escuro afetado por sistema de culturas. Rev Bras Ciênc Solo 15:135–140

Pérombelon MCM (2002) Potato diseases caused by soft rot erwinias: an overview of pathogenesis. Plant Pathol 51:1–12

Ragassi CF, Favarin JL, Shiraishi FA, Moita AW, Sako H, Melo PCT (2009) Efeito da descompactação profunda de solo na produção da cultura da batata. Horticultura Brasileira 27:484–489. doi:10.1590/S0102-05362009000400014

Rizzo LTB (2000) Indicadores da resiliência do latossolo vermelho escuro cultivado com citros e eucalipto em Itapetininga-SP: recuperação de um solo degradado pela compactação. Dissertation, University of São Paulo, São Paulo, São Paulo State, Brazil

Sanchez PA (1976) Properties and management of soil in the tropics. North Carolina State University, Raleigh

Secor GA, Gudmestad NC (2001) Disease of unknown cause: pink eye. In: Stevenson R, Loria R, Franc GD, Weingartner DP (eds) Compendium of potato diseases, 2nd edn. APS Press, Saint Paul

Silva IR, Mendonça ES (2007) Matéria orgânica do solo. In: Novais RF, Alvarez VH, Barros NF, Fontes RLF, Cantarutti RB, Neves JCL (eds) Fertilidade do solo, 1st edn. Sociedade Brasileira de Ciência do Solo, Viçosa

Silva I, Mielniczuk J (1997a) Ação do sistema radicular de plantas na formação e estabilização de agregados do solo. Rev Bras Ciênc Solo 21:113–117

Silva I, Mielniczuk J (1997b) Avaliação do estado de agregação do solo afetado pelo uso agrícola. Rev Bras Ciênc Solo 21:313–331

Sojka RE, Westermann DT, Kincaid DC, McCann IR, Halderson JL, Thornton M (1993) Zone-subsoiling effects on potato yield and grade. Am Potato J 70:475–484

Stalham MA, Allen EJ, Rosenfeld AB, Herry FX (2007) Effects of soil compaction in potato (*Solanum tuberosum*) crops. J Agric Sci 145:295–312. doi:10.1017/S0021859607006867

Stolf R, Fernandes J, Furlani Neto V (1983) Penetrômetro de impacto- IAA/ Planalsucar – STOLF (Recomendações para seu uso). STAB – açúcar, álcool e subprodutos 1:18–23

Thornton M, Stark J, Hopkins BG, Thornton RE (2008) Selecting and preparing the planting site. In: Johnson DA (ed) Potato health management, 2nd edn. The American Phytopathological Society, Saint Paul

Tisdall JM, Oades JM (1979) Stabilization of soil aggregates by the root systems of ryegrass. Aust J Soil Res 17:429–441

Wale S, Platt HW, Cattlin N (2008) Diseases, pests and disorders of potatoes: a colour handbook. Manson Publishing, London

Webster CC, Wilson PN (1980) Agriculture in the tropics. Longman Group, London

Wrege MS, Herter FG, Pereira AS, Caramori PH, Gonçalves SL, Braga HJ, Pandolfo C, Matzenauber R, Camargo MPB, Brunini O, Steinmetz S, Reisser Junior C, Ferreira JSA, Sans LMA (2004) Caracterização climática das regiões produtoras de batata no Brasil. Embrapa Clima Temperado, Pelotas

Young IM, Bengough AG, Mackenzie CJ, Dickson JW (1993) Differences in potato development (*Solanum tuberosum* cv. Maris Piper) in zero and conventional traffic treatments are related to soil physical conditions and radiation interception. Soil Tillage Res 26:341–359

Chapter 25
Developing Integrated Pest Management for Potato: Experiences and Lessons from Two Distinct Potato Production Systems of Peru

Jürgen Kroschel, Norma Mujica, Jesus Alcazar, Veronica Canedo, and Octavio Zegarra

Abstract Potato (*Solanum tuberosum* L.) is cultivated in diverse agroecosystems, which may harbor different insect pests; accordingly, potato farmers need to have appropriate site-specific pest control solutions. We developed Integrated Pest Management (IPM) strategies for potato production systems of the high Andes and the coast of Peru. This required considering all economically important pests and developing technological innovations to replace farmers' pesticide applications with equal efficacy. Examples are the use of plastic barriers that effectively prevent infestations of migrating Andean potato weevils (*Premnotrypes* spp.), the use of attract-and-kill for managing potato tuber moths [*Phthorimaea operculella* (Zeller), *Symmetrischema tangolias* (Gyen.)], or the rational use of insecticides to control flea beetles (*Epitrix* spp.) in the Andean highlands, or the leafminer fly [*Liriomyza huidobrensis* (Blanchard)] and the bud midge [*Prodiplosis longifila* (Gagne)] in the coastal lowlands. Moreover, the resilience of potato agroecosystems can be increased through augmentation strategies for natural enemies at the field level and inoculative biological control to recuperate species lost through the intensive use of pesticides. Potato IPM showed clear economic and ecological benefits at pilot sites. Strong public-private partnerships will be crucial for technology delivery, and well-trained field advisors are required to support the specific needs of farmers to adopt IPM.

J. Kroschel (✉) • N. Mujica • J. Alcazar • V. Canedo • O. Zegarra
Global Program – Integrated Crop and System Research,
International Potato Center (CIP) P.O. Box 1558, Lima 12, Peru
e-mail: j.kroschel@cgiar.org

Z. He et al. (eds.), *Sustainable Potato Production: Global Case Studies*,
DOI 10.1007/978-94-007-4104-1_25, © Springer Science+Business Media B.V. 2012

25.1 Introduction

The origin of the potato (*Solanum tuberosum* L.) is the High Andes in South America. Its global distribution began about 300 years ago: first to Europe and then to other parts of the world. Today, potato is grown in over 140 countries, more than 100 of them in tropical and subtropical zones, with China being the largest producer. Losses due to insect pests, which occur during the potato cropping and storage period, are often very high; a 16% of loss was estimated, worldwide (Oerke et al. 1994). Locally, if not routinely controlled, reductions in tuber yield and quality can be between 30% and 70% for various insect pests (Raman and Radcliffe 1992). Potato producing farmers in developing countries worldwide have to contend with about 20 insect pests. With some exceptions, a minimum of 2–4 pests often reach pest status requiring the application of control methods. Many pests have evolved in the center of origin of the potato, and hence farmers in the Andean region are confronted by a higher number of pest species than farmers in Africa or Asia. Some species such as the potato tuber moth [*Phthorimaea operculella* (Zeller)] and the leafminer fly [*Liriomyza huidobrensis* (Blanchard)] have become invasive and occur today as serious pests in many tropical and subtropical regions.

In developing countries, the one-sided use of often highly toxic insecticides is a common pest management practice. Farmers receive inadequate training on Integrated Pest Management (IPM) and there is often poor availability and quality of protection equipment and no collection services for used pesticide containers (Orozco et al. 2009; Arica et al. 2006). Serious health threats of pesticides to farmers, consumers and the environment have increased the interest in the search for safer control alternatives through the development of IPM. In addition, a rapid build-up of pest resistance to pesticides has frequently limited their use in potato (Llanderal-Cazares et al. 1996). However, farmers have often been reluctant to adopt IPM for many reasons, including the following: (i) IPM strategies are mainly targeted towards, or available against, single pest species and do not consider all pests in a specific agroecosystem (Trumble 1998; Horne and Page 2008); (ii) Promising IPM technologies are not available on the market or are too expensive; (iii) IPM technologies are not as immediately effective as pesticides, which are relatively cheap, and they are too complex to apply. Farmers can more easily rely on simple and proven pesticide applications (Daxl et al. 1994). Regular insecticide applications are encouraged, even if no pests have been observed, to avoid any risks; (iv) IPM is not understood or applied as a system approach; for this, experienced IPM field advisors would be crucial to support farmers in their IPM decision making. A survey by Horne et al. (1999) showed that the IPM adoption rates of Australian potato growers varied from 0% to 100% depending on the presence or absence of an IPM advisor. Similar observations have been made by Alston and Reading (1998).

By applying a holistic framework for the development of IPM, the objective of our research was to develop practical, economic and ecological solutions to pest management, applicable to resource-poor farmers in the Andean highlands and coastal region of Peru.

25.2 Pest Problems and Management in Andean Highland and Coastal Lowland Potato Production Systems of Peru

25.2.1 *Andean Highlands*

In the Peruvian Andean highlands, potato is mainly produced by small-scale farmers at altitudes between 2,800 and 4,200 m from October to May under rain-fed conditions with an annual precipitation of some 600 mm. Potato cultivation is severely constrained by many pest problems, but farmers' main response is to control infestations of Andean potato weevils by applying hazardous class Ia and Ib (e.g., carbofuran, metamidophos) insecticides (Orozco et al. 2009). Andean potato weevils belong to the genera *Premmnotrypes, Rhigopsidius,* and *Phyrdenus*. With 12 species, the genus *Premnotrypes* is the most important and the most widely distributed. *P. suturicallus* Kuschel occurs mainly in the central highlands, *P. vorax* (Hustache) in the northern and *P. latithorax* (Pierce) in the southern highlands of Peru (Alcazar and Cisneros 1999). While Andean potato weevils represent the only major biotic insect pest problem at altitudes above 3,800 m, in inner Andean valleys, other important pests are the potato tuber moth (*P. operculella*) and the Andean potato tuber moth [*Symmetrischema tangolias* (Gyen.)]; both pests are native to South America and are mainly controlled by farmers in potato storage (Keller 2003). Occasional pests are flea beetles (*Epitrix* spp.), infestations of which are kept under control by insecticide applications for the Andean potato weevil.

25.2.2 *Coastal Lowlands*

The Peruvian coastal region, at altitudes below 500 m, is characterized by an extremely arid subtropical desert climate, with a precipitation of less than 50 mm yearly. Here, the Cañete Valley is one of the most intensive cropping regions, covering some 23,000 ha, which are mostly cultivated by small-to-medium-scale farmers, who produce for the local and export markets. Among the many agricultural and horticultural crops produced in this valley, the potato is an important crop during the winter cropping season. The leafminer fly (*L. huidobrensis*) is a serious pest on potato and many other vegetable crops, which farmers try to control by the frequent application (8–13 times per season) of mostly highly hazardous pesticides. Insecticides represent the highest input costs (an average of US$600/ha) followed by fertilizers, fungicides, and manure. Without control, yields are commonly reduced by more than 50% (Cisneros and Mujica 1997). The frequent insecticide applications have caused severe secondary infestations by the white mite [*Poliphagotarsonemus latus* (Banks)] and the bud midge [*Prodiplosis longifila* (Gagne)]. Depending on the climatic conditions, both pests can seriously affect tuber yield and quality and need to be addressed in an IPM program.

25.3 Framework for IPM Research and Development in Potato

The International Potato Center (CIP) is investigating safer alternatives to replace toxic chemicals, and anticipates a widespread impact on poverty reduction, food security, human health and environmental protection through the development and adoption of IPM in potato production. In its global research program, a holistic working framework for potato pest management research and development (research outputs) and its application by farmers (research outcomes) was developed (Fig. 25.1).

In this framework, ecosystem research constitutes the basis, aiming at understanding the specific characteristics of potato agroecosystems, the influence of farmers' crop management interventions on pests and natural enemies, as well as the interactions and relationships among these groups of insects. An inventory of insect communities in potato agroecosystems is being built up with reference material in CIP's entomological museum. Investigations into the efficacy of the functional diversity in natural pest control are related to landscape fragmentation and farmers' practices. This research reveals species important in balancing pest problems and self-regulating agroecosystems, and contributes in the long term to the assessment of changes in the stability of systems by identifying relevant bioindicators (Mujica and Kroschel 2011). Pest population ecology and biological studies investigate pest population dynamics influenced by climatic (abiotic), biotic, and external factors such as cultural practices (rotation systems, pest-infested seed,

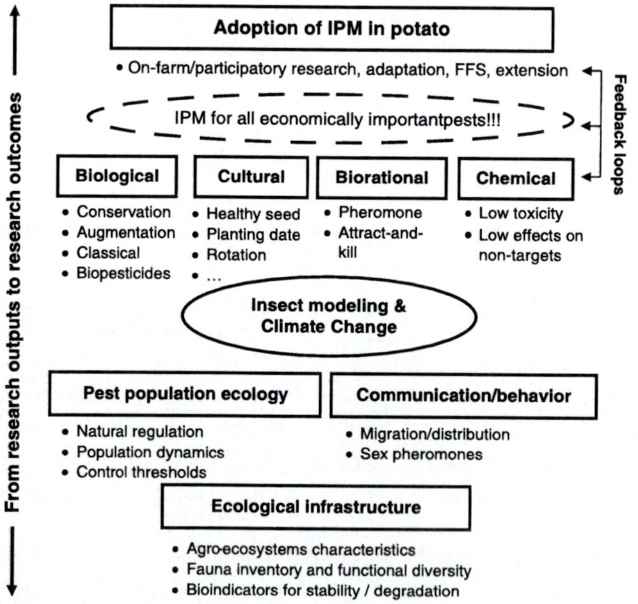

Fig. 25.1 Framework for integrated pest management (IPM) research and development (research outputs) and its application by farmers (research outcomes)

change in sowing date, etc.) applied by farmers (Keller 2003; Kroschel and Koch 1994). This is supported by the development and application of phenology modeling to predict pest population development in different agroecological zones, as well as critical infestation periods of pests (Kroschel and Sporleder 2006; Sporleder et al. 2004; Keller 2003). In combination with Geographic Information Systems (GIS), phenology models make it possible to carry out pest risk assessments under the current and future climate conditions (Kroschel et al. 2011a; Sporleder et al. 2008). Modeling requires insect life-table data, which are collected in autoecological temperature studies at constant and fluctuating temperatures for which insect rearing systems are established and maintained. Control thresholds are defined for targeted IPM interventions (Kroschel 1995). The assessment of intraspecific communication within pest species (pheromones) as well as that of the pest migration capacities and spatial dispersal is fundamental for developing innovative systems for effective pest management (e.g. attract-and-kill, physical control) (Rios and Kroschel 2011; Kroschel and Zegarra 2010; Kroschel et al. 2009). We are aiming at developing environmentally sound pest-management strategies based on manipulating the self-regulating capacity of agro-ecosystems through conservation (adaptation of cultural practices) and augmentation (inoculative/inundative) strategies for pests' natural antagonists, and biopesticide approaches (entomopathogens). In order to achieve this, an overall understanding and analysis of the antagonists' biology and ecology is essential, including assessments of their efficacy and the development of systems for formulation and application. At farm level, achieving a compatibility of control methods for different economic pests is important. Interdisciplinary and participatory research is conducted to learn lessons with farmers for adapting innovative technologies to farmers' situations. Ideally, farm communities are involved, which is often the case in the Andes. For the purpose of up-scaling IPM technologies and outreach, appropriate extension materials and training courses for farmers and extension staff are developed and conducted in collaboration with national programs (Kroschel et al. 2011b).

25.4 Experiences and Innovations in Potato IPM for the Andean Highlands

25.4.1 Sustaining the Resilience of Potato Agroecosystems

Maintaining and promoting the natural antagonistic potential is a major element of IPM to stabilize agroecosystems and reduce the incidence of pests. For controlling potato pests, a large number of entomophagous species in families Aphidiidae, Syrphidae, Carabidae, Coccinellidae and Chrysopidae are important natural limiting factors in Europe (Hassan 1989). In developing countries, little information is available as yet on existing beneficial insects present in the wide range of agroecosystems in which the potato is produced. In potato production in the Republic of

Yemen, where there is little or no use of insecticides, there are good indications that due to a large number of predators (25) and several parasitoids, the agro-ecosystem functions in a self-regulating manner, keeping aphids (*Myzus persicea* Sulz., *Macrosiphum euphorbiae* Thomas) or the cutworm *Agrotis ipsilon* Hfn. under the control threshold (Kroschel 1995).

The landscape structure and plant diversity play an important role in the conservation and enhancement of natural enemies to keep pest populations under the control threshold. Agroecosystems with a higher structural diversity showed, in most cases, an increased interaction between pests and natural enemies, and provided a higher potential for self-regulation (Tscharntke et al. 2005). Natural vegetation can effectively support beneficial arthropod species (Gurr et al. 2005; Landis et al. 2000) and arable weeds provide important resources for natural enemies such as nectar, pollen and alternative preys and hosts. A possible way to reintroduce insect biodiversity into large-scale monocultures is by establishing vegetation-diverse field margins and/or hedgerows, which may serve as biological corridors (Altieri 1999). Agricultural practices including insecticide application act as disturbances, reducing natural enemy populations and species richness (Croft 1990). Understanding the impacts of common agricultural practices on natural enemies is critical in developing ecologically based pest management (Landis et al. 2000).

For the Andean highlands of Peru, several beneficial insect predators and parasitoids that could play an important role in the control of potato pests have been reported. These include predators in families Carabidae, Coccinelidae, Nabidae, Lygaeidae, Chrysopidae, and Syrphidae (Cisneros 1995). The carabids *Harpalus turmalinus* Van Emden and *Notiobia schmusei* Van Emden have been reported as predators of Andean potato weevils (Loza and Bravo 2001; Cisneros 1995); and recently, Kroschel et al. (2009) reported high numbers of carabids of the genera Blennidus, Metius, Pelmatellus, Incagonum, and *Notiobia peruviana* (Dej.) in the central highlands of Peru that greatly affect the Andean potato weevil. *Apanteles subandinus* Blanchard, *Dolichogenidea gelechiidivoris* (Marsh), *Copidosoma koehleri* Blanchard, and *Incamyia cuzcensis* Townsend are parasitoids of the potato tuber moths; *Gonia peruviana* Townsend and *Patelloa robusta* (Wied.) are larval parasitoids of cutworms (Cisneros 1995).

As the potato is often a pesticide-intensive crop, it is very difficult to gather information on the natural control of potato pests in relation to the natural habitat. To better understand and exploit the potential of habitat management for the promotion and natural conservation of beneficial insects, we started in-depth ecological studies in the Andes of the Mantaro Valley as well as in the Canete Valley in the central coastal region of Peru (see Sect. 25.5.1) to develop practical recommendations for IPM.

25.4.1.1 The Impact of Insecticides

To explore the effects of insecticide applications on potato pests and natural enemies, ten potato fields were equally divided into insecticide-treated (farmers' practice)

Table 25.1 Insect numbers of three functional groups sampled with different evaluation methods in ten insecticide-treated (*I*) and insecticide-untreated fields (*C*) in potato cropping systems of Huasahuasi and the Mantaro Valley, Peru

		Huasahuasi			Mantaro valley		
Functional group	Methods of evaluation	I	C	Chi^2	I	C	Chi^2
Phytophagous	Plant evaluation	193	615	***	1,989	4,133	***
	Sweeping net	25	21	ns	1,331	2,655	***
	Pitfall trap	1,350	2,844	***	1,468	3,957	***
Parasitoids	Plant evaluation + sweeping net	35	45	ns	120	225	***
	Plant evaluation	68	131	***	55	112	***
	Plant evaluation[a]	26	50	ns	35	84	***
Predators	Sweeping net	0	0		40	41	ns
	Pitfall trap[b]	1,654	2,654	***	871	752	**
	Pitfall trap[a]	335	453	***	162	197	ns
TOTAL		3,686	6,813	***	6,071	12,156	***

***$P \leq 0.001$; **$P \leq 0.01$; ns = non significant
[a]Araneae
[b]Mainly carabidae and staphylinidae

Production Systems of Peru and insecticide-untreated field plots (>1,500 m^2) in an altitude gradient between 2,800 and 3,850 m of the Mantaro Valley and the region of Huasahuasi, in the central highlands of Peru (Kroschel and Cañedo 2009). Active and passive evaluation methods were used to monitor arthropods. Phytophagous insects were the most numerous functional groups, accounting for 75.4% of total arthropods. Predatory insects and parasitoids represented 23% and 1.4% of the total insect population, respectively (Table 25.1).

Of 36 identified phytophagous insects, 23 species are phytophagous on potato. Among 23 predator species, ground-dwelling predators were the most numerous (21.7%); plant-inhabiting predatory species represented only 1.3% of the total insect population. Although a total of 16 parasitoids were identified, parasitism of potato tuber moths was low (0.4%). Insecticide treatments significantly reduced about 50% of the arthropod population compared to untreated fields (Table 25.1).

In fields not treated with insecticides, tuber damage by Andean potato weevils and potato tuber moths ranged between 74–95% and 5–35%, respectively. Very striking were the higher infestations with flea beetles (*Epitrix yanazara* Bechyné), with up to 70% of potato foliar damage causing yield losses of up to 72% (Fig. 25.2, data are shown for the Mantaro Valley only). Insecticide applications were not always effective in controlling Andean potato weevil tuber infestations ranging between 20% and 70%.

25.4.1.2 The Impact of the Structural Diversity

To identify and to understand plant-beneficial insect interactions and the contribution of plant biodiversity to natural biological control, an inventory of the flora and of

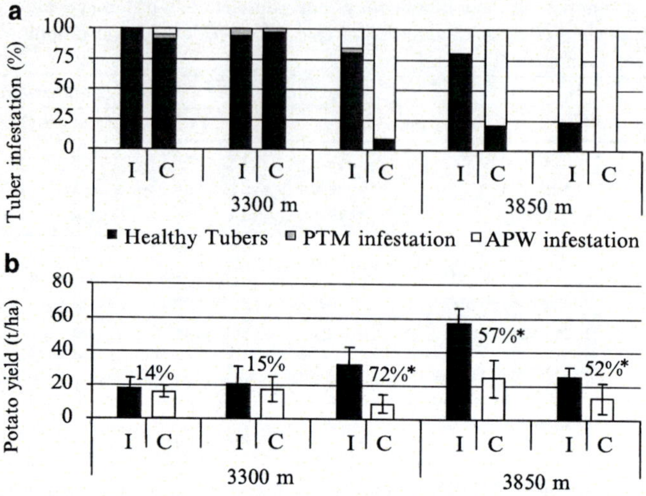

Fig. 25.2 Tuber infestation by the potato tuber moth complex, *Phthorimaea operculella* and *Symmetrischem tangolias*, and the Andean potato weevil, *Premnotrypes suturicallus* (**a**), and total potato yield (**b**) in five insecticide-treated (I) and non-treated fields (C) in the potato growing regions of the Mantaro Valley. *P≤0.05

arthropods (pests and natural enemies) was made in complex and simple structured landscapes in potato cropping systems of the central Andean highlands at two altitudes (3,250 and 3,850 m a.s.l.). The vegetation composition of grasses, herbs, shrubs and trees was assessed on representative areas as well as arable weeds in potato. Active and passive evaluation methods were used to monitor arthropods.

A total number of 122 plant species comprising grasses (22), herbs (79), shrubs (16) and trees (5) were identified, the distribution of which was highly influenced by altitude: 51% and 42% of the plants occurred only at 3,250 and 3,850 m a.s.l, respectively, with 7% species common in potato systems at both altitudes. There was a clearly higher plant diversity of herbs and shrubs at 3,250 m a.s.l. with a high species diversity of the families Asteraceae, Fabaceae and Brassicaceae. Brassicaceae species (7) were found at 3,250 m a.s.l. only. In contrast, trees were more diverse (5 species) at 3,850 m (e.g., *Alnus acuminate* Kunth) than at 3,250 m a.s.l. (2 species, e.g., *Eucalyptus globules* Labill.) (Cañedo et al. 2010).

A total number of 38 weed species were identified, 61% and 29% occurred at 3,250 and 3,850 m a.s.l., respectively, with only 1% found at both altitudes. The most representative families were Poaceae at 3,850 m, and Brassicaceae and Asteraceae at 3,250 m a.s.l. In complex landscapes, *Veronica persicae* Poiret was predominant during the whole potato cropping period; in contrast, in simple landscapes, *Amaranthus hybridus* L., *Brassica rapa* subsp. *campestris* L., and *Chenopodium murale* L. were most common in potato at 3,250 m a.s.l.

Thirty-nine plant species were identified as important sources of food for predators, parasitoids and pollinators of 11 insect families. *Senecio rudbeckiaefolius*

Meyen & Walp., distributed at both altitudes, hosted the highest number of parasitoids (7) and predators (13); however, because of their high abundance, plants of the families Asteraceae (e.g., *Baccharis penthandii*, 4 parasitoids and 3 predators at 3,250 m or *Bidens andicola* Kunth, 3 parasitoids and 3 predators at 3,850 m a.s.l.) and Brassicaceae (e.g., *B. rapa* subsp. *campestris*, 1 parasitoid and 3 predators at 3,250 m) are more important as entomophilous plants attractive to insects.

Phytophagous insects had a significantly higher abundance in simple than in complex landscapes at both altitudes. *P. suturicallus* dominated in the simple landscape at 3,850 m a.s.l., while *E. yanazara* was dominant at 3,250 m a.s.l. in the simple landscape. The abundance of parasitoids (3.4%) was very low, but we found 99 species from 10 families; the families Tachinidae (39) and Ichneumonidae (30) were the most representative ones. Predators (17.2% of all insects collected) comprised 75 species; carabids were dominant in the complex landscape at both altitudes. The main carabid species were *Pelmatellus columbianus* (Reiche), *Pelmatellus* sp., *Blennidus mateui* Straneo and *Blennidus* sp. at 3,850 m a.s.l. and *Incagonum* sp. (near *chilense*) at 3,250 m a.s.l. in complex landscapes; unlike *Notiobia* sp., which was more abundant in simple structured landscapes at 3,250 m a.s.l. At an altitude of 3,250 m, the higher diversity of plants, especially of the families Asteraceae and Brassicaceae, in the surrounding vegetation of potato fields as well as within fields, supported the abundance and diversity of natural enemies more than the vegetation of shrubs and trees forming the complex landscape structure (Cañedo et al. 2010).

25.4.2 The Development of Technological Innovations

Technological innovations are important to replace insecticide applications by other means of control that are equally or more efficacious at the same or lower costs for farmers. In our research several opportunities for innovations were identified, which could be developed and tested under various agroecological conditions of the Andean highlands.

25.4.2.1 Plastic Barriers for Andean Potato Weevil Management

Andean potato weevils are flightless and the knowledge of the migratory behavior of weevils, which walk long distances from the previous year's potato fields to new potato fields, supported our studies to develop simple physical barriers to prevent the migration of these weevils to potato fields (Kroschel et al. 2009). This was further supported by our findings that potato fields are the main source for new weevil infestation and that weevils do not occur – or only in negligible numbers – on fallows (Rios and Kroschel 2011).

In most cases, participatory research with farmers starts after technologies have been developed and tested for many years under laboratory and controlled experimental field conditions. Hence, feedback is seldom received from the end users at the start of the research process regarding the technology already under development; and this has probably often been the cause for low adoption rates of innovations. In our participatory research of testing plastic barriers for Andean potato weevil management, we included farmers from two Andean villages in the evaluation of the technology. Experiments under controlled conditions were conducted merely to prove that adult weevils are unable to climb plastic barriers, and to establish the appropriate height for practical field evaluation. The technology, whereby plastic sheets are fixed to wooden stakes and installed around potato fields (10 cm below and 40–50 cm above soil surface), has been evaluated over several seasons. It proved more effective than several insecticide applications by farmers, where potato is grown in rotation with other crops or after fallow. Ideally, to fully prevent migration a plastic barrier needs to be installed at the time of sowing, or even earlier, but in any case no later than the end of October or beginning of November. In cases where farmers grow potato after potato, plastic barriers prevent new migration, although the potato fields are already a source of infestation. But under these conditions, plastic barriers also proved to be an advantage, since only one initial weevil population needed chemical control, which made it possible to reduce the insecticide applications to only one application (Kroschel et al. 2009). In the High Andes above 3,800 m a.s.l., where Andean potato weevils are the only economic biotic constraint, potato production without the use of insecticides seems to become possible with this simple technology, especially in potato rotational cropping systems (see Sect. 25.4.3). Further, plastic barriers also proved to be effective in controlling other weevil species, such as *P. solaniperda* Kuschel, *P. latithorax* and *P. vorax* in potato agroecologies of Peru, Bolivia, and Ecuador.

25.4.2.2 Attract-and-Kill for Field and Storage Management of the Potato Tuber Moth Complex

Sex pheromones have been identified for the potato tuber moths, *P. operculella* and *S. tangolias*. Since then, they have been mainly deployed for monitoring and studying the flight activity of males, but for practical and economic reasons they have not been used in managing the pests. Attract-and-kill was an innovation successfully developed for controlling other lepidopteran pests (Lösel et al. 2000), a strategy consisting of an insecticide-pheromone co-formulation whereby the male moths are attracted by the pheromone and killed through contact with the insecticide. Considering the small sizes of potato fields in the Andes and in other developing countries, we adapted this technology to be used for the two potato tuber moth species. Attract-and-kill resulted in a 100% kill of both moth species after 3 and 4 days under controlled conditions (Kroschel and Zegarra 2010). In field experiments, at a droplet volume of 100 μl and a droplet density of one drop per 4 m^2, equal to 2,500 droplets ha^{-1}, immediately 1 day after the attract-and-kill treatment, daily catches of male moths of either *P. operculella* or *S. tangolias* were drastically reduced (>95%) in the pheromone water traps. In contrast, catches continued to be high in the pheromone water traps set up in the respective control fields. Male catches generally

Table 25.2 Efficacy of attract-and-kill to reduce *Phthorimaea operculella* and *Symmetrischema tangolias* male population in two potato agroecosystems of Peru as monitored with pheromone-baited water traps

Mean number of male catches in pheromone water traps (numbers)*				
	P. operculella		*S. tangolias*	
	Ancash	Huancayo	Ancash	Huancayo
Treatment	2008	2009	2008	2009
Control	8,329 (7.84) a	3,712 (4.71) a	7,786 (9.91) a	8,564 (14.79) a
Attract-and-kill	150 (0.44) b	259 (0.64) b	202 (0.98) b	1,008 (6.92) b
Efficacy %	98	93	98	90

*For each location and year, mean of five replications for the attract-and-kill treatments and the untreated control with field sizes between 2,000–4,000 m^2. Means followed by different letters are significantly different according to the Tukey's HSD-Test at $P \leq 0.05$

Table 25.3 Efficacy of attract-and-kill to control the potato tuber moth species *Phthorimaea operculella* and *Symmetrischema tangolias* in potato storage

	P. operculella		*S. tangolias*	
	Tuber infestation (%)			
Treatment	Infestation rate*	Infestation intensity	Infestation rate	Infestation intensity
Control	89.3 (5.65) a	39.9 (6.59)a	82.4 (8.61) a	66.0 (12.03) a
Attract-and-kill	12.2 (4.03) b	4.8 (27.1) b	10.5 (10.25) b	10.0 (3.69) b
Efficacy %	86.3	88.0	87.3	84.8

*Mean of three experiments with four replications of each treatment. Means followed by different letters are significantly different according to the Tukey's HSD-Test at $P \leq 0.05$

remained very low over a long period of time after application (>70 days). At the final evaluation, the total male populations of both *P. operculella* and *S. tangolias* were reduced between 90% and 98%, respectively, compared to untreated potato fields (Table 25.2).

Attract-and-kill also proved highly efficacious in reducing the infestation by *P. operculella* and *S. tangolias* in potato stores. After storage period of 75 days, the infestation rate and intensity in potato tubers was reduced between 84% and 88% (Table 25.3).

25.4.2.3 Biological Control

Re-naturalization of Lost Biological Control Agents

For decades the frequent use of insecticides has been the farmers' main practice for managing potato pests in the Andean highlands (Ewell et al. 1990). Considering all study sites, insecticide treatments reduced approximately 50% of the arthropod population compared to non-treated fields within one crop cycle (see Sect. 25.4.1.1). Since the potato is the most widely grown crop, especially at altitudes above

3,800 m, the regular use of insecticides may have caused long-term negative effects on natural enemy populations, especially where no effective recovery from unsprayed fields or hedgerows was given; however, no earlier reference base line studies exist that could provide more evidence for this assumption.

Long-term field studies in the United Kingdom in a range of arable crops found few adverse long-term effects of pesticides on non-target organisms including insects, spiders, earthworms and soil microbes (Young et al. 2001). Here, the application of broad-spectrum insecticides resulted in declines in the number of many non-target arthropods, but these usually recovered within the same growing season. Less temporary effects were seldom noted and affected only soil-dwelling collembolans (springtails). Numbers of these organisms remained comparatively low in treated plots up to 2 years after application (Devine and Furlong 2007). Apart from the persistence of the insecticide, the degree to which affected populations can recover is also dependent upon the recruitment of new individuals from unaffected populations, which often permit the rapid recovery of species in insecticide-treated fields (Jepson and Thacker 1990).

From our research we learned that the landscape structure in the High Andes does not provide effective niches for natural enemies to recuperate, and annual plants (weeds) surrounding potato fields were more supportive than species-poor hedgerows. Further, pest-specific parasitoids, e.g. those of the potato tuber moth complex, can probably not build up sufficient populations due to the use of broad-spectrum insecticide applications, which was also the limiting factor after the introduction of parasitoids for the classical biological control of *P. operculella* in Australia (Horne and Page 2008). The introduction of plastic barriers as an effective technology to control the Andean potato weevil gives the opportunity to introduce a new IPM program in the Andes, which may fully replace – or will at least drastically reduce – insecticide interventions. This could be accompanied by augmentation strategies for natural parasitoids of the potato tuber moth complex, whose populations were reduced or partly lost through the intensive use of pesticides, to further increase the resilience of potato agroecosystems. For this purpose, CIP maintains and supplies to national programs parasitoids such as *Copidosoma koehleri* Blanchard, *Apanteles subandinus* Blanchard, and *Orgilus lepidus* Muesebeck, which have co-evolved with the potato tuber moth.

Bacillus thuringiensis for Low-Cost Storage Management of the Potato Tuber Moth Complex

Bacillus thuringiensis subspec. *kurstaki* (*Btk*) was known to be highly effective against *P. operculella,* but it has hardly been used by small-scale potato farmers because of its relatively high cost. A very low proportion of *Btk*, 40 g *Btk* mixed with 960 g fine sand dust containing quartz provided effective storage control for 1 ton of stored potato in the Republic of Yemen (Kroschel and Koch 1994). Alternatively, *P. operculella* granulovirus (*Phop*GV) was produced nationally in Peru and Bolivia using low-cost facilities for propagation. A dust formulation, produced by selecting and grinding *Phop*GV-infected larvae mixed with ordinary talcum, has been used at the rate of 5 kg per ton of stored potatoes (20 infected larvae per kilogram of talcum). Research showed that *Phop*GV reduced damage in stores by 91% and 78%, 30 and 60 days after application (Raman and Alcázar 1990). However, the pest situation has changed in the High Andes and *S. tangolias* has become the most prevalent species in potato stores (Keller 2003) for which *Phop*GV is not effective; it has not been possible to identify a *S. tangolias*-specific granulovirus. Winters and Fano (1997) pointed out that the market potential of a bioinsecticide directed to only one species of the potato tuber moth complex may limit its potential use and adoption. We found that the commercial product Dipel2X reformulated in magnesium silicate (15 g/1 kg talcum) effectively protects tubers against both tuber moth species, which makes the product highly competitive compared to the current *Phop*GV-product or to chemical pesticides. Since farmers have adopted the use of a dusted *Phop*GV formulation to treat potatoes prior to storage, the probability for adoption of a specific *Btk*-talcum formulation, which controls all species, is high. However, for effective management of potato tuber moths in rustic stores, storage hygiene is also very important. In particular, physical protection is needed to prevent new moths from entering storage facilities and thereby prevent the infestation of tubers through young unprotected potato sprouts.

Potential of Entomopathogens

Many studies have been conducted to assess the pathogenicity and practical use of entomopathogenic fungi for the control of Andean potato weevils (Kaya et al. 2009). Kühne (2007) demonstrated that *Beauveria bassiana* (Balsamo) Vuillmen kills *P. suturicallus* under laboratory conditions but is not effective when applied in the field, mainly due to the low temperatures at high elevations. In contrast, Cisneros and Vera (2001) obtained good results applying the fungus against weevil larvae under potato tuber storage conditions. However, Winters and Fano (1997) concluded that the benefits of using a biocontrol product based on *B. bassiana* for Andean potato weevil storage control are low, and farmers are not expected to purchase the product in sufficient quantities to allow a market to develop.

Alcázar and Kaya (2003) isolated an undescribed entomopathogenic nematode in the genus *Heterorhabditis*, designated as strain Alcázar-1, from last instar *P. suturicallus* larvae in soil from a commercial potato storage at 2,750 m a.s.l. Parsa et al. (2006) studied the biological activity and demonstrated its high efficacy to infect all stages of *P. suturicallus* and to protect potato tubers from infestation by neonate larvae; further, it showed adaptability to low temperatures. In field experiments, applied in suspension at a rate of 50 infective juveniles/cm^2, the entomopathogenic nematode significantly reduced tuber damage and plant infestation by Andean potato weevil larvae compared to the control; tuber damage was reduced by 41.4% and the larval infestation by 53.2%, respectively (Alcazar et al. 2007). To commercialize entomopathogenic nematodes requires investments in large-scale production facilities, which may not yet be economically viable for small-scale farmers; however, small-holder potato production systems of the high Andes could profit from a wider use of nematode-based bioinsecticides in high value crops, e.g., asparagus (*Asparagus officinalis* L.) along the Peruvian coast.

25.4.3 Ecological and Economic Benefits of IPM

With regard to the altitude where potato is produced and the temperature-dependent related abundance of insect pests, our studies concluded with two IPM approaches for the Andean Highlands. At altitudes >3,800 m, where Andean potato weevils are the only insect pest problem, the use of plastic barriers may constitute the main IPM technology. In contrast, in inner Andean valleys such as the Mantaro Valley (3,200–3,500 m a.s.l.), in addition to Andean potato weevils potato tuber moths and flea beetles may cause economic damage and losses. Based on our ecological studies different strategies and technological innovations need to be integrated in a pest management program. We evaluated the ecological and economic benefits of the two IPM systems versus farmers' practice.

With the participation of 40 farmers (i.e., 40 individual on-farm field evaluations) of the villages Ñuñunhuayo (3,800 m a.s.l.) and Aymara (3,900 m a.s.l.), the use of plastic barriers proved effective in reducing Andean potato weevil tuber infestation. Compared to farmers' practice using 3–4 insecticide applications per season and a tuber infestation of 18.8%, the barriers reduced the infestation by 68% to a total of 6% at potato harvest (Table 25.4).

The EIQ (Environmental Impact Quotient) is a method for measuring the impact of pesticides on the environment. EIQ values ranging from 0 to 20 indicate a low environmental impact; from 20.1 to 40, medium; and >40 high environmental impact (Mazlan and Munford 2005). Because no insecticides were applied in IPM, the environmental impact (EI) of insect pest management was accordingly zero, while farmers' practice resulted in an EI of 122.7. Finally, IPM resulted in an excellent investment with mean increase in income of about US$600 for farmers of two Andean villages.

In the Mantaro Valley, IPM for potato was tested against farmers' practice and an untreated control at four locations. The IPM system included the installation of plastic barriers to control Andean potato weevils, the application of attract-and-kill to control the two potato tuber moth species *P. operculella* and *S. tangolias*, intercropping of potato with faba bean (*Vicia faba* L.) and the establishment of field borders (*Brassica rapa* var. *campestris*) to increase the overall functional diversity of natural enemies to reduce the incidence of other pests such as flea beetles or aphids (*M. persicae*, *M. euphorbiae*). Further, one insecticide application (beta-cyfluthrin) was considered for flea beetle control. In contrast, farmers' practice included four insecticide applications (3x carbofuran, 1x methamidophos).

Plant damage caused by flea beetles was significantly reduced in the IPM treated fields compared to farmers' practice and untreated control (Table 25.5). Tuber damage caused by Andean potato weevils was similar in farmers' practice (1.9%) and

Table 25.4 Potato tuber infestation by the Andean potato weevil, estimated net production benefits (US$/ha) and environmental impact quotient (EIQ/ha) for plastic barriers (IPM) and insecticide treated farmers' fields (farmers' practice) in the two Andean villages Ñuñunhuayo and Aymara (N=20 for each treatment at each community)

	Farmer's practice	IPM
Potato pest		
Andean Potato weevil (Premnotrypes suturicallus)		
Tuber damage (%)	18.82 ± 13.8 a*	6.0 ± 6.5 b
Environmental impact		
N sprays/season	3.0 ± 1.1	0
Total pesticide (L or kg/ha/season)	5.1 ± 3.5	0
EI/ha/season	122.7 ± 86.7	0
Economic analysis		
Cost of component of control		
Insecticides (US$)	184.7 ± 128.5	0
Insecticide applications (US$)	65.2 ± 23.6	0
Plastic barriers (US$)	–	57.0
Plastic barriers installation (US$)	–	14.5
Total (US$)	249.9	71.5
Partial budget		
Yield t/ha	13.9 ± 10.2 a	15.2 ± 10.9 a
Damage at harvest (%)	18.8 ± 13.8 a	6.0 ± 6.5 b
Value/t (US$)	161.7	175.7
Total Benefit (US$)	2248.5 ± 1563.1	2671.5 ± 1952.1
Control costs (US$)	249.9	71.5
Net income** (US$)	1998.6	2600.0
Increase in income (US$)		601.4

*Means followed by different letters are significantly different according to the LSD-test at $P \leq 0.05$

**Does not include costs for seed, fertilizer, fungicides, labor for field preparation and weeding, etc.

IPM (4.7%) plots, but significantly reduced compared to the untreated control (31.3%). Attract-and-kill significantly reduced the flight activity of both potato tuber moth species as well as the infestation of potato plants and tubers at harvest. In the IPM system, one application of beta-cyfluthrin (EIQ=31.57) was made, while farmers used three applications of carbofuran (EIQ=50.67) and one application of methamidophos (EIQ=36.83). Therefore, IPM showed high ecological benefits with an EI of only 8.3 compared to farmers' practice with an EI of 129.5.

In all IPM plots, there was a positive interaction between the growth of potato and faba bean, demonstrated by land equivalent ratio (LER) values of 1.39–2.52. However, potato intercropping with faba bean or the weed *B. rapa* var. *campestris* failed to increase the biological control of pests and the number of parasitoids and predators. Hence, yield calculations were made for the potato area only (Table 25.5).

The higher pesticide costs increased the total potato production costs in farmers' practice plots. However, due to insignificant higher yields and lower tuber infestations, farmers' practice and IPM plots showed no differences in total net income.

Table 25.5 Efficacy of IPM versus farmers' practice in potato production systems of the Mantaro Valley, Peru

	Farmer's practice	IPM	Untreated control
Potato pest			
Flea bettle (*Epitrix yanazara*)			
Adult infestation (adults/plant)	11.6 ± 5.6 a*	7.6 ± 3.6 b	12.6 ± 5.9 b
Foliar damage (%)	4.2 ± 2.3 a	5.1 ± 2.9 a	14.3 ± 9.1 b
Andean Potato weevil (*Premnotrypes suturicallus*)			
Tuber damage (%)	1.9 ± 2.4 a	4.7 ± 5.5 a	31.3 ± 20.4 b
Potato tuber moth (*Phthorimaea operculella*)			
Flight activity (males/trap/day)		1.8 ± 0.8 a	8.8 ± 2.8 b
Andean Potato tuber moth (*Symmetrischema tangolias*)			
Flight activity (males/trap/day)		0.2 ± 5.2 a	91.5 ± 35.9 b
Both PTM species			
Infestation rate of potato plants (%)		13.0 ± 4.5 a	73.3 ± 6.9 b
Infestation intensity of potato plants (mines/plant)		1.5 ± 5.3 a	4.6 ± 2.7 b
Tuber infestation rate (%)		8 ± 2.1 a	20 ± 6.3 b
Ecological impact			
Foliage-inhabiting phytophagous (insects counts/evaluation date)	82 ± 71.4 ab	67.6 ± 61.6 a	94.7 ± 64.5 b
Foliage-inhabiting predators (insects counts/evaluation date)	3.5 ± 4.7 a	3.3 ± 2.9 a	4.0 ± 4.8 a
Foliage-inhabiting parasitoids (insects counts/evaluation date)	3.7 ± 2.8 a	4.6 ± 7.0 a	5.7 ± 6.2 a
Environmental impact			
N° sprays/season	4	1	–
Total pesticide (L or Kg/ha/season)	13.5	0.5	–
EI/ha/season	129.5	8.3	–
Economic analysis			
Cost of components of control			
Insecticides (US$)	485.5	20.9	–
Insecticide applications (US$)	130.9	10.9	–
Plastic barriers (US$)	–	57.0	–
Plastic barriers installation (US$)	–	14.5	–
Attract-and-kill (US$)	–	30.0	–
Attract-and-kill application (US$)	–	5.4	–
Total (US$)	616.4	138.8	–
Partial budget			
Yield t/ha	33.0 ± 8.8 a	30.9 ± 6.0 a	18.1 ± 5.8 b
Damage at harvest (%)	1.9 ± 2.1 a	4.7 ± 5.0 a	33.9 ± 21.0 b
Value/t (US$)	180.0	180.0	180.0
Total Benefit (US$)	5887.3 ± 1610	5415.8 ± 1149.5	2699.8 ± 966
Control costs (US$)	616.4	138.8	–
Net income** (US$)	5270.9	5277.0	2699.8
Increase in income (US$)		6.1	

*Means followed by different letters are significant different according to the LSD-test at $P \leq 0.05$
**Does not include costs for seed, fertilizer, fungicides, labor for field preparation and weeding, etc.

In contrast, clear differences of both systems with the untreated control occurred, demonstrating the overall need for an effective pest management system in potato. It has to be considered that calculations do not include health related costs of the use of highly toxic pesticides, which are another important social cost factor.

25.5 Experiences and Innovations in Potato IPM for the Coastal Lowlands

25.5.1 Sustaining the Resilience of Potato Agroecosystems

The leafminer fly (*L. huidobrensis*) is regulated by numerous hymenoptera parasitoids, some of which were successfully used in biological control programs (Murphy and LaSalle 1999). Likewise, along the Peruvian coast a high diversity of 42 parasitoids were identified on *L. huidobrensis*, and the most abundant species are the ectoparasitoid *Diglyphus websteri* (Crawford) (Eulophidae) and the endoparasitoids *Halticoptera arduine* (Walter) (Pteromalidae) and *Chrysocharis caribea* Boucek (Eulophidae) (Mujica and Kroschel 2011). For the bud midge (*Prodiplosis longifila*) no efficient parasitoids have been detected. Only the parasitoid *Synopeas* sp. (Platygasteridae) has been recovered from asparagus in the north coast with a parasitism that hardly exceeds 20% (Cisneros 1995). Because both the leafminer fly and the bud midge pupate in the soil, ground-dwelling predators play an important role in the regulation of populations of these insects. Generalist predators are regarded as an indispensable complement to specialized parasitoids, buffering outbreaks of pests, when populations of parasitoids and oligophagus predators are still low (Riechert and Lawrence 1997). Only few studies on soil dwelling predators in agroecosystems of the central coast of Peru have been conducted, showing the dominance of spiders (Araneae), ground beetles (Carabidae, 4 spp.), rove beetles (Staphylinidae, 16 spp,) and earwigs (Labiduridae, e.g. *Labiduria riparia,* and Anisolabidae) in potato and other agricultural crops (Schuller and Sanchez 2003; Rondon and Vergara 2003).

25.5.1.1 The Impact of Insecticides

The effect of broad-spectrum (farmers' practice) and selective insecticide applications in potato in comparison to untreated fields was studied on the leafminer infestation and the abundance and diversity of its natural enemies. Selective insecticides significantly reduced leafminer fly larvae infestation (75%), compared to the use of broad-spectrum insecticides (35%) (Fig. 25.3a). Populations of the two endoparasitoids *H. arduine* and *C. caribea* were greatly reduced by >80% by either kind of insecticide. In the arthropod community, Isopods were the most sensitive to the applications of insecticides and were reduced by 58.7% and 89.9% by

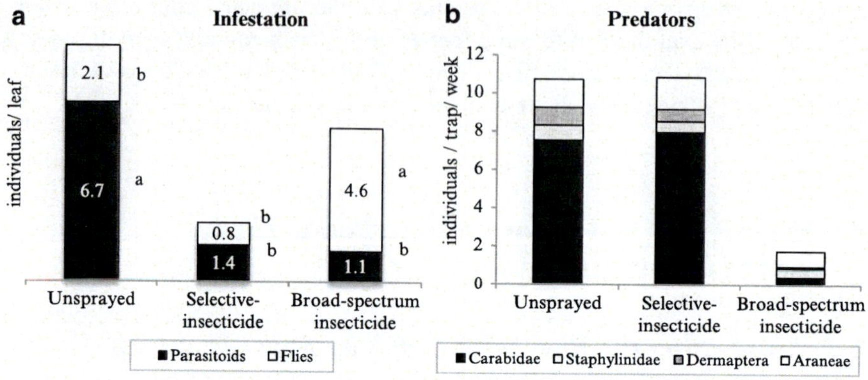

Fig. 25.3 Effect of selective and broad-spectrum (farmers' practice) insecticides on (**a**) leafminer infestation and parasitism, and on (**b**) the abundance of ground-dwelling arthropods and spiders, in the Cañete Valley, Peru

broad-spectrum and selective insecticides, respectively. Species of the orders Coleoptera and Hymenoptera were most abundant in non-treated (46.6% and 22.3%) and selective insecticide (50.8% and 15.6%) treated plots; however, they were greatly reduced by 86% and 78% through the application of broad-spectrum insecticides. In contrast, in farmers' fields Collembolla (64.3%) and Diptera (16.8%) were the most abundant orders. Broad-spectrum insecticides also significantly reduced the density of ground-beetles (Carabidae) (95%), earwigs (Labiduridae and Anisolabidae) (90%) and spiders (Aranea) (49.2%) (Fig. 25.3b), but no differences were found between non-insecticide and selective insecticide treated plots. The density of rove beetles was not affected by any treatment. The use of selective insecticides constitutes a particularly effective conservation biological control strategy, because it combines direct targeted reduction in pest numbers with predator conservation.

25.5.1.2 The Impact of the Structural Diversity

The landscape structure is a key factor for maintaining the abundance, community structure, and effectiveness of natural enemies in agricultural landscapes. The lower Cañete Valley (Ihuanco and Cañete) was characterized based on vegetation diversity and percent of arable land. The region of Cañete represents 81.6% of arable land associated to large-scale monoculture and a structurally simple landscape; instead in Ihuanco, arable land covers only 20.9% and most of the non-cropped and structurally complex landscape was made up of temporary herbaceous vegetation (72.3%). Although no differences in the vegetation diversity were found between the two landscapes, a higher percentage of field margins and a higher abundance of trees and shrubs were found in the complex landscape.

Leafminer infestation in potato and parasitism of leafminer fly larvae were not affected by the landscape structure, although a higher abundance and diversity of leafminer fly parasitoids was found in the complex landscape. Neither did the landscape structure affect the relative abundance of the species *H. arduine* and *D. websteri,* which are the most important parasitoid species, representing 72.4% and 17.9% of the total parasitoid population, respectively. The arthropod community composition was also similar between landscapes. Hexapoda was the dominant taxa in complex (81%) and simple (90%) landscapes, but Isopoda was more abundant in the complex landscape. By contrast, the insect community structure varied greatly between the two agricultural landscapes. Hymenoptera, for instance, was more abundant in the complex (72%) than in the simple (17%) landscape. Ground-dwelling predators were primarily ground beetles (Carabidae), rove beetles (Staphylinidae), earwigs (Labiduridae) and spiders (Aranea) representing 60.2%, 9.5%, 6.6% and 23.7%, respectively. Densities of each group of predators differed significantly between the two types of landscapes. In the complex landscape, Aranea and Carabidae were the most abundant predators, with 42.8% and 38.5% respectively. In the simple landscape Carabidae (71.8%) was the dominant predator group. The Staphylinidae family included 14 species and was the most species-rich family compared to Carabidae with six and one species for the Labiduridae and Anisolabidae families. The dominant species was the carabid *Pterostichus* sp. with 59.6% of total predators captured, and abundances of 71.3% and 37.7% in the simple and complex landscape, respectively. The Aranea order was represented by 10 families and 10 morphospecies. In general, for ground-dwelling predators and for all four predator families, species diversity (richness, Shannon and Simpson index) was higher in the complex than in the simple landscape, but a higher activity and abundance of Carabidae and Dermaptera were observed in the simple landscape.

25.5.1.3 The Impact of Weeds

In the central Peruvian coastal area, we assessed leafminer fly-parasitoid relationships in arable weeds. Twenty-two weed species of 14 plant families were identified as hosts of 11 leafminer fly species. *Liriomyza sabaziae* Spencer (58.7%) was the dominant species followed by *L. huidobrensis* (13.6%); however, only *L. huidobrensis* is regarded as an economically important pest, concluding that weeds associated to potato are mainly a source of leafminer flies with non-economic importance. The weeds *Galinsoga parviflora* Cav. (host of *L. sabaziae*), *Malva parviflora* L. [host of *Calycomyza malvae* (Burgess)], and *Commelina fasciculata* R&P [host of *Liriomyza commelina* (Frost)] are non-hosts of *L. huidobrensis*, but leafminer flies infesting these species shared the highest number of common parasitoids. On average, parasitism of leafminer flies in weeds reached 48.9% by a total of 50 hymenoptera parasitoids, of which *Chrysocharis caribea* Boucek, *C. flacilla*, *Halticoptera arduine*, *Diglyphus begini* (Ashmead) and *Diglyphus websteri* were the most important. In potato, in contrast, *L. huidobrensis* was parasitized by five parasitoid species only. Considering parasitoid richness, abundance and diversity, *L. sabaziae, Cerodontha dorsalis* (Loew), *L. commelina and C. malvae* were the main parasitoid sources for leafminer fly. A rational management of arable weeds associated with potato could therefore support the augmentation of parasitoids and natural control of *L. huidobrensis*. We tested this hypothesis by establishing intercropping strips of the weed *G. parviflora* in potato (see Sect. 25.5.2.3). However, weeds can also be sources for other insect pests and diseases of potato which need to be taken into consideration as suggested by Altieri (1999) and Kroschel (1995).

25.5.2 The Development of Technological Innovations

25.5.2.1 Trapping Devices

The strong attraction of leafminer flies to yellow surfaces was reported in the early 1980s, and, covered with a sticky substance, these were used for trapping and monitoring adults. CIP tested, optimized and adapted the use of yellow sticky traps for the mass trapping of leafminer flies in potato. Fixed traps (50×50 cm) were set up at plant emergence, and after becoming saturated with flies they were replaced at flowering stage. After flowering, the flight activity of the leafminer fly decreased and captures were substantially reduced. With the use of 80 fixed yellow sticky traps per hectare, about 5 million insects were recorded in one potato season representing 91.9% adults of leafminer flies and only 8.1% parasitoids, respectively. Highest captures (80%) were made from potato hilling to potato flowering, which were the recommended stages to use traps. In an evaluation of three trap densities of 60, 80 and 100 traps/ha with two replacements during the cropping season, an average of 1.15, 1.60 and 1.81 millions of adults were trapped, respectively (Mujica et al. 2000). A further development was the use of mobile yellow sticky traps (4×1 m) that are passed over the plant canopy covering four or more rows at a time. Farmers adopted and modified them according to their ingenuity. Comparing the efficacy of a combination of 95 fixed traps/ha (one change/cropping season) and 9 mobile trap passages with chemical leafminer fly control, a cumulative capture of 7.28 million adults/ha led to a reduction of 66.7% of the control costs (US$66.7/ha) compared to chemical control (US$200/ha) only with a total of 6 insecticide applications per season. Farmers' use of yellow sticky traps reduced leafminer fly adult populations effectively and made insecticide sprayings unnecessary to control adults, which reduced costs and did not destroy natural enemies at an early stage of plant growth (Mujica et al. 2000). An impact evaluation of the use of trapping devices in the Cañete valley showed that 90% of farmers reduced insecticide applications. The greatest economic benefits (US$200/ha) were observed when farmers used the combination of fixed and mobile traps. Trapping can replace the use of insecticides; however, this might not be fully effective to prevent the development and mining of larvae in the potato foliage as well as yield reductions, especially if the control by natural enemies is limited. In this case, applications of selective insecticides become necessary.

25.5.2.2 Selective Insecticides

When insecticide applications are unavoidable they should be used as selectively as possible. Several studies have been conducted on the effect of different groups of insecticides on leafminer larvae and their parasitoids. Cyromazine is an effective and selective larvicide with no effects on adults of leafminer flies or parasitoids. However, applied before and after oviposition of the parasitoid *H. arduine* it produced high mortality of leafminer larvae (>70%), but did not allow the hatching of parasitoids or development of neonate parasitoid larvae. When cyromazine was applied to leafminer larvae already parasitized by *H. arduine* (larvae already developed), leafminer fly larvae were killed but parasitoid larvae survived. Abamectine, which is a selective but expensive product, deriving from a soil fungus, affects the feeding behavior and oviposition of leafminer fly adults and also kills first instar larvae of the leafminer fly; its effect decreases as larvae grow. It showed no detrimental effects on populations of leafminer parasitoids (CIP 2004). Under field conditions, adding vegetable oil to abamectin increased its efficacy and reduced treatment costs by 73.5% and 74.2% as compared to the use of abamectin alone (Mujica et al. 2001). Imidacloprid was not effective against leafminer fly adults, but killed the parasitoid *H. arduine*. Pyrethroids (cypermethrine; cyfluthrine and b-cyfluthrine) affected both leafminer fly adults and parasitoids (CIP 2004). Other translaminar compounds (bensultap, cartap, chlorfluazuron, spinosad) significantly affected the leafminer larvae development. Parasitoids, parasitism and species richness were affected by the application of most pesticides; ectoparasitoids being the most susceptible.

25.5.2.3 Intercropping System

An increasing complexity of a landscape or cropping system can enhance the functional biodiversity in agroecosystems and can therefore positively contribute to the natural regulation of pests. Potato, as host plant of *Liriomyza huidobrensis,* was intercropped with conservation strips of maize (*Zea mays* L.) as host of *Liriomyza graminivora* (Korytkowski), or the weed *Galinsoga parviflora* as host of *L. sabaziae*, both in complex (Cañete Valley) and simple (Ihuanco Valley) landscapes. The structurally complex landscape was associated with a reduced leafminer larvae infestation and leaf damage and also with higher parasitism (Fig. 25.4). In the potato-maize intercropping system a minor larvae infestation, leaf damage and increased parasitism were observed compared to the potato-weed intercropping system. Intercropping strips with the weed *G. parviflora* or maize contributed to a diversification at the landscape scale, but differences in infestation and parasitism were more evident in the simple than in the complex landscape. Therefore, potato pest management will benefit from intercropping systems mostly in structurally poor agricultural landscapes, in which the leafminer fly is a main pest problem.

Ground-dwelling predators were dominated by the carabid *Pterostichus* sp. followed by rove beetles and earwigs. Although the seasonal activity pattern of these beneficial species varied spatially and temporally, they consistently preferred potato as compared to the intercropping strips. Predators responded differently to intercropping systems and the landscape context, showing that these variables did not influence all species equally, and interacting communities were made up of species with different spatial strategies. Biological diversity and ecological functions are

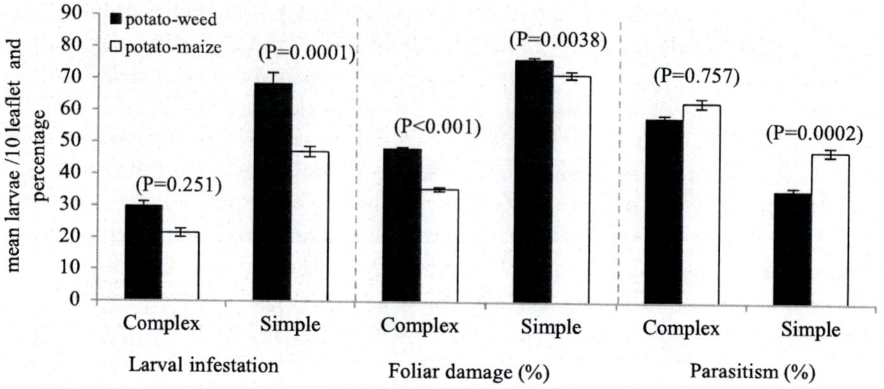

Fig. 25.4 Mean variation of *Liriomyza huidobrensis* larvae infestation, foliar damage, and parasitism affected by two intercropping systems in complex and simple structured landscapes in the Cañete Valley, Peru

not only affected by the overall landscape characteristics but also by the cropping systems, and must be considered to improve agro-ecosystems for biodiversity conservation and biological control of the leafminer fly (Schmidt et al. 2004). According to our results, natural control could benefit from early planting of conservation strips (at least 30 days before potato planting) to allow colonization by natural enemies, which later should be disturbed to facilitate the movement of natural enemies towards the crop.

25.5.2.4 Biological Control Using Entomopathogens

In 2003, natural epizootics of the Hypocreales fungus *Isaria fumosorosea* (Wize) were found in *L. huidobrensis* adult populations. Bioassays proved the pathogenicity against adults but not to other life stages. In preliminary field experiments, the application of 1.29×10^{12} conidia ml^{-1} of the isolate CIP-LMF II reduced the leafminer fly adult population by 68.1% within 8 days compared to population increases by 144% in untreated controls. In order to assess the efficacy of the fungus to cause mortality on a known adult population of *L. huidobrensis*, semi-field experiments with potatoes grown in screen houses (4.8 × 6.0 m) covered with fine nylon mesh were conducted. The screen houses were artificially infested with 750 adults (400 females and 350 males). Ten days after 4 and 2 fungal applications of 7.38×10^{12} conidia ml^{-1} at 2-days and 5-days intervals, an Abott mortality of 47.4% and 40.1% as well as a fungal infection of 21% and 13% of adults was observed. However, the reduction of the initial adult population was not sufficient to reduce leafminer fly leaf infestation, and hence the potato yields in all fungal treatments and the control were not significantly different.

We concluded that the biocontrol of adults would need complementary biological control efforts that control other life stages of the leafminer fly. Entomopathogenic nematodes of the families Steinernematidae and Heterorhabditidae have been considered as potential control agents against leafminers in recent years. Nematodes enter infested plants via leafminer oviposition holes and feeding sites (Harris et al. 1990) attacking the larvae through natural openings (Poinar 1990). Lebeck et al. (1993) found that all larval instars of *Liriomyza trifolii* (Burgess) were susceptible to *Steinernema carpocapsae* (Filipjev); *Heterorhabditis bacteriophora* Poinar also showed good efficacy in controlling this pest (Olthof and Broadbent 1991). We assessed 42 strains of entomopathogenic nematodes against *L. huidobrensis* and found large differences in their efficacy. The *Heterorhabditis* sp. isolate 3712 showed the highest efficacy to induce larvae mortality between 90–100% within a temperature range of 15–25°C. Its application on infested leaves of potted potato in a concentration of 10,000, 5,000 and 2,500 IJs/ml induced larvae mortality by 83%, 70%, and 55%, respectively. Applied in field experiments on leafminer fly infested faba beans, a mortality of larvae in the lower and middle part of the plants of 69% and 48.3% was achieved. In the upper leaves no mortality of larvae was indicated. The preliminary results are promising but before entomopathogenic nematodes can be considered in an integrated pest management program for the control of *L. huidobrensis,* further field evaluations, including a combination with the entomopathogenic fungus *I. fumosorosea*, are needed.

25.5.3 Ecological and Economic Benefits of IPM

The ecological, environmental, and economic benefits of an IPM program based on the use of action thresholds, trapping devices, and selective insecticides was evaluated for the potato crop in comparison to farmers' practice in the Cañete Valley of the central coast of Peru. The aim was to evaluate and combine IPM strategies profitable for farmers, but with fewer environmental and human risks.

During the potato season, the leafminer fly *L. huidobrensis* and the bud midge *P. longifila* were the most important pests. Mobile traps utilized during the vegetative growth of potato made it possible to reduce the flight activity of the leafminer fly in IPM fields (Table 25.6); further, the larvae infestation of both pests as well as the leafminer foliar damage was significantly reduced through IPM.

The arthropod diversity in sweeping net samples was similar in both pest management systems. In general, the abundance of phytophagous (66.3%, 12 spp.) was higher than that of predators and parasitoids (12.6%, 26 spp.). In farmers' practice fields the higher abundance of parasitoids was due to a higher leafminer larvae infestation. In IPM fields, the total number of plant-inhabiting predators, ground beetles (*Pterostichus* sp.) and earwigs (*Labiduria riparia*), was significantly higher than in farmers' fields. On average, double the amount of pesticides per season were applied by farmers (16.7 sprays) compared to IPM (8.8 sprays). However, the total amount of pesticides used per season was higher in IPM fields due to the specific control of the bud midge with the insecticide chlorpiriphos (2.5%). Still, the applied IPM strategy reduced the environmental impact compared to farmers' practice by 70.3%, from 137.4/ha to 40.8/ha, respectively, and gained a 35% higher marketable potato yield. Partial budget analysis showed a clear net profit for IPM, with an average increase in income of US$1410 per farmer.

Table 25.6 Efficacy of IPM versus farmers' practice for potato production systems of coastal lowlands in the Cañete Valley, Peru

	Farmer's practice	IPM
Potato pest		
Leafminer fly (*Liriomyza huidobrensis*)		
Adult infestation (adults/trap/week)	314.0±62.5 a*	164.7±42.8 b
Larvae infestation (larvae/leaflet)	1.7±0.15 a	0.73±0.35 b
Foliar damage (%)	20.3±4.03 a	10.3±2.29 b
Bud midge (*Prodiplosis longifila*)		
Infested buds (%) (51 days after planting)	18.9±1.72 a	10.0±6.01 b
Ecological impact		
Foliage-inhabiting arthropods (number/evaluation)	276.0±105.7 a	448.3±216.1 a
Soil-inhabiting predators (captured/trap/week)	2.67±0.29 a	5.36±0.62 b
Parasitoids (number/leaflet)	1.7±0.35 a	0.7±0.15 b
Parasitism (%)	66±4.5 a	61±4.4 a
Environmental impact		
N° sprays/season	16.7±2.23 a	8.8±0.79 b
Total pesticide (L or K/ha/season)	14.3±4.37 a	26.2±2.04 b
EI/ha/season	137.4±19.72 a	40.8±3.98 b
Economic analysis		
Cost of components of control		
Pesticides (US$)	336.3	177.5
Pesticide applications (US$)	79.7	46.6
Trapping device (US$)	–	20.14
Scouting (US$)	–	83.2
Total (US$)	415.9	327.5
Partial budget		
Marketable yield (t/ha)	14.25	21.97
Value /t (US$)	171.3	171.3
Total Benefit (US$)	2441.2	3763.3
Control costs (US$)	415.9	327.5
Net income** (US$)	2025.2	3435.8
Increase in income (US$)		1410.6

*Means followed by different letters are significant different according to the LSD-test at $P \leq 0.05$
**Does not include costs for seed, fertilizer, fungicides, labor for field preparation and weeding, etc.

25.6 IPM Support Tools

25.6.1 *Insect Simulation Modeling and Climate Change Effects on Pests*

Insect life table data, developed under a wide range of temperatures, give good predictions for the best temperature conditions that insects require for optimal growth and development. Based on such life table data, temperature-based phenology models are useful tools to gain an understanding of how temperature affects pest population growth potentials in different agroecologies. CIP develops temperature-based phenology models for potato pests; e.g., for *P. operculella*, which has been validated both in the laboratory and the field (Sporleder et al. 2004; Keller 2003). By linking this model to Geographic Information Systems (GIS) the pest developmental potential can be visualized for potato regions worldwide in order to support decision-making for appropriate pest management practices (Sporleder et al. 2008).

Global Warming will affect pest abundance and severity in agricultural and horticultural crops, and will most likely increase the yield and quality losses caused by pests. Insect population growth potentials are mainly temperature-driven, and a rise in temperature may either increase or decrease insect development rates and related crop damages depending on the insect species' optimum temperature range. We used the process-based climatic phenology model for *P. operculella* and applied three risk indices (establishment index, generation index, and activity index) in a geographic information system (GIS) environment to map and quantify changes for climate-change scenarios of the year 2050 based on downscaled climate-change data of the A1B scenario from the WorldClim database (Kroschel et al. 2011a). All applications and simulations were made using the Insect Life Cycle Modeling (ILCYM) software recently developed by The International Potato Center, Lima, Peru (Sporleder et al. 2011). The study concludes that the *P. operculella* damage potential will progressively increase in all regions where the pest already prevails today, with an excessive increase in warmer cropping regions of the tropics and subtropics. A range expansion into tropical temperate mountainous regions with a moderate increase of its damage potential is also predicted; i.e., that in Bolivia, Ecuador and Peru 44,322, 9,569, and 39,511 ha of potato will be under new risk of infestation. This information is important to prepare farmers for such change and to develop adaptation strategies to climate change.

25.7 Technology Transfer

The development of IPM calls for an agroecosystem-/site-specific approach and the most practical research outputs (see CIP's IPM research framework, Fig. 25.1) need to be combined to define an IPM program that is the most relevant to farmers' conditions and can have a high rate of adoption. Embedded into the best cultural practices of potato cropping for managing the different potato insect pests by the use of high quality pest-free seed, adequate crop rotation, optimal planting and harvest dates, best practices of weeding and hilling, etc. (Kühne 2007; Kroschel 1995), according to the recent research results, an IPM program for the Andean highlands would ideally consist of new technologies and strategies (i) to inhibit the migration of Andean potato weevils into potato fields by using plastic barriers; (ii) to scout the occurrence of *Epitrix* spp. and eventually control by one single application of low-toxic insecticides, and (iii) to apply attract-and-kill at a rate of 2,500 droplets/ha at the pre-flowering stage to reduce potato tuber moth tuber infestations (Fig. 25.5a) (Kroschel et al. 2011b). Additional control efforts would be needed for pest management in potato stores by using talcum-*Btk* or attract-and-kill.

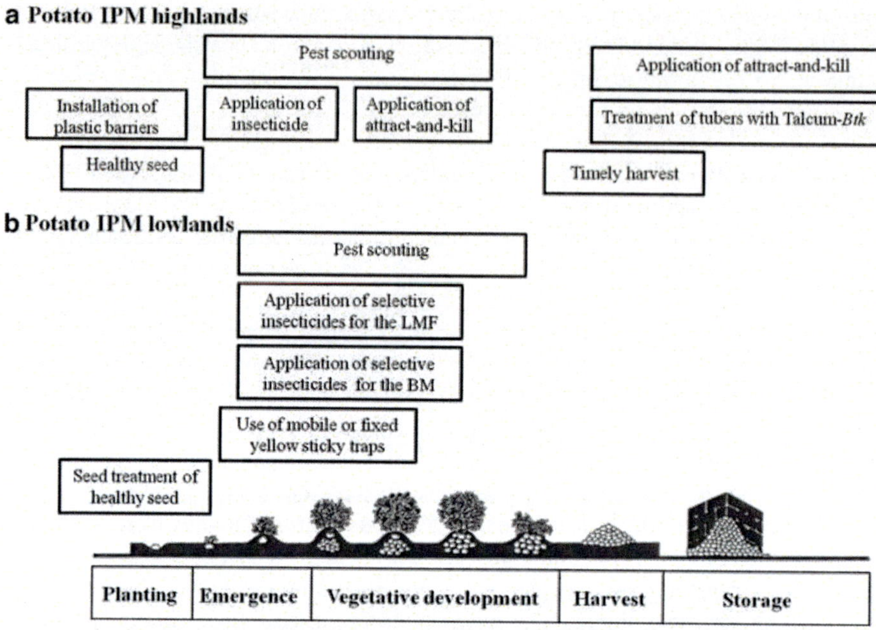

Fig. 25.5 Potato growth stages and most relevant IPM practices for potato croping systems of the Andean highlands (**a**) and the lowlands (coast) (**b**) of Peru. These technologies have to be embedded into the best cultural practices for potato which are not included in this sketch. *Application on insecticides might be needed to control flea beetle (*Epitrix* spp.) at specific locations. LMF = leafminer fly (*Liriomyza huidobrensis*), BM = bud midge (*Prodiplosis longifila*)

Likewise, for lowland potato production in the coastal region of Peru, the IPM program would be based on (i) seed treatments, (ii) use of mobile or fixed yellow sticky traps, and (iii) pest scouting for both leafminer fly and bud midge, and the target-specific application of low-toxic, selective insecticides. Mobile and yellow stick traps have been already successfully introduced and are widely adopted by farmers (Fonseca and Mujica 1997; Mujica et al. 2000). Potato storage is not practiced in the lowlands and quality seed derives from highland production regions (Fig. 25.5b). Both IPM systems would also profit from intercropping of potato with other crops, or from establishing field borders of naturally occurring weeds, to increase the floristic diversity and to support natural biological control.

Before the development of plastic barriers, which inhibit the migration of adult weevils to potato fields, insecticide applications were the only direct control method for Andean potato weevil adults to prevent oviposition and tuber damage by larvae. Therefore, previous IPM efforts in the Andes were based on the training of farmers to better understand the weevils' biology and to use cultural practices appropriately to reduce weevil infestation sources (Ortiz et al. 1996); however, this has achieved a low level of adoption (Ortiz et al. 2009). A better adoption of the proposed new potato IPM program could profit from the high cost-efficacy of plastic barriers and the accessibility to plastic in the local markets. Nevertheless, the wider use and application of all three new IPM technologies – plastic barriers, attract-and-kill, and talcum-*Btk* – can be achieved only if governmental and non-governmental organizations, as well as the private sector, are interested in the implementation and commercialization of the new products. Training courses have been initiated and the new technologies were well perceived by agricultural extension workers and farmers. Organic potato growers, in particular, have shown immediate interest in adopting plastic barriers with the support of non-governmental organizations. The interest in IPM differs from one village to another, and often depends on how well the farming communities are organized internally. Complementary demonstration trials at the village level will be important so that farmers' groups can jointly experiment for themselves and gain confidence in the new technologies. Currently, partnerships are being built up with the private sector to introduce and make available plastic in adequate sizes (40 cm wide) and quality (durability of at least 2–3 years), to reformulate existing *B. thuringiensis* products with talcum to register and commercialize a low-cost biocontrol product, as well as to register attract-and-kill in Peru. Main limiting factors and constraints for implementing IPM in Peru are the non-existent governmental extension service as well as the heavy competition of agrochemical products to which farmers have easy access through a strong network of pesticide sellers.

Generally, the implementation of IPM is knowledge-demanding and farmers would profit from strong support of experienced advisors with an entomological background, who can adequately assist farmers in their decision making throughout the cropping period. This kind of extension service would require a private IPM potato consultancy. A successful example is given by Horne and Page (2008) for potato production in Australia, which can serve as a model for other countries.

25.8 Lessons Learned and Conclusions

It is often argued that the international agricultural research centers (IARC) should prioritize research on global or regional plant health problems; in the case of CIP, giving priority to research on regional pests like the Andean potato weevils or globally invasive pests such as potato tuber moths or the leafminer fly. National agricultural research should then adapt technologies to their specific locations. Our lessons from the Andean highlands and coastal lowland potato production systems clearly demonstrated that a system approach is required to develop IPM. This needs to consider all economically important pests in an agroecosystem, i.e., globally minor but locally important pests to be addressed by farmers. In the Andean highlands flea beetles can become more important pests if farmers adopt non-chemical pest management strategies for Andean potato weevils. Developing and applying a system approach in IPM as outlined in CIP's framework for IPM research and development (Fig. 25.1) requires not only sufficient human capacity but also long-term funding, which in most cases cannot be afforded by national programs.

Research efforts on globally important pests have already generated a lot of information for their management on which basis new and more effective technologies could be developed (such as plastic barriers, attract-and-kill). In contrast, basic information is often missing for local pests such as flea beetles or the bud midge. Therefore, it was not possible to include alternative control methods in our IPM program for these pests, which have to be managed for the time being by the rational use of insecticides. Likewise, it has not so far been possible to adapt highly effective technologies for two potato tuber moth species, such as attract-and-kill, for the control of the Guatemalan tuber moth *Tecia solanivora* (Povolny), an important potato pest in Colombia and Ecuador.

In our research, agroecological studies were conducted to better understand the impact of farmers' pest management and of the landscape structure on the abundance and diversity of pests and natural enemies. This revealed important information on how long-term pesticide use might have reduced arthropod populations or how natural biological control could be augmented in the two agroecosystems. Further, such information may be relevant for monitoring changes over time.

On-farm field trials to test IPM versus farmers' conventional pest management practices demonstrated the environmental and economic viability of IPM programs for both potato cropping systems. Evaluation methods such as ecological assessments of foliage and soil inhabitant arthropods and parasitoids, the use of the environmental impact quotient (EIQ) and partial budgeting proved most practical in assessing the ecological, environmental, and economic impact of the systems under evaluation.

Acknowledgements The authors gratefully acknowledge the financial support for this research that was provided by the Regional Fund for Agricultural Technology (FONTAGRO), Washington, DC, for the project "Developing and use of ecological approaches in pest management for enhancing sustainable potato production of resource-poor farmers in Andean regions of Bolivia, Ecuador, and Peru", and the Federal Ministry of Cooperation and Development (BMZ), Germany, for the project "Tackling *Liriomyza* leafmining flies: invasive pests of global proportions". Finally, we are thankful to the valuable comments on our manuscript provided by Dr. Oscar Ortiz.

References

Alcazar J, Cisneros F (1999) Taxonomy and bionomics of the Andean potato weevil complex: *Premnotrypes* spp. and related genera. In: Impact on a changing world. Program Report 1997–1998, International Potato Center (CIP), Peru, pp 141–151

Alcazar J, Kaya H (2003) Hallazgo de un nemátodo nativo del género Heterorhabditis sp., parásito del gorgojo de los Andes *Premnotrypes suturicallus* en Huasahuasi – Junín. Resumen de la XLV Convención Nacional de Entomología Ayacucho, Peru

Alcazar J, Kroschel J, Kaya H (2007) Evaluation of the efficacy of an indigenous Peruvian entomopathogenic nematode *Heterorhabditis* sp. to control the Andean Potato weevil *Premnotrypes suturicallus* Kuschel under field conditions. In: XVI international plant protection congress, 15–18 Oct 2007. Congress proceedings vol II. Glasgow, Scotland, pp 544–545

Alston DG, Reading ME (1998) Factors influencing adoption and educational outreach of integrated pest management. J Ext 36(3)

Altieri M (1999) The ecological role of biodiversity in agroecosystems. Agr Ecosyst Environ 74:19–31

Arica D, Kroschel J, Forbes G, Saint Pere K (2006) Persistent Organic Pollutants (POPs) and Hazardous Pesticides in Andean Farming Communities in Peru. Final report to the Canada POPs Fund, World Bank. International Potato Center, Lima, Peru, 73pp

Cañedo V, Rojas J, Alvarado J, Kroschel J (2010) Efecto de la composición del paisaje sobre las plagas y enemigos naturales de la papa en la sierra central del Perú. In: Proceedings of the XXIV congress of the Latin-American Potato Association (ALAP), 23–28 May 2010, Cusco, Peru, pp 300–301

CIP (2004) Research on selective and nonselective insecticide compounds against *Liriomyza huidobrensis* and their hymenopteros parasitoids under laboratory and field conditions. Compendium of activity reports from 1997 to 2003. International Potato Center, Peru

Cisneros F (1995) Control de plagas agrícolas. Editorial Impresión, Lima

Cisneros F, Mujica N (1997) Developing IPM components for leafminer fly in the Cañete Valley of Peru. In: International Potato Center (ed) Program report 1995–1996, Lima, Peru, pp 177–184

Cisneros F, Vera A (2001) Mass producing *Beauveria brongniartii* inoculum, an economical, farm-level method. In: Scientist and farmer: partners in research for the 21st century. Program report, 1999–2000, International Potato Center, Lima, Peru, pp 155–160

Croft BA (1990) Arthropod biological control agents and pesticides. Wiley, New York

Daxl R, von Kayserlingk N, Klein-Koch C, Link R, Waibel H (1994) Integrated pest management: guidelines. Deutsche Gesellschaft für Technische Zusammenarbeit (GTZ) GmbH, Schriftenreihe der GTZ No. 249, Eschborn, Germany, 119pp

Devine GJ, Furlong MJ (2007) Insecticide use: contexts and ecological consequences. Agric Human Values 24(3):281–306

Ewell PT, Fano H, Raman KV, Alcazar J, Palacios M, Carhuamaca J (1990) Farmer management of potato insect pests in Peru. International Potato Center, Lima, 87pp

Fonseca C, Mujica N (1997) Impacto del control etologico sobre la mosca minadora (*Liriomyza huidobrensis*) en el valle de Cañete. Internal report, International Potato Center, Lima, Peru

Gurr G, Wratten S, Tylianakis J, Kean J, Keller M (2005) Providing plant foods for natural enemies in farming systems: balancing practicalities and theory. In: Wackers FL, van Rijn PCJ, Bruin J (eds) Plant provided food for carnivorous insects: a protective mutualism and its applications. Cambridge University Press, New York

Harris M, Begley J, Warkenstein D (1990) *Liriomyza trifolii* (Diptera: Agromyzidae) suppression with foliar applications of *Steinernema carpocapsae* (Rhabditida: Steinernema) and abemactin. J Econ Entomol 83:2380–2384

Hassan SA (1989) Vorstellungen der IOBC-Arbeitsgruppe "Pflanzenschutzmittel und Nutzorganismen" zur Erfassung der Nebenwirkungen von Pflanzenschutzmitteln auf Nutzlinge. Gesunde Pflanze 8:295–302

Horne PA, Page J (2008) IPM dealing with potato tuber moth (PTM) and all other pests in Australian potato crops. In: Kroschel J, Lacey L (eds) Integrated pest management for the potato tuber moth – a potato pest of global proportion. Tropical agriculture 20, Advances in crop research 10. Margraf Publishers, Weikersheim, Germany, pp 111–117

Horne PA, Rae JE, Henderson AP, Spooner-Hart R (1999) Awareness and adoption of IPM by Australian potato growers. Plant Protect Q 14:139–142

Jepson P, Thacker J (1990) Analysis of the spatial component of pesticide side-effects on non-target invertebrate populations and its relevance to hazard analysis. Funct Ecol 4:349–355

Kaya HK, Alcázar J, Parsa S, Kroschel J (2009) Microbial control of the Andean potato weevil complex. In: Tennant P, Benkeblia N (eds) Potato II. Fruit, vegetable and cereal science and biotechnology 3(Special issue 1):39–45

Keller S (2003) Integrated pest management of the potato tuber moth in cropping systems of different agroecological zones. In: Kroschel J (ed) Tropical Agriculture 11, Advances in Crop Research 1, Margraf Publisher, Weikersheim, Germany, 156pp

Keller M, Baker G (2002) Impact of conservation biological control practices on natural enemies: a case study of diamondback moth and its parasitoids. In: USDA Forest Service (ed)

First international symposium on biological control of arthropods, USDA Forest Service, Washington, USA, pp 215–219

Kroschel J (1995) Integrated pest management in potato production in the Republic of Yemen with special reference to the integrated biological control of the potato tuber moth (*Phthorimaea operculella* Zeller). Tropical Agriculture 8, Margraf Verlag, Weikersheim, Germany, 227 pp

Kroschel J, Zegarra O (2010) Attract-and-kill: a new strategy for the management of the potato tuber moths *Phthorimaea operculella* (Zeller) and *Symmetrischema tangolias* (Gyen) in potato – laboratory experiments towards optimizing pheromone and insecticide concentration. Pest Manag Sci 66:490–496

Kroschel J, Cañedo V (2009) How do insecticides affect potato yield and ecosystem resilience to manage potato pests? An ecological assessment from the central highlands of Peru. In: 15th triennial symposium of the International Society for Tropical Root Crops, 2–6 Nov 2009, International Potato Center, Lima, Peru, pp 9–22

Kroschel J, Koch W (1994) Studies on the population dynamics of the potato tuber moth (*Phthorimaea operculella* Zeller) (Lepidoptera: Gelechiidae) in the Republic of Yemen. J App Ent 118:327–341

Kroschel J, Koch W (1996) Studies on the use of chemicals, botanicals and biological control in the management of the potato tuber moth in potato stores. Crop Prot 15(2):197–203

Kroschel J, Sporleder M (2006) Ecological approaches to integrated pest mangement of the potato tuber moth, *Phthorimaea operculella* (Zeller) (Lepidoptera: Gelechidae). In: Proceedings of the 45th annual Washington State potato conference, 7–9 Feb 2006, Moses Lake, Washington, pp 85–94

Kroschel J, Alcazar J, Pomar P (2009) Potential of plastic barriers to control Andean potato weevil *Premnotrypes suturicallus* Kuschel. Crop Prot 28:466–476

Kroschel J, Sporleder M, Tonnang HE, Juarez H, Carhuapoma P, Gonzales JC, Simon R (2011a) Predicting climate change caused changes in global temperature on potato tuber moth *Phthorimaea operculella* (Zeller) distribution and abundance using phenology modeling and GIS mapping. J Agric Forest Meteorology (Special Issue) (under revision)

Kroschel J, Cañedo V, Alcazar J, Miethbauer T (2011b) Manejo de plagas de la papa en la región andina del Perú – Guía de capacitación. International Potato Center, Lima, 85pp

Kühne M (2007) The Andean potato weevil *Premnotrypes suturicallus* ecology and interactions with the entomopathogenic fungus *Beauveria bassiana*. Ph.D. thesis, Georg-August-University, Göttingen, Germany, 177pp

Landis D, Wratten S, Gurr G (2000) Habitat management to conserve natural enemies of arthropod pests in agriculture. Annu Rev Entomol 45:175–201

LeBeck L, Gaugler R, Kaya H, Hara A, Jhonson M (1993) Host stage suitability of the leafminer *Liriomyza trifolii* (Diptera: Agromyzidae) to the entomopathogenic nematode *Steinernema carpocapsae* (Rhabditida: Steinernematidae). J Invertebr Pathol 62:58–63

Llanderal-Cazares C, Lagunes-Tejada A, Carillo-Sanchez JL, Sosa-Moss C, Vera-Graziano J, Bravo-Mojica H (1996) Susceptibility of *Phthorimaea operculella* (Zeller) to insecticides. J Entomol Sci 31:420–426

Lösel P, Penners G, Potting R, Ebbinghaus D, Elbert A, Scherkenbeck J (2000) Laboratory and field experiments towards the development of an attract and kill strategy for the control of the codling moth, *Cydia pomonella*. Entomol Exp Appl 95:39–46

Loza AL, Bravo R (2001) Poblaciones de carabidos (Coleoptera) en agroecosistemas del altiplano peruano. Rev Per Ent 42:79–87

Mazlan N, Munford J (2005) Insecticide use in cabbage pest management in the Cameron Highlands, Malaysia. Crop Prot 24(1):31–39

Mujica N, Kroschel J (2011) Leafminer fly (Diptera: Agromyzidae) occurrence, distribution and parasitoid associations in field and vegetable crops along the Peruvian coast. Environ Entomol 40(2):217–230

Mujica N, Fonseca C, Suarez F, Fabian F, Marchena M, Cisneros F (2000) Reducción del uso de insecticidas en el control de la mosca minadora, *Liriomyza huidobrensis* Blanchard, por medio del uso de técnicas etológicas. In: Arning I, Lizárraga A (eds) Control Etológico.

Uso de feromonas, trampas de colores y luz para el control de plagas en la agricultura sostenible. Red de Acción en Alternativas al Uso de Agroquímicos (RAAA), Lima, Peru

Mujica N, Pravatiner M, Cisneros F (2001) Effectiveness of abamectin and plant-oil mixtures on eggs and larvae of the leafminer fly, *Liriomyza huidobrensis* Blanchard. In: International Potato Center (ed) Scientist and farmer: partners in research for the 21st century. Program report 1999–2000. CIP, Lima, Peru

Murphy ST, La Salle J (1999) Balancing biological control strategies in IPM of New World invasive *Liriomyza* in field vegetable crops. Biocontrol News INF 20(3):91–104

Oerke EC, Dehne HW, Schönbeck F, Weber A (1994) Crop production and crop protection. Estimated losses in major food and cash crops. Elsevier Science, Amsterdam, p 808

Olthof T, Broadbent B (1991) Control of chrysanthemums leafminer, *Liriomyza trifolii* with the entomopathogenic nematode *Heterorhabditis heliothidis*. Nematologica 36:379

Orozco FA, Cole DC, Forbes G, Kroschel J, Wanigaratne S, Arica D (2009) Monitoring adherence to the International Code of Conduct – highly hazardous pesticides in central Andean agriculture and farmers' rights to health. J Occup Env Health 15(3):255–268

Ortiz O, Alcazar J, Catalan W, Villano W, Cerna V, Fano H, Walker T (1996) Economic impact of IPM practices on the Andean potato weevil in Peru. In: Walker T, Crissman C (eds) Case studies of the economic impact of CIP-related technologies. International Potato Center, Lima, 157pp

Ortiz O, Kroschel J, Alcázar J, Orrego R, Pradel W (2009) Evaluating dissemination and impact of IPM: lessons from case studies of potato and sweetpotato IPM in Peru and other Latin American countries. In: Peshin R, Dhawan AK (eds) Integrated pest management: dissemination and impact. Springer, Dordecht, pp 419–434

Parsa S, Alcazar J, Salazar J, Kaya H (2006) An indigenous Peruvian entomopathogenic nematode for suppression of the Andean potato weevil. Biol Control 39:171–178

Poinar G (1990) Biology and taxonomy of Steinernematidae and Heterorhabditidae. In: Gaugler R, Kaya H (eds) Entomophathogenic nematodes in biological control. Boca Raton, Florida, pp 23–61

Raman KV, Alcazar J (1990) Peruvian virus knocks Potato tuber moth. In paper presented at the 74th annual meeting of Potato Association of America, Québec, Canada

Raman K, Radcliffe EB (1992) Pest aspects of potato production. Part 2. Insect pests. In: Harris PM (ed) The potato crop: the scientific basis for improvement, 2nd edn. Chapman and Hall, London, pp 476–506

Riechert SE, Lawrence K (1997) Test for predation effects of single versus multiple species of generalist predators: spiders and their insect prey. Entomol Exp Appl 84:147–155

Rios A, Kroschel J (2011) Evaluation and implications of Andean potato weevil infestation sources for its management in the Andean region. J Appl Entomol 135(10):738–748

Rondon SI, Vergara C (2003) Effects of insecticide applications in the population dynamic of *Labiduria riparia* Phallas (Dermaptera: Labiduridae) in four cultivars of sweet potato. Rev Per Entomol 43:121–127

Schmidt MH, Thies C, Tscharntke T (2004) The landscape context of arthropod biological control. In: Gurr GM, Wratten SD, Altieri MA (eds) Ecological engineering for pest management: advances in habitat manipulation for arthropods. CSIRO Press, Collingwood

Schuller S, Sanchez G (2003) Los artrópodos del suelo depredadores en agroecosistemas de maíz en el valle de Chancay, Lima, Perú. Rev Per Entomol 43:47–58

Sporleder M, Kroschel J, Gutierrez Quispe MR, Lagnaoui A (2004) A temperature-based simulation model for the potato tuberworm, *Phthorimaea operculella* Zeller (Lepidoptera; Gelechiidae). Environ Entomol 33:477–486

Sporleder M, Simon R, Juarez H, Kroschel J (2008) Regional and seasonal forecasting of the potato tuber moth using a temperature-driven phenology model linked with geographic information systems. In: Kroschel J, Lacey L (eds) Integrated pest management for the potato tuber moth – a potato pest of global proportion. Tropical Agriculture 20, Advances in Crop Research 10. Margraf Publishers, Weikersheim, Germany

Sporleder M, Simon R, Gonzales J, Chavez D, Juarez H, Mendiburu F, Kroschel J (2011) ILCYM: a software for developing insect pest phenology models based on temperature and for pest risk

mapping (English version). Manual for ILCYM (Insect Life Cycle Modeling), version 2.0. International Potato Center (CIP), Lima, Peru, 78 pp
Trumble JT (1998) IPM: Overcoming conflicts in adoption. Integr Pest Manage Rev 3:195–207
Tscharntke T, Klein A, Kruess A, Steffan-Dewenter I, Thies C (2005) Landscape perspectives on agricultural intensification and biodiversity – ecosystem service management. Ecol Lett 8:857–874
Winters P, Fano H (1997) The economics of biological control in Peruvian potato production. Social Science Department, Working paper No. 1997-7. International Potato Center (CIP), Lima, Peru
Young JEB, Griffin MJ, Alford DV, Ogilvy SE (2001) Reducing agrochemical use on the arable farm: the TALISMAN and SCARAB projects. DEFRA, London, 416pp

Part VIII
Improvement of Potato Production Systems by Organic Amendments: Italian and Egyptian Reports

Chapter 26
Soil Fertility Management in Organic Potato: The Role of Green Manure and Amendment Applications

Stefano Canali, Corrado Ciaccia, and Fabio Tittarelli

Abstract During the last decade in the European Union, the organic food and farming (OFF) sector has grown considerably. The potato (*Solanum tuberosum* L.) crop plays an important role in organic farming systems, being one of the most highly demanded products on the market for organic produce. In this chapter, the role of green manure and organic amendments application for soil fertility management in agro-ecosystems based on organically managed potato crop is discussed in the light of the most relevant scientific literature. Moreover, as a case study, the results of a field experiment designed to evaluate the combined effects of green manure and organic amendment applications on organic potato yield, nitrogen (N) use efficiency and soil mineral N dynamic are presented. Our results indicated that legume green manure management and the recycling of organic materials may provide a valid alternative to the conventional synthetic fertilizer-based management system to sustain potato yield without enhancing potential environmental risks due to N leaching. Our study demonstrated that ecofunctional intensification of potato-based organically managed cropping systems is achievable through the exploitation of the combined effect of legume green manure with organic amendments application.

26.1 Introduction

During the last decade, in the European Union, the organic food and farming (OFF) sector has grown considerably and great interest has been directed to its capacity to provide safe, quality food, while preserving the environment and addressing the

---

S. Canali (✉) • C. Ciaccia • F. Tittarelli
Research Centre for the Soil-Plant System, Research Council for Agriculture,
Via Navicella 2, I-00184 Rome, Italy
e-mail: stefano.canali@entecra.it

Z. He et al. (eds.), *Sustainable Potato Production: Global Case Studies*,
DOI 10.1007/978-94-007-4104-1_26,

Table 26.1 Area under organic potato production, percentage of organic potato in total organic and total potato production in selected EU (European Union) and extra EU countries (years 2007–2008)

Country		Area under organic potato(ha)	Organic potatoes in total organic production(%)	Organic potatoes in total potato production(%)
EU countries	Denmark	1,268	0.84	3.08
	Germany	8,150	0.90	2.96
	Italy	1,014	0.10	1.41
	Netherland	1,271	2.52	0.79
	Poland	1,861	0.59	0.33
	Romania	143	0.10	0.01
	United Kingdom	3,270	0.44	2.35
Extra EU countries	Azerbaijan	251	1.18	0.37
	Canada	447	0.07	0.28
	Morocco	50	1.45	0.08
	South Africa	398	0.91	0.69
	Turkey	136	0.12	0.09
	Ukraine	79	0.03	0.01
	United States	3,348	0.17	0.73

socio-cultural and economic requirements of farmers and society (European Commission 2010). The total organic area in EU-27 continues to show an upward trend and the fully converted and under conversion area was 7,764,722 ha in 2008, revealing an increase of 7.4% relative to 2007 (Eurostat 2010).

The potato (*Solanum tuberosum* L.) crop plays an important role in organic farming systems, being one of the most highly demanded products on the market for organic produce (Tamm et al. 2004). According to the data reported by FAO and organic-world.net statistics, in many countries, organic potato cultivation represents an important crop within organic agriculture (Table 26.1).

According to the International Federation of Organic Agricultural Movements (IFOAM), basic principles for soil health and quality management in organic farming rely on the returns of microbial, plant, and animal organic material to the soil. Cultivation techniques should contribute to increasing soil biological activity and nutrient inputs and must be applied in a way that does not harm soil, water, and biodiversity (IFOAM 2010). Accumulation and decomposition of soil organic matter are, therefore, at the basis of fundamental life-promoting processes representing a key issue of organic production. With a similar approach, the current European Regulation for organic food and farming (EC Regulation 834/2007) established that soil fertility management is based on multi-annual crop rotation, recycling organic materials, choice of appropriate – locally adapted – plant species, crop varieties and cultivation techniques. Additional inputs for fertility management (i.e. fertilizers and soil conditioners) should be used only if they are compatible with the objectives and principles of organic production (i.e. the use of chemically synthesized inputs is strictly limited).

Under this conceptual-regulative framework, wastes and by-products of plant and animal origin should be recycled to return nutrients and organic carbon (C) to soils. Organic amendments (i.e. manure and compost) are thus a useful tool for soil fertility management and long-term sustainability of crop production. Indeed, many studies have noted the benefits of organic sources in improving soil structure, water holding capacity, root development and soil microbial activity (Sommerfeldt et al. 1988; Gilley and Risse 2000; Grandy et al. 2002; Lynch et al. 2008).

Similarly, utilisation of cover crops within cropping systems is consistent with the main organic principles, and they are introduced in the rotation with the primary aim of increasing soil organic matter content and/or nutritive elements availability to crops (Doran and Smith 1991; Lu et al. 2000; Dabney et al. 2001; Lenzi et al. 2009).

In this chapter, the role of green manure and organic amendment applications for soil fertility management in agroecosystems based on organically managed potato crop is discussed, in light of the most relevant scientific literature. Moreover, as a case study, results from a field experiment designed to evaluate the combined effects of green manure and organic amendment applications on organic potato yield, nitrogen (N) use efficiency and soil mineral N dynamic are presented.

26.2 Organic Potato Research in Europe

In accordance with the importance of the crop, a large number of studies about organically grown potatoes have been carried out in the last several years (i.e. Mirabelli et al. 2005; Speiser et al. 2006; Flier et al. 2007; Hagman et al. 2009). In more depth, American Journal of Potato Research and Potato Research published 63 papers on organic potato between 2001 and 2010. Furthermore, the bibliographic research carried out by the Springerlink search engine showed that among all the found papers dealing with organic potatoes cultivation, 27 (43%) were referred to European experiences, 7 (11%) dealt with control of pests and diseases, 6 (10%) had potato breeding as main subject and 6 (10%) were related to agronomical strategies. Likewise, looking for the keywords "organic farming" and "potato" between 2000 and 2011 in the SCIRUS search engine, 15 papers were found. Fourteen of them focused on researches carried out in the EU: among these, 5 referred to cultivation techniques, 2 to pest and disease control and 2 to plant breeding for organic agriculture.

As far as soil fertility management is specifically concerned, a number of recent studies were performed by several authors with particular attention to N management (i.e. Döring et al. 2005; Haase et al. 2007a, b). Tuber yield response was indeed dependent on the rate at which N was released from precedings crops residues (Stockdale et al. 1992; Köpke 1995; van Delden 2001) and highly responsive to N fertilization, cover crop and manure treatments (Reents and Möller 2000; Bélanger et al. 2001; Sincik et al. 2008; Zelalem et al. 2009). At the same time, efficient use of N fertilizer was found to be essential to increase the economic return of the crop and minimize potentially negative effects on water and air quality (Harris 1992). Thus, the typical organic potato production was characterized by extended rotations involving leguminous crops as green manures and/or organic amendments utilization (Lynch et al. 2008).

26.3 Green Manure Utilisation in Potato-Based Crop Rotations

Cover crops help to maintain soil organic matter, improve soil health, prevent and slow erosion and assist in nutrient management, enhancing their availability (Lu et al. 2000; Dabney et al. 2001; Lenzi et al. 2009). They can also contribute to weed management (Creamer et al. 1996; Teasdale 1996; Lu et al. 2000; Davis 2010), increase water infiltration (Dabney et al. 2001), maintain or increase populations of beneficial fungi (Galvez et al. 1995), and help with the management of insect pests, diseases and nematodes (Mojtahedi et al. 1991; Johnson et al. 1992; Tillman et al. 2004; Larkin et al. 2010). Cover crop utilisation and management are hence an important research lode for organic agriculture.

As a function of the different types of services provided by cover crops and in relation to the agro-ecosystem characteristics in which they are implemented, a range of cover crop families and species have been introduced in organic potato agro-ecosystems. In general, non-leguminous cover crops (sunflower; crucifer; cereals, such as rye and barley) are beneficial because they generate organic matter, compete with weeds and help in preventing soil erosion. They may be utilized as a N catch crop when planted either before and after potato, optimizing N utilisation during the whole rotation system (i.e. rye; Evanylo 1991; Reents and Möller 2000; Larkin and Griffin 2007). In particular, planted after cash crops (i.e. potato), when the soil is still warm and microbes are releasing nitrates, they capture N that otherwise might be leached from the soil (Jégo et al. 2008). Moreover, some non-leguminous cover crops, such as winter rye, ryegrass, brassicas and buckwheat, have also been shown to reduce soil-borne diseases when used in rotation with potatoes (Edwards 1986; Boydson and Hang 1995).

On the other hand, leguminous crops can be planted as full season cover crops with a cereal nurse crop (e.g.: small red clover undersown in oats, barley or wheat) or as the sole cover crop in the year before potatoes (Odland and Sheehan 1957; Schmidt et al. 1999). They are commonly terminated as green manures, and it has been estimated that the effect of a preceding leguminous cover crop was equivalent to the application of N fertilisers for a total of about 20–150 kg N ha^{-1} (Doran and Smith 1991; Ledgard 2001). Indeed, it is generally accepted that the release of N from decomposing green manure residues may be well-timed with plant uptake, possibly increasing N uptake efficiency and crop yield while reducing N leaching losses (Bath et al. 2006). For these reasons, they are particularly useful when preceding potato, which require high N levels (Sincik et al. 2008).

26.4 Organic Amendment Utilisation in Potato Crops

Potato rotations are often characterized by low levels of soil organic matter and consequently exhibit a poor soil physical condition. This is attributable to relatively low organic C inputs and the sandy soil types generally associated with potato production, which have a limited capability to retain organic C (Carter et al. 2003). Mallory and Porter (2007), in a long-term experiment carried out in Maine (USA), found that soil management based on the addition of organic amendments enhanced potato yield and reduced year-to-year variability of those yields. Consequently, the use of organic amendments is a common feature in intensive potato production systems (e.g., Gagnon et al. 2001; Grandy et al. 2002). Crops grown on soil that received organic amendments have been shown to have access to greater soil moisture and to be more resistant to weed and insect pressure (Gallandt et al. 1998; Liebig and Doran 1999; Lotter et al. 2003; Alyokhin et al. 2005). On the other hand, Lynch et al. (2008), evaluating the nutritive effects of N applied by organic amendments to an organic potato field, observed that tuber yield was decreased by the application of a swine manure-sawdust compost with a C/N ratio of 22 and attributed this effect to N net immobilization. In fact, the C/N ratio is generally considered to be a useful, if only approximate, guide to likely net mineralization, and it is generally accepted that easily decomposable organic materials, characterized by a C/N ratio below 20, release N on decomposition, but that material with a C/N above 20, immobilizes N temporarily (Whitmore 2007). Organic amendment use in consolidated production practices, therefore, requires some knowledge of N rate of release. In particular, for compost, a large fraction of total N (>90%) is not easily available for plant uptake (Amlinger et al. 2003). The greatest fraction is bound to the organic N-pool, while mineral, readily plant available N represents less than 2% of the total N content (Day and Shaw 2001). Indeed, according to the literature, in the first year after compost addition to the soil, available N is less than one fifth of the total N applied (Hadas and Portnoy 1994; Coutinho et al. 2006; Zhang et al. 2006). Therefore, in the short-term, compost does not fulfill the N needs of crops, so its use as the sole source of N for crops is not recommended. In their study on different composts as N source on a crop succession including potato and catch crops, Passoni and Borin (2009) found that crop response and N uptake were scarcely affected by compost fertilization. On the other hand, Carter et al. (2004), in their experiment aiming to test the influence of compost application on a potato rotation, found an increase in tuber yield, even above the maximum yield obtained with N application. This "non-nitrogen" compost yield effect was proposed to be related to the slight, but significant, improvement in soil water-holding capacity.

Similarly, farmyard manure (FYM) in organic farming systems plays a very important role for crop nutrition and the maintenance of soil fertility (Mäder et al. 2002). However, different authors reported the use of manure in organic agriculture hampered the optimization of more than one nutrient in terms of the nutrition of the potato crop (Dewes and Hünsche 1998; Shepherd et al. 2002). In particular, they showed the low potential of manure to increase plant available N and tuber N uptake. Indeed, Stein-Bachinger and Werner (1997) stated that N from FYM is usually not readily available in the season of application. Consequently, the utilisation of organic amendments can give the best performance on crop productivity when they are not utilised as the only input resources in soil fertility management strategies for the potato crop. Thus, combined applications of manure/compost with cover crop/green manures could potentially enhance the effectiveness of organic amendment fertility and improve yields and soil fertility, while at the same time reducing the risk of nutrient (i.e. N) leaching (Singer et al. 2008; Cambardella et al. 2010).

26.5 Effects of Green Manure-Soil Amendment Interactions on N Availability for Organic Potatoes

As is already well-known, restricted availability of any one of the needed nutrients will result in growth reductions that may also reduce crop quality and yield. To avoid this risk, nutrients must be available from the soil in amounts that meet the minimum requirements for the whole plant. This requirement is much higher than just the nutrients removed from the harvested yield. All the plant parts require nutrients at specific times during plant growth and development. Nutrients such as N and phosphorus (P) often move beyond the bounds of the agricultural field because the management practices used fail to achieve good congruence between nutrient supply and crop nutrient demand (van Noordwijk and Cadisch 2002). Soil fertility management should be optimized to supply nutrient requirements at the appropriate time and at sufficient levels, to support healthy plant growth. In this context, timing of organic amendment (e.g. compost) application within the rotation may contribute to achieve these results (Willson et al. 2001). In this regard, agro-ecosystem management strategies should be balanced to obtain high short-term efficiency as well as maximizing the cumulative crop yield response over time (Dobermann 2007).

Concerning N, synchronizing its release with plant requirements is indeed important with the double aim to promote yield and limit N leaching (Bath 2000). Potato crops show a relatively low ability to take up available soil mineral nitrogen (SMN), (Tyler et al. 1983; Dilz 1987). Since it is impossible to accurately predict the total crop N requirements and soil mineral N supply during the growing season, in conventional farming splitting of N fertilizer application is a suitable approach to better match N need and supply (Vos and MacKerron 2000; Goffart et al. 2008). With reference to organically managed cropping systems, where the use of high soluble/easy mineralisable inputs for fertility management is not promoted (EC Regulation 834/2007), Sikora and Enkiri (2000) observed that to optimize the use of nutrient sources, to enhance N supply in relation to crop demand and to achieve maximum crop yields, the combination of different N sources as cover crop residues, manures and composts may be an effective approach. Accordingly, Nyiraneza and Snapp (2007) reported that potato N uptake and mineralization of N from organic sources could be synchronized if a mixture of different residue qualities are used, including low N (high C/N ratio, such as compost) and high N (low C/N ratio, such as legume green manure) tissues.

In light of the above reported consideration, and despite the potential of organic amendments and legume green manure cover crops in managing soil N fertility, only a few studies have investigated the combined effect of green manure and different types of amendment applications on organic potato yield and environmental impact (Bath et al. 2006; Nyiraneza and Snapp 2007). Moreover, none were carried out under Mediterranean conditions.

26.6 Organic Potato Under Mediterranean Conditions: A Case Study

As already mentioned above, the primary challenge in organic potato systems, and perhaps even more so in general organic agriculture, is synchronizing nutrient release from organic sources, particularly N, with crop requirements. In this context, a field experiment was carried out in Tuscany (Central Italy) with the objective to assess the contribution of farmyard manure and compost utilised in combination with a green manure legume cover crop (*Trifolium subterraneum* L.) to potato crop nutrition, evaluating potato yield, N uptake and use efficiency (Canali et al. 2010). In the experiment, SMN dynamic was also studied to evaluate the potential impact on the environment.

Climate in the area is typical Mediterranean; the monthly mean minimum (January) and maximum (July) temperatures are 9°C and 20°C respectively. Rainfall (average 900 mm year^{-1}) is unevenly distributed during the year, being concentrated mainly in the winter months.

The experimental field, according to a split-plot layout was divided into two main strips, representing different management systems, in which a subterranean clover (*Trifolium subterraneum* L.) green manure was cultivated (GM+) or not (GM–). Within each strip, elementary plots, which received farmyard manure (FYM) or green compost (C) at three different rates, corresponding to an amount of 0, 50 and 100 kg N ha^{-1} (0, 50 and 100), were randomly distributed (Fig. 26.1). N applied to the soil by the green manure accounted for about 20 kg ha^{-1}. Compost was produced starting from green (garden) residues collected in the area. The compost heavy metals concentration complied with the European Regulation on organic farming (EC 834/2007) and the Italian regulation on organic fertilizers and amendments (Legislative Decree 217/2006). Cattle farmyard manure was obtained from an organic animal farm located close to the experimental site. The main characteristics of the two amendments used are reported in Table 26.2.

During the cropping cycle and at harvest, potato yield, total and above-ground biomass, and total N content (Bremner and Mulvaney 1982) were determined, allowing the calculation of Total N uptake (N content x biomass dry matter). On the basis of these measurements, the following parameters were calculated:

- harvest index (HI) as the ratio of the tuber yield to total biomass (Jennings 1964);
- N harvest index (NHI) as the ratio of the tuber N uptake to total N (Montemurro 2009);
- N utilization efficiency (NUE) as the ratio of tuber yield to total N uptake (Montemurro 2009).

Fig. 26.1 Farmyard manure (FYM) and Compost (C) application to the plots in the experimental site

Table 26.2 Main chemical characteristics of organic amendments (on dry matter basis) (Canali et al. 2010)

	FYM	C
Total C[a] (g kg^{-1})	386	285
Total N[b] (g kg^{-1})	23	12
C/N	17	24
pH	8.9	7.9
Ashes[c] (g kg^{-1})	228	480
Cd[d] (mg kg^{-1})	0.1	0.3
Hg[d] (mg kg^{-1})	<1.5	<1.0
Cu[d] (mg kg^{-1})	20.1	22.3
Zn[d] (mg kg^{-1})	117.9	128.9
Ni[d] (mg kg^{-1})	7.5	22.3
Pb[d] (mg kg^{-1})	5.0	76.9
Cr VI[d] (mg kg^{-1})	<0.5	<0.5

All analytical data are reported as the mean of three replicate determinations

FYM cattle farmyard manure, *C* green wastes compost (crop residues, pruning materials and lawn mowing)

[a]Springer and Klee (1954)

[b]Kjeldahl method (Bremner and Mulvaney 1982)

[c]residue to weight loss at 400°C

[d]ICP-AES after incineration of compost samples at 400°C for 24 h and elemental extraction in acidic environment

Table 26.3 Mean effects of management system, type of amendment and dose on tuber yield, above-ground biomass, total biomass, harvest index, and N indexes on potato (Canali et al. 2010)

	Tuber yield(t ha^{-1})	Aboveground biomass(t ha^{-1})	Total biomass (t ha^{-1})	HI(%)	NHI(%)	NUE
Management system						
GM+	6.80 a	2.90 a	9.70 a	70.4	65.1 b	222 b
GM–	5.55 b	1.78 b	7.33 b	73.5	68.9 a	269 a
Amendment						
FYM	6.86 a	2.16	9.02	74.5 a	71.6 a	263 a
C	5.48 b	2.53	8.01	69.5 b	62.4 b	227 b
Dose						
0	5.27 b	2.33	7.60	69.5	70.1	225 b
50	5.70 b	2.24	7.94	72.0	65.0	249 a
100	7.55 a	2.45	10.0	74.4	66.0	263 a
Means	6.17	2.35	8.51	72.0	67.0	246

The mean values in each column followed by a different letter are significantly different according to LSD and DMRT (two and more than two comparison. respectively) at the $P \leq 0.05$ probability level *GM+* rotation including green manure, *GM–* rotation without green manure, *FYM* farmyard manure, *C* compost, *HI* harvest index (tuber yield/total biomass), *NHI* N harvest index (tuber N uptake/total N uptake), *NUE* N utilization efficiency (tuber yield/total N uptake)

At the same sampling times SMN was measured: NO_3^--N and NH_4^+-N were extracted by 2 M KCl and determined by continuous flowing system (Henriksen and Selmer-Olsen 1970; Krom 1980).

Results were analyzed using univariate analysis of variance (ANOVA) considering the management system (GM+ and GM–), amendment (FYM and C) and dose (0, 50, and 100 kg ha^{-1}) as fixed factors. Means comparison was carried out according to the Least Square Difference (LSD) test and the Duncan Multiple Range Test (DMRT), both at $P \leq 0.05$ probability level, for two and more than two comparisons, respectively.

26.6.1 *Combining Green Manure and Organic Amendments: Effects on Yield and N Dynamics*

The GM+ system presented significantly higher values of above-ground biomass (62.9%) and total biomass (32.3%) relative to the GM–. Similarly, the total potato yield showed an increase of 22.5 and 25.1% of the GM+ treatment with respect the GM– treatment and of the FYM in comparison with C, respectively (Table 26.3). These results are in accordance with the study of Sincik et al. (2008) in which the responses of potato to green manure cover crops, combined with different N fertilization rates, was evaluated. In their experiment, they reported that green manure legume cover crops resulted in a 35% increase on tuber yield compared with potatoes following winter wheat, when no N fertilizer was applied.

These results showed an increase of 43.3 and 16.9% for potato production and N use efficiency for the highest dose of organic amendments relative to the unfertilised control (Table 26.3). Tuber yield also showed significant differences between the amendment with higher value for FYM than the C treatment. This achievement is probably related to the greater capability of the FYM to release mineral N in accordance with the lower C/N ratio with respect to C (Table 26.2). These results seemed to be in discordance with findings of Willekens et al. (2008) who, in their 4-year rotation comparison trial between farmyard manure and compost, found significantly higher tuber yield in treatments with compost addition. However, they explained this achievement as due to the higher N input by manure which promoted the plant growth and the subsequent leaf blight infection (*Phytophtora infestans;* Möller and Reents 2007).

As far as the HI is concerned, results obtained in the whole experiment showed a high average value (72.0%), similar to the results of Neele (1990) and higher than those reported by Mussaddak (2007), which ranged from 42% to 56% and from 50% to 63% for spring and fall potato, respectively. No significant differences were observed for the amendment dose and the management system treatments, whereas, HI was significantly higher for FYM than for C. Significant increase of both NUE and NHI indices was found in GM− treatment with respect to GM+, showing that there was a higher efficiency in translocation of the absorbed N in the yield components under lower N levels in the soil, which occurred when clover green manure was not previously cultivated. Similarly, the FYM showed a significant increase in NUE and NHI relative to the C treatment, which followed the same pattern regarding tuber yield. No significant differences were observed between the amendment dose treatments for NHI. Meanwhile, NUE showed similar values for the 50 and 100 doses, significantly higher than control (0 dose). This findings suggested the lack of differences in translocation ability when different doses of amendments were applied (NHI), while the absence of differences of NUE in the 50 and 100 kg amendment treatments may reflect a poor crop use of added N at the highest dose.

SMN content showed a similar trend in all doses, with an increase between 6 and 34 DAP, probably due to organic materials mineralization, followed by a decrease from 34 to 82 DAP (harvest) (Fig. 26.2). As reported by Paré et al. (1995) and Jowkin and Schoenau (1998), progressive SMN depletion can be attributable to the increasing plant N uptake along the potato cropping cycle. At the end of potato cropping cycle a slight and significant difference in SMN (about 10 kg N ha^{-1}) was found among the three amendment dose treatments, with the control (dose 0) having the lowest value. However, at the end of the cropping cycle the combination of clover green manure and amendment applications did not increase the SMN (no significant interaction between the two factors, data not showed). These results demonstrated that the combination of legume green manure with amendment applications did not contribute to increasing the potential environmental risks due to N leaching.

Average values of above-ground biomass, tubers and whole plant N uptake measured throughout the potato cropping cycle are reported in Fig. 26.3. In accordance with the low tuber yield obtained (average of 6.17 t ha^{-1}respect to the Italian average marketable yield for the conventional crop estimated in 23.6 t ha^{-1}; Table 26.3) a low level of tuber N uptake at harvest was found, ranging between 11 for GM− and

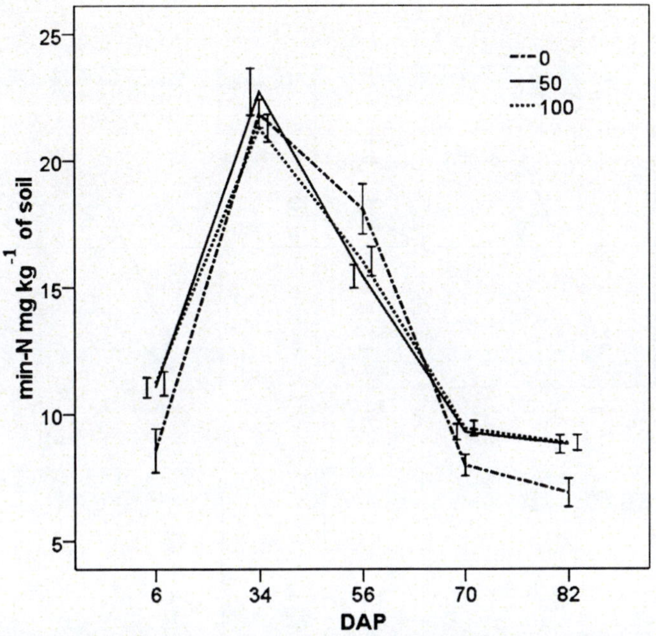

Fig. 26.2 Mineral soil N divided per dose (0, 50, 100 kg ha^{-1}). Data are obtained considering the mean of system and amendment treatments. (Note: *DAP* days after planting; bars represent the confidence interval at the $P \leq 0.05$ probability level) (Canali et al. 2010)

20 kg ha^{-1} for GM+ (significant difference). Similarly, the above-ground biomass and tubers N uptake showed the same trend in all systems and amendment treatments. In particular, above-ground biomass N were 10 and 5 kg ha^{-1} for GM+ and GM– and 9 and 6 kg ha^{-1} for C and FYM, respectively. Considering the whole cropping cycle, total N uptake showed a similar trend in all the system x amendment combinations (Fig. 26.3a–d), resulting in a consistent increase of N uptake over time. At harvest the GM+ treatments showed significantly higher values with respect to the GM– systems (33% higher), explainable through the higher amount of N supplied by clover in GM+. This result was in accordance with the findings of Ten Holte and van Keulen (1989), who carried out an experiment to evaluate the responses of potato and sugarbeet to different levels of N fertilization and different green manures. Their results showed that N supplied by green manure to the potato crop became available over time, matching the potato needs during its cropping cycle. Looking to the whole cropping cycle, in all system x amendment combinations, up to 56 days after planting (DAP), above-ground biomass N was higher than tuber N uptake. Afterwards, the above-ground biomass N decreased and tuber uptake showed an opposite trend. N uptake by the above-ground biomass and tubers became approximately equal at 70 DAP. This pattern was due to N translocation from above to below ground biomass, and at harvest, the average N contained into the above-ground biomass and tubers was respectively 33% and 67% of total N uptake by the crop. This is in accordance with the study by Alva et al. (2002)

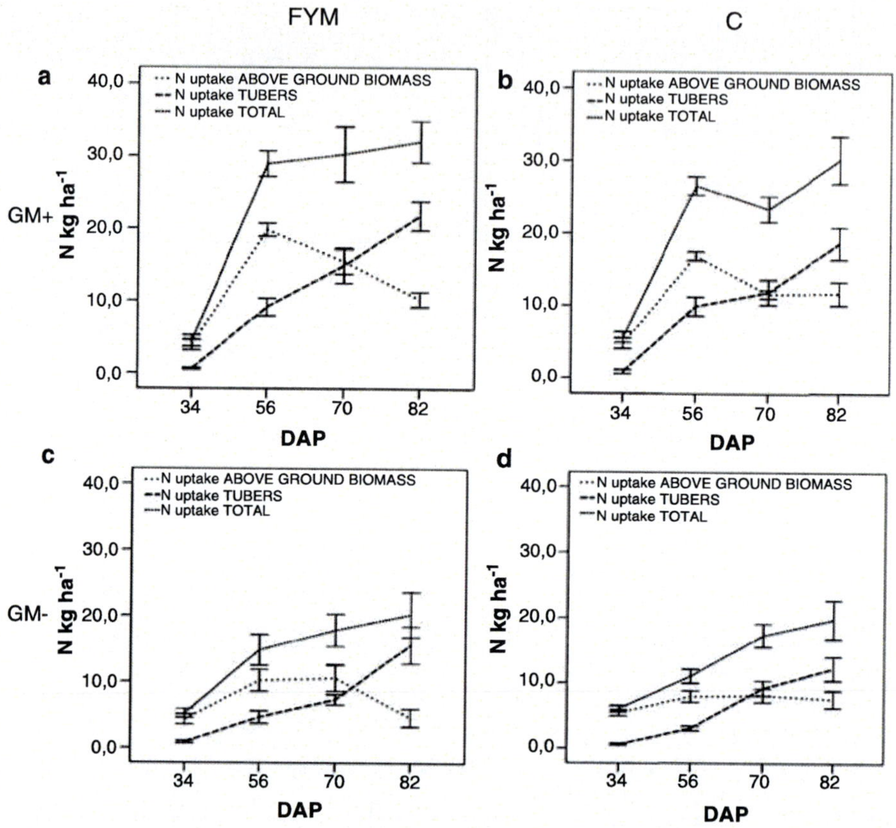

Fig. 26.3 Total N uptake, tubers N uptake and above-ground N uptake of potato as affected by amendment and system treatments. (Note: *DAP* days after planting, *FYM* farmyard manure, *C* compost, *GM+* green manure system, *GM−* no green manure. Bars represent the confidence interval at the $P \leq 0.05$ probability level) (Canali et al. 2010)

regarding N accumulation and partitioning in potato, in which an increase in tuber weight was recorded during 60–100 DAP, whereas the above-ground biomass decreased rapidly in the second half of the cropping cycle. Comparing the different system x amendment combinations, the results underlined the absence of influence of green manure and organic amendment on potato N nutrition physiology, indicating the lack of potential synergic effects of N deriving from green manure and organic soil amendments on N uptake plant physiology.

26.6.2 Conclusions

Typically, potato above-ground biomass is recycled into soil and the nutritive element applied to soil by this technique represents a valuable contribution to the N nutrition of the next crops, especially in organically managed cropping system, where N may become a limiting factor for crop nutrition. In confirmation of this, in the above reported experience, the results showed that the legume green manure management and the recycling of organic materials could represent valid alternatives to the conventional – synthetic fertilizers based – management to sustain potato yield. In particular, the combination of different sources of N seems to enhance the crop performance providing higher availability of the nutrient during the cropping cycle. Simultaneously, this approach did not enhance potential environmental risks due to N leaching. Thus our study demonstrated that ecofunctional intensification of organically managed cropping systems based on potato is achievable through the exploitation of the combined effect of legume green manure with organic amendments application.

References

Alva AK, Hodges T, Boydston RA, Collins HP (2002) Dry matter and nitrogen accumulations and partitioning in two potato cultivars. J Plant Nutr 25(8):1621–1630

Alyokhin A, Porter G, Groden E, Drummond F (2005) Colorado potato beetle response toil amendments: a case in support of the mineral balance hypothesis? Agric Ecosyst Environ 109:234–244

Amlinger F, Götz B, Dreher P, Geszti J, Weissteiner C (2003) Nitrogen in biowaste and yard waste compost: dynamics of mobilization and availability-a review. Eur J Soil Bid 39:W7–U6

Bath B (2000) Matching the availability of N mineralized for crops with the N-demand of field vegetables. Doctoral thesis, Swedish University of agricultural science, Upssala, Sweden

Bath B, Ekbladh G, Ascard J, Olsson K, Andersson B (2006) Yield and nitrogen uptake in organic potato production with green manures as pre-crop and the effect of supplementary fertilization with fermented slurry. Biol Agric Hortic 24:135–148

Bélanger G, Walsh JR, Richards JE, Milburn PH, Ziadi N (2001) Tuber growth and biomass partitioning of two potato cultivars grown under different N fertilization rates with and without irrigation. Am J Potato Res 78:100–117

Boydson RA, Hang A (1995) Rapeseed (*Brassica napus*) green manure crop suppresses weeds in potato (*Solanum tuberosum*). Weed Technol 9(4):669–675

Bremner JM, Mulvaney CS (1982) Nitrogen total. In: Page AL (ed) Methods of soil analysis. Part 2, vol 9, 2nd edn, Agronomy Monograph. ASA and SSSA, Madison, pp 595–624

Cambardella CA, Moorman TB, Singer JW (2010) Soil nitrogen response to coupling cover crops with manure injection. Nutr Cycl Agroecosyst 87:383–393

Canali S, Ciaccia C, Antichi D, Barberi P, Montemurro F, Tittarelli F (2010) Interactions between green manure and amendment type and rate: effects on organic potato and soil mineral N dynamic. J Food Agric Environ 8:537–543

Carter MR, Angers DA, Gregorich EG, Bolinder MA (2003) Charcterizing organic matter retention for surface soils in eastern Canada using density and particle size fractions. Can J Soil Sci 83:11–23

Carter MR, Sanderson JB, MacLeod JA (2004) Influence of compost on the physical properties and organic matter fractions of a fine sandy loam throughout the cycle of a potato rotation. Can J Soil Sci 84(2):211–218

Coutinho J, Dalla Costa L, Borin M, Battilani A, Doleïal F, Bízic J, Mazurczyk W, Plauborg F (2006) One single value for maximum N application from organic residues: is it technically and environmentally sound?. In: Proceedings of the IX ESA Congress, Warsaw, Poland

Creamer NG, Bennett MA, Stinner BR, Cardina J, Regnier EE (1996) Mechanisms of weed suppression in cover crop-based production systems. HortSci 31:410–413

Dabney SM, Delgado JA, Reeves DW (2001) Using winter cover crops to improve soil quality and water quality. Commun Soil Sci Plant Anal 32:1221–1250

Davis AS (2010) Cover-crop roller–crimper contributes to weed management in no-till soybean. Weed Sci 58:300–309

Day M, Shaw K (2001) Biological, chemical, and physical processes of composting. In: Stoffella P, Kahn B (eds) Compost utilization in horticultural cropping systems. Pub. Lewis Publishers, Boca Raton

Dewes T, Hünsche E (1998) Composition and microbial degradability in the soil of farmyard manure from ecologically-managed farms. Biol Agric Hortic 16:251–268

Dilz K (1987) Efficiency of uptake and utilization of fertilizer nitrogen by plants. In: Jenkinson DS, Smith KA (eds) Nitrogen efficiency in agricultural soils. Elsevier, London/New York, pp 1–26

Dobermann A (2007) Nutrient use efficiency: measurement and management. In: Proceedings of the IFA workshop on fertilizer best management practices, Brussels, Belgium, 7–9 Mar 2007. Int Fertil Ind Assoc, Paris

Doran JW, Smith MS (1991) Role of cover crops in nitrogen cycling. In: Hargrove WL (ed) Cover crops for clean water. Soil and Water Conservation Society, Ankeny, pp 85–90

Döring TF, Brandt M, Heß J, Finckh MR, Saucke H (2005) Effects of straw mulch on soil nitrate dynamics, weeds, yield and soil erosion in organically grown potatoes. Field Crop Res 94(2–3):238–249

EC Regulation 834/2007 of the Council, 28th June 2007. Official Journal of European Communities, L 189/1 of 20th July 2007

Edwards L (1986) Spring interseeding of winter rye with cover crops. J Soil Water Conserv 41(3):190–191

European Commission (2010) An analysis for the EU organic sector. Directorate General for Agriculture and Rural Development http://ec.europa.eu/agriculture/analysis/markets/organic_2010_en.pdf. Accessed 29 Aug 2010

Eurostat (2010) Statistics on focus, Agriculture and fishering http://epp.eurostat.ec.europa.eu/cache/ITY_OFFPUB/KS-SF-10-010/EN/KS-SF-10-010-EN.PDF. Accessed 20 Jan 2011

Evanylo GK (1991) Rye nitrogen cycling for corn and potato production. In: Hargrove WL (ed) Cover crops for clean water. Soil and Water Conservation Society, Ankeny, pp 101–103

FAO http://faostat.fao.org. Accessed 22 Feb 2011

Flier WG, Kroon LPNM, Hermansen A, van Raaij HMG, Speiser B, Tamm L, Fuchs JG, Lambion J, Razzaghian J, Andrivon D, Wilcockson S (2007) Genetic structure and pathogenicity of populations of Phytophthora infestans from organic potato crops in France, Norway, Switzerland and the United Kingdom. Plant Pathol 56(4):562–572

Gagnon B, Lalande R, Fahmy SH (2001) Organic matter and aggregation in a degraded potato soil as affected by raw and composted pulp residue. Biol Fertil Soil 34:441–447

Gallandt ER, Mallory EB, Alford AR, Drummond FA, Groden E, Liebman M, Marra MC, McBurnie JC, Porter GA (1998) Comparison of alternative pest and soil management strategies for Maine potato production systems. Am J Altern Agric 13:146–161

Galvez L, Douds DD Jr, Wagoner P, Longnecker LR, Drinkwater LE, Janke RR (1995) An over-wintering cover crop increases inoculum of VAM fungi in agricultural soil. Am J Altern Agric 10:152–156

Gilley JE, Risse LM (2000) Runoff and soil loss as affected by the application of manure. Trans Am Soc Agric Eng 43:1583–1588

Goffart JP, Olivier M, Frankinet M (2008) Potato crop nitrogen status assessment to improve N fertilization management and efficiency: past–present–future. Potato Res 51:353–381

Grandy AS, Porter GA, Erich MS (2002) Organic amendment and rotation crop effects on the recovery of soil organic matter and aggregation in potato cropping systems. Soil Sci Soc Am J 66:1311–1319

Haase T, Schüler T, Heß J (2007a) The effect of different N and K sources on tuber nutrient uptake, total and graded yield of potatoes (*Solanum tuberosum* L.) for processing. Eur J Agron 26:187–197

Haase T, Schüler C, Piepho HPP, Thöni H, Heß J (2007b) The effect of preceding crop and pre-sprouting on crop growth, N use and tuber yield of maincrop potatoes for processing under conditions of N stress. J Agron Crop Sci 193:270–291

Hadas A, Portnoy R (1994) Nitrogen and carbon mineralization rates of composted manures incubated in soil. Environ Qual 23:1184–1189

Hagman JE, Mårtensson A, Grandin U (2009) Cultivation practices and potato cultivars suitable for organic potato production. Potato Res 52(4):319–330

Harris PM (1992) Mineral nutrition. In: Harris PM (ed) The potato crop. Chapman & Hall, London, pp 162–213

Henriksen A, Selmer-Olsen AR (1970) Automatic methods for determining nitrate and nitrite in water and soil extracts. Analyst 95:514–518

IFOAM (2010) IFOAM basic standard. www.ifoam.org. Accessed 22 Feb 2011

Jégo G, Martinéz M, Antiqüedad I, Launay M, Sanchez-Pérez JM, Justes E (2008) Evaluation of the impact of various agricultural practices on nitrate leaching under the root zone of potato and sugar beet using STICS soil-crop model. Sci Total Environ 394:207–221

Jennings PR (1964) Plant type as a rice breeding objective. Crop Sci 4:13–15

Johnson AW, Golden AM, Auld DL, Sumner DR (1992) Effects of rapeseed and vetch as green manure crops and fallow on nematodes and soil-borne pathogens. J Nematol 24:117–126

Jowkin V, Schoenau JJ (1998) Impact of tillage and landscape position on nitrogen availability and yield of spring wheat in the Brown soil zone in southwestern Saskatchewan. Can J Soil Sci 78(3):563–572

Köpke U (1995) Nutrient management in organic farming systems – the case of nitrogen. Biol Agric Hortic 11(1–4):15–29

Krom MD (1980) Spectrophotometric determination of ammonia: a study of a modified Berthelot reaction using salicylate and dichloroisocyanurate. Analyst 105:305–316

Larkin RP, Griffin TS (2007) Control of soilborne potato diseases using Brassica green manures. Crop Prot 26(7):1067–1077

Larkin RP, Griffin TS, Honeycutt CW (2010) Rotation and cover crop effects on soilborne potato diseases, tuber yield, and soil microbial communities. Plant Dis 94(12):1491–1502

Ledgard SF (2001) Nitrogen cycling in low input legume-based agriculture, with emphasis on legume/grass pastures. Plant Soil 228:43–59

Legislative Decree 217 (2006) Revisione della disciplina in materia di fertilizzanti. Off J Ital Repub n. 141 of 20th June. Annex n. 152

Lenzi A, Antichi D, Bigongiali F, Mazzoncini M, Migliorini P, Tesi R (2009) Effect of different cover crops on organic tomato production. Renew Agric Food Syst 24(2):92–101

Liebig MA, Doran JW (1999) Impact of organic production practices on soil quality indicators. J Environ Qual 28:1601–1609

Lotter DW, Seidel R, Liebhardt W (2003) The performance of organic and conventional cropping systems in an extreme climate year. Am J Altern Agric 18:146–154

Lu Y, Watkins KB, Teasdale JR, Abdul-Baki AA (2000) Cover crop in sustainable production. Food Rev Int 16(2):121–157

Lynch HD, Zheng Z, Zebarth BJ, Martin RC (2008) Organic amendment effects on tuber yield, plant N uptake and soil mineral N under organic potato production. Renew Agric Food Syst 23(3):250–259

Mäder P, Fliebach A, Dubois D, Gunst J, Fried P, Niggli U (2002) Soil fertility and biodiversity in organic farming. Science 296:1694–1697

Mallory EB, Porter GA (2007) Potato yield stability under contrasting soil management strategies. Agron J 99:501–510

Mirabelli C, Colla G, Fiorillo A, Cardarelli M, Rouphael Y, Paolini R (2005) The effect of mechanical weed control technique and irrigation method on yield, tuber quality and weed suppression in organic potato, vol 684, Acta Horticulturae. International Society for Horticultural Science (ISHS), Leuven, pp 127–133

Mojtahedi H, Santo GS, Hang AN, Wilson JH (1991) Suppression of root-knot nematode populations with selected rapeseed cultivars as green manure. J Nematol 23:170–174

Möller K, Reents HJ (2007) Impact of agronomic strategies (seed tuber pre-sprouting, cultivar choice) to control late blight (*Phytophthora infestans*) on tuber growth and yield in organic potato (*Solanum tuberosum* L.) crops. Potato Res 50(1):15–29

Montemurro F (2009) Different nitrogen fertilization sources, soil tillage, and crop rotations in winter wheat: effect on yield, quality, and nitrogen utilization. J Plant Nutr 32(1):1–18

Mussaddak J (2007) Efficiency of nitrogen fertilizer for potato under fertigation utilizing a nitrogen tracer technique. Commun Soil Sci Plant Anal 38:2401–2422

Neele AEF (1990) Study on the inheritance of potato tuber yield by means of harvest index component and its consequences for choice of parental material. Euphytica 48:159–166

Nyiraneza J, Snapp S (2007) Integrated management of inorganic and organic nitrogen and efficiency in potato systems. Soil Sci Soc Am J 71(5):1508–1515

Odland TE, Sheehan JE (1957) The effect of redtop and red clover on yields of following crops of potatoes. Am Potato J 34:282–284

Organic-world.net http://www.organic-world.net. Accessed 22 Feb 2011

Paré T, Gregorich EG, Ellert BH (1995) Comparison of soil nitrate extracted by potassium chloride and adsorbed on an anion exchange membrane in situ. Commun Soil Sci Plant Anal 26(5–6):883–898

Passoni M, Borin M (2009) Effects of different composts on soil nitrogen balance and dynamics in a biennial crop succession. Compost Sci Util 17(2):108–116

Reents HJ, Möller K (2000) Effects of different green manure catch crops grown after peas on nitrate dynamics in soils and on yield and quality of subsequent potatoes and wheat. In: Alfodi T et al (eds) Proceedings of the 13th international IFOAM scientific conference, IFOAM, Zürich, pp 73–76

Schmidt H, Philipps L, Welsh JP, von Fragstein P (1999) Legume breaks in stockless organic farming rotations: nitrogen accumulation and influence on the following crops. Biol Agric Hortic 17:159–170

Shepherd M, Philipps L, Jackson L, Bhogal A (2002) The nutrient content of cattle manures from organic holdings in England. Biol Agric Hortic 20:229–242

Sikora LJ, Enkiri NK (2000) Efficiency of compost-fertilizers blends compared with fertilizer alone. Soil Sci 165:444–451

Sincik M, Metin Turan Z, Tanju Goksoy A (2008) Responses of potato (*Solanum tuberosum* L.) to green manure cover crops and nitrogen fertilization rates. Am J potato Res 85:150–158

Singer JW, Cambardella CA, Moorman TB (2008) Enhancing nutrient cycling by coupling cover crops with manure injection. Agron J 100(6):1735–1739

Sommerfeldt TG, Chang C, Entz T (1988) Long-term annual manure applications increase soil organic matter and nitrogen, and decrease carbon to nitrogen ratio. Soil Sci Soc Am J 52:1668–1672

Speiser B, Tamm L, Amsler T, Lambion J, Bertrand C, Hermansen A, Ruissen MA, Haaland P, Zarb J, Santos J, Shotton P, Wilcockson S, Juntharathep P, Ghorbani R, Leifert C (2006) Improvement of late blight management in organic potato production systems in Europe: field tests with more resistant potato varieties and copper based fungicides. Biol Agric Hortic 23(4):393–412

Springer U, Klee J (1954) Prüfung der Leistungsfähigkeit von einigen wichtigeren Verfahren zur Bestimmung des Kohlemstoffs mittels Chromschwefelsäure sowie Vorschlag einer neuen Schnellmethode. Z Pflanzenernähr Dang Bodenk 64:1

Stein-bachinger K, Werner W (1997) Effects of manure on crop yield and quality in an organic agricultural system. Biol Agric Hortic 14:221–235

Stockdale EA, McKinlay RG, Rees RM (1992) Soil nitrogen management and interactions with pests and diseases in organic farming. Asp Appl Biol 30:387–392

Tamm L, Smit AB, Hospers M, Janssens SRM, Buurma JS, Molgaard JP, Laerke PE, Hansen HH, Hermans A, Bodker L, Bertrand C, Lambion L, Finckh MR, van Lammerts CE, Ruissen T, Nielsen BJ, Solberg S, Speiser B, Wolfe MS, Philipps S, Wilcoxon SJ, Leifert C (2004) Assessment of the socio-economic impact of late blight and state of the art management in European organic potato production systems. http://orgprints.org/2936. Accessed 1 Feb 2011

Teasdale JR (1996) Contribution of cover crops to weed management in sustainable agricultural systems. J Prod Agric 9:475–479

Ten Holte L, van Keulen H (1989) Effects of white clover and red clover as a green crop on growth, yield and nitrogen of sugar beet and potatoes. Dev Plant Soil Sci 37:16–24

Tillman G, Schomberg H, Phatak S, Mullinix B, Lachnicht S, Timper P, Olson D (2004) Influence of cover crops on insect pests and predators in conservation tillage cotton. J Econ Entomol 97(4):1217–1232

Tyler KB, Broadbent FE, Bishop JC (1983) Efficiency of nitrogen uptake by potatoes. Am Potato J 60:261–269

van Delden A (2001) Yield and growth components of potato and wheat under organic nitrogen management. Agron J 93:1370–1385

van Noordwijk M, Cadisch G (2002) Access and excess problems in plant nutrition. Plant Soil 247:25–40

Vos J, MacKerron DKL (2000) Basic concepts of the management of supply of nitrogen and water in potato production. In: Haverkort AJ, MacKerron DKL (eds) Management of nitrogen and water in potato production. Wageningen, Wageningen, the Netherlands, pp 15–33

Whitmore AP (2007) Determination of the mineralization of nitrogen from composted chicken manure as affected by temperature. Nutr Cycl Agroecosyst 77:225–232

Willekens K, De Vliegher A, Vandecasteele B, Carlier L (2008) Effect of compost versus animal manure fertilization on crop development, yield and nitrogen residue in the organic cultivation of potatoes. In: Proceedings of the 16th IFOAM organic world congress, Modena, Italy, 16–20 June 2008

Willson TC, Paul EA, Harwood RR (2001) Biologically active soil organic matter fractions in sustainable cropping systems. Appl Soil Ecol 16:63–76

Zelalem A, Tekalign T, Nigussie D (2009) Response of potato (*Solanum tuberosum* L.) to different rates of nitrogen and phosphorus fertilization on vertisols at Debre Berhan, in the central highlands of Ethiopia. Afr J Plant Sci 3(2):16–24

Zhang M, Heaney D, Henriquez B, Solberg E, Bittner E (2006) A four year study on influence of biosolids/MSW compost application in less productive soils in Alberta: nutrient dynamics. Compost Sci Util 14(1):68–80

Chapter 27
Effect of the Application of Humic Substances on Yield, Quality, and Nutrient Content of Potato Tubers in Egypt

Ahmed A. Mosa

Abstract Modern potato cultivation systems in the arid climate of Egypt require the use of non-traditional strategies to improve water use efficiency and soil nutrient supplies. This chapter reviews a few studies conducted in Egypt that used fertigation to apply humic substances (HS) to potato fields. When applied to soil, HS tend to increase moisture retention in the root zone, and therefore can increase irrigation efficiency. Moreover, HS can increase the nutrient content of soil and the nutrient supply potential, which is reflected in increased fertilizer use efficiency. In addition, application of HS can play a considerable role in increasing plant resistance against common potato diseases. Based on these effects, we conclude that application of HS to potato production systems can increase both quantitative and qualitative characteristics of tubers, and can also improve soil fertility and quality.

27.1 Introduction

Egypt has an area of about one million square kilometers. The total agricultural land in Egypt amounts to nearly 3.5 million ha, accounting for around 3.5% of the total land area; however, this area is primarily considered to be virgin desert, with a sandy soil texture that is not optimal for crop production. The imbalance between a growing population and available agricultural land has led to a shortage in food supplies; therefore, the implementation of management practices that can increase crop productivity is of great interest in an area where the amount of land available for cultivation is continuously declining as a result of urbanization (Adriansen 2009). Land reclamation in the Egyptian context means converting desert areas to agricultural land and rural settlements. This is not only done by extending the water canals

A.A. Mosa (✉)
Soils Department, Faculty of Agriculture, Mansoura University, Mansoura, Egypt
e-mail: ahmedmosa@mans.edu.eg

Z. He et al. (eds.), *Sustainable Potato Production: Global Case Studies*,
DOI 10.1007/978-94-007-4104-1_27, © Springer Science+Business Media B.V. 2012

from existing agricultural land into the desert, but also by working with the soil, ploughing in manure in order to enhance its fertility, and finally by providing the infrastructure for making new villages (Ibrahim and Ibrahim 2003). Agronomic productivity in the newly reclaimed soil in Egypt is limited by low water holding capacities, high infiltration rates, high evaporation rates, low fertility levels, very low organic matter content, and excessive deep percolation losses, all of which may induce low water use efficiency (Al-Omran et al. 2004). The arid climate of Egypt is characterized by high evaporation rates (1,500–2,400 mm $year^{-1}$) and scant precipitation (5–200 mm $year^{-1}$) (Robaa 2008), which leaves the River Nile as the predominant fresh water supply. Consequently, effective management practices should therefore be developed and implemented in order to simultaneously improve the fertility and quality of Egyptian soils, maximize water use efficiency, and increase crop production in order to meet the food demands of the growing population.

Several approaches have been investigated to increase the water use efficiency and maintain the productivity in the sandy soils in Egypt. One method, the combined application of soil conditioners and mineral fertilizers, is increasingly gaining recognition as an appropriate means for increasing fertility in depleted soils, particularly in arid regions (Vanlauwe et al. 2010). A more traditional and common approach for improving the water holding capacity and fertility of sandy soils is to incorporate organic residues (e.g. green manure, farmyard manure, crop residues) into the plough layer in order to increase the soil organic matter content; however, large amounts of residue must be applied to significantly improve the water holding capacity and fertility of sandy soils. In addition, organic residues can be a source of both plant and human pathogens, and weed seeds.

It is well known that biogenic wastes, including sewage sludge, can be a good source of plant nutrients, and these types of materials show promise as organic fertilizer sources for sandy soils (Ahmed et al. 2003). However, the application and utilization of some types of organic waste has the potential to cause environmental problems, such as the introduction of pathogenic bacteria and parasites, and soil loading of toxic organic compounds and heavy metals (Badawy 2003; El-Motaium and Abo El-Seoud 2007). Furthermore, sandy soils are prone to leaching; therefore, application of organic wastes to soils with a high sand content presents a risk of groundwater contamination (Abdel-Shafy et al. 2008).

Previous studies have indicated that surface mulching with bitumen emulsions, particularly when using hydrophobic material prepared from local Egyptian bitumen, can improve fertility of sandy soil. Bitumen emulsion has been shown to protect soil against wind and water erosion, reduce evaporation, increase the preserved moisture below the mulch layer, modify soil temperature, increase plant growth and nutrient uptake, and stimulate soil biological activity (El-Hady 1999; El-Hady et al. 2008). However mulching with bitumen can be expensive, and application to a virgin soil will cause a lot of environmental problems (Muratova et al. 2003).

Due to the expense and insufficient longevity of synthetic polymers as soil conditioners, natural soil amendments offer promise as an alternative material to improve the chemical and physical properties of sandy soils (Al-Omran et al. 2004).

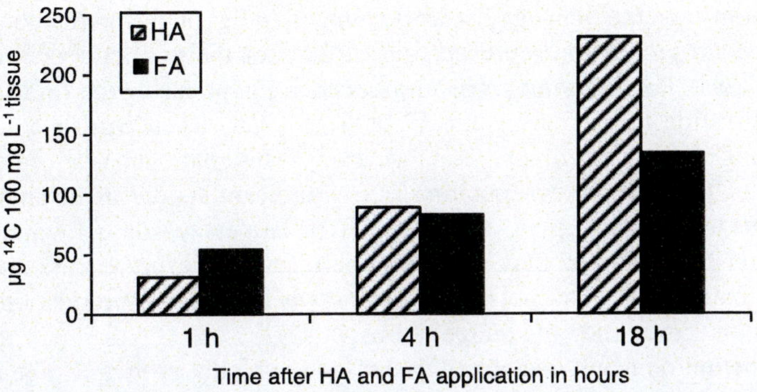

Fig. 27.1 Uptake of ^{14}C-humic acid (HA) and fulvic acid (FA) by pea roots after 1, 4 and 18 h (Modified from Vaughan 1986)

The use of natural deposits may provide a cost-effective means of increasing soil productivity, especially in the areas where these materials are abundant. The application of tafla (particular type of clay deposits formed in Upper Egypt) and other clay deposits has been shown to improve the water holding capacity of sandy soils and increase crop yield (Thabet et al. 1994). Nevertheless, salt accumulation at the soil surface has been attributed to the application of tafla or other clay deposits, especially with low water quality types, which characterized with high salinity levels (Al-Omran et al. 2005).

Of the different natural amendments that have been investigated for use as soil conditioners, humic substances (HS) are among the most effective (Piccolo et al. 1997). The term HS is a general descriptor for a family of organic molecules composed and includes humic acids (HA), fulvic acids (FA), and humin (Stevenson 1982). Early research on the effect of HS on plant growth relied on color changes in plant tissue to serve as an indicator of HS uptake (Prat 1963). More recently, ^{14}C-labeled HS has been used to investigate plant uptake (Vaughan and Ord 1981). In his intensive work, Vaughan (1986), excised roots (25–35 mm long) from 2-day old peas (*Pisum sativum*), and found that HA and FA concentration in roots increased with time (Fig. 27.1), thus indicating that HS could be directly absorbed by plants.

Due to the numerous reported benefits of HS in promoting sustainable agricultural systems, intensive investigation should be conducted to determine the role of HS in maintaining water holding capacity and fertility in sandy soils. In general, sandy soil requires frequent, albeit light, irrigation applications near the cultivated plant, and drip irrigation can be an effective means of meeting these requirements. Consequently, the application of HS by fertigation through drip irrigation systems could be an efficient technique in order to obtain maximum crop yield and quality in sandy soils, especially for economic crops.

Potato (*Solanum tuberosum* L.) is a major food crop in many countries, ranking fourth among the world's various agricultural products in production volume, following wheat, rice and corn (Fabeiro et al. 2001). Thus, it is an important crop in the global economy and the fight against world hunger. In Egypt, the amount of land allocated for potato production represents about 20% of the total cultivated area (Kabeil et al. 2008a). Potato farming was introduced to Egypt during the 1800s, and large scale cultivation began during the First World War, when British colonial officials encouraged its production to feed their troops (International Year of Potato 2008). Export of Egyptian grown potatoes is a significant source of revenue, as potato represents over half of Egypt's net exports. The harvest season in Egypt coincides with winter in the United States and Europe; therefore, exporters are able to internationally market Egyptian potatoes to areas where seasonal production has ceased due to snow cover and cold temperatures.

Potato production on newly reclaimed desert land is highly profitable, and crop yields in reclaimed areas are greater those obtained in the traditional production areas of the Nile Delta and Nile Valley. Increased yield in reclaimed soil is primarily due to a reduction in most common diseases, especially Brown Rot disease, which negatively affects potato production in the Nile Delta and Nile Valley (Kabeil et al. 2008b). Farms on newly reclaimed land tend to be owned by large corporations involved in potato export and processing, and are of greater size and more technologically advanced than farms in traditional production areas. In addition, large corporate farms have sufficient resources to purchase certified seeds, utilize crop rotation practices, and install modern irrigation systems, all of which can help reduce the occurrence of disease and increase crop yields (Pautsch and Abdelrahman 1998).

To obtain maximum tuber yield on sandy, low-fertility, native Egyptian soil, large amounts of mineral fertilizer are normally applied to potato fields. However, modern crop production systems require efficient, sustainable, and environmentally sound fertilizer management practices. Consequently, the adequate rates, appropriate sources and efficient methods of application are important strategies for maintaining the nutrient supply potential of soils (Fageria and Baligar 2005). On the other hand, the agricultural practices in the Nile Delta and valley had been changed under the completion of the Aswan High Dam in 1964. Perennial, furrow irrigation replaced basin flooding, and multiple cropping replaced the single crop per year resulting in shorter fallow periods (Lenney et al. 1996). Therefore, soil fertility declined due to intensive cultivation practices coupled with a lack of systematic nutrient replacement and a loss of the alluvium deposits (Metz 1991). For this reason, modern strategies should be taken in order to increase soil organic matter content, and to compensate the continuous depletion of nutrients. In this chapter we will focus on some case studies of research done in Egypt in order to evaluate the effects of HS application on potato growth and yield.

27.2 Characteristics of Humic Substances

Characterized by long carbon chains with numerous active radicals, such as phenols and other aromatics, HS has an advantage over other natural and synthetic soil amendments as a longer-term soil conditioner. This is primarily due to the refractory

Table 27.1 Elemental composition and atomic and E_4/E_6 ratios of humic acids extracted from different compost materials (Modified from Taha and Modaihsh (2003))

	Elements concentration(%)				Atomic ratio			
Source of HA	C	H	N	O	C/N	C/H	O/H	E_4/E_6
Khaldia	53.68	5.40	5.65	35.27	11.08	0.83	0.41	6.50
Sanbest	50.55	4.45	4.90	40.10	12.04	0.95	0.56	7.90
Yanbost	46.91	4.16	6.34	42.59	8.63	0.94	0.64	5.70
Al-Kharj	44.16	4.41	7.49	43.94	6.88	0.83	0.62	6.40
Al-Enzy	44.09	5.42	4.01	46.48	12.83	0.68	0.54	5.30
Bostan	51.23	6.38	7.63	34.76	7.83	0.67	0.34	4.60

nature of chemical structure of HS, which provides more resistance to microbial attack than simpler, less condensed compounds. Based on published and unpublished data, characteristics of HS introduced it to be an effective source for increasing water and nutrients supply potentials of soil, and stimulating plant growth and yield.

Taha (1985) carried out some physical and chemical measurements on HA isolated from an Egyptian alluvial soil at the Experimental Station of Mansoura University. Humic acid was extracted and purified according to the method reported by Sonbol and El-arquan (1977). The elemental analysis indicated that the extracted soil HA contained 51.43% C, 3.55% H, 3.37% N, 0.37% S and 41.28% O. These data revealed that the extracted HA seems to be in a humified state as a result of the climatic conditions of Egypt, which characterized with little precipitation and high temperature (Robaa 2008). The visible spectra of the isolated HA indicated that it had a red-brown color, which give an indication that it is in a more oxidized state.

Results also showed that HA contents of COOH, phenolic OH and total carbonyl groups were relatively high; however, the alcoholic OH groups content were low. Infra Red (IR) spectroscopy of the studied HA showed strong absorption band at 3,600–3,100 cm^{-1} (H-bonded OH groups); 2,970–2,840 cm^{-1} (aliphatic C-H stretching); 1,700 cm^{-1} (carboxyl and carbonyl of carboxylic acids); 1,600 cm^{-1} (C=C of breathing bands COO^- and quinone groups); and 1,400 cm^{-1} (OH deformation of aliphatic C-H and COO^- groups). The IR spectra of HA-ions complexes showed increasing absorption bands for COO^- at 1,600 and 1,400 cm^{-1}, which explained to be caused by bonding of the ions to the carboxylate or phenolic OH groups.

Another investigation carried out by Taha and Modaihsh (2003) to study some physicochemical characteristics of HAs extracted from different compost materials. The used commercial compost materials were Khaldia (Slaughterhouse wastes compost), Sanbest (Slaughterhouse wastes and crop residues compost), Yanbost (crop residues compost), Al-Kharj (farmyard manure compost), Al-Enzy (farmyard manure and crop residues compost) and Bostan (sludge compost). Humic acid was extracted from different compost materials using 0.1 M NaOH according to Kononova (1966). The obtained data showed that carbon, hydrogen and nitrogen of the studied HAs were affected by the origin of the HAs (Table 27.1). The C/H and O/H values for HA extracted from Yanbost compost were higher than other HAs. Humic acids extracted from Sanbost and Al-Enzy composts have the lowest value of E_4/E_6 ratio indicating a higher degree of aromatic condensation and low aliphatic structure. The IR spectra of the studied HAs showed a broad similarity among the different HAs. The intensities of the absorption band vary slightly among different HAs. They differed mainly in the ratios of the number of functional groups, and the degree of polymerization. On the other hand, HA extracted from sludge (Bostan) contained the highest percentage of aliphatic carbon, associated with polysaccharides structures.

Another unpublished data obtained from the Fertilizers Development Center, El-Delta Fertilizers Plant, Egypt demonstrated the chemical analysis of HS extracted from composted crop residues. Humic substances were extracted using KOH 0.1 M (1:7 w/v). Potassium hydroxide was used instead of other conventional sodium-extractants to increase the benefit of the product from its potassium content. Humic substances fractionation was carried out according to Kononova (1966). Data revealed that HS product contained 14.8% humic acid and 3.5% fulvic acid. The Cation Exchange Capacity (CEC) of HS was 440 meq/100 g. The pH value of HS product was 7.7 and the EC value was 0.97 dSm^{-1}. Macronutrient concentrations were 3.9, 0.13 and 3.22% for N, P and K, respectively; however micronutrients concentrations were 248, 436 and 216 mg kg^{-1} for Zn, Fe and Mn, respectively.

The aforementioned characteristics of HS, which carried out on different origins of HS in Egypt, showed a high ability toward water and nutrients binding as a result of their functional groups. On the other hand, the chemical analysis of HS showed an appropriate content of different plant nutrients. Therefore, HS application to the Egyptian soils could be an effective source for maintaining soil fertility parameters, and promoting plant growth and yield.

27.3 Using HS as a Treatment for Potato Diseases

Early blight, which caused by the fungal pathogen *Alternaria solani* is one of the most common diseases affecting potato plants (Waals et al. 2004; Pasche et al. 2005). In the Noubareia region of Egypt, severe infection with late and early blight diseases has been recorded (El-Gamal et al. 2007). Early blight can affect both leaves and tubers of potato, forming lesions and leading to decreases in potato yield if left uncontrolled. Fungicide application is the predominant means employed to control blight in potato production systems; however, intensive use of agrochemicals in conventional cropping systems has caused irreversible effects on soil and water ecosystems, including pollution of surface and fresh water reserves, and endangering food safety. To circumvent these issues, there have been recent efforts directed at the development of new approaches for controlling plant diseases that are effective, reliable, and safe for the environment.

Abd-El-Kareem (2007) reported that bean plants that had been treated with HA had higher resistance against root rot and Alternaria leaf spot, and improved bean yields. Also, sulfur (S) has been used in organic farming systems to control plant diseases, and can be used as a preventive fungicide (Scherm and Savelle 2001). Sulfur prevents fungal spores from germinating; therefore, it must be applied before the disease develops for effective results. It is well known that soil microorganisms are a crucial factor in the soil S cycle. However, the low organic matter content in sandy soils will affect the efficiency of S in such soil conditions as the low organic matter content will leads to less microbial growth and activity, which would decrease S cycling rates.

Abd-El-Kareem et al. (2009) believed that the integrated applications of HA and S could be an effective remedy for early blight disease of potato plants. For this purpose, greenhouse and field experiments were carried out in El-Nubareia district, Behera Governorate of Egypt, which has a characteristic sandy loam textured soil. For the greenhouse portion of the experiment the effects of different concentrations of HA and S on the severity of early blight infection of potato plants was evaluated. Potato plants (cv. Nigola) were grown in plastic pots (30 cm diameter) under greenhouse conditions (23–25°C) and experimental treatments included the application of HA at rates of 6.0 and 8.0 mL L^{-1}, the application of S at rates of 3 and 4 gL^{-1}, and the integrated application of both HA and S (HA+S). Depending on the particular treatment received, potato plants at the 4–5 compound leaf stage were sprayed with HA, S, or HA followed by S application 3 days later. After 5 days of treatment, plants were inoculated with early blight by spraying potato plants with spore suspensions (10^6 spores ml^{-1}) of *A. solani*. The severity of early blight was based on the percent of the leaf area infected, and rated on a scale of 0–4, according to method of Cohen et al. (1991). The obtained results indicated that all of the HA, S, and HA+S treatments significantly reduced the disease severity ($p<0.05$). The highest reduction in disease severity was obtained with the integrated application of HA at 8 mL L^{-1} and S at 4 gL^{-1}, which was effective for reducing blight severity by 89%, as compared to control treatment. While single treatments of HA and S reduced early blight severity by 54–60%, respectively. Thus, the combined effect of the two treatments was notably improved over only HA or S application alone.

Plant chitinases are induced as a result of the presence of pathogenic infections, as well as by abiotic agents (Yun et al. 1997). *In vitro* studies have shown that chitinases possess antifungal activity and cause lysis of hyphal tips (Leah et al. 1991). Consequently, an increase in plant chitinase activity could serve as a mean of protecting growing potato plants from early blight and other fungal pathogens. Results from this study showed that all of the foliar applications under investigation increased the chitinase activity of potato plants as compared with the control treatment. The chitinase activity of the control treatment was 2.5 units mL^{-1}. Although the conventional treatment of S as a preventive fungicide in organic farming systems increased chitinase activity by up to 64%, the application of HA at both 6 and 8 mL L^{-1} resulted in chitinase activity that was 124% higher than the control. Therefore, the application of HA was more effective at increasing plant chitinase activity than S. Furthermore, the combined application of HA+S did not record a significant increase in chitinase activity as compared with the application of HA alone.

The same treatments in pot experiments were applied under field conditions to study their effect against early blight disease as compared with the application of Ridomil-plus fungicide at 2 gL^{-1}. Field experiments were conducted under natural infection in plots (4×10 m) each comprised of 8 rows (40 holes/row) in a randomized complete block design with 3 replicates (plots) for each treatment. Results from the field studies show clearly that foliar application of all HA and S treatments significantly reduced the severity of early blight during both growing seasons. The most effective treatments were the integrated treatments of HA+S, which reduced the severity of early blight to more than 90% as compared with 72% with Ridomil-plus fungicide.

The role of HA in mitigating early blight disease may be due to an enhancement of the natural resistance by plants against diseases and pests. Experimental results from the greenhouse portion of this study indicated that the highest increase in chitinase activity was obtained from HA application. Β-1,3-glucanases and chitinases are responsible for the hydrolysis of Β-1,3-glucan and chitin, respectively, which are the major components of fungal cell walls (Mohammadi et al. 2001). On the other hand, HA has been reported to stimulate plant growth by increasing the rate of cell division, optimizing the plant uptake of nutrients and water (Delgado et al. 2002), and stimulating soil microorganisms (Garcia et al. 2004). Thus, while the specific mechanism by which HA increases disease resistance is not clear, this study does indicate that application of HA and a combination of HA and S can decrease the severity of early blight on potato plants and increase the yield of potato tubers produced.

27.4 Effect of Humic Substances Application on Potato Tuber Yield

The role of HS in stimulating the growth and yield of plants has been attributed to both direct and indirect effects. It has been demonstrated that HS can directly affect plant growth by inducing an increase in the absorptive surface area of roots via an ordered remodeling of the root morphology (Schmidt et al. 2007). In addition, HS has been shown to stimulate the proliferation of lateral roots, along with the activation of plasmalemma and vacuolar H^+-ATPases and tonoplast H^+-PPase (Zandonadi et al. 2007). Furthermore, it has recently been shown that HS can interact with root organic acid exudates to influence the root area, primary root length, the number of lateral roots, and lateral root density (Canellas et al. 2008). There have also been reports revealing that HS are able to enhance the respiration rate and increase the permeability of cell membranes in higher plants (Vaughan and Malcom 1985; Samson and Visser 1989). In addition to their role in stimulating enzyme activity and hormone-like activity (Piccolo et al. 1992; Nardi et al. 1994), a number of studies have demonstrated that the two main fractions of HS found in soil, HA and FA, can assist in controlling plant diseases by inhibiting the growth of some soil-borne phytopathogenic fungi (Loffredo et al. 2007).

Indirectly, HS can improve both yield and quality characteristics of crops by enhancing soil enzyme activity and promoting the growth and activity of microorganisms in the rhizosphere. Increased microbial numbers following HS application is primarily due to the role of HS in creating soil conditions that favor microbial replication (Sellamuthu and Govindaswamy 2003). Improved aggregate stability is a cornerstone for success in the reclamation of sandy soils, as it is a crucial factor involved in protection against soil erosion. Fortun et al. (1989) reported that when HS extracted from farmyard manure was applied to soil, that soil aggregation was improved and aggregate stability was increased, as compared to soil that received bulk farmyard manure, even when higher rates of manure than HS were applied.

Several experiments carried out in Egypt to investigate the effect of HS on potato tubers yield. Selim et al. (2009a) carried out a field experiment to examine the effect HS application through fertigation system on increasing water use efficiency of potato grown under sandy soil conditions. The arid climate of Egypt, which marked with high evaporation, requires a modification for drip irrigation system to decrease evaporation losses. Therefore, authors compared the efficiency of subsurface drip irrigation system with the conventional surface irrigation system. The experiment was carried out in the winter growing season between November and February, when the average temperature was 17.4°C, and the average precipitation was 20.3 mm. The soil was sandy in texture (*Entisol-Typic Torripsamments*), with a field capacity of 10.7%, pH of 8.4, and a total $CaCO_3$ content of 5.36%. The soil was not classified as saline (EC=0.31 dSm^{-1}), and the water used for irrigation was considered to be acceptable quality, with a salinity value of 0.43 dSm^{-1}. Results obtained from this study indicated that HS increased tubers yield under both surface and subsurface drip irrigation systems. Humic substances application at rate of 120 kg ha^{-1} was more efficient than 60 kg ha^{-1} on stimulating tubers yield quantity. Humic substances application through surface drip irrigation system led to increase potato tubers yield over the control treatment by 2.82% and 29.10%; however, this increment was 4.43% and 17.98% in case of subsurface drip irrigation system following the application of HS at rates of 60 and 120 kg ha^{-1}, respectively.

The rapid growth of population of Egypt and the rapid urbanization are causing an imbalance between water supply and demand. There are two ways to overcome this problem, the first of which is by increasing the efficiency with which current water needs are met (e.g. more crop per drop), and secondly by increasing the use of non-conventional water resources, such as saline agriculture drainage water/brackish groundwater, reclaimed wastewater, and the conjunctive use of surface and groundwater (Ragab et al. 2005). Ezzat et al. (2009) conducted a study in order to evaluate the effects of the application of HS on water use efficiency, potato growth, and tuber yield in the North Nile Delta of Egypt under deficit irrigation conditions. Irrigation water was applied at rates of 2,000, 4,000 and 6,000 m^3 ha^{-1} by a drip irrigation system for two growing seasons. The optimum irrigation requirement of potato is 6,000 m^3 ha^{-1}.Potassium humate (K-humate) served as the source of HS, and it was applied as powder additives beside potato seeds. The soil was classified as a clay loam in texture. The highest marketable yield quantity was obtained from the irrigation rate of 4,000 m^3 ha^{-1} with K-humate application (30.1 Mg ha^{-1}). However, the obtained yield of the irrigation requirement of 6,000 m^3 ha^{-1} was 26.9 Mg ha^{-1}. This means that it could be possible to decrease the irrigation quantity to about 67% from the recommended irrigation requirement in case of K-humate application. The irrigation rate of 2,000 m^3 ha^{-1} was associated with the lowest marketable yield quantity (18.2 m^3 ha^{-1}). However, K-humate application increased

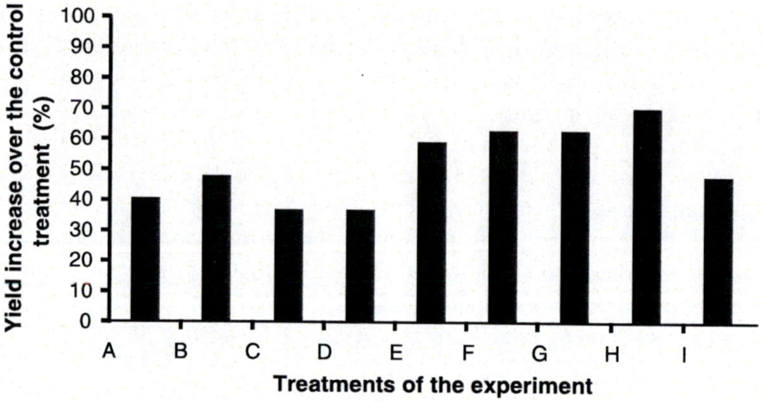

Fig. 27.2 Yield increase in potato yield over the control treatments with different remediation treatments of humid substances (HA) and sulfur (S): (A) HA at 6 mL L^{-1}, (B) HA at 8 mL L^{-1}, (C) S at 3 gL^{-1}, (D) S at 4 gL^{-1}, (E) HA at 6 mL L^{-1}+S at 3 gL^{-1}, (F) HA at 6 mL L^{-1}+S at 4 gL^{-1}, (G) HA at 8 mL L^{-1}+S at 3 gL^{-1}, (H) HA at 8 mL L^{-1}+S at 4 gL^{-1} and (I) redomil-plus at 2 gL-1

the marketable yield quantity by about 33%. On the other hand, the lowest irrigation treatment (2,000 m^3 ha^{-1}) was associated with the highest yield of potatoes that were considered to be unmarketable. Thus, application of K-humate led to an increase in the quantity of marketable tubers and improved tuber quality as compared to plots that did not receive K-humate. These results were attributed to increased membrane permeability of plants, which would, promote greater nutrient uptake, and accelerate the net rate of photosynthesis by increasing the concentration of photosynthetic pigments in the plant leaves (Zhang et al. 2003).

Potato plants are attacked by several plant pathogens causing serious diseases during the growing season, this consisted approximately 19% of crop loss (El-Mougy 2009). Early blight, caused by *Alternaria solani* is a very common disease of potato, and is found in most potato growing areas. Values in the literature for measuring crop losses due to early blight vary enormously from 5% to 78% (Waals et al. 2004; Pasche et al. 2005).

Abd-El-Kareem et al. (2009) carried out field experiments in El-Nubareia district, Behera Governorate, of Egypt to examine the effect of the combined application between HA and Sulfur (S) on early blight disease of potato plants. Treatments of the experiments were the application of HA at concentrations of 0.6 and 0.8 mL L^{-1} and S at concentrations of 3 and 4 gL^{-1} solely or in combination as compared with the fungicidal application of Redomil-plus at concentration of 2 gL^{-1}. The control treatment was left untreated. Data presented in Fig. 27.2 illustrated that HA and S treatments either alone or in combination increased potato yield as compared with the control treatment (without spraying). The combined application of HA at 8 mL L^{-1}+S at 4 gL^{-1} was the most efficient treatment for increasing potato yield (70.4% increase over the control treatment). On the other hand the conventional fungicidal remediation of early blight disease by Redomil-plus at concentration of 2 gL^{-1} led to an increase

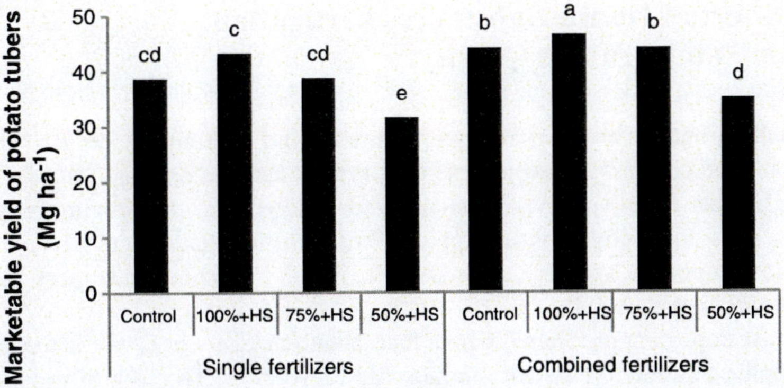

Fig. 27.3 The effect of fertigation with HS and different rates of single and combined NPK fertilizers on the marketable yield of potato tubers (Mg ha^{-1})

in potato yield by 48.1% over the control treatment. Therefore, the combined application of HA and S could be considered as an efficient method for early blight disease of potato. This method is low in cost and friendly with the environment.

A theory behind why fertigation has become the state of the art in plant nutrition in arid environments is that nutrients can be applied in the correct dosage and at the required time appropriate for each specific growth stage (Hebbar et al. 2004; Badr et al. 2010). However, the irrigated agricultural areas are particularly susceptible to groundwater pollution because irrigated crops are abundantly fertilized and the sandy soils are not able to protect plant nutrients against leaching (Vázquez et al. 2006).

Selim et al. (2009b) conducted a study in to evaluate the effect of co-application of HS with either single nutrient or mixed NPK fertilizers via a drip irrigation system to potato planted in a sandy soil. The experimented soil was sandy in texture (*Entisol-Typic Torripsamments*) with a calcium carbonate content of 6%. Treatments of the experiment were assigned in a split plot design with 3 replicates. Fertilization was applied by fertigation through a drip irrigation system, and two forms of mineral fertilizers (single and combined fertilizers) were the main treatments. Sub treatments were the application of HS with 50%, 75% or 100% of the recommended mineral fertilizer rate with HS application in addition to the control treatment (100% of the recommended mineral fertilizer without HS application). Results indicated that application of HS through the fertigation system significantly increased the total marketable yield of potato tubers (Fig. 27.3). The highest marketable yield of potato tubers was obtained from the co-application of HS with 100% of the recommended fertilization rate of NPK. This increment in potato yield is primarily attributed to the enhancement of fertilizer use efficiency, which decreased the leaching of nutrients from the rooting zone, and increased plant nutrient uptake.

27.5 Effect of Humic Substances Fertigation on Potato Tubers Quality

Good-quality potatoes are firm, relatively smooth, and without any defects, sprouts and unfavorable colors. However, these factors may vary according to the degree of maturity, harvest time, variety, and storage conditions. The first quality judgment made by a consumer is by its visual appearance. In this context tubers size and color are the most important appearances attributes, which influence consumers' acceptability (Nourian et al. 2003).

The most important problems, which face potato growers in Egypt is the lack of water supplies and the low fertility in most land resources. The lack of water availability is one of the most important constraints to potato yield, and an adequate water supply is required from planting until maturity. The primary effect of drought or water stress on potato is yield and size reduction. Water deficit during early plant growth can increase the occurrence of spindled tubers, which is more noticeable in oval than in round tuber varieties (El-Ghamry and El-Shikha 2004). Furthermore, growth during drought conditions followed by irrigation may result in tuber cracking or tubers with "hollow hearts"; therefore, water supply and scheduling have important impacts on potato growth, yield and tuber quality (Lutaladio et al. 2009). On the other hand, because most of soils in Egypt are characterized by low fertility standards, Egyptian potato growers typically add large amounts of mineral fertilizers in order to obtain maximum crop yield, particularly when planting in sandy soils. However, modern agricultural production practices require efficient, sustainable, and environmentally sound fertilizer management practices. Consequently, important strategies for maintaining the nutrients supply potential of soils require that fertilization should be conducted at adequate rates, with appropriate sources, and by efficient methods of application.

Tuber size is rated as the most important characteristic of potato quality according to the grower's preference. This finding was reported by Govinden (2006) after his investigation about potato tuber characteristics preferred by growers. He also mentioned that more than two-thirds of the growers preferred large or extra-large tubers. In this respect, Mahmoud and Hafez (2010) demonstrated that humic acid application through drip irrigation system increased potassium fertilizer use efficiency as a result of decreasing its leaching under sandy soils conditions. This was associated with increasing tubers weight and diameter (Figs. 27.4 and 27.5). Potassium fertilization treatments were applied as K-sulfate at rates of 80, 160 and 240 kg ha^{-1}. Humic acid (HA) treatments were added through drip irrigation systems in 5-equal portions at rates of 0, 2.5 and 5 kg ha^{-1}. The experimental soil was sandy in texture, with a field capacity of 16.5%, EC of 1.7 dSm^{-1}, and pH of 8.2.

According to Johnson (2003), potassium has two roles in the functioning of plant cells. First, it has an irreplaceable part to play in the activation of enzymes which are fundamental to metabolic processes, especially the production of proteins and sugars. Thereafter, it has a vital role in a process called carbohydrate metabolism. This process converts simple sugar to more complex sugar and starch. Second, potassium is

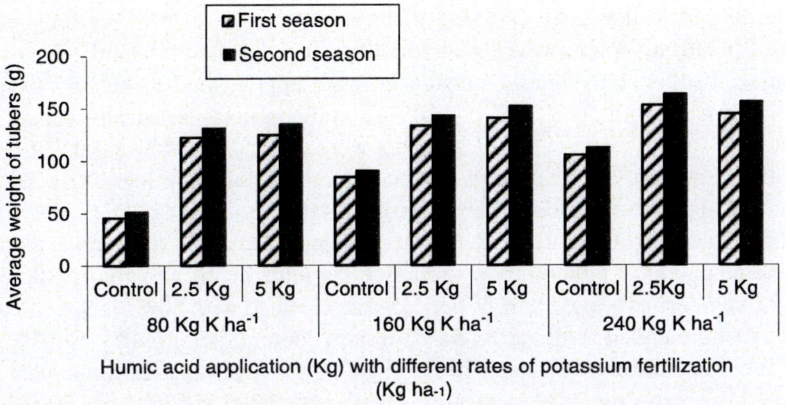

Fig. 27.4 Average tuber weight as a result of HA application with different potassium fertilization levels

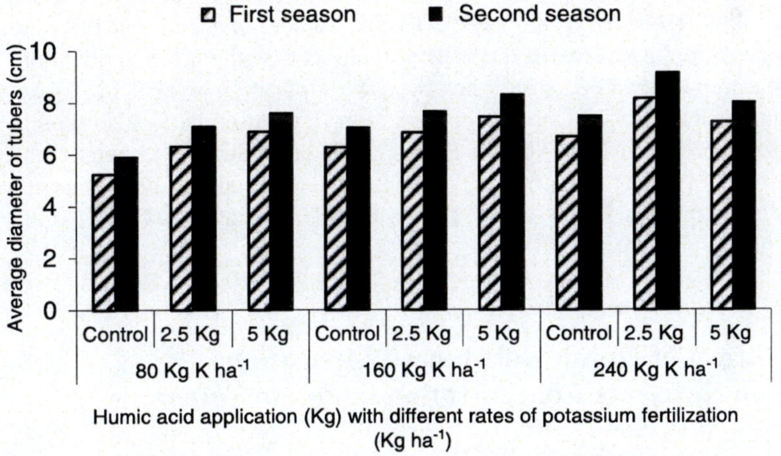

Fig. 27.5 Average tuber diameter (cm) as a result of HA application with different potassium fertilization levels

the "plant-preferred" ion for maintaining the water content in plant cells as it creates conditions that cause water to move into the cell (osmosis) through the porous cell wall. These two important functions are responsible for increasing tubers size and its starch content.

Specific gravity has been used as an important criterion of potato quality because of its close relationship to dry matter content and the rapidly of its measurement (Schark et al. 1956). High specific gravity of potato, which characterized by high starch content, is better suited for baking, frying, mashing and chipping. However, low specific gravity, which characterized by low starch content, is more suited for boiling and canning (Robinson et al. 2006).

According to Selim et al. (2009b), there was no significant effect on specific gravity of potato tubers, however; starch content increased significantly due to HS application. In this study humic substances were applied at 120 kg ha^{-1} through fertigation system with 50%, 75% or 100% from the mineral fertilization dose. The increment of starch content in potato tubers were 1.37%, 16.65% and 3.19% over the control treatment (application of the recommended dose of mineral fertilization without HS application) with mineral fertilization rates of 50%, 75% and 100% from the mineral fertilization dose. Results obtained from this study revealed also that protein content of potato tubers increased significantly as a result of HS application through fertigation system. When HS were applied with 50%, 75% and 100% from the recommended fertilization dose, protein content increased by 14.17% and 5.48% over the control treatment with 75% and 100% from the recommended fertilization dose, respectively. However, the fertilization level of 50% from the recommended dose reduced protein content by 6.68% less than the control treatment.

According to Selim et al. (2009a) starch content of potato tubers increased due to HS application through surface and subsurface drip irrigation systems. Mean values of starch content in potato tubers were 13.01%, 13.98% and 14.61% with HS levels of 0, 60 and 120 kg ha^{-1}, respectively. Authors revealed that there was no significant difference between the control treatment (without HS application) and the application level of 60 kg ha^{-1}; however, the application level of 120 kg ha^{-1} led to a significant increase in starch concentration in potato tubers. Also there was a significant increase in Total Soluble Solids (TSS) concentration in potato tubers as a result of HS application through fertigation system. Total Soluble Solids concentrations were 5.10%, 5.24% and 5.47% with HS application levels of 0, 60 and 120 kg ha^{-1}.

27.6 Effect of Humic Substances Fertigation on Nutrients Concentration in Potato Tubers

Potassium (K) is a crucial element for optimal potato production, which is unlike the specific nutrient requirements of most other vegetable crops. Potassium is important to potato, as it strengthens stems and thus helps to prevent lodging, increases tuber yield, size and quality, increases specific gravity and starch content, and improves fry color and storage quality (Ibrahim et al. 1987; Omran et al. 1991). Furthermore, K can reduce the susceptibility of potato to black spot bruise, decreased the occurrence of darkening after cooking, and lower tuber sugar content. It also allows the crops to adapt to environmental stress viz. salinity stress (Akram et al. 2009), water stress (Kanai et al. 2011), and promotes tolerance of plants against insect infection and increases resistance to fungal disease (Kettlewell et al. 1990; Menzies et al. 1992).

Sandy soils are characterized by low clay content and small buffer capacity and the application of K fertilizers can result in localized increases in K concentration in the soil solution, as there is not sufficient clay mineral surface area to bind ionic K^+.

Under these conditions, soluble K can be leached through the soil profile by rainfall or irrigation water. In arid and semi-arid regions, the leaching of K is enhanced by the presence of calcite and gypsum (Jalali and Rowell 2003).

The primary aim of the study done by Mahmoud and Hafez in 2010 was to maximize the productivity of potato by increasing K use efficiency under sandy soil conditions. The application of HA beside different potassium fertilization levels led to increase potassium concentration in potato tubers. The highest concentration of K in potato tubers was associated with the application of K-fertilizer rate of 240 kg ha^{-1} with HA at 5 kg ha^{-1}. However, the lowest concentration was associated with the application level of 80 kg ha^{-1}. The mean values of K concentration in potato tubers as affected by HA application were 1.4%, 1.66% and 1.58% with 0, 2.5 and 5 kg ha^{-1}, respectively. The application of HA likely decreased K^+ leaching due to the influence of functional groups commonly present in HA, including carboxyl, phenol and hydroxyl, which contributed in K^+ binding by HA (Wang and Huang 2001). Furthermore, HA could have had a stimulating effect on plant physiological properties, thereby increasing K uptake. According to Samson and Visser (1989), HA can induce an increase in the permeability of biomembranes for electrolytes, resulting in increased uptake of K. Humic acid application not only increased K concentration in potato tubers, but also increase N and P concentrations. The mean values of N concentrations were 1.35%, 1.73% and 1.86%; however, P concentrations were 0.66, 0.77 and 0.75 with the application of HA at rates of 0, 2.5 and 5 kg ha^{-1}, respectively.

Magnesium (Mg) is a very important element in potato nutrition system. It is a vital constituent in the chlorophyll molecule that regulates photosynthesis. In addition, it acts as an activator for many enzyme systems involved in carbohydrate metabolism and synthesis of nucleic acids and helps in translocation of sugar. High rates of K and ammonium-N (NH_4^+-N) application reduce the uptake of Mg. It is evident that high concentration of these cations in soil solution interferes with Mg^{2+} uptake by plants (Marschner 1995). The fertilization programs used in Egypt rarely include Mg supplementation due to sufficient levels of Mg present for potato production in these soils; however, under these conditions, the presence of an excess of fertilizer K and N will lead to a decrease in Mg uptake by potato plants. Moreover, Mg has a pronounced role in the activation of nitrate reductase, which is one step in the pathway responsible for nitrate assimilation into amino acid compounds (Morgan et al. 1972). Therefore, Mg has an indirect effect on alleviating NO_3^- accumulation in potato tubers, and increasing tuber protein content (Herraera and Johnson 1997). In addition, the chelating power of HA in soil could increase plant Mg uptake by protecting Mg^{2+} ions from leaching, forming insoluble Mg-carbonates, or competing with K^+ or NH_4^+ ions for plant uptake.

Awad and El-Ghamry (2007) conducted a study in order to investigate the effects of HA on the activity of soil microorganisms and Mg availability to potato grown in an alluvial soil. The experimented soil was clayey in texture (*Clayey, Superactive, Mesic, Vertic Xerofluvents*) with a pH value of 7.8, and calcium carbonate content of 3.48%. Potato pieces (cv. Spunta) were planted in the fall of 2005 and 2006. Potato plants were fertilized with mineral N, P and K fertilizers at rates of 425,

80 and 200 kg ha^{-1} as a control treatment. Humic acid and Mg were applied alone or in combination with the aforementioned fertilization doses. Humic acid at rate of 100 mL (1:500 w:v) was applied to soil directly beside growing potato plants at 28, 44, and 60 days after planting. While Mg was applied in the form of magnesium sulphate (0.5%) as foliar spraying at 7 and 9 weeks after planting. Cultivation practices for potato production were carried out according to the recommendations of the Egyptian Ministry of Agriculture (Egyptian Agricultural Research Center 2003). The obtained results from this study indicated that HA application also influenced the NO_3^- concentration of potato tubers, revealing that HA application either alone or in combination with Mg, resulted in a reduction in NO_3^- concentration in potato tubers as compared with the control treatment. The mean values of NO_3^- concentrations in potato tubers were 67, 60, 36 and 45 mg kg^{-1} with the treatments of control, humic acid, Mg and (humic acid+Mg), respectively. As mentioned before, Mg has a role in stimulating nitrate reductase, and would therefore increase the rate of NO_3^- assimilation into amino acids in potato tubers. On the other hand, it has been shown that HA can have an inhibitory effect on the activity of urease enzyme, which could potentially decrease the efficiency of NH_3 utilization by potato plants. This is confirmed with results obtained by Thorn and Mikita (1992), using ^{15}N and ^{13}C NMR techniques. They detected that ^{15}N-labeled ammonia was incorporated into humic acid in the laboratory incubation and that the average N content of humic acid increased from 0.88% to 3.17%. It was also found that the concentration of nutrients in both potato tubers was affected by application of HA, likely by directly increasing plant nutrient uptake, and by increasing the nutrient availability in the root zone. The most efficient treatment was the combination of HA, and Mg, showed a synergistic effect. It led to increase N, P and K concentration in potato tubers more than 20% as compared with the control treatment. The binding power of HA can protect NO_3^- ions from leaching following irrigation. This binding power is also able to play a definite role in liberating the fixed K by expanding silicate clays (Tan 1978). Humic acid can increase P availability by complexing with soil minerals and forming stable organo-mineral compounds, decreasing P fixation as apatite and other mineral-phosphates, and allowing P to remain exchangeable for plant uptake (Seyedbagheri 2010).

Because the transport of micronutrients to the plant roots occurs *via* diffusion, low soil moisture content will reduce micronutrient uptake (Hu and Schmidhalter 2005). In a deficit irrigation experiment, Ezzat et al. (2009) examined the role of HS application on mitigating the harmful effect of water stress. Deficit irrigation treatments (2,000 and 4,000 m^3 ha^{-1}) were compared with the optimum irrigation requirement of potato plants (6,000 m^3 ha^{-1}). Results illustrated in Fig. 27.6 revealed that HS application led to increase micronutrients concentration in potato tubers. Humic substances structure presents a variety of potential sites for binding of trace metals. Binding could be occurred through: (1) a water bridge; (2) electrostatic attraction to a charged COO- group; (3) formation of coordinate linkages and ring structures; and (4) formation of chelate structures, such as those with COO- and phenolic OH- site combinations (Shenker and Chen 2005).

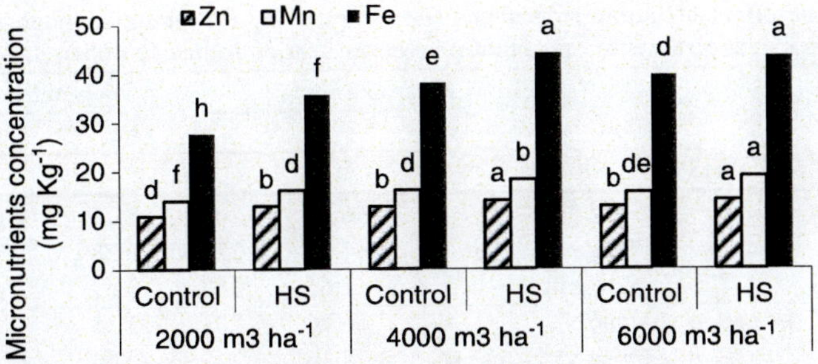

Fig. 27.6 Effect of HS application under different irrigation levels on micronutrients concentration in potato tubers (mg kg^{-1})

27.7 Effect of Humic Substances Fertigation on Soil Fertility After Harvesting

Most of soils of Egypt are considered sandy soils, which characterized with low fertility limits. On the other hand, extensive and frequent cropping under the conditions of an unsustainable irrigation water management and of improper agricultural practices in the Nile Valley and in the Delta have resulted in a depletion and a deficiency in many nutrient elements. This situation has been exacerbated after the construction of the High Dam, which sharply decreased the annual additions of the fertile sediments to the soils. Consequently, all Egyptian soils are poor in their content of organic matter, total nitrogen, and other nutritive elements (Hussein 2011). Humic substances properties (e.g. chelation, mineralization, buffer effect, clay mineral-organic interaction, plant nutrients content and cation exchange) encourage potato growers to use it a conditioner in the sustainable cultivation system (Seyedbagheri 2010). According to Selim et al. (2009a) the application of HS by fertigation increased the level of macro- and micronutrients that were retained in soil after potato harvesting ($p<0.05$), as shown in Table 27.2, and was likely due to an improvement of the nutrient supply potentials of those sandy soils (Suganya and Sivasamy 2006).

27.8 Conclusion

Due to resistant against microbial attack, numerous active radicals and high content of plant nutrients, HS could be considered as an efficient soil conditioner in the modern potato cultivation systems. According to the aforementioned case studies, HS application led to increase water and nutrients supply potentials of soils. Consequently, it could be possible to decrease amounts of irrigation water and mineral fertilizers application. This will lead to provide huge amounts of irrigation

Table 27.2 The effect of humic substances (HS) applied by fertigation through surface and subsurface drip irrigation systems on soil nutrient concentration following potato harvest

		Macronutrients			Micronutrients		
		N	P	K	Fe	Mn	Zn
Irrigation method and HS application rate		mg kg^{-1}					
Surface drip irrigation	None	47	6.67	200	3.15	1.16	0.93
	60 kg ha^{-1}	48	7.25	226	3.38	1.30	1.11
	120 kg ha^{-1}	50	8.65	253	3.67	1.74	1.25
Subsurface drip irrigation	None	47	6.76	203	3.35	1.27	0.96
	60 kg ha^{-1}	49	8.24	245	3.46	1.5	1.13
	120 kg ha^{-1}	53	9.22	256	3.87	1.87	1.28
Mean values as affected by HS application rate							
	None	47^{b}	6.72^{b}	201^{c}	3.25^{b}	1.22^{c}	0.94^{c}
	60 kg ha^{-1}	48^{b}	7.75^{b}	235^{b}	3.42^{b}	1.40^{b}	1.12^{b}
	120 kg ha^{-1}	52^{a}	8.94^{a}	254^{a}	3.77^{a}	1.81^{a}	1.27^{a}

Mean values followed by the same letter within treatments are not significantly different ($p<0.05$) according Duncan's multiple range test (Duncan 1955)

water, protecting environmental resources from the excessive mineral fertilization and decreasing the cultivation costs and labor. Furthermore, application of HS led to enhance potato quality characteristics without significant reduction in tubers yield. Therefore, it is recommended to add humic substances through drip irrigation systems in the newly reclaimed sandy soils. Also it is suggested to use HS as organic conditioner in the alluvial soils of the Nile Delta and Valley. Taking into account the motivation of the induced resistant of potato plants due to HS spraying, future studies should be undertaken in order to investigate the role of HS spraying against common potato diseases. This was evident in the enhancement of chitinase activity following HS spraying.

References

Abd-El-Kareem F (2007) Induced resistance in bean plants against root rot and Alternaria leaf spot diseases using biotic and abiotic inducers under field conditions. Res J Agric Biol Sci 3:767–774

Abd-El-Kareem F, Abd-El- Latif FM, Fotouh YO (2009) Integrated treatments between humic acid and sulfur for controlling early blight disease of potato plants under field infection. Res J Agric Biol Sci 5:1039–1045

Abdel-Shafy HI, Guindi KA, Tawfik NS (2008) Groundwater contamination as affected by long-term sewage irrigation in Egypt. I. In: Baz A et al (eds) Efficient management of wastewater. Springer, Berlin/Heidelberg, pp 53–63

Adriansen HK (2009) Land reclamation in Egypt: a study of life in the new lands. Geoforum 40:664–674

Ahmed AH, Rehan MG, Shaheien AH, Awad SS (2003) Effect of sewage sludge, phosphorine inoculation, and cultivars on growth, yield, and elemental composition of pea under newly reclaimed soil. Mansoura Univ J Agric Sci 28:6465–6485

Akram MS, Ashraf M, Akram NA (2009) Effectiveness of potassium sulfate in mitigating salt-induced adverse effects on different physio-biochemical attributes in sunflower (*Helianthus annuus* L.). Flora 204:471–483

Al-Omran AM, Falatah AM, Sheta AS, Al-Harbi AR (2004) Clay deposits for water management of sandy soils. Arid Land Res Manag 18:171–183

Al-Omran AM, Sheta AS, Falatah AM, Al-Harbi AR (2005) Effect of drip irrigation on squash (*Cucurbita pepo*) yield and water-use efficiency in sandy calcareous soils amended with clay deposits. Agric Water Manag 73:43–55

Awad EM, El-Ghamry AM (2007) Effect of humic acid, effective microorganisms and, magnesium on potatoes in clayey soil. J Agric Sci Mansoura Univ 32:7629–7639

Badawy SH (2003) Comparison of Cd, Cu, Ni, and Zn partitions in soils of long term fertility experiments receiving sewage sludge. Mansoura Univ J Agric Sci 28:703–718

Badr MA, Abou Hussein SD, El-Tohamy WA, Gruda N (2010) Nutrient uptake and yield of tomato under various methods of fertilizer application and levels of fertigation in arid lands. Gesunde Pflanz 62:11–19

Canellas LP, Teixeira LRL, Dobbss LB, Silva CA, Medici LO, Zandonadi DB, Fraçana AR (2008) Humic acids crossinteractions with root and organic acids. Ann Appl Biol 153:157–166

Cohen Y, Gisi U, Mosinger E (1991) Systemic resistance of potato plants against *Phytophthora infestans* induced by unsaturated fatty acids. Physiol Mol Plant Pathol 38:255–263

Delgado A, Madrid A, Kassem S, Andreu L, Campillo M (2002) Phosphorus fertilizer recovery from calcareous soils amended with humic and fulvic acids. Plant Soil 245:277–286

Duncan DB (1955) Multiple range and multiple F test. Biometrics 11:11–44

Egyptian Agricultural Research Center (2003) Cultivation and production of potato. In Arabic. Ministry of Agriculture & Land Reclamation, Egypt

El-Gamal G, Abd-El-Kareem NF, Fotouh YO, El- Mougy SN (2007) Induction of systemic resistance in potato plants against late and early blight diseases using chemical inducers under greenhouse and field conditions. Res J Agric Biol Sci 3:73–81

El-Ghamry AM, El-Shikha DM (2004) Effects of different irrigation systems and nitrogen fertilizer sources on potato growth and yield. J Agric Sci Mansoura Univ 29:6393–6410

El-Hady OA (1999) Twenty five years of research with asphalt emulsions for soils reclamation, conditioning, fertilization, remediation and conservation. In: The 3rd international conference on petroleum and environment. Fac Petroleum and Mining Eng. Suez Canal Univ. EPRI, VEA and ISA, Cairo, 4–6 Dec, pp 295–300

El-Hady OA, Ghaly NF, Wanas Sh E (2008) Hydrophobic-hydrophilic combination for sandy soil conditioning and plantation. 1-Growth response and water and fertilizers use efficiency by casuarina trees. Am-Eurasian J Agric Environ Sci 4:332–342

El-Motaium RA, Abo El-Seoud MA (2007) Irradiated sewage sludge for production of fennel plants in sandy soil. Nutr Cycl Agroecosyst 78:133–142

El-Mougy NS (2009) Effect of some essential oils for limiting early blight (*Alternaria solani*) development in potato field. J Plant Prot Res 49:57–62

Ezzat AS, Saif Eldeen UM, Abd El-Hameed AM (2009) Effect of irrigation water quantity, antitranspirant and humic acid on growth, yield, nutrients content and water use efficiency of potato (*Solanum tuberosum* L.). J Agric Sci Mansoura Univ 34:11585–11603

Fabeiro C, de Santa M, Olalla F, de Juan JA (2001) Yield and size of deficit irrigated potatoes. Agric Water Manag 48:255–266

Fageria NK, Baligar VC (2005) Enhancing nitrogen use efficiency in crop plants. Adv Agron 88:97–185

Fortun A, Fortun C, Ortega C (1989) Effect of farmyard manure and its humic fraction on the aggregate stability of sandy-loam soil. J Soil Sci 40:293–298

Garcia JC, Plaza C, Senesim N, Brunetti G, Polo A (2004) Effects of sewage sludge amendment on humic acids and microbiological properties of a semiarid Mediterranean soil. Biol Fertil Soil 39:320–328

Govinden M (2006) Potato tuber characteristics preferred by growers. MSIRI occasional report no. 34

Hebbar SS, Ramachandrappa BK, Nanjappa HV, Prabhakar M (2004) Studies on NPK drip fertigation in field grown tomato (*Lycopersicon esculentum* Mill.). Eur J Agron 21:117–127
Herraera R, Johnson C (1997) Involvement of magnesium and ATP in the regulation of nitrate reductase activity in *Sinapis alba*. Phytochemistry 44:11–16
Hu Y, Schmidhalter U (2005) Drought and salinity: a comparison of their effects on mineral nutrition of plants. J Plant Nutr Soil Sci 168:541–549
Hussein IA (2011) Desertification process in Egypt. In: Brauch HG et al (eds) Coping with global environmental change, disasters and security, hexagon series on human and environmental security and peace. Springer, Berlin/Heidelberg
Ibrahim FN, Ibrahim B (2003) Egypt: an economic geography. I.B. Tauris, Bodmin
Ibrahim SA, El-Zawily AI, Zayed EA (1987) Effect of NPK level and ratio on growth, yield and chemical composition of potato plants grown in sandy loam soil. Egypt J Soil Sci 27(2):131–144
International Year of Potato. New Lights on a hidden treasure (2008) An end-of- year review. Food and Agriculture Organization of the United Nation, Rome
Jalali M, Rowell DL (2003) The role of calcite and gypsum in the leaching of potassium in a sandy soil. Exp Agric 39:379–394
Johnson AE (2003) Understanding potassium and its use in agriculture. European Fertilizer Manufacturers Association, Brussels
Kabeil SS, Amer MA, Matarand SM, El-Masry MH (2008a) *In planta* biological control of potato brown rot disease in Egypt. World J Agric Sci 4:803–810
Kabeil SS, Lashin SM, El-Masry MH, El-Saadani MA, Abd-Elgawad MM, Aboul-Einean AM (2008b) Potato brown rot disease in Egypt: current status and prospects. Am-Eurasian J Agric Environ Sci 4:44–54
Kanai S, Moghaieb RE, El-Shemy HA, Panigrahi R, Mohapatra PK, Ito J, Nguyen NT, Saneoka H, Fujita K (2011) Potassium deficiency affects water status and photosynthetic rate of the vegetative sink in green house tomato prior to its effects on source activity. Plant Sci 180:368–374
Kettlewell PS, Bayley GL, Domleo RL (1990) Evaluation of late season foliar application of potassium chloride for disease control in winter wheat. J Fertil Issue 7:17–23
Kononova MM (1966) Soil organic matter, 2nd edn. Pergamon Press, Oxford
Leah R, Tommerup H, Svendsen I, Mundy J (1991) Biochemical and molecular characterization of three barley seed proteins with antifungal properties. J Biol Chem 266:1564–1573
Lenney MP, Woodcock CE, Collins JB, Hamdi H (1996) The status of agricultural lands in Egypt: the use of multitemporal NDVI features derived from landsat TM. Remote Sens Environ 56:8–20
Loffredo E, Berloco M, Casulli F, Senesi N (2007) In vitro assessment of the inhibition of humic substances on the growth of two strains of *Fusarium oxysporum*. Biol Fertil Soils 43:759–769
Lutaladio N, Ortiz O, Haverkort A, Caldiz D (2009) Sustainable potato production, International Year of the Potato. FAO, Rome
Mahmoud AR, Hafez MM (2010) Increasing productivity of potato plants (*Solanum tuberosum*, L.) by using potassium fertilizer and humic acid application. Int J Acad Res 2:83–88
Marschner H (1995) Mineral nutrition of higher plants, 2nd edn. Academic, London
Menzies J, Bowen P, Ehret D, Glass ADM (1992) Foliar applications of potassium silicate reduce severity of powdery mildew on cucumber, muskmelon, and zucchini squash. J Am Soc Hortic Sci 117:902–905
Metz HC (1991) Egypt. A country study. Library of Congress, Washington
Mohammadi M, Roohparvar R, Torabi M (2001) Induced chitinase activity in resistant wheat leaves inoculated with an incompatible race of *Puccinia striiformis* f. sp. *tritici*, the causal agent of yellow rust disease. Mycopathologia 154:119–126
Morgan MA, Jackson WA, Volk RJ (1972) Nitrate absorption and assimilation in ryegrass as influenced by calcium and magnesium. Plant Physiol 50:485–490
Muratova A, Hübner T, Narula N, Wand H, Turkovskaya O, Kuschk P, Jahn R, Merbach W (2003) Rhizosphere microflora of plants used for the phytoremediation of bitumen-contaminated soil. Microbiol Res 158:151–161

Nardi S, Panuccio MR, Abenavoli MR, Muscolo A (1994) Auxin-like effect of humic substances extracted from faeces of *Allolobophora caliginosa* and *A. rosea*. Soil Biol Biochem 26:1341–1346
Nourian F, Ramaswam HS, Kushalappa AC (2003) Kinetics of quality change associated with potatoes stored at different temperatures. Lebensm Wiss U Technol 36:49–56
Omran MS, Taysee M, El-Shinnawi MM, El-Sayed MM (1991) Effect of macro– and micronutrients application on yield and nutrients content of potatoes. Egypt J Soil Sci 31:27–42
Pasche JS, Piche LM, Gudmestad NC (2005) Effect of the F129L Mutation in *Alternaria solani* on fungicides affecting mitochondrial respiration. Plant Dis 89:269–278
Pautsch GR, Abdelrahman AH (1998) Effects of Egyptian economic reforms: the horticultural sector. Food Policy 23:199–210
Piccolo A, Nardi S, Concheri G (1992) Structural characteristics of humic substances as related to nitrate uptake and growth regulation in plant systems. Soil Biol Biochem 24:373–380
Piccolo A, Pietramellara G, Mbagwu JSC (1997) Use of humic substances as soil conditioner to increase aggregate stability. Geoderma 75:267–277
Prat S (1963) Permeability of plant tissues to humic acids. Biol Plant 5:279–283
Ragab R, Malash N, Abdel Gawad G, Arslan A, Ghaibeh A (2005) A holistic generic integrated approach for irrigation, crop and field management. 1. The SALTMED model and its calibration using field data from Egypt and Syria. Agric Water Manag 78:67–88
Robaa SM (2008) Evaluation of sunshine duration from cloud data in Egypt. Energy 33:785–795
Robinson JG, Thompson A, Preston D (2006) Potatoes from garden to table. NDSU Extension Service, North Dakota State University, Fargo
Samson G, Visser SA (1989) Surface-active effects of humic acids on potato cell membrane properties. Soil Biol Biochem 21:343–347
Schark AE, Peterson CE, Carlin F (1956) The influence of variety on the specific gravity-mealiness relationships of potatoes. Am Potato J 33:79–83
Scherm H, Savelle AT (2001) Control of peach scab with reduced midseason fungicide programs. Plant Dis 85:706–712
Schmidt W, Santi S, Pinton R, Varanini Z (2007) Water-extractable humic substances alter root development and epidermal cell pattern in *Arabidopsis*. Plant Soil 300:259–267
Selim EM, Mosa AA, El-Ghamry AM (2009a) Evaluation of humic substances fertigation through surface and subsurface drip irrigation systems on potato grown under Egyptian sandy soil conditions. Agric Water Manag 96:1218–1222
Selim EM, El-Neklawy Soad AS, El-Ashry M (2009b) Beneficial effects of humic substances fertigation on soil fertility to potato grown on sandy soil. Aust J Basic Appl Sci 3:4351–4358
Sellamuthu KM, Govindaswamy M (2003) Effect of fertiliser and humic acid on rhizosphere microorganisms and soil enzymes at an early stage of sugarcane growth. Sugar Technol 5:273–274
Seyedbagheri M (2010) Influence of humic products on soil health and potato production. Potato Res 53:341–349
Shenker M, Chen Y (2005) Increasing iron availability to crops: fertilizers, organo-fertilizers, and biological approaches. Soil Sci Plant Nutr 51:1–17
Sonbol HA, El-arquan MY (1977) Dynamics of mineralization and humification of rice straw. J Agric Sci Mansoura Univ 2:136–143
Stevenson FJ (1982) Humus chemistry: genesis, composition, reactions. Wiley-Inter-Science, New York
Suganya S, Sivasamy R (2006) Moisture retention and cation exchange capacity of sandy soil as influenced by soil additives. J Appl Sci Res 2:949–951
Taha AA (1985) Study of physic-chemical properties of humus and its complexes. PhD thesis, Fac Agri Mansoura Univ, Egypt
Taha AA, Modaihsh AS (2003) Chemical and spectroscopic measurements on the humic acid extracted from some organic composts. J Agric Sci Mansoura Univ 28:5073–5082
Tan KH (1978) Effects of humic and fulvic acids on release of fixed potassium. Geoderma 21:67–74
Thabet EMA, Abdallah AAG, Mohammed ARAG (1994) Productivity of onion grown in reclaimed sandy soil using Tafla as affected by water regimes and nitrogen levels. Ann Agric Sci 39:337–344

Thorn KA, Mikita MA (1992) Ammonia fixation by humic substances: a nitrogen-15 and carbon-13 NMR study. Sci Total Environ 113:67–87

Vanlauwe B, Bationo A, Chianu J, Giller K, Merckx R, Mokwunye U, Ohiokpehai O, Pypers P, Tabo R, Shepherd K (2010) Integrated soil fertility management operational definition and consequences for implementation and dissemination. Outlook Agric 39:17–24

Vaughan D (1986) Effetto delle sostanze umiche sui processi metabolic delle piante. In: Burns RG, Dell'Agnola G, Miele S, Nardi S, Savoini G, Schnitzer M, Sequi P, Vaughan D, Visser SA (eds) Sostanze Umiche effetti sul terreno e sulle piante. Ramo Editoriale degli Agricoltori, Roma, pp 59–81

Vaughan D, Malcom RE (1985) Influence of humic substances on growth and physiological processes. In: Vaughan D, Malcom RE (eds) Soil organic matter and biological activity. Kluwer, Dordrecht, pp 37–76

Vaughan D, Ord BG (1981) Uptake and incorporation of 14C-labelled soil organic matter by roots of *Pisum sativum* L. J Exp Bot 32:679–687

Vázquez N, Pardo A, Suso ML, Quemada M (2006) Drainage and nitrate leaching under processing tomato growth with drip irrigation and plastic mulching. Agric Ecosyst Environ 112:313–323

Waals JE, Korsten L, Slippers B (2004) Genetic diversity among *Alternaria solani* isolates from potatoes in South Africa. Plant Dis 88:959–964

Wang FL, Huang PM (2001) Effects of organic matter on the rate of potassium adsorption by soils. Can J Soil Sci 81:325–330

Yun D, Bressan RA, Hasegawa P (1997) Plant antifungal proteins. Hortic Rev 14:39–87

Zandonadi DB, Canellas LP, Fraçana AR (2007) Indolacetic and humic acids induce lateral root development through a concerted plasmalemma and tonoplast H^+ pumps activation. Planta 225:1583–1595

Zhang X, Ervin EH, Schmidt RE (2003) Physiological effects of liquid applications of a seaweed extract and a humic acid on creeping bentgrass. J Am Soc Hortic Sci 128(4):492–496

Chapter 28
Evaluation of Residual Pesticides and Heavy Metals Levels in Conventionally and Organically Farmed Potato Tubers in Egypt

Sameeh A. Mansour

Abstract Egypt is one of the top 10 potato exporting countries in the world. The increasing demand of food safety has stimulated research regarding the risk associated with consumption of foods contaminated by pesticides and heavy metals. Egypt, like many other countries, is moving towards production of organic crops along with traditional ones. This chapter examines and discusses contamination levels of pesticides and heavy metals in potato tubers in Egypt. As there is a widespread belief that organic agriculture products are safer and healthier than conventional foods, this chapter comparatively analyzes the data of pesticides and heavy metals in conventionally and organically farmed potato tubers produced by the chapter author and his colleagues. Information in this chapter shows it is difficult to come to conclusions, but what should be made clear to the consumer is that "organic" does not automatically equal "safe". More research efforts are needed to evaluate and assure sustainable production of high quality and safe potato from both conventional and organic farming practices.

28.1 Introduction

Food safety is a major public concern worldwide and food consumption has been identified as the major pathway for human exposure to certain environmental contaminants, accounting for >90% of intake compared to inhalation or dermal routes of exposure (Fries 1995). About 30% of human cancers are caused by low exposure

S.A. Mansour (✉)
Pesticide Chemistry Department, Environmental Toxicology Research Unit (ETRU), National Research Centre, Dokki, Cairo, Egypt
e-mail: samansour@hotmail.com

Z. He et al. (eds.), *Sustainable Potato Production: Global Case Studies*, DOI 10.1007/978-94-007-4104-1_28, © Springer Science+Business Media B.V. 2012

to initiating carcinogenic contaminants in the diet (Tricker and Preussmann 1990). During the last decades, the increasing demand of food safety has stimulated research regarding the risk associated with consumption of foods contaminated by pesticides, heavy metals and/or toxins (D'Mello 2003).

Pesticide usages have contributed to dramatic increases in crop yields and in the quantity and variety of the diet. Also, they have helped to limit the spread of certain diseases. Thus, pesticides cannot be totally abolished from agricultural practice, without causing famine. On the other hand, pesticide usages may cause injury to human health and the environment. The range of these adverse health effects include acute and persistent injury to the nervous system, lung damage, injury to the reproductive organs, and dysfunction of the immune and endocrine systems, birth defects, and cancer (Mansour 2004, 2009). Heavy metals are also among the major contaminants of food supply and may be considered as one of the most important problems to the environment (Zaidi et al. 2005). Dietary intake of heavy metals poses risk to animals and human health. Heavy metals such as Cd and Pb have been shown to have carcinogenic effects (Trichopoulos 1997), and high concentrations of Cu, Cd and Pb in fruits and vegetables were related to high prevalence of upper gastrointestinal cancer (Turkdogan et al. 2002). Heavy metal contamination may occur due to irrigation with contaminated water, the addition of fertilizers and metal-based pesticides, industrial emissions, transportation, harvesting process, storage and/or sale. It is well known that plants take up metals by absorbing them from contaminated soils as well as from deposits on parts of the plants exposed to the air from polluted environments (Dach and Starmans 2005; Sharma et al. 2008).

Potato is a major industrial crop in Egypt and is one of the main food crops grown mainly in delta, and middle Egypt. Its production is concentrated in certain Governorates; namely: Beheira, Menufia, Gharbia, Giza, Dagahlia and Minia (Medany 2006). The total cultivated area of potato is 89,000 ha (ca. 374,000 feddan), which produced about two million tons, with an average yield of 10.34 tons/feddan, and a total potato export of 296,000 tons in 2005 (Abdrabbo et al. 2010). In the meantime, Egypt, like many other countries, is moving towards production of organic crops along with traditional ones. Thus, this chapter reviews and discusses research in Egypt on the residual levels of pesticides and heavy metals in potato tubers with the focus on comparative analysis of the author's data on conventionally and organically farmed potato tubers. Information in this chapter may be helpful in potato production and safety monitoring and management in Egypt and other regions.

28.2 Pesticide Contamination in Potato

Several investigators (e.g., Tricker and Preussmann 1990; Zaidi et al. 2005; Bhanti and Taneja 2005, 2007; Peris et al. 2007; Sharma et al. 2008) have reported that residues of organochlorine pesticides (OCPs), organophosphorus pesticides (OPPs) and certain heavy metals in foods at levels currently regarded as safe adversely affect human health. Most of the OCPs have been banned for decades in many parts of the world, including

Egypt, but their residues still appear as pollutants in food as well as in the environment. Their occurrence and long-range transport at local, regional and global scales has been recently explored (Harner et al. 2006). Frequent detection of these compounds in different environmental compartments could provide information about to what extend the threats posed by these xenobiotics may exist (Dogheim et al. 1999).

Occurrence of this group of long persistent compounds in potato tubers collected from local markets in different locations in Greater Cairo has been evaluated by several investigators. The total concentrations of organochlorine pesticide (OCP) residues recorded were 0.107 ppm (Dogheim et al. 1996a), 1.313 ppm (Abou-Arab et al. 1998), and 0.692 ppm (Soliman 1999). Such values are comparable with those found by Mansour et al. (2009a): 0.035–1.131 ppm in conventionally farmed potato and 0.037–0.573 ppm in organically farmed potato. Also, the occurrence of this long persistent group of compounds in vegetables from other countries has been reported either in traditionally-farmed vegetables (Zawiyah et al. 2007) or organically-farmed ones (Zohair et al. 2006).

Leafy and root vegetables such as potato plants are liable to accumulate OPP residues in the fruits (Dogheim et al. 2002, 2004). As a matter of fact, the occurrence of OPPs (e.g., malathion, dimethoate, chlorpyrifos, pirimiphos-methyl and fenitrothion) in potato tubers has been long recognized by Egyptian workers (Table 28.1). The highest contamination frequency with insecticide methamidophos was found to occur in 58.3% and 50.0% of the samples analyzed from conventional and organic potatoes, respectively; 11.1% and 8.3% of the samples contained levels higher than the maximum residue limits (MRLs) (Mansour et al. 2009a) (Fig. 28.1). Such findings coincided with the results of Dogheim et al. (2004) who reported that the most analyzed samples of vegetables and fruits contained methamidophos, although at levels below the MRLs. They attributed methamidophos contamination to spray drifts from neighboring crops such as cotton. Also, in Brazilian monitoring programs in selected vegetables and fruits, methamidophos was ranked second after chlorpyrifos among the ten most frequently detected pesticides (Caldas et al. 2006).

It is difficult to draw a clear pattern for time-contamination relationship of OCP residues in potato from currently available data. In light of the data presented in Table 28.1, potato samples collected from different locations and various sampling times throughout 1991–2007 years were found contaminated with some OCPs, such as HCHs and DDTs. Analyses concerned with monitoring of OPP residues revealed the presence of some insecticides (e.g. malathion, pirimiphos-methyl, dimethoate, fenitrothion and profenofos). Salim (2006) determined residues of OCPs in three kinds of organically farmed vegetables (green onion, beetroot, potatoes), as well as in the corresponding soils, from samples collected in Sadat City, Egypt. Results indicated that ∑OCP residues in the soil ranged between 9.94 and 10.82 $\mu g\ kg^{-1}$. Organic vegetables showed detectable residues ranging from 3.49 to 5.61 $\mu g\ kg^{-1}$, levels which are si gnificantly below the MRLs (Codex 2006). In the case of potato, most of the estimated OCPs were found in soil at concentration levels higher than those in potato tubers grown in this soil. An exception, endrin was found to be 0.93 $\mu g\ kg^{-1}$ in the tubers while undetected in the soil. Also, endosulfan in the tubers recorded 0.64 $\mu g\ kg^{-1}$ corresponding to 0.23 $\mu g\ kg^{-1}$ in the soil. As a general result,

Table 28.1 Overall pesticide contamination results for potato tubers based on monitoring studies conducted by Egyptian investigators during 1991–2007

Sampling year	# samples	Detected pesticides	Mean residues (mg kg^{-1})	Reference
1991/1992	54	Total HCH	0.053	Dogheim et al. 1996a
		Cyclodienes	0.036	
		Total DDT	0.054	
		Chlorpyrifos	0.028	
		Dimethoate	0.011	
		Malathion	0.212	
		Pirimiphos-methyl	0.507	
		Fenitrothion	2.220	
		Parathion-methyl	0.465	
		Profenofos	0.508	
1994	9	Total HCH	0.004	Dogheim et al. 1996b
		Total DDT	0.004	
1996	42	Dicofol	0.050	Dogheim et al. 2001
1997	39	None	–	Dogheim et al. 2002
1997	50	HCH	0.024	Abou-Arab et al. 1998
		Lindane	0.850	
		Dieldrin	0.001	
		Total DDT	0.840	
		Malathion	0.864	
		Dimethoate	0.961	
1997/1998	18	Fenitrothion	0.035	Abbassy 2001
		Malathion	0.028	
1998	50	HCH	0.014	Soliman 1999
		Lindane	0.141	
		Total DDT	0.537	
		Dimethoate	0.082	
		Pirimiphos-me	0.066	
		Malathion	0.601	
2006/2007	36	Many[a]	0.835[b]	Mansour et al. 2009a
			1.144[c]	
			0.527[d]	

[a] A number of OC & OP pesticides were detected in the analyzed samples (for their names, please refer to Fig. 28.1)
[b] Total pesticide residues in potato samples, generally
[c] Total pesticide residues in potato samples from conventional farming
[d] Total pesticide residues in potato samples from organic farming

total pesticidal contamination level amounted to 3.51 μg kg^{-1} in potato versus 9.94 μg kg^{-1} in soil; i.e. 35.3%. In agreement with Salim (2006), Mansour et al. (2009a) reported the presence of detectable residues of DDTs, lindane, dieldrin, heptachlor and endrin in organically farmed potatoes (Table 28.2). However, there were big differences between the estimated concentration levels in both cases. Organic potato tubers collected from the market (Mansour et al. 2009a) contained levels higher than those collected from a specified organic farm (Salim 2006).

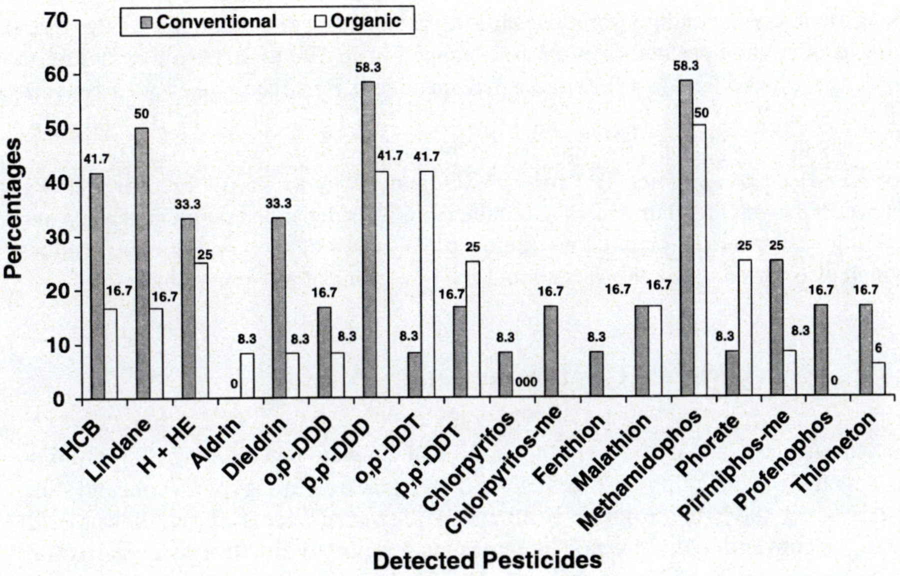

Fig. 28.1 Percentage of contaminated samples of different types of potato tubers collected from Egyptian local markets during 2006/2007 with respect to detected pesticides in the analyzed samples (n = 36 samples for each type). H+HE: heptachlor & heptachlor epoxide (Adapted from Mansour et al. (2009a))

Table 28.2 Comparison of organochlorine pesticide (OCP) contamination levels (μg kg^{-1}) in organic potato samples collected from different locations and conditions

Pesticide	Salim 2006[a]	Mansour et al. 2009a[b]	X – times[c]
HCB	0.0	157.0	–
Total DDTs	1.78	25.0	14
Lindane	0.60	38.0	63
Dieldrin	0.30	7.0	23
Hepttachlor	0.59	81.0	137
Endosulfan	0.24	0.0	–
Total pesticides	3.51	308.0	87.7

[a]Potato samples collected from an organic farm located at Sadat City, Egypt
[b]Organic potato samples collected from Egyptian local markets
[c]X –times = column 3/column 2

According to the data presented in Table 28.2, the total contamination level by OCP residues in the market samples (308 μg kg^{-1}; Mansour et al. 2009a) reached 87.7 times those from an organic farm (3.51 μg kg^{-1}; Salim 2006). Such findings may give an indication that a result of a single supervised farm may not reflect the market which receives products from different sources.

In comparison, Zohair et al. (2006) analyzed PAHs, PCBs, and OCPs in three carrot and four potato varieties obtained from organic farms in England, UK, as well as in their corresponding soils. Results revealed that concentrations of the three classes of studied pollutants in the soils ranged from 590 to 2,301 μg kg^{-1}, 3.56 to 9.61 μg kg^{-1}, and 52.2 to 478.0 μg kg^{-1}, respectively. Residue uptake for carrot varieties was 8.42–40.1 μg kg^{-1} for ∑PAHs, 0.83–2.68 μg kg^{-1} for ∑PCBs, and 8.09–133.0 μg kg^{-1} for ∑OCPs. Residue uptake for potato varieties was 8.50–33.68 μg kg^{-1} for ∑PAHs, 1.04–2.68 μg kg^{-1} for ∑PCBs, and 10.36–88.36 μg kg^{-1} for ∑OCPs. The authors concluded that residue uptake from soils depended on plant variety, and peeling either carrots or potatoes removed 52–100% of the contaminant residues which also varied with crop variety and the properties of the contaminants.

28.3 Heavy Metals Contamination in Potato

Metals such as Cd, Cu, Zn, Fe, Mn, Cr, Ni and Co were detected in different kinds of vegetables, including potatoes, from local markets in Egypt at concentration levels below the MLs (Soliman et al. 1997). However, excess of the allowable Pb level in conventionally farmed potatoes was reported by several investigators (e.g., Soliman et al. 1997; Dogheim et al. 2004; Mansour et al. 2009a), as well as in organically farmed potatoes (Salim 2006; Mansour et al. 2009a).

Data presented in Table 28.3 show concentration levels of heavy metals in potato samples collected from different locations and times from the Egyptian markets. For comparison, Zn, Cu, Cd and Pb, as examples, were found 0.96, 7.66, 0.009 and 0.65 mg kg^{-1}, respectively (Soliman et al. 1997); compared to 33.72, 3.91, 0.094 and 0.047 mg kg^{-1}, respectively (Radwan and Salama 2006); and 0.584, 0.772, 0.025 and 0.337 mg kg^{-1}, respectively (Mansour et al. 2009a). Differences of results in such studies can be referred to many factors, such as time, location and farming conditions.

28.4 Comparison of Contamination Levels Between Conventional and Organic Potatoes

According to a recent survey, 70% of consumers said that they purchased organic foods to avoid pesticides (Whole Foods Market 2005). Clearly, one of the drivers of the organic food industry is the differentiation between organic foods and conventional foods with respect to pesticide use and perceived food residues. However, only a small number of studies have looked at specific differences between pesticide residues on organic and conventional foods (Winter and Davis 2006).

The literature (Gonzalez et al. 2003, 2005; Trewavas 2004; Salim 2006; Zohair et al. 2006; Mansour et al. 2009a) offers evidence regarding the occurrence of pesticides and heavy metals in organically farmed crops. These studies report these contaminants present generally at levels below the allowed by Codex Alimentarius. Mansour et al. (2009a, b) reported actual comparisons between both farming types of potatoes and cucumbers. This research was performed within a monitoring

Table 28.3 Concentration of heavy metals in potato samples collected from different locations in Egypt

Sampling year	Mean concentration (mg kg^{-1})									Reference
	Zn	Cu	Mn	Fe	Cd	Cr	Pb	Ni	Co	
1997	0.96	7.66	4.17	12.44	0.009	0.004	0.65	0.002	0.008	Soliman et al. 1997
2005	33.72	3.91	nd	nd	0.094	nd	0.047	nd[a]	nd	Radwan and Salama 2006
2006/2007	0.584	0.772	0.654	29.876	0.025	0.758	0.337	0.022	0.032	Mansour et al. 2009a

[a] nd-not determined

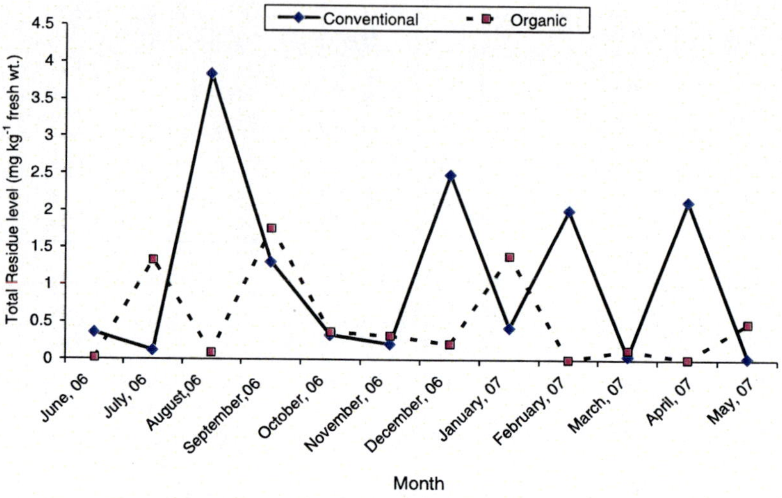

Fig. 28.2 Monthly total pesticide residue levels in conventionally and organically produced potatoes, based on samples collected over a year (June 2006–May 2007) from local markets in Egypt (Adapted from Mansour et al. (2009a))

program of pesticide residues and heavy metals in samples collected from the Egyptian markets. The general contamination pattern is discussed below.

Whereas both conventional and organic potatoes contained the same kinds of pesticides and their residues, the percentage of contaminated samples was generally higher in conventional potatoes than organic ones (Mansour et al. 2009a). However, the percentage of samples contaminated by aldrin, o,p′-DDT, and phorate were higher in organic potatoes (8.3%, 41.7%, and 25%, respectively), compared to 0%, 8.3%, and 8.3%, respectively, in the conventional samples (Fig. 28.1). Levels of pesticide residues in conventional potatoes were noticeably raised in several months (e.g., August, December, February and April), the highest peak for organic potatoes was reached in September (Fig. 28.2). Such situation created seasonal variation with respect to levels of total pesticide residues in both types of samples compared. For instance, total pesticide contamination level in conventional potatoes underwent the following arrangement: winter>summer>fall>spring; those for organic potatoes were: fall>summer>winter>spring (Table 28.4). Generally, percentage of violated samples (i.e., those contained insecticide residue levels exceeding MRLs – Codex 2006) was greater among conventional potatoes than organic ones. Fluctuation of pesticide contamination levels among the monthly analyzed samples throughout a complete year may refer to the pressure of pesticide quantity used on the crop of different three production cycles (e.g., summer, winter, and 'nili').

As with the case of pesticide contamination, both conventional and organic potatoes were found to contain the same kinds of heavy metals but contamination levels were generally higher in conventional potatoes than organic ones. Metals concentrations in conventional potatoes were noticeably raised in several months

Table 28.4 Seasonal variation of pesticide residues in potato tubers produced from conventional (C) and organic (O) farming, based on samples collected from the Egyptian local markets during 2006/2007

	Pesticide residues (ppm)					
	OCPs[a]		OPPs[b]		Total	
Season	C	O	C	O	C	O
Summer	1.027	0.413	0.409	0.129	1.436	0.542
Fall	0.545	0.573	0.213	0.252	0.758	0.825
Winter	1.131	0.037	0.513	0.496	1.644	0.533
Spring	0.035	0.210	0.702	ND[c]	0.737	0.210

Adapted from Mansour et al. (2009a)
[a]OCPs-Organochlorine pesticides
[b]OPPs-Organophosphorus pesticides
[c]ND: not detected

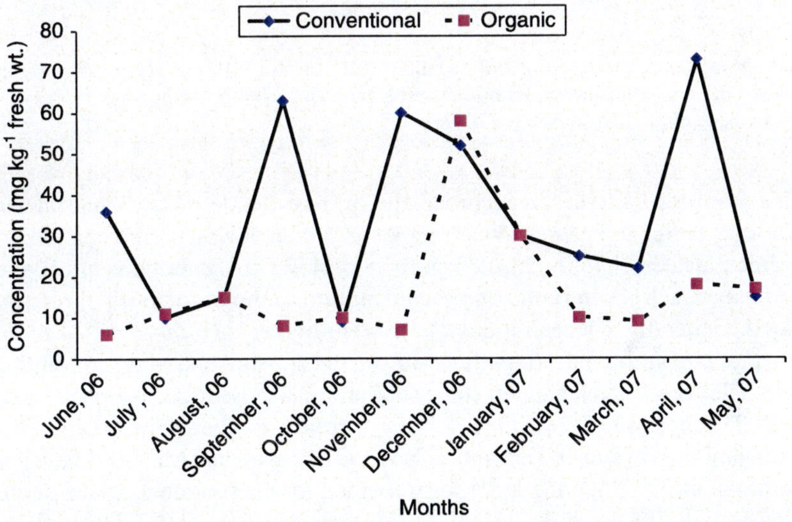

Fig. 28.3 Monthly total concentration levels of heavy metals in conventionally and organically produced potatoes, based on samples collected over a year (June 2006–May 2007) from local markets in Egypt (Adapted from Mansour et al. (2009a))

(e.g., September, November and April), the highest peak for organic potatoes was reached in December (Fig. 28.3). Such situation created seasonal variation with respect to levels of heavy metals in both types of samples compared. For instance, iron (Fe) was found at concentrations relatively higher than the other measured metals and except for the winter season samples, organic potatoes contained lower iron concentrations than the conventional potatoes throughout the other seasons (Mansour et al. 2009a). For total heavy metals content, both conventional and organic potato samples contained nearly equal concentrations (33.64 and 32.96 mg kg^{-1}).

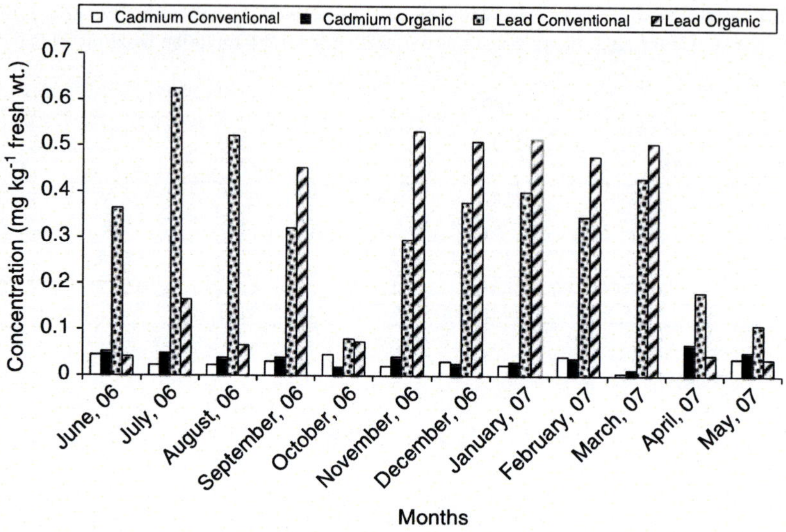

Fig. 28.4 Mean concentration levels of cadmium (Cd) and lead (Pb) in potato tuber samples collected from conventional and organic farming products. N = 36 samples for each type of potatoes (Adapted from Mansour et al. (2009a))

Samples from the fall season recorded the highest heavy metals concentration level for conventional potatoes (44.72 mg kg^{-1}), but the lowest contamination level for organic potatoes (7.75 mg kg^{-1}). Figure 28.4 shows concentrations of cadmium (Cd) and lead (Pb) in monthly collected potato samples of both farm types compared. Generally, concentrations of Pb were higher than those of Cd in both potato types. In some months (e.g., June, July, August, April and May) conventional potato samples contained higher Pb concentrations than those found in organic potato samples. The opposite was found for the samples collected in other months (e.g., September, November, December, January, February and March). The highest concentration of Pb (0.624 mg kg^{-1}) was recorded for conventional potato samples collected in July 2006, while the lowest Pb concentration (0.037 mg kg^{-1}) was observed with organic potato in May 2007 (Mansour et al. 2009a).

28.5 Risk/Safety Assessment by Bioassay

Based on results of their monitoring studies, Mansour et al. (2009a) estimated risk/safety attributed to dietary intake of pesticides and heavy metals by potatoes by comparing residue levels they found with the WHO-ADIs (World Health Organization – Acceptable Daily Intake). The results revealed that only phorate insecticide residues either in conventional or organic potatoes may pose risks to human health. None of the studied heavy metals were estimated to cause dietary intake risks to human health.

This study (Mansour et al. 2009a) shed light on the problem of multi toxicants in potatoes, and led Mansour and Gad (2010) to investigate a novel bioassay method using *Daphnia magna* Straus to assess risk posed by multi-contaminants (e.g., pesticides and heavy metals) in cucumber fruits and potato tubers. Based on the estimated lethal time to 50% mortality (LT50) in daphnids, Mansour and Gad (2010) suggested a classification to categorize toxic hazards in six definite ratings. Either samples of cucumbers (from conventional, greenhouse and organic farming) or potatoes (from conventional and organic farming) were evaluated for toxic hazard of the mixture of pesticide residues, the mixture of heavy metals, and the mixture of both.

For potato samples, Mansour and Gad (2010) categorized toxic hazards in six definite ratings: Extremely Toxic: ET (LT50=<1 h) for 11.1%; Very Toxic: VT (LT50=1 to <3 h) for 50.0%; Highly Toxic: HT (LT50=3 to <12 h) for 13.9%; Moderately Toxic: MT (LT50=12 to <24 h) for 11.1%; Slightly Toxic: ST (LT50=24–48 h) for 0.0%; and Practically Non-Toxic: NT (LT50=>48 h) for 13.9% of the samples bioassayed. The results of LT50 values and the corresponding toxicity ratings with respect to concentration levels of total pesticide residues either in conventionally or organically farmed potatoes are presented in Table 28.5a. Similar findings concerned with total heavy metals are presented in Table 28.5b. The LT50 values and their respective hazard rating in *D. magna* exposed to (pesticide residues+heavy metals) contaminating potato tuber samples (conventional & organic) are shown in Table 28.5c. It was observed that conventional potato samples of February and April of 2007 as well as organic potato samples of August of 2006 were considered as "extremely toxic; ET". The rest of the bioassayed samples appeared to be "very toxic; VT"; where their LT50 values ranged between (1 - < 3 h). Finally, Mansour and Gad (2010) pointed out that the investigated bioassay method, compared to chemical methods (e.g., gas chromatography and atomic absorption), is simple, easy to be applied by laboratory technicians, time saving and needs no sophisticated equipments or fine reagents. Moreover, it evaluates toxicity of all the mixture constituents and their metabolites and/or degradation products "in one shot". Thus, Mansour and Gad (2010) encouraged researchers to investigate the use of daphnids as an alternative test model organism instead of or in conjunction with vertebrate organisms.

28.6 Conclusion

Generally speaking, organic agricultural products may contain toxic residues of pesticides and heavy metals at levels lower than those in conventional products and mostly below the threshold limits. However, there is a growing awareness that residues of certain pesticides and heavy metals in foods at levels currently regarded as safe adversely affect human health. This led Egypt, like many other countries, to move towards production of organic crops along with traditional ones. However, most monitoring programs directed towards vegetables and fruits do not consider the types of farming production with respect to their chemical contaminants

Table 28.5 Lethal time for 50% mortality (LT50 values) and their respective hazard rating in *Daphnia magna* exposed to pesticide residues and heavy metals contaminating conventional (C) and organic (O) potato tuber samples collected from Egyptian local market

(A) Total pesticide residues

	Total pesticide residues in the bioassay solution (ppm)		LT50 (h)		Toxicity rating	
Sampling time	C	O	C	O	C	O
August, 06	3.84	0.09	11.02	16.68	HT	MT
September, 06	1.73	1.78	438.17	6.58	NT	HT
November, 06	0.21	0.32	21.05	14.94	MT	MT
January, 07	0.43	1.39	>1,000	6.40	NT	HT
February, 07	2.01	0.0	78.58	>1,000	NT	NT
April, 07	2.13	0.0	16.76	299.57	MT	NT

(B) Total heavy metal

	Total heavy metals concentration in the bioassay solution (ppm)		LT50 (h)		Toxicity rating	
Sampling time	C	O	C	O	C	O
August, 06	0.06	0.06	3.11	2.77	HT	VT
September, 06	0.26	0.03	1.65	2.56	VT	VT
November, 06	0.25	0.03	1.43	6.72	VT	HT
January, 07	0.10	0.12	1.65	2.02	VT	VT
February, 07	0.10	0.05	2.27	2.74	VT	VT
April, 07	0.32	0.09	0.09	2.11	ET	VT

(C) Mixtures pesticides and heavy metals residues

	C		O	
Sampling time	LT50 (h)	Toxicity rating	LT50 (h)	Toxicity rating
August, 06	1.09	VT	0.98	ET
September, 06	1.60	VT	1.77	VT
November, 06	1.93	VT	1.58	VT
January, 07	1.78	VT	2.24	VT
February, 07	0.45	ET	2.61	VT
April, 07	0.04	ET	1.92	VT

Adapted from Mansour and Gad (2010)
Toxicity rating (based on LT50 values): *ET* extremely toxic (<1 h), *VT* very toxic (1 to <3 h) *HT* highly toxic (3 to <12 h), *MT* moderately toxic (12 to <24 h), *ST* slightly toxic (24–48 h), *NT* practically non-toxic (>48 h)

(e.g., pesticide residues and heavy metals). Information examined in this chapter shows that organic potatoes do contain toxic residues of pesticides and heavy metals, but in most cases, their levels are lower than those in conventional products and mostly below the threshold limits. However, it is premature to conclude that either food system is superior to the other with respect to safety based on currently available data. Whereas there is a widespread belief that organic agriculture products are safer and healthier than conventional foods, more research is needed to achieve a scientifically-rigorous conclusion.

References

Abbassy MS (2001) Pesticide residues in selected vegetables and fruits in Alexandria city, Egypt, 1997–1998. Bull Environ Contam Toxicol 67:225–232

Abdrabbo MAA, Khalil AA, Hassanien MKK, Abou-Hadid AF (2010) Sensitivity of potato yield to climate change. J Appl Sci Res 6(6):751–755

Abou-Arab AAK, Soliman KM, Naguib Kh (1998) Pesticide residue contents in Egyptian vegetables and fruits and removal by washing. Bull Nutr Inst Cairo Egypt 18(1):117–138

Bhanti M, Taneja A (2005) Monitoring of organochlorine pesticide residues in summer and winter vegetables from Agra, India – a case study. Environ Monit Assess 110:341–346

Bhanti M, Taneja A (2007) Contamination of vegetables of different seasons with organophosphorous pesticides and related health assessment in northern India. Chemosphere 69:63–68

Caldas ED, Boon PE, Tressou J (2006) Probabilistic assessment of the cumulative acute exposure to organophosphorus and carbamate insecticides in the Brazilian diet. Toxicology 222:132–142
Caldas ED, Boon PE, Tressou J (2006) Probabilistic assessment of the cumulative acute exposure to organophosphorus and carbamate insecticides in the Brazilian diet. Toxicology 222:132–142

Codex (2006) Codex Alimentarius Commission, Joint FAO/WHO Food Standards Programme, Pesticide Residues in Food. Geneva, Switzerland

D'Mello JPF (2003) Food safety: contaminants and toxins. CABI Publishing, Wallingford/Cambridge

Dach J, Starmans D (2005) Heavy metals balance in Polish and Dutch agronomy: actual state and previsions for the future review. Agric Ecosyst Environ 107(1):309–316

Dogheim SM, El-Zarka M, Gad Alla SA, El-Saied S, Salama EY, Ayoub MM, Fahmy SM (1996a) Monitoring of pesticide residues in human milk, soil, water and food samples collected from Kafr El-Zayat governorate. J Assoc Off Anal Chem 79:111–116

Dogheim SM, Gad Alla SA, El-Syes SAM, Almaz MM, Salama EY (1996b) Organochlorine and Organophosphorus pesticide residues in food from Egyptian local markets. J Assoc Off Anal Chem 79:949–952

Dogheim SM, Gad Alla SA, El-Marsafy AM (1999) Monitoring of pesticide residues in Egyptian fruits and vegetables in 1995. J Assoc Off Anal Chem 82:955–984

Dogheim SM, Gad Alla SA, El-Marsafy AM (2001) Monitoring of pesticide residues in Egyptian fruits and vegetables in 1996. J Assoc Off Anal Chem 84(2):519–531

Dogheim SM, El-Marsafy AM, Salama EY, Gad Alla SA, Nabil YM (2002) Monitoring of pesticide residues in Egyptian fruits and vegetables during 1997. Food Addit Contam 19(11):1015–1027

Dogheim SM, Ashraf EMM, Gad Alla SA, Khorshid MA, Fahmy SM (2004) Pesticides and heavy metals levels in Egyptian leafy vegetables and some aromatic medicinal plants. Food Addit Contam 21(4):323–330

Fries GF (1995) A review of the significance of animal food products as potential pathways of human exposures to dioxins. J Anim Sci 73(6):1639–1650

Gonzalez M, Miglioranza KSB, Aizpún de Moreno JE, Moreno VJ (2003) Occurrence and distribution of organochlorine pesticides (OCPs) in tomato (Lycopersicon esculentum) crops from organic production. J Agric Food Chem 51(5):1353–1359

Gonzalez M, Miglioranza KSB, Aizpún de Moreno JE, Moreno VJ (2005) Evaluation of conventionally and organically produced vegetables for high lipophilic organochlorine pesticide (OCP) residues. Food Chem Toxicol 43:261–269

Harner T, Pozo K, Gouin T, Macdonald A-M, Hung H, Cainey J, Peters A (2006) Global pilot study for persistent organic pollutants (POPs) using PUF disk passive air samplers. Environ Pollut 144:445–452

Mansour SA (2004) Pesticide exposure – Egyptian scene. Toxicology 198:91–115

Mansour SA (2009) Persistent organic pollutants (POPs) in Africa: Egyptian scenario. Hum Exp Toxicol 28(9):531–566

Mansour SA, Gad MF (2010) Risk assessment of pesticides and heavy metals contaminants in vegetables: a novel bioassay method using *Daphnia magna* Straus. Food Chem Toxicol 48:377–389

Mansour SA, Belal MH, Abou-Arab AK, Ashour HM, Gad MF (2009a) Evaluation of some pollutant levels in conventionally and organically farmed potato tubers and their risks to human health. Food Chem Toxicol 47:615–624

Mansour SA, Belal MH, Abou-Arab AK, Gad MF (2009b) Monitoring of pesticides and heavy metals in cucumber fruits produced from different farming systems. Chemosphere 75:601–609

Medany M (2006) Assessment of the impact of climate change and adaptation on potato production. Central Lab for Agricultural Climate, Agricultural Research Center, Ministry of Agriculture & Land Reclamation, Egypt. Final report, 15pp

Peris M, Micó C, Recatalá L, Sánchez R, Sánchez J (2007) Heavy metal contents in horticultural crops of a representative area of the European Mediterranean region. Sci Total Environ 378:42–48

Radwan MA, Salama AK (2006) Market basket survey for some heavy metals in Egyptian fruits and vegetables. Food Chem Toxicol 44:1273–1278

Salim A (2006) Evaluation of heavy metal contents and organochlorine pesticides (OCPs) residues in Egyptian organically-farmed vegetables. J Agric Sci Mansoura Univ 31(3):1601–1612

Sharma RK, Agrawal M, Marshall FM (2008) Heavy metal (Cu, Zn, Cd and Pb) contamination of vegetables in urban India: a case study in Varanasi. Environ Pollut 154:254–263

Soliman KM (1999) Changes in concentration of some pesticide residues in potatoes during washing and cooking. J Agric Sci Mansoura Univ 24(5):2503–2511

Soliman KH, Abou-Arab AAK, Badawy A, Naguib KH (1997) Heavy metal contamination levels in Egyptian vegetables, fruits and elimination by washing procedure. Bull Nutr Inst Cairo Egypt 17(2):120–146

Trewavas A (2004) A critical assessment of organic farming and food assertions with particular respect to the UK and the potential environmental benefits of no-till agriculture. Crop Prot 23:757–781

Trichopoulos D (1997) Epidemiology of cancer. In: DeVita VT (ed) Cancer: principles and practice of oncology. Lippincott Company, Philadelphia, pp 231–257

Tricker AR, Preussmann R (1990) Chemical food contaminants in the initiation of cancer. Proc Nutr Soc 49:133–144

Turkdogan MK, Kilicel F, Kara K, Tuncer I (2002) Heavy metals in soil, vegetables and fruits in the endemic upper gastrointestinal cancer region of Turkey. Environ Toxicol Pharmacol 13:175–179

Whole Foods Market (2005) Whole foods market organic trend tracker. Whole Foods Market, Austin

Winter CK, Davis SF (2006) Quality and safety comparisons of organic and conventional foods. J Food Sci 71(9):117–124

Zaidi MI, Asrar A, Mansoor A, Farooqui MA (2005) The heavy metal concentrations along roadside trees of Quetta and its effects on public health. J Appl Sci 5(4):708–711

Zawiyah S, Che Man YB, Nazimah SAH, Chin CK, Tsukamoto I, Hamanyza AH, Norhaizan I (2007) Determination of organochlorine and pyrethroid pesticides in fruit and vegetables using SAX/PSA clean-up column. Food Chem 102:98–103

Zohair A, Salim A, Soyibo AA, Beck AJ (2006) Residues of polycyclic aromatic hydrocarbons (PAHs), polychlorinated biphenyls (PCBs) and organochlorine pesticides in organically-farmed vegetables. Chemosphere 63:541–553

Part IX
Improving Potato Yields in the Tropical Highlands of Africa

Chapter 29
Optimization of Late Blight and Bacterial Wilt Management in Potato Production Systems in the Highland Tropics of Africa

Modesto Olanya, Richard Nyankanga, Peter Ojiambo, Berga Lemaga, Rogers Kakuhenzire, and Dominic Fontem

Abstract Late blight and bacterial wilt are two formidable disease constraints on potato and account for significant losses in Sub-Saharan Africa (SSA). In this chapter, various management techniques for late blight and bacterial wilt diseases are highlighted with examples drawn from diverse research. The modified disease management approaches include resistant cultivars, reduced fungicide applications, disease monitoring based on field scouting, cultural practices, post-harvest management and farmer training. Deployment of cultivars with resistance genes and quantitative resistance in addition to fungicide use has contributed significantly to sustained late blight management in tropical Africa. Similarly, cultural practices such as date of planting, disease-free tubers, roguing and bio-rational approaches (plant-derived extracts and phosphoric acid) have been used to a lesser degree. Disease monitoring and weather-based predictions in relation to fungicide applications have been utilized in conjunction with host-plant resistance. Similarly, bacterial wilt has been successfully managed through non-chemical means which include

M. Olanya (✉)
New England Plant, Soil and Water Laboratory, USDA-Agricultural Research Service, Orono, ME 04469, USA
e-mail: modesto.olanya@ars.usda.gov

R. Nyankanga
Department of Plant Science and Crop Protection, Department of Plant Pathology, University of Nairobi, P.O Box 30197–00100, Nairobi, Kenya

P. Ojiambo
Department of Plant Pathology, North Carolina State University, Raleigh, NC 27695, USA

B. Lemaga • R. Kakuhenzire
Sub-Saharan Africa region, International Potato Centre, Sub-Saharan Africa Liasion Office, P. O. Box 22274, Kampala, Uganda

D. Fontem
Department of Plant Protection, D.N. Fontem, University Dschang, P.O. Box 208, Dschang, Cameroon

Z. He et al. (eds.), *Sustainable Potato Production: Global Case Studies*, DOI 10.1007/978-94-007-4104-1_29, © Springer Science+Business Media B.V. 2012

crop rotation, sanitation (removal of wilted plants, destruction of crop residues), and minimum post-emergence cultivation of potatoes. Small-scale seed plot technique, non-diseased tubers, soil amendments and less susceptible cultivars have been important components for integrated management of bacterial wilt disease. The holistic approach for control of late blight and bacterial wilt ultimately lead to increased potato production and better economic returns in the diverse potato production region of SSA.

29.1 Introduction

Potato (*Solanum tuberosum* L.) production in Africa as a whole has increased tremendously, from 1.3 million tons in the early 1960s to 16.7 million tons in 2007 (FAO-CIP 1995; Scott et al. 2000). Potato consumption is increasing in the developing countries and accounts for more than a third of the global harvest. The potato's ease of cultivation and high energy content has made it a valuable cash and food crop for millions of farmers in Africa. The crop has grown in importance as food in rapidly growing urban areas. Potato is the fastest growing major crop in the great lakes area of Uganda, Rwanda, Burundi, and Democratic Republic of Congo (East and Central Africa), where it has increased by about 250% since the mid 1990s (FAO-CIP 1995). In these production areas, potatoes play an important role as both cash and subsistence crop for highland farmers.

Broadly, there are three major potato production systems in tropical Africa. In the densely populated, high-potential highland regions (1,800–2,750 m) of eastern and central Africa, potatoes are grown by small-scale farmers (0.5–2 ha) for the fresh produce market and for home consumption. In these production regions, tuber yields range from 5 to 20 tons ha^{-1}, with average production of 8–10 tons ha^{-1}. In the highland tropics of East and Central Africa, inputs such as seeds, fertilizers, fungicides and some insecticides are utilized to a varying degree (Olanya et al. 2001a; Namanda et al. 2004; Nyankanga et al. 2004). The rates for chemical fertilizers and fungicide use vary widely among potato farmers, countries and production regions, but are generally low because the high costs of pesticides are prohibitive to many farmers (Ojiambo et al. 2001; Olanya et al. 2001b; Mukalazi et al. 2001; Muchiri et al. 2009). However, supply of some farm inputs such as seed is often unreliable or chemicals for pest control are used incorrectly. Seed inputs are primarily obtained from informal, local sources where production practices are often minimally regulated and the average seed quality is low. Late blight, bacterial wilt, and potato viruses are the most important biotic constraints contributing to low yield and significant losses.

In West Africa, potato is principally produced in Cameroon and Nigeria, where production of potatoes is an important small holder crop in higher elevations of the Cameroon highlands and the Jos Plateau, respectively (Fontem and Aighewi 1993; Fontem et al. 2001). Elsewhere in West Africa, potato is grown on a very limited scale as a high-value vegetable crop, and usually under irrigation production. In southern Africa region, potato is primarily produced in the Republic of South Africa.

In this region, potato is grown on a relatively large scale in this modern farming, high input system. Irrigation is an important component of potato production, where good-quality seed and other high inputs are used intensively. The average yield is relatively high and ranges from 15 to over 25 tons ha^{-1} in South Africa. Potato production in most developing countries is characterized by low yields (Scott et al. 2000).

Many factors other than disease cause yield instability in Africa. These include low soil fertility, water stress and frost. Although potato is the principal food security crop in the tropical highlands of sub-Saharan Africa, its sustained production is constrained by major pests and diseases, particularly late blight (LB), bacterial wilt (BW) and viruses. Late blight caused by *Phytophthora infestans* (mont.) de Bary occurs throughout the main potato growing regions in the tropical highlands and is by far the most important disease constraint impacting the crop (Fontem et al. 2005; Olanya et al. 2001a, b; Nyankanga et al. 2004). Whereas late blight can be controlled by fungicides, bacterial wilt lacks economically feasible chemical control strategies, resulting in both yield and quality losses.

Bacterial wilt (*Ralstonia solanacearum* (Smith) Yabuuchi) is also a major production impediment at lower elevations of less than 800 m above sea level (m.a.s.l) and higher elevations greater than 1,000 m above sea level (Kinyua et al. 2001, 2005). The disease is also important particularly in farming practices where solanaceous crops such as tomatoes, egg plants, and pepper is rotated with potato (Katafire et al. 2005). Early blight, caused by *Alternaria solani* occurs towards the end of the cropping seasons and in the drier production areas. Generally, the disease causes less damage on potato compared to late blight and bacterial wilt. Black scurf (*Rhizoctonia solani*) and black leg (*Erwinia spp.*) have also been reported as minor constraints to potato production.

Potato viruses similarly pose significant production constraints in various regions. The major potato viruses reported or documented in tropical Africa include: Potato leaf-roll virus (PLRV), Potato virus Y, and Potato viruses A, S, X, as well as alfalfa mosaic virus. The aphid-transmitted PLRV and PVY cause the most losses, especially in seed stocks retained by farmers and in situations where seed lots are obtained through the informal seed trade. Aphid-borne viruses are known to spread in rustic diffused light-storage facilities. PVX is also widespread and its incidence is high throughout all potato-growing areas, often reaching over 90% in some varieties.

29.2 Occurrence of Late Blight and Bacterial Wilt

29.2.1 *Late Blight Occurrence and Economic Importance*

Late blight is the most important potato disease in the tropical region of Africa. The disease has been reported in the potato growing countries of Sub-Saharan Africa ranging from East, Central Africa, and West Africa to Southern Africa. Disease distribution and severity is mainly confined to the highland regions

(1,800–3,000 masl) and on potato and tomato hosts. Variation in yield losses caused by late blight have been documented in several countries and have been shown to range from 30% to 75% on susceptible cultivars (Olanya et al. 2001a). Under experimental conditions at research stations, potato yield losses attributed to late blight range from 2.7% to 47% in Ethiopia (Bekele and Hiskias 1996). At Kalengyere Research Station in Uganda, yield losses of up to 20% have been reported in a season with low late blight pressure (Mukalazi et al. 2001). Similarly, yield increases in excess of 20% have been recorded under on-farm situations in Kiambu, Kenya (Njuguna et al. 1998). The disease is less significant in the drier areas of southern Africa. Generally, there is limited research on the importance of tuber blight in potato production in Africa (Nyankanga et al. 2007).

The primary sources of inoculum have not been adequately documented. However, the continuous cropping of potato and tomato during the year ensures inoculum presence year-around in tropical Africa. Potato is produced in the highland tropics in multiple cycles, with variations in the quantity planted generally following rainfall and climatic patterns and prevailing commodity prices. The low incidence of tuber blight and the lack of evidence for potato seed-borne infection suggests that tuber blight is not a significant source of primary inoculum in the tropics. Although late blight occurs mainly on potato and tomato, infection of wild hosts belonging to the *Solanum* spp. such as black nightshade (*Solanum nigrum* L.), hairy nightshade (*Solanum sarrachoides* L.), common morning glory (*Pharbitis purpurea* L) or asteraceous species by *P. infestans* may occur (Olanya et al. 2009; Fontem et al. 2004, 2005; Natrass and Ryan 1951). These hosts overlap geographically with potato and provide additional opportunities for pathogen survival and reproduction. Wild *solanaceous* hosts such as pear melon and wild tomato found in humid areas of Ecuador have been implicated in pathogen perpetuation during drier periods of the cropping cycle.

Population studies of *P. infestans* in Sub-Saharan Africa (SSA) have been conducted on isolates from Uganda, Kenya and S. Africa. Mating type tests with A1 tester isolates coupled with DNA analysis revealed that the fungal isolates from Uganda, Kenya and S. Africa are of US 1 clonal lineage, A1 mating type (McLeod et al. 2001; Mukalazi et al. 2001; Ochwo et al. 2002). Variation and lack of consistency in oospore production (10% selfing, 24% mating, and 15% non-oospore producers) have been detected among the isolates from Uganda and Kenya (Mukalazi et al. 2001; Vega-Sanchez et al. 2000). Similarly, variability in metalaxyl sensitivity has been detected among some isolates of *P. infestans* (Mukalazi et al. 2001).

29.2.2 Bacterial Wilt Occurrence and Economic Importance

Bacterial wilt caused by *Ralstonia solanacearum* is the second most important constraint to potato production in the tropical highlands of Africa (Kinyua et al. 2005).

Table 29.1 Presence of *Ralstonia solanacearum* races/biovars and importance in selected countries in some East and central Africa

Country	Races (biovars)	% crop losses*
Burundi	3 (2-A)	30–50
Ethiopia	2 (2-A)	45
Kenya	1, 3 (2-A, 2-T)	50–70 (seed)
Rwanda	3 (2-A)	-
Uganda	1, 3 (2-A)	26–100

*The crop losses resulting from bacterial wilt infections are for potato (*Solanum tuberosum* L.)

The important hosts for this pathogen include species in the family Solanaceae (potato, tomato, egg plant, bitter tomato, tobacco, black night shade, pseudo-apple), Musaceae (banana), Papilionaceae (ground nut) and Zingerbaraceae (ginger) (Hayward 1991). Bacterial wilt is principally soil-borne and primarily disseminated by infected seed tubers (Hayward 1991). The pathogen occurs in many tropical and sub-tropical potato production regions of Uganda, Ethiopia, Kenya, Madagascar, Rwanda, Burundi, Nigeria and Cameroon (Fontem and N'tchorere 1999; Hayward 1991; Boucher et al. 1992; Kinyua et al. 2005).

Two races and 3 biovars of bacterial wilt are known to occur on potato: Race1 biovar 3 and biovar 1, and Race 3 biovar 2 have been reported in the highland tropics (Priou et al. 2005). Race 1 occurs exclusively in tropical lowlands or sub tropical regions where soil never freezes (Denny and Hayward 2001). Race 3 biovar 2 was reported to occur in the same geographic region where late blight is endemic in potato production regions of Uganda, Kenya, Ethiopia, Rwanda, Burundi and Madagascar (Lemaga et al. 2005; Olanya et al. 2001a). Whereas bacterial wilt has been reported in Mali in West Africa (Coulibaly et al. 2002), no identification of the race or biovar has been published. Race 1 (biovars 1, 3, 4), is pathogenic on potato and on a broad range of hosts and are restricted to tropical areas, whereas Race 3, which has a narrow host range (potato and tomato) and a lower optimum temperature, occurs in cool upland areas of tropical regions and warm temperate areas. Due to this low temperature optimum, Race 3 or biovar 2 is the causal organism of bacterial wilt on potato and tomato in some African countries, such as Burundi, Egypt, Kenya, Libya, Reunion, South Africa, and Zambia (CABI/EPPO 1999). According to Elphinstone (2005), *R. solanacearum* Race 1 have been reported in at least 21 countries in Africa; biovar 2 Race 3 has been reported on potato in Burundi, Egypt, Ethiopia, Kenya, Libya, Reunion, Rwanda, South Africa, Tanzania and Uganda. Isolates collected from potato in Kenya, Nigeria and Cameroon were found to have the biovar 2T phenotype. In east and central Africa, races 1 and 3 and biovar 2T have been reported (Table 29.1). Research work conducted in Kenya and in Uganda showed that bacterial wilt disease is extensively disseminated by use of infected seed lots, lack of proper crop rotation and the increase in propagules of *R solanacearum* populations in infested soil (Ateka et al. 2001; Lemaga et al. 2001a, b; Katafire et al. 2005; Kinyua et al. 2005; Smith et al. 2005) (Table 29.1).

29.3 Management of Potato Diseases in the Highland Tropics of Africa

29.3.1 Host Resistance (Resistant Cultivars) and the Management of Late Blight

In the tropical highlands of Africa, host plant resistance is still the most economical and sustainable option for the management of potato late blight. Much of this is driven by the fact that year round availability of disease inoculum would dictate several fungicide sprays for adequate disease control. The use of fungicides for late blight control is, however, limited by the cost and potential detrimental impact on the environment (Hallberg 1989). Many of the traditional potato cultivars grown in the tropical highlands in Africa have very low to moderate levels of resistance to potato late blight.

Foliar resistance to late blight is governed by both qualitative and quantitative genes (Gees and Hohl 1988). Like in many other pathosystems, qualitative resistance for late blight is readily overcome by evolution of new pathotypes that are virulent and more aggressive to the existing qualitative genes (Fry 2008). Thus, several breeding efforts have been designed to breed for quantitative resistance. In the tropical highlands of Africa, the International Potato Center (CIP) in conjunction with the national potato programs in sub-Saharan Africa has undertaken research that has led to the development of genotypes that express horizontal (Landeo et al. 1997) and vertical resistance. The genotypes have been evaluated in multi-location trials in Cameroon, Uganda, Kenya and Ethiopia. Within these countries, some of these genotypes have been selected for adaptability to local conditions and for superior tuber yield and quality (Olanya et al. 2006). For example, at the Kenya regional center for CIP, the cultivars Asante and Tigoni (Table 29.2) were released from a CIP population bred for horizontal resistance free from Resistant (R) genes (Olanya et al. 2006). Similar efforts have led to the release of the cultivars Victoria and Rutuku in Uganda (Hakiza et al. 2000), and Genet and Awash in Ethiopia. The varieties Asante, Tigoni, Victoria and Awash have quantitative resistance to potato late blight, whereas Rutuku possesses qualitative resistance with one or two major *R*-genes. Differences in disease severity (AUDPC) have been detected among these potato cultivars in the tropical highlands and the relative ranking of genotypes have been reported to be consistent across years and locations (Olanya et al. 2006). Specifically, cultivars with major gene resistance (Cruza 148 and Rutuku) consistently show higher levels of resistance than varieties with horizontal resistance. However, cultivars with horizontal resistance (Tigoni, Asante, Genet) consistently had greater yield in presence of high disease pressure and may be more suitable in regions in the tropical highlands where late blight is known to be a common occurrence (Table 29.2).

One aspect of host resistance to late blight that has received little attention is tuber resistance. Unlike foliar resistance, the genetics of tuber blight resistance have

Table 29.2 Late blight severity (area under disease progress curve) and yield of potato cultivars evaluated for disease development at two locations in Kenya in 2001 and 2002

		Loreto		Kabete	
Year	Variety*	AUDPC (% disease days)	Yield (Mg ha^{-1})	AUDPC (% disease days)	Yield (Mg ha^{-1})
2001	Tigoni	703.4 b	17.8 a	23.8 b	41.9 a
	Asante	834.2 b	13.1 ab	18.8 b	45.1 a
	Nyayo	902.1 b	11.8 ab	322.6 a	32.5 b
	Rutuku	323.2 c	15.2 ab	42.4 b	28.7 b
	Genet	334.8 c	11.8 ab	57.9 b	43.1 a
	Kerr's Pink	2248.1 a	2.8 c	427.7 a	16.9 c
	Cruza 148	396.2 c	14.5 ab	42.2 b	35.4 ab
	Awash	980.8 b	9.5 bc	58.9 b	30.6 b
2002	Tigoni	2016.4 c	12.1 b	385.7 b	23.4 ab
	Asante	1799.1 cd	10.8 b	237.3 bc	23.8 ab
	Nyayo	1356.8 de	19.5 a	925.8 a	30.7 a
	Rutuku	193.3 f	13.9 b	36.9 c	12.1 bc
	Genet	952.2 e	12.0 b	378.6 b	23.1 ab
	Kerr's Pink	3086.0 a	3.5 c	1078.8 a	10.0 c
	Cruza 148	1067.5 e	10.4 b	305.8 bc	14.1 bc
	Awash	2612.5 b	5.8 bc	863.1 a	12.5 bc

*Asante, Tigoni, Victoria and Awash have quantitative resistance, whereas Rutuku and Cruza 148 possess qualitative resistance. Nyayo is a local cultivar whose inheritance to late blight is not known. Disease or yield values followed by the same letter are not significantly different at $\alpha = 0.05$ (Olanya et al. 2006)

not been studied extensively. In an effort to develop a model for tuber blight infection, a series of studies were conducted in the tropical highlands in Kenya to determine important weather and foliar resistance variables that influence tuber infection (Nyankanga et al. 2007). Soil temperature, precipitation, tuber depth, and foliar resistance were negatively correlated with incidence of tuber blight. Regression models using these variables had low predictive ability ($0.40 < R^2 < 0.46$) for the incidence of tuber blight. This model has been validated using data collected in the US (Nyankanga et al. 2011) but its utility under the tropical highlands in Africa has not yet been evaluated. From that research work, the cultivar Tigoni had consistently recorded lower incidence of tuber blight on potato (Nyankanga et al. 2011).

29.3.2 Management of Late Blight with Fungicides

Fungicide application strategies, timing, rates and mixtures are important for the management of potato late blight. Due to the continuous cultivation of potato that ensures abundant *P. infestans* inoculum for most of the year, disease control by use of fungicides is common in many agro-ecosystems. In the tropics of Africa, contact

Table 29.3 Effect of fungicide mixtures on late blight (*Phytophthora infestans*) disease levels and tuber yield at Tigoni, Kenya in 1999–2000

Fungicide mixtures (1999–2000)	Application rate (ha^{-1})	Disease severity (%)*	Total tuber yield ($Mg\ ha^{-1}$)**
Untreated control	–	98.5	5.5
Fenamidone+mancozeb	0.9 L	56.3	5.5
	1.0 L	55	15.5
	1.1 L	53.8	11.3
	LSD (.05)	2.7	18.4
Propamocarb HCL+mancozeb	2 L	40.0	18.7
	3 L	33.1	22.4
	4 L	26.8	23.6
	LSD (.05)	6.1	3.5
Mancozeb	1.5 kg	66.5	14.1
	2.5 kg	63.0	16.7
	LSD (.05)	3.6	2.7
Metalaxyl	2.5 kg	8.5	36.0
LSD (0.05)		21.5	6.3

*Disease severity refers to % leaf area diseased and **total tuber yield
Disease or yield values differ based on $LSD_{0.05}$ values (Muchiri et al. 2009)

(mancozeb), systemic (metalaxyl, propamocarb HCL) and protectant fungicides have been used routinely for the management of potato late blight. For example, mancozeb and a premix of metalaxyl+mancozeb have been widely used for late blight management in many tropical countries of Africa (Fontem et al. 2005; Ojiambo et al. 2001; Olanya et al. 2001b; Nyankanga et al. 2004). Published reports from various parts of the world have documented genetic changes in populations of *P. infestans* as more virulent and metalaxyl resistant strains became common (Davidse et al. 1981; Dowley and O'Sullivan 1981). In the tropical highlands of Africa, the presence of new clonal lineages (US 8, A2 mating type) has not been documented. However, various studies have shown that isolates of *P. infestans* are insensitive to metalaxyl particularly in tropical Africa where there is excessive use of metalaxyl (Ojiambo et al. 2001; Mukalazi et al. 2001; Fontem et al. 2005).

In field studies conducted on fungicide efficacy in the highlands of Kenya, late blight incidence and severity varied with the timing, frequency, rates and mixtures of fungicides used (Muchiri et al. 2009). In studies where mixtures of fungicide compounds consisting of fenamidone+mancozeb and propamocarb HCL+mancozeb at various rates, and application of metalaxyl and mancozeb were evaluated for control of late blight (US-1 genotype), disease levels differed significantly among treatments (Table 29.3). Propamocarb HCL+mancozeb significantly ($P<0.05$) reduced foliar blight compared with mancozeb and the untreated control at the Tigoni location. Late blight severity was also significantly lower when mixtures of propamocarb HCL+mancozeb were applied at the rate of 4 $L\,ha^{-1}$ at Tigoni (Table 29.3).

Similar results were obtained with identical set of treatments during the 2000–2001 cropping cycle. All fungicide treatments resulted in significantly lower late blight severity irrespective of the application rate. The application of metalaxyl resulted in

Table 29.4 Effect of fungicide mixtures on late blight severity (*Phytophthora infestans*) and tuber yield at Tigoni, Kenya in 2000–2001

Fungicide mixtures (1999–2000)	Application rate (ha^{-1})	Disease severity (%)*	Total tuber yield ($Mg\ ha^{-1}$)**
Untreated control	–	81.3	12.0
Fenamidone+mancozeb	0.9 L	13.8	16.9
	1.0 L	9.4	18.1
	1.1 L	8.8	20.1
	LSD (.05)	5.7	2.2
Propamocarb HCL+mancozeb	2 L	10.6	18.3
	3 L	5.4	18.8
	4 L	4.5	20.1
	LSD (.05)	5.9	1.8
Mancozeb	1.5 kg	63.5	14.9
	2.5 kg	51.4	15.7
	LSD (.05)	12.4	0.8
Metalaxyl	2.5 kg	8.0	21.9
LSD (0.05)		12.5	2.9

*Disease severity refers to % leaf area diseased and **total tuber yield in tons/ha. Disease or yield values differ based on $LSD_{0.05}$ values (Muchiri et al. 2009)

the lowest disease severity compared to all other treatments. In fungicide treatments where Propamocarb HCL+mancozeb were applied, moderate disease levels were recorded, whereas plants treated with mancozeb alone had higher late blight severity (Table 29.4). In both cropping years, propamocarb HCL+mancozeb and mancozeb alone resulted in significantly ($P<0.05$) greater yield than the untreated control. In 1999–2000, total yield ranged from 5.5 to 37 $t\,ha^{-1}$ whereas in 2000–2001, yield ranged from 12 to 21.9 $t\,ha^{-1}$.

The fungicide mixtures were effective in control of late blight compared with the application of mancozeb alone or the use on the untreated control suggesting that mixtures can work equally well for late blight management. Even though none of the fungicide mixtures or mancozeb were as effective as metalaxyl, metalaxyl should not be used excessively due to potential development of fungicide resistance. Elsewhere, mixtures of sytemic and protectant fungicides have also been documented for their effectiveness in the control of other oomycetes (Samoucha and Gisi 1987).

In addition to the use of mixtures of fungicidal compounds, combinations of protectant and systemic fungicides are frequently used for management of potato late blight in the highlands of East and Central Africa (Ojiambo et al. 2001; Mukalazi et al. 2001; Namanda et al. 2004), where repeated applications of 4–5 times per season is conducted for late blight management. Elsewhere in other parts of tropical Africa such as the highlands of Cameroon (West Africa), late blight is controlled by the intensive application of fungicides (Fontem et al. 2005). In such a potato production region, a premix formulation with copper oxide (Ridomil Plus) or with mancozeb (Ridomil MZ) is utilized for late blight control (Fontem et al. 2005). In general, fungicide efficacy on late blight in tropical weather conditions has been one of the rationales mentioned by farmers in using the product.

Table 29.5 Comparison of late blight disease and tuber yield at different planting dates at Kalengyere, Uganda in experiments conducted from 2002 to 2004

Cropping season	Planting dates	Final disease severity	Actual rAUDPC*	Potato yield (control plot) (Mg ha^{-1})	Potato yield in fungicide treated plots (Mg ha^{-1})
Season 1	1 Mar.	63.4	27.1	15.7	26.5
	22 Mar.	38.1	20.4	14.0	26.9
	13 Apr.	40.2	19.2	13.2	30.3
Season 2	1 Sep.	59.8	18.8	28.4	32.8
	22 Sep.	70.3	30.1	24.7	34.7
	13 Oct.	68.6	31.9	16.0	30.8
	3 Nov.	69.4	39.4	17.2	30.4

*Data represents average of three cropping years, *rAUDPC* relative area under disease progress curves (AUDPC/number of days disease was assessed)

29.3.3 *Cultural Practices (Use of Disease-Free Seed Tubers, Manipulation of Date of Planting, Sanitation/Removal of Diseased Plants)*

Cultural measures for late blight management in tropical Africa have been used in various ways. Given that potato culls or residue, volunteer plants, and infected tomato plants are sources of inoculum, their elimination will greatly reduce late blight risk. Various cultural practices have been shown to be effective strategies for late blight management in the Kenyan highlands (Nyankanga et al. 2004). Hilling of potato plants, dehaulming or vine kill to minimize tuber borne infections are most often used in tropical Africa. However, the removal of volunteer potatoes or infected tomato plants, or the spot treatment of pockets of disease development is not readily practiced by farmers. There is, however, a paucity of quantitative data concerning the contribution of these sources of inoculm to disease epidemiology in tropical Africa. There are virtually very limited use of containment policy or regulatory procedures that deal with late blight. In general, specific tolerance levels for late blight and visual inspections have proven inadequate for disease management in potato.

The use of cultural practices such as staggering of date of planting during the cropping cycle or manipulation of potato planting sequences in response to favorable environments for late blight have been documented as a late blight control strategies (Table 29.5). Manipulation of planting date, planting density and inter-cropping are often used as a disease escape strategy and to reduce build up of air-borne inoculums during the period of crop growth. In experiments conducted from 2002 to 2004, the effects of planting dates of potato cultivars on the development of late blight and potato tuber yield were evaluated at Kalengyere, Uganda. Potato cultivars with different levels of resistance to late blight were planted in two cropping seasons of each year. Contact and systemic fungicides were applied periodically to vary the levels of late blight disease in field plots. The results of this study showed that late blight severity varied among cultivars, planting dates and years.

Disease severity levels ranged from 38% to 70% among cultivars and years. Similarly, potato tuber yield ranged from 13.2 to 38 ton ha^{-1} across cultivars (Victoria, NAKPOT4 and NAKPOT5) in the untreated (control) plots, whereas in fungicide treated plots, tuber yield were in the range of 26.5–34.7 ton ha^{-1} during both cropping seasons. The cropping seasons and planting dates impacted tuber yield loss and the tuber yield loss attributed to late blight generally ranged from 40% to 57% and 13–49% during the first and second cropping seasons, respectively. Other cultural and cropping practices such as seed selection and removal of diseased plants have also been used (Olanya et al. 2010).

29.3.4 *Integrated Techniques for Late Blight Management (Host Resistance + Fungicides, Weather-Based Diseased Monitoring, Calendar-Based Fungicide Applications)*

Integrated disease management has been used in many host-pathosystems in diverse crops and regions of the world due to cost reductions, environmental and health stewardships and economic gains from the practice (Namanda et al. 2004). Integrated management of potato late blight is perhaps one of the best management options available for this devastating disease, particularly in growing regions where resources are limited (Namanda et al. 2004; Olanya et al. 2010). In tropical Africa, combinations of host resistance, fungicide applications, cultural measures, seed management, bio-rational approaches, and post-harvest management have been utilized to varying degrees. In research studies conducted to assess the effectiveness of fungicide application and host resistance for late blight control, it was shown that fungicide applications considerably reduced late blight in cultivars with high levels of late blight with corresponding increases in tuber yield compared to untreated plots or susceptible cultivars (Namanda et al. 2004). Disease scouting and weather monitoring prior to first fungicide applications resulted in significant economic gains compared to other strategies for late blight control. Similar strategies such as use of IPM, weekly, bi-weekly applications, and untreated control have been utilized in assessing the effectiveness of integrated late blight management. Strategies involving various combinations of resistant cultivars with potato cropping dates and fungicide management strategies also resulted in significant gains in disease control (Kankwatsa et al. 2002).

29.3.5 *Bio-Rational Approaches – Plant-Derived Extracts (Tagetes Minuta), Phosphoric Acids*

Amino butyric acid (BABA) and fosetyl–Al, foliarly applied at early stages of crop growth, can increase the resistance of potato foliage and tubers to late blight (Andreu et al. 2006). Phosphoric acid has been reported to inhibit sporulation of *P. infestans*

Table 29.6 Effect of plant extract (Stinging nettle), phosphoric acid and fungicide combinations on late blight severity (%) at Tigoni location in 2008 and 2009

2008	Disease severity (%)		rAUDPC (% disease days)*	
Treatment	Desiree	Tigoni	Desiree	Tigoni
Metalaxyl	10.2	3.9	13.0	4.0
Mancozeb	12.8	5.87	15.5	7.5
Phosphate	9.0	4.4	11.3	5.6
Stinging nettle (extract)	15.1	8.0	18.4	9.8
Metalaxyl+Phosphite	6.9	3.5	8.9	4.6
Mancozeb+phosphate	10.2	4.4	12.8	4.5
Mancozeb+extract	–	–	–	–
Untreated	33.6	13.6	39.4	16.5
Means	44.5	20.0	12.34	18.2
LSD (0.05)	2.7	1.7	6.6	3.2
2009	**Disease severity (%)**		**rAUDPC (% disease days)**	
Treatment	Desiree	Tigoni	Desiree	Tigoni
Metalaxyl	10.9	5.5	11.2	5.6
Mancozeb	27.6	12.5	27.1	12.6
Phosphate	10	5.9	10.3	5.8
Stinging nettle (extract)	28.4	12.7	28.2	13.1
Metalaxyl+Phosphite	8.8	4.7	9.1	4.8
Mancozeb+phosphate	13.8	7.1	14.1	7.1
Mancozeb+extract	24.8	11.1	25.2	11.4
Untreated	50	22.8	50.2	22.8
Means	44	20.0	12.34	18.2
LSD (0.05)	2.4	2.1	2.6	2.3

**rAUDPC* relative area under disease progress curves (AUDPC/number of days disease was assessed), Nyankanga et al. unpublished data

and phosphoric acid may have some fungitoxic activity after pathogen infection (Schwinn and Margot 1991). In experiments involving the use of plant extract from stinging nettle (bio-pesticide) for control of plant diseases, the extract was reported to be effective on *Phomopsis thea* of tea and also inhibited some oomycetes (Onyiango et al. 2005). In experiments conducted to evaluate the efficacy of phosphoric acid and extract of stinging nettle alone or in combination with other fungicides for late blight control in Kenya, alternating metalaxyl with phosphate fungicide had the most suppressive effect on relative area under disease progress curve (R.AUDPC) and percentage disease severity (Table 29.6). The application of phosphoric acid by itself resulted in significantly ($P<0.05$) higher tuber yield compared to metalaxyl treatment at the Tigoni location. The suppressive effect of plant extract from stinging nettle on late blight was relatively moderate when compared to the untreated control in 2008 and 2009.

A similar trend was obtained when treatment effects at Marimba location were compared. During 2008, suppressive effects for all treatments were greater on the cultivar Tigoni than on Desiree (Table 29.7). The lowest amount of late blight severity and AUDPC were recorded when combinations of metalaxyl+phosphate

Table 29.7 Effect of plant extract (Stinging nettle), phosphoric acid and fungicide combinations on late blight severity (%) at Marimba location in 2008 and 2009

2008	Disease severity (%)		rAUDPC (% disease days)*	
Treatment	Desiree	Tigoni	Desiree	Tigoni
Metalaxyl	10.4	4.4	16.4	5.5
Mancozeb	14.0	6.3	17.1	11.1
Phosphate	10.2	4.6	16.2	5.8
Stinging nettle (extract)	17.8	9.1	21.5	14.3
Metalaxyl+Phosphite	7.4	3.7	12.7	4.7
Mancozeb+phosphate	10.8	5.4	16.4	6.9
Mancozeb+extract	–	–	–	–
Untreated	40.1	18.0	49.1	24.3
Means	44.5	20.0	12.34	18.2
LSD (0.05)	2.5	2.3	8.4	5.2
2009	Disease severity (%)		rAUDPC (% disease days)	
Treatment	Desiree	Tigoni	Desiree	Tigoni
Metalaxyl	12.1	6.2	12.4	6.3
Mancozeb	29.1	13.3	24.7	12.9
Phosphate	11	6.2	11.4	6.4
Stinging nettle (extract)	30.5	13.7	30.4	13.9
Metalaxyl+Phosphite	9.6	5.3	10	5.5
Mancozeb+phosphate	15.88	8.1	16.4	8.1
Mancozeb+extract	29	12.8	28.9	12.1
Untreated	53.2	24.5	52.5	24.1
Means	44.	20.0	12.34	18.2
LSD (0.05)	2.8	2.2	3.1	2.7

**rAUDPC* relative area under disease progress curves (AUDPC/number of days disease was assessed), Nyankanga et al. unpublished data

were used and the greatest amount of disease was obtained in the untreated control. Similarly, disease severity (%) and AUDPC were lowest in treatment consisting of metalaxyl+phosphate. With the exception of the untreated control, the extract from stinging nettle was the least effective (Table 29.7).

29.4 Bacterial Wilt Management

29.4.1 Crop Rotation Practices

Potato cultivation in most tropical highlands in SSA is characterized by continuous planting of the crop on same piece of land sometimes with no crop rotation for several consecutive cropping seasons. Where crop rotation is practiced, the rotation cycles are too short to break the disease cycle. Additionally, volunteer potatoes are usually nurtured in rotation crops, taking the semblance of an intercrop, as a source

Table 29.8 The effect of one season rotations with various crops on bacterial wilt incidence at Kachwekano, Uganda in 1999

Treatment	Bacterial wilt incidence (%)	Total yield (Mg ha^{-1})	Market yield (Mg ha^{-1})	Market yield increase (%)	Total yield market (%)
Potato-onions-potato	12.4 (3.5b)*	2.04a	2.00a	80	98
Potato-peas-potato	5.0 (2.2b)	1.86a	1.85a	67	99
Potato-cabbage-potato	6.9 (2.61b)	1.77a	1.76a	59	99
Potato-sweetpotato-potato	3.8 (1.57b)	1.64ab	1.64a	48	100
Potato-millet-potato	3.2 (1.7b)	1.67ab	1.67a	50.5	100
Potato-carrots-potato	11.0 (3.3b)	2.04a	2.04a	84	100
Potato-beans-potato	7.4 (2.2b)	1.81a	1.81a	63.1	100
Potato-potato-potato	62.2 (7.8a)	1.24b	1.11b	–	90

*Values in parentheses are Square root transformed. Means followed by the same letter are not significantly different ($P<0.05$). Source: Lemaga et al. (2001a)

of the most needed food during a critical period of food shortage, thus, perpetuating the bacterial wilt pathogen. It is often difficult to find suitable crops that do not serve as symptomless hosts for *R. solanacearum* and best fitted in the rotation and farming systems of the highland tropics. Whereas potato is grown by most farmers, an equally high number of farmers grow beans that are known symptomless carriers of the BW pathogen. Farm survey data from southwestern Uganda revealed that 25% or more of the farmers planted beans after potato during the previous two cropping seasons and that 10–15% of the farmers plant potato on the same piece of land after just one season of crop rotation. Data also showed that 20–24% of the farmers either fallow their land or grow sorghum after potato. Crop rotation with sorghum or maize or establishment of fallow would be helpful in reducing or eradicating BW in previously infected fields, primarily if volunteer potatoes are destroyed and the dominant vegetation is composed of grasses.

Crop rotation experiments conducted in Uganda showed that a one season rotation of potato with cereals, pulses, vegetables and root crops in mildly infested fields (15–20%), significantly ($P<0.05$) reduced bacterial wilt incidence and significantly increased fresh tuber yield compared to plots continuously planted to potato (Table 29.8). Similarly, a two-cropping season (one-year) rotation with cereals and beans in a previous highly infested field (>90% incidence) reduced BW incidence by 50% compared to 80% BW incidence with continuous potato cropping (Lemaga et al. 2001a).

A two-year crop rotation study in Uganda involving wheat, peas and fallow for two cropping seasons after potato rendered a previously BW infested field free of the pathogen beyond detectable limits (Kakuhenzire et al. 2000). In Rwanda a two-year crop rotation also resulted in the production of BW-free seed tubers (Van der Zaag 1986). Crop rotation and fallowing, however, are beneficial if ground keepers are removed, volunteer potatoes destroyed and solanaceous crops are not planted before potato is planted again. There are also many symptomless hosts of *R. solanacearum*, including beans and other herbaceous weeds (Tusiime et al. 1997) and these should not be allowed to flourish in a rotation crop or fallow. Cereals and grasses have been reported to suppress bacterial wilt and inoculum build-up in the soil (Lemaga et al. 2001a). Some plant extracts of herbaceous species such as *Clotalaria, Tephrosia,* wild cabbage (*Brassica integrifolia*) and cissus (*Cissus araliodes*) suppress in-vitro growth of *R. solanacearum*. Soils amended with vegetable matter of *Clotalaria sp. and Tephrosia* sp. had low levels of BW severity and higher tuber yield than the controls.

29.4.2 Diverse Cultural Measures for Bacterial Wilt Management

Cultural practices that can limit the introduction, population increase or dissemination of bacterial wilt on potato include: destruction of crop residues and waste tubers, timely weed management and minimum post-emergence cultivation, removal of wilted plants, selection of healthy (non-diseased) plants, disinfecting farm tools, shoes and equipment used in potato fields, and reducing or preventing run-off from neighboring fields.

Haulms cut from infected potato and discarded in the field can be a source of inoculums or perpetuation of the BW pathogen in the soil. Disposal of such infected haulm can spread the disease to neighboring fields. Therefore, haulm from BW-infected potato plants and infected tubers should be destroyed in dumping pit where the bacteria can be destroyed through natural decomposition processes. Such dumping or refuse sites should be decontaminated and used exclusively for that purpose. Any volunteer potato that may grow at such a site should be destroyed with a systemic herbicide.

Some common weeds such as *Ageratum conyzoides, Amaranthus sp. Bidens pilosa, Oxalis latifolia, Polygonum napalense, Rumex abyssinicum* and *Tagites minuta* have been shown to be symptomless hosts of *R. solanacearum* (Tusiime et al. 1996, 1997). Such weeds should be removed from potato fields, particularly after potato cultivation. Improved fallows that would suppress such weeds are more appropriate than natural fallow systems having mixed plant species where some plants maybe symptomless carriers of *R. solanacearum* pathogen. It should be noted that in such areas, post-emergence cultivation should be done with care to minimize or prevent root damage. Wounded roots of potato plants can create entry points for the BW pathogen. Planting practices that utilized potato ridges can be optimized instead of instead of planting in flat fields and reconstituting ridges after crop emergence (Kinyua et al. 2005).

Harvest practices often leave some residual tubers below the soil in potato fields. Assessment of field soils after harvest conducted in Uganda revealed that 600–1,000 kg of potatoes per hectare may be recovered after the completion of harvest operation. Removal of residual tubers should be a routine practice that can greatly reduce the number of volunteer potato plants and potential *R. solanacearum* hosts for pathogen perpetuation. Volunteer potatoes that inevitably germinate, should be destroyed by rouging or spraying with a systemic herbicide such as glyphosate.

Table 29.9 Incidence of latent bacterial wilt infection in plants grown from seed from positive selection, farmer practice, small seed plot technique and improved seed in Uganda during 2010

Seed production practice	Sample size	Number of infected samples	Infected samples within production practice (%)	Total infected samples (%)
Farmer selection	25	9	36.4	8.5
Positive selection	22	3	14.1	2.9
Small seed plot	20	1	7.5	1.4
Improved seed	40	8	19.0	7.1
Total	107	21	19.9	19.9

Planting seed pieces is a principal source for dissemination of BW in many crops (Hayward 1991). Cropping practices that reduce the use of bacterial wilt infected seed can reduce pathogen dissemination in potato fields. The rouging of diseased/wilted plants can reduce the sources of infection and spread of the BW pathogen in the field, resulting in seed potato with minimum or no latent BW infection. However, the best practice would be to use healthy, certified seed if available. In the absence of this, small scale potato farmers can reduce BW in potato seed by selecting and marking healthy, vigorous and non-diseased potato plants for seed (positive selection). This practice reduces latent infection in seed tubers and minimizes or eliminates the incidence of BW in the progeny tubers. In general, seed health testing or detection of BW in seed by use of sensitive serological techniques (ELISA) should be done for seed lots. The assessment of latent BW infection in various seed potato production practices in Uganda showed that positive selection and small-scale seed plot techniques produced more healthy seed potatoes (Table 29.9). Both positive selection and small seed plot technique, however, are appropriate for small scale farmers planting less than 500 kg of seed per cropping season that is a common phenomenon in most tropical highlands.

The bacterial wilt pathogen can be spread among potato fields on contaminated farm tools and machinery. Hoes, cutlasses, and sickles used in BW infected field can spread the disease to a healthy field if used without being disinfected. Similarly, contaminated field shoes and boots can disseminate the disease in adhering soil unless they are decontaminated between potato fields. Farm machinery such as tractors and mowers should be thoroughly cleaned with a disinfectant after every farm operation to reduce or prevent pathogen dissemination.

29.4.3 Use of Healthy (Disease-Free Tubes) and Regulatory Measures

Various national potato programs in eastern and central Africa (ECA) have been producing and availing seed potato to farmers every year following strict quality standards, though in limited quantities. Quality seed as a tool to bacterial wilt management in the ECA region is hampered by the low seed potato production capacity.

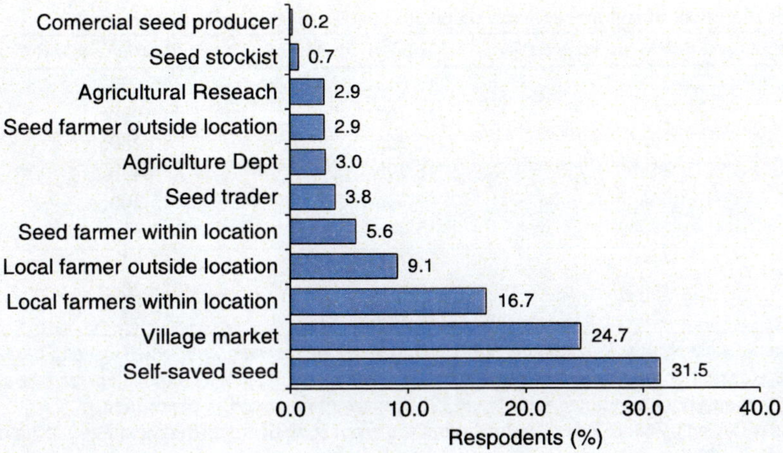

Fig. 29.1 Major seed potato sources among farmers in southwestern Uganda during 2009 cropping season. A sample size of 588 farmers responded

The quantity of mini-tubers produced on a regional scale in order to generate disease-free seed is still very low. Consequently, national programs usually bulk seed tubers for several field generations before adequate quantities are accumulated, increasing possibilities of acquiring bacterial and viral infections. Since 2000, all the major potato growing countries in ECA, with the exception of Rwanda, have been producing less than 150 t of basic seed for distribution to farmers annually. This means that most of the seed potato planted by farmers is from informal sources with high probability of disease infection. A 2009 farm family survey data in Uganda revealed that the majority of farmers plant seed potato obtained from sources with unknown health status. More than 30% of the farmers plant seed retained from previous harvest, whereas about 6% obtain quality seed from the agriculture department and agricultural research stations (Fig. 29.1).

Therefore to improve the supply of quality potato seed to farmers, the quantity of mini-tubers produced per annum in the potato growing areas of the highland tropics should be increased to reduce the number of field generations before seed reaches the ware potato farmer. This is being gradually achieved in some SSA countries, partly through adoption of aeroponics technology for mini-tuber production and optimizing seed potato production and distribution (Olanya et al. 2010). For example by using the aeroponics technology, Kenya produced 792, 656 mini-tubers in a single year compared to 39,791 mini-tubers produced by conventional methods during the same period. Increased seed supply, however; is likely to be hampered by inadequate implementation of seed quality regulations. Among the countries in the East and Central Africa, only Kenya has a functional seed potato inspection service. In other countries, the potato seed supply system is largely informal and national programs provide less than 1% of the national seed requirement.

Table 29.10 Effect of soil amendments on potato yields over three growing seasons

Treatment*	BW incidence (%)	Total yield (Mg ha^{-1})	Market yield (Mg ha^{-1})
S	28ab**	18.01a	16.63b
L	30ab	16.76bc	14.93bc
S+PK	19c	22.35a	20.280a
L+PK	26ab	18.79b	16.44b
S+P	26a	17.13bc	15.20bc
L+P	27ab	15.43c	13.31c
NPK	23ab	18.75b	16.92b
NP	16bc	16.61bc	15.20bc
Control	32a	11.24d	9.69d

*N nitrogen, P phosphorus, K potassium, *BW* bacterial wilt, *L Leucaena diversifolia* (L.), and *S Sesbania sesban* L. (*Sesbania* and *Leucaena* were used to supply organic matter equivalent of 100 kg of N ha^{-1} either singly or combined with P and PK in this soil amendment experiment)
**Means followed by same letters within columns are not statistically different at $P<0.05$ (source: Lemaga et al. 2005)

The impact of quality seed to reduce bacterial wilt is illustrated by a study conducted in Uganda for three consecutive cropping seasons, involving an improved package that consisted of bacterial wilt–free seed of a less susceptible variety, row planting with hilling at planting and minimum post-emergence cultivation that reduced BW incidence by 56% and increased marketable tuber yield by 29% compared to farmer practice (Lemaga 2000). The highest contribution to reducing wilting incidence and increasing marketable tuber yield was largely attributed to BW-free seed.

29.4.4 Soil Health and Soil Amendments

One of the principal components of integrated bacterial wilt management besides disease-free seed is BW-free soil. However, in tropical highlands, characterized by continuous potato cropping and seed tuber recycling, clean seed is limited and BW-free fields are not prevalent, the latter being exacerbated by land shortage. Continuous potato planting also depletes the soil of essential plant nutrients where malnourished plants are more susceptible to pest and disease attack than crops in rich soils (Katafire et al. 2005). In an experiment conducted in southwestern Uganda involving amending soil with either organic (*Leuceana* sp. and *Sesbania* sp.) or inorganic soil amendments or their mixtures resulted in high reduction in BW incidence, severity and increase in total and marketable tuber yield (Lemaga et al. 2001b). The benefits of soil fertility enhancement on reduction of BW incidence, severity and yield increase were more manifested where both organic and inorganic amendments were combined than where either of the nutrient sources was used (Table 29.10). Therefore, to get the full benefit of soil amendment, both organic and inorganic sources of plant nutrients should be combined to synergize each other (Lemaga et al. 2001b).

29.4.5 *Tolerance to Bacterial Wilt Disease and Use of Less Susceptible Cultivars*

Conventional breeding for resistance to BW has not developed stable varieties largely because of high interaction between cultivars, pathogen strains and the environment (Tusiime et al. 1996; French and de Lindo 1982). There are cultivars, for example Cruza 148, that are tolerant to BW, and could be grown in soil in fields infested with BW, however; their seeds are likely to transfer the disease latently to probably disease-free areas (Kakuhenzire et al. 2000). Previous research in Uganda during the 1990s revealed that cultivars with high levels of tolerance to BW showed some plasticity to pathogen attack over seasons and locations (Tusiime et al. 1996). In another research, it was shown that cultivars with high levels of tolerance to BW had high susceptibility to late blight (Kakuhenzire et al. 2000). Therefore, in tropical highlands and mid-altitude elevations, there is need to obtain a balance between BW and late blight resistance and cultivar acceptability in the market if such cultivars are to be widely adopted. Similar experiments conducted in Kenya also showed that cultivars had varying reactions to bacterial wilt in field experiments.

29.4.6 *Integrating Various Bacterial Wilt Management Techniques*

Individual components of BW management will not help to prevent or reduce the impact of disease in most crops prone to *R. solanacearum* infection, especially in fields that are already infested (Kinyua et al. 2001). It is more prudent to use disease-free seed as planting materials in BW-infested soils. However, among all BW management techniques, disease-free soil combined with healthy planting material and sanitation will delay or totally prevent infection of a potato crop particularly when other disease predisposing factors are operating at a low level. The efficacy of BW control technique also depends on the BW race common in the region because of differences in the host range and pathogen resilience (Priou et al. 1999; Hayward 1991). The various techniques for BW control have been weighted and shown to be effective for control of the disease and furthermore, they offer potential for the eradication of *R. solanacearum*.

29.5 Conclusions and Recommendations

Late blight and bacterial wilt remains the most formidable constraints to potato production and accounts for significant losses in tropical Africa. Although late blight can be effectively controlled through the use of fungicides, the economic costs associated with chemical use are prohibitive for the resource constrained farmers in those regions. Moreover, the deleterious impacts of chemical use to human health and environment make this a less attractive technique for control of potato diseases. Management efforts for late blight management have, therefore, been focused on host-plant resistance and integrated approaches. The combined approaches for late blight management that have been highlighted in this chapter include: resistant cultivars, reduced fungicide application, disease monitoring/prediction based on field scouting and weather criteria, bio-rational approaches, post-harvest management, regulatory techniques and farmer participatory training in holistic late blight management techniques.

The spread of bacterial wilt has continued to increase with high demand for potato forcing farmers to have short crop rotation cycles before potato is returned to the same piece of land and growing the crop beyond where it was traditionally grown. Bacterial wilt incidence and spread are further exacerbated by a warming environment due to climate change. Inadequate availability of quality seed, decreasing farm family land holdings and poverty are likely to impact on future incidence, severity and potato crop loss due to BW. Appearance of BW at high altitudes in the tropics where the disease has not been traditionally known, probably due to global warming and poor farm practices will reduce zones where disease-free seed can be produced.

To sustain potato production in tropical highlands in sub-Saharan Africa, and reduce bacterial wilt occurrence, new approaches to disease management and seed potato supply systems have to be developed and existing ones improved. One of the critical factors that should be addressed to improve the supply of quality seed potato is to reduce the number of unregulated field multiplication cycles before ware potato farmers can access seed due to tremendous seed degeneration that may occur. Seed potato production technologies such as: tissue culture and mini-tuber production in solid media, solid-free mini-tuber production technique in aeroponics, and private sector participation in seed production and multiplication should be optimized in order to reduce the costs of quality and quantity of potato seed. To improve quality seed potato supply as a tool for managing diseases such as LB, BW and potato viruses, harmonization of regulations and policies across the sub-Saharan Africa region can greatly facilitate seed trade and improve potato production in the region.

References

Andreu AB, Guevara MG, Wolski EA, Daleo GR, Caldiz DO (2006) Enhancement of natural disease resistance in potatoes by chemicals. Pest Manag Sci 62:162–170

Ateka EM, Mwang'ombe AW, Kimenju JW (2001) Reaction of potato cultivars to *R. solanacearum* in Kenya. Afr Crop Sci J 9:251–256

Bekele K, Hiskias Y (1996) Tuber yield loss assessment of potato cultivars with different levels of resistance to late blight. In: Proceedings of the 3rd annual conference of Crop Protection Society of Ethiopia, 18–19 May 1996. Addis Ababa, Ethiopia, pp 149–152

Boucher CA, Gough CL, Arlat M (1992) Molecular genetics of pathogenicity determinants of *Pseudomonas solanacearum* with special emphasis on *hrp* genes. Annu Rev Phytopathol 30:443–461

CABI/EPPO (1999) Distribution maps of plant diseases. Map No. 785, CAB International, Wallingford, UK

Coulibaly M, Dembele D, Vanderhofstadt B (2002) Unite de production de plants de pomme de terre au Mali. Soc. International, Amatevi et IER

Davidse LC, Looijen L, Turkensteen LT, van der wal D (1981) Occurrence of metalaxyl resistant strains of *Phytophthora infestans* in Dutch potato fields. Eur J Plant Pathol 87:65–68

Denny TP, Hayward AC (2001) Ralstonia. In: Schaad NW et al (eds) Laboratory guide for the identification of plant pathogenic bacteria, 3rd edn. APS Press, St. Paul, pp 151–174, 373pp

Dowley LJ, O'Sullivan E (1981) Metalaxyl-resistant strains of *Phytophthora infestans* (Mont.) de bary in Ireland. Potato Res 24:417–421

Elphinstone JG (2005) The current bacterial wilt situation: a global overview. In: Allen C, Prior P, Hayward AC (eds) Bacterial wilt disease and the *Ralstonia Solanacearum* species complex. APS Press, St. Paul, pp 9–28, 510pp

FAO-CIP (1995) Potato in the 1990's: situation and prospects for the world potato economy. Food and Agriculture Organization (FAO), Rome, 39pp

Fontem DA, Aighewi B (1993) Effect of fungicides on late blight control and yield loss of potato in the western highlands of Cameroon. Int J Pest Manag 39:152–155

Fontem DA, Gumedzoe YD, Olanya M (2001) Quantitative effects of late blight on potato yield in Cameroon. Afr Crop Sci Conf Proc 5:449–453

Fontem DA, Olanya OM, Njualem BF (2004) Reaction of certain solanaceous and asteraceous plant species to inoculation with *Phytophthora infestans* in Cameroon. J Phytopathol 154:331–336

Fontem DA, Olanya OM, Tsopmbeng GR, Owona MAP (2005) Pathogenicity and metalaxyl sensitivity of *Phytophthora infestans* isolates obtained from garden huckleberry, potato and tomato in Cameroon. Crop Prot 24:449–456

French ER, de Lindo L (1982) Resistance to *Pseudomonas solanacearum* in potato: Specificity and temperature sensitivity. Phytopathology 72:1408–1412

Fry W (2008) *Phytophthora infestans*: the plant (and *R* gene) destroyer. Mol Plant Pathol 9:385–402

Gees R, Hohl HR (1988) Cytological comparison of specific (R3) and general resistance to late blight in potato tissue. Phytopathology 78:350

Hakiza JJ, Turyamureeba G, Kakuhenzire RM, Odongo B, Kanzikwera R, Adipala E (2000) Potato and sweet potato improvement in Uganda. A historical perspective. Afr Potato Assoc Conf Proc 5:47–58

Hallberg GR (1989) Pesticide pollution of ground water in the humid United States. Agric Ecosyst Environ 26:299–367

Hayward AC (1991) Biology and epidemiology of bacterial wilt caused by *Pseudomonas solanacearum*. Annu Rev Phytopathol 29:65–87

Kakuhenzire R, Hakiza JJ, Lemaga B, Tusiime G, Alacho F (2000) Past, present and future strategies for integrated management of potato bacterial wilt in Uganda. Afr Crop Sci Proc 5:353–359

Kankwatsa P, Adipala E, Hakiza JJ, Olanya M, Kidanemariam HM (2002) Effect of integrating planting time, fungicide application and host resistance on potato late blight development in south-western Uganda. J Phytopathol 150:248–257

Katafire M, Adipala E, Lemaga B, Olanya M, El-Bedewy R, Ewell P (2005) Management of bacterial wilt of potato using one-season rotation crops in south-western Uganda. In: Allen C, Prior P, Hayward AC (eds) Bacterial wilt disease and the *Ralstonia Solanacearum* species complex. APS Press, St. Paul, pp 197–204, 510pp

Kinyua ZM, Smith JJ, Lung'aho C, Olanya M, Priou S (2001) On-farm success and challenges of producing bacterial wilt-free tubers in seed plots in Kenya. Afr Crop Sci J 9:279–285

Kinyua ZM, Olanya M, Smith JJ, El-Bedewy R, Kihara SN, Kakuhenzire RK, Crissman C, Lemaga B (2005) Seed plot technique: empowerment of farmers in production of bacterial wilt-free seed potato in Kenya and Uganda. In: Allen C, Prior P, Hayward AC (eds) Bacterial wilt disease and the *Ralstonia Solanacearum* species complex. APS Press, St. Paul, pp 167–175, 510pp

Landeo JA, Gastelo M, Forbes G, Zapata JL, Flores FJ (1997) Developing horizontal resistance to late blight in potato. In: International potato center program report, 1995–1996, Lima, Peru, pp 122–125

Lemaga B (2000) Integrated control of potato bacterial wilt in Kabale district, southwestern Uganda. In: Scientist and farmer. Partners in research for the 21st century. Program report 1999–2000. International Potato Center, pp 129–141

Lemaga B, Kanzikwera R, Kakuhenzire R, Hakiza JJ, Manzi G (2001a) The effect of crop rotation on bacterial wilt incidence and potato tuber yield. Afr Crop Sci J 9:257–266

Lemaga B, Siriri D, Ebanyat P (2001b) Effect of soil amendments on bacterial wilt incidence and yield of potato in south western Uganda. Afr Crop Sci J 9:267–278

Lemaga B, Kakuhenzire R, Bekele K, Ewell PT, Priou S (2005) Integrated control of potato bacterial wilt in Eastern Africa: the experience of African highlands initiative. In: Allen C, Prior P, Hayward AC (eds) Bacterial wilt disease and the *Ralstonia Solanacearum* species complex. APS Press, St. Paul, pp 145–158, 510pp

McLeod A, Denman S, Denner FDN (2001) Characterization of South African isolates of *Phytophthora infestans*. Plant Dis 85:287–291

Muchiri FN, Narla RD, Olanya OM, Nyankanga RO, Ariga ES (2009) Efficacy of fungicide mixtures for the management of *Phytophthora infestans* (US-1) on potato. Phytoprotection 90:19–29

Mukalazi J, Adipala E, Sengooba T, Hakiza JJ, Olanya M, Kidanemariam HM (2001) Metalaxyl resistance, mating type and pathogenicity of *Phytophthora infestans* in Uganda. Crop Prot 20:379–388

Namanda S, Olanya OM, Adipala E, Hakiza JJ, El-Bedewy R, Baghsari AS, Ewell P (2004) Fungicide application and host resistance for potato late blight management: benefits assessment from on-farm studies in S.W. Uganda. Crop Prot 23:1075–1083

Natrass RM, Ryan M (1951) New hosts of *Phytophthora infestans* in Kenya. Nature 168:85–86

Njuguna JGM, Oduor GI, Njenga DN (1998) Rate of adoption of new late blight resistant potato cultivars by farmers in Kiambu district. In: Farrell G, Kibata GN (eds) Proceedings of the 2nd biennial crop protection conference, 16–17 Semptember 1998, Nairobi, Kenya, pp 196–202

Nyankanga RO, Wein HC, Olanya OM, Ojiambo PS (2004) Farmers' cultural practices and management of potato late blight in Kenya Highlands: implications for development of integrated disease management. Int J Pest Manag 50:135–144

Nyankanga RO, Wien HC, Olanya MO, Ojiambo PS (2007) Relationship between late blight (*Phytophthora infestans*) of potato on tuber and foliage, as affected by disease severity on foliage, cultivar resistance, and atmospheric and soil variables. Can J Plant Pathol 29:372–387

Nyankanga RO, Olanya OM, Ojiambo PS, Wien HC, Honeycutt CW, Kirk WW (2011) Validation of tuber blight (*Phytophthora infestans*) prediction model. Crop Prot 30:547–553

Ochwo MKN, Kamoun S, Adipala E, Rubaihayo PR, Lamour K, Olanya M (2002) Genetic diversity of *Phytophthora infestans* (Mont.) de Bary in Eastern and Western Highlands of Uganda. J Phytopathol 150:541–542

Ojiambo PS, Namanda S, Olanya OM, El-Bedewy R, Hakiza JJ, Adipala E, Forbes GA (2001) Impact of fungicide application and late blight development on potato growth parameters and yield in tropical highlands of Kenya and Uganda. Afr Crop Sci J 9:225–233

Olanya OM, Adipala E, Hakiza JJ, Kedera JC, Ojiambo PS, Mukalazi J, Forbes G, Nelson R (2001a) Epidemiology and population dynamics of *Phytophthora infestans* in sub-Saharan Africa: progress and constraints. Afr Crop Sci J 9:185–193

Olanya OM, El-Bedewy R, Ojiambo PS, Ewell PT, Hakiza JJ (2001b) Relationships of fungicide application to late blight development and potato growth parameters in the tropical highlands of Uganda and Kenya. In: International Potato Center, 2001, Scientist and Farmer partners in research for the 21st century. Program report, 1990–2000, Lima, Peru, pp 77–86

Olanya OM, Ojiambo PS, Nyankanga RO (2006) Dynamics of development of late blight (*Phytophthora infestans*) in potato, and comparative resistance of cultivars in the highland tropics. Can J Plant Pathol 28:84–94

Olanya OM, Plant AB, Larkin RP, Honeycutt CW (2009) Infection potential of hairy nightshade (*Solanum sarrachoides*) by *Phytophthora infestans* and late blight implications of the alternate host. J Phytopathol 157:427–437

Olanya M, Nelson R, Hakiza J, Ewell P, El-Bedewy R, Kakuhenzire R, Namanda S, Kasheija I, Wagoire W, Ngombe B, Musoke C (2010) Comparative assessment of pest management practices in potato production at Farmer Field Schools. Food Secur 2:327–341

Onyiango MO, Khare KB, Achwanya OS, Otieno W (2005) Efficacy of synthetic and non-synthetic fungicides against stem canker disease of tea. Afr Crop Sci J 13:209–214

Priou S, Aley P, Chujoy E, Lemaga B, French E (1999) Integrated control of bacterial wilt of potato. CIP Slide Training Series IV-3. International Potato Centre, Lima, Peru, 30pp

Priou S, Aley P, Gutarra L (2005) Assessment of resistance to bacterial wilt in CIP advanced potato clones. In: Allen C, Prior P, Hayward AC (eds) Bacterial wilt disease and the *Ralstonia Solanacearum* species complex. APS Press, St. Paul, pp 261–268, 510pp

Samoucha Y, Gisi U (1987) Systemicity and persistence of cymoxanil in mixture with oxadixyl and mancozeb against *Phytophthora infestans* and *Plasmopara viticola.* Crop Prot 6:393–398

Schwinn FJ, Margot P (1991) Control with chemicals. In: Ingram DS, Williams PH (ed). Advances in Plant Pathology pp. 225–265. Academic Press, San Diego, CA, USA

Scott GJ, Rosegrant MW, Ringler C (2000) Global projections for root and tuber crops to the year 2020. Food Policy 25:561–597

Smith JJ, Murimi ZK, Offord LC, Clayton S, Mienie N, Gouws NR, Priou S, Olanya M, Simons S, Saddler GS (2005) Processes in the development of a biocontrol agent against bacterial wilt. In: Allen C, Prior P, Hayward AC (eds) Bacterial wilt disease and the *Ralstonia Solanacearum* species complex. APS Press, St. Paul, pp 61–72, 510pp

Tusiime G, Adipala E, Opio F, Bhagsari AS (1996) Occurrence of *Pseudomonas solanacearum* latent infection in potato tubers and weeds in highland Uganda. Afr J Plant Prot 6:108–118

Tusiime G, Adipala E, Opio F, Bhagsari AS (1997) Towards developing an integrated control package for potato bacterial wilt in lowland Uganda. Afr Crop Sci Conf Proc 3:989–994

Van der Zaag P (1986) Potato production under *Pseudomonas solanacearum* conditions: sources and management of planting material. In: Persley (ed) Disease in Asia and South Pacific: proceedings of an international workshop, Los Banos Philippines, Oct 1985. Aust Centre Int Agric Res Proc 13:84–88

Vega-Sanchez ME, Erselius LJ, Rodriguez AM, Bastidas O, Hohl HR, Ojiambo PS, Mukalazi J, Vermeulen T, Fry WE, Forbes GA (2000) Host adaptation to potato and tomato within the US-1 clonal lineage of *Phytophhtora infestans* in Uganda and Kenya. Plant Pathol 49:531–539

Index

Z. He et al. (eds.), *Sustainable Potato Production: Global Case Studies*, DOI 10.1007/978-94-007-4104-1, © Springer Science+Business Media B.V. 2012